The Caledonide Geology
of Scandinavia

The Caledonide Geology of Scandinavia

edited by

R.A. Gayer

Department of Geology, University of Wales,
Cardiff, UK

Graham & Trotman

A member of the Kluwer Academic Publishers Group
LONDON/DORDRECHT/BOSTON

First published in 1989 by

Graham and Trotman Limited
Sterling House
66 Wilton Road
London
SWIV 1DE
UK

Graham and Trotman Inc.
Kluwer Academic Publishers Group
101 Philip Drive
Assinippi Park
Norwell, MA 02061
USA

British Library Cataloguing in Publication Data
The Caledonide geology of Scandinavia.   1. Scandinavia. Caledonian   orogenic regions   I. Gayer, R.A. (Rodney A.)   551.7′2

Library of Congress Cataloging in Publication Data
The Caledonide geology of Scandinavia. "Originally presented at a conference held at University College Cardiff, 22nd–23rd September 1987."   Bibliography: p.   1. Geology—Scandinavia—Congresses.   2. Geology—Greenland—Congresses.   I. Gayer, Rodney A. II. University   College, Cardiff.   QE277.C34 1989   554.9   88-21387

ISBN 1 85333 067 1

© R. A. Gayer, 1989

Photoset by Interprint Ltd. Malta
Printed in Great Britain by Dotesios Printers Ltd, Trowbridge, Wiltshire

Contents

Introduction

This book represents an edited selection of papers and discussion originally presented at a conference held at University College Cardiff, 22–23 September 1987. The conference and this book are dedicated to the memory of Dr. Sven Føyn, former Director of Norges Geologiske Undersøkelse, who with Professor J. G. C. Anderson in 1962 established the Norwegian Caledonides research group in the Department of Geology at University College Cardiff. The conference thus celebrated the 25th anniversary of this research group and was attended by some 75 delegates, mainly from British and Scandinavian institutions.

The 25 papers reproduced in this book present new and original research material on the Scandinavian Caledonides and on the Caledonides of East Greenland. The Caledonides of the North Atlantic Region are represented by superbly exposed sections in Scandinavia and the Arctic, where much recent active research has rapidly expanded our knowledge of this major continental collision orogenic belt. The deep erosion that has occurred since the Mid-Devonian has exposed sections through all levels of the orogen, rarely seen in modern active mountain belts. These have allowed insight into the fundamental tectonic processes associated with the evolution of an oceanic domain, from its initiation, through its spreading and closure histories, to its final destruction associated with continental collision. The rift phase of this development, in the late Precambrian, stretched and thinned the continental crust of Baltica and Laurentia, allowing the development of major extensional sedimentary basins. The post-rift phase of ocean spreading resulted in the two continents drifting apart and the formation of the major Iapetus Ocean, with sedimentary basins forming on its passive continental margins. The various stages of ocean closure during the Ordovician and early Silurian, with subduction of oceanic lithosphere generating island arc magmatism, arc-related basins and obduction of ophiolites, provides the first evidence for contractional deformation in the mountain belt. Oblique closure of Iapetus may well have resulted in strike-slip displacement of terranes along the oceanic margins. The final stages of closure resulted in underthrusting of Baltica beneath Laurentia, with the development of high-pressure metamorphism in the deeply buried crust. The Baltica margin contracted and thickened with the development of major, far-travelled allochthonous thrust sheets and foreland basins that are so characteristic of this and other continental-collision mountain belts. Gravitational collapse of the isostatically uplifted orogen eventually thinned the crust, allowing it to regain its original thickness with the development of major Devonian sedimentary basins controlled by extensional shear zones and faults. The Palaeozoic rocks, which record this complex history of oceanic evolution, form the basement that partly controlled the development of younger basins on the continental shelf bordering the modern North Atlantic. These basins are sites for vigorous hydrocarbon exploration.

The selection of papers in this book aims to present modern work and ideas on a broad spectrum of this complex Caledonian history. The material has been grouped into seven sections covering the areas of Finnmarkian Geology (isotopic dating and early Caledonian events), Regional Geology (structure and lithostratigraphy), Igneous Geology, Metamorphic Geology, Palaeontology and Biostratigraphy, Devonian Geology and East Greenland Caledonian Geology.

In order to understand the complex plate interactions developed during the closure of Iapetus, it is important to determine the precise age of deformation, metamorphism and igneous activity along the mountain belt. A critical area has been in northernmost Norway, in the county of Finnmark where an early phase of orogenesis (the Finnmarkian orogeny) has been dated between 540 and 490 Ma. In the section on **Finnmarkian Geology**, various approaches towards understanding the Finnmarkian Orogen are investigated. In the first paper *R. B. Pedersen, G. R. Dunning* and *B. Robins* (Bergen and Toronto) present two zircon ages of 531 ± 2 Ma and 523 ± 2 Ma from nepheline syenite pegmatites emplaced during an extensional phase at the close of the intrusion of the synorogenic Seiland Igneous Province. Dates from this province have previously been used to define the extent of the early Caledonian Finnmarkian Orogeny and the new dates suggest that deformation had ceased by mid to late Cambrian times, older than the Ordovician obduction of ophiolites further south in Scandinavia. *R. D. Dallmeyer, A. Reuter, N. Clauer* and *N. Liewig* (Athens, Georgia and Strasbourg) discuss K–Ar and $^{40}Ar/^{39}Ar$ ages of 440 Ma for illites developed during low-grade metamorphism of the Lower Allochthon and Autochthon in Finnmark and conclude that these Baltic margin sediments have only been affected by the late Caledonian collisional event, dismissing previous suggestions of a Finnmarkian history. *R. E. Binns* (Trondheim) proposes a correlation between the Helliskogen/Skibøtn Nappe Complexes of Troms, which contain an upper Ordovician/Silurian fauna, with the Kalak Nappe Complex of Finnmark, suggesting no early Caledonian deformation in these units and implying erroneous previous dates. In contrast, *D. Tietzsch-Tyler* (Dublin) presents structural and stratigraphic evidence for an early Caledonian deformation in Central Norway and, in a tectonic model for its development, suggests that this might correlate with Finnmarkian Orogeny. *C. Townsend* and *R. A. Gayer* (Cambridge and Cardiff) assess the evidence for timing of orogenesis in Finnmark and conclude that only in the Kalak Nappe Complex of the Middle Allochthon is any early Caledonian deformation recorded and that this can be divided into a Pre-Mid-Cambrian event and an Early Ordovician thermal event. Neither equates in time to the previously defined Finnmarkian Orogeny, although the latter appears to correlate with events farther south in Scandinavia.

The largest section of the book deals with papers describing **Regional Geology** and it is in this section that detailed new work on structural and stratigraphic relationships is presented from various parts of the mountain belt. In the first paper of the section, *R. O. Greiling* (Heidelberg) describes the Middle Allochthon of Västerbotten, northern Sweden, where tectonic windows through the Upper Allochthon (Seve Nappe) show that the Middle Allochthon has a similar lithostratigraphy to that of the Stalon Nappe Complex of the eastern Caledonian margin but with a more ductile deformation and metamorphosed to a higher grade following thrust emplacement. These relations are explained by suggesting that the window rocks were initially subducted beneath a colliding western plate but were later accreted to the base of the western plate and thrust with it. The thrust geometry of the windows, described as antiformal stacks, agrees with this model. The Middle Allochthon of the Caledonian margin in northern Sweden is described by *R. O. Greiling* and *R. Kumpulainen* (Heidelberg and Stockholm) who record two distinct metasedimentary units separated by a thick zone of mylonites interpreted as a lateral thrust ramp. Turbidites in the northern unit were derived from an unidentified igneous source to the east and cannot be correlated with other sequences in the Middle Allochthon. In another paper dealing with the northern Swedish Caledonides, *L. Hansen* (Uppsala) describes down-to-the-west normal faults cutting the autochthonous Cambrian sediments in the tunnel sections of the Vietas Hydropower Station, but themselves being truncated by the basal decollement of the Lower Allochthon. It is not clear whether the extensional faults relate to an earlier phase of compression/transcurrence or whether they represent initial synsedimentary rifting structures. Farther to the north in Troms, northern Norway, *M. W. Anderson* (Southampton) analyses the structure and metamorphic grade of a series of tectonic windows through the Upper Allochthon. Assuming a constant metamorphic gradient increasing westwards towards the Iapetus suture, the Rombak Window has been transported at least 45 km as an antiformal stack and the Caledonian thrust front must have lain a further 120 km east-southeast of its present erosional position. The final two papers of the section describe aspects of the regional geology of Finnmark. *C. Townsend, A. H. N. Rice* and *A. Mackay* (Cambridge and Cardiff) extend the knowledge of thin-skinned thrusting of the Lower Allochthon (Gaissa Thrust Belt) to the west of Porsangerfjord and recognize a new stratigraphic group (Ai'roai'vi Group) in an antiformal stack at the trailing branch line of the Gaissa Thrust Belt. The structure is interpreted as an inversion of an original extensional basin, and the Gaissa Thrust Belt has been shortened by at least 65 km in this area. The overlying Middle Allochthon (Kalak Nappe Complex) also shows evidence of extensional faults being re-used as thrusts. *R. A. Gayer* and *A. H. N. Rice* (Cardiff) describe the tectonic controls of deposition in the Finnmark Caledonides. They recognize three sedimentary basins active during the late Riphean–Tremadoc that were controlled by pre- and syn-rifting extensional faults that were later re-activated as thrusts, lateral ramps and a major dextral transcurrent fault (Trollfjord–Komagelv Fault) during the main phase of Caledonian deformation in Finnmark.

The three papers in the third section of the book provide three different approaches to the application of **Igneous Geology**. In the first, *T. F. Emmett* (Cambridge) uses the geochemistry and country rock relations of a set of alkalic and tholeiitic dykes emplaced 900–700 Ma ago into hot orthogneisses of the Jotun Nappe of SW Norway, to argue for crustal thinning and intrusion during early Iapetus rifting of Laurentia/Baltica. In the second, *A. P. Boyle* (Liverpool) presents evidence from the geochemistry of major elements and REE (rare-earth elements) that the Sulitjelma Ophiolite and overlying volcanic rocks in Nordland formed in a marginal ensialic back-arc basin above a west-dipping subduction zone on the Laurentian rather than the Baltica margin. Hence, during this stage of ocean closure Iapetus lay to the east and the entire Sulitjelma sequence represents an allochthonous terrane, thrust onto the Baltic margin during continental collision. In the third example, *J. M. Leaver, M. C. Bennett* and *B. Robins* (Bristol and Bergen) describe nodules from a swarm of alkali olivine dykes intruded during a phase of extension of the Seiland Igneous Province of Finnmark. The nodules, representing xenoliths from the underlying lithospheric mantle at a depth of 45–55 km, indicate that, although both depleted and enriched mantle heterogeneities existed, the prevalent mantle source for the nodules was undepleted, with major implications for the nature and thickness of the mid-Cambrian lithosphere beneath the Seiland Igneous Province.

The **Metamorphic Geology** of an area is directly related to its tectonic evolution. The first paper of this section, by *A. H. N. Rice, R. E. Bevins, D. Robinson* and *D. Roberts* (Cardiff, Bristol and Trondheim), uses low-grade metamorphism in the Finnmark Caledonides. The Lower Allochthon, on metamorphic grounds, is shown to consist of the Gaissa Thrust Belt, the Laksefjord Nappe Complex, the Komagfjord Antiformal Stack and the Barents Sea Caledonides, each deriving their metamorphism from contact beneath the overriding *hot* Middle Allochthon (Kalak Nappe Complex). There is no major sole thrust between the Gaissa Thrust Belt and the East Finnmark Autochthon, suggesting the development of a passive roof duplex in the area. *A. J. Barker* (Southampton) reviews the large amount of new metamorphic data for the Caledonian nappes of north central Scandinavia, where peak metamorphism was achieved during early D2, related to Laurentia/Baltica collision, and late D2 shearing produced inverted isograds. Thrusts developed in D3, emplacing the major nappes and significant post-thrusting retrogression can be documented. Regionally, the metamorphism increases from sub-greenschist facies in the imbricates of the Lower Allochthon at the foreland, to upper amphibolite facies in the highest allochthonous units where maximum temperatures reached 530–650°C and maximum pressures 6–9 kb. Early Caledonian metamorphism is only locally preserved in lenses of eclogite facies, partially retrogressed to amphibolite facies. *H. L. H. van Roermund* (Utrecht) describes two types of such high-pressure metamorphism in ultramafic rocks within the Seve Nappes. The first is interpreted as subcontinental Baltic upper mantle with initially high-pressure assemblages that were later retrogressed, whilst the second was originally lower-pressure ophiolitic ultramafic rocks that were raised to high-pressure conditions by deep crustal ductile imbrication during early Caledonian crustal thickening and later retrogressed in Scandian thrusting. Eclogites are also the subject of the final paper in the section, where *I. Bryhni* (Oslo) discusses the origin of the world's largest eclogite occurrence in the Western Gneiss Region of SW Norway. Here the high-pressure metamorphism is thought to have resulted from late Caledonian continental obduction of Laurentia over Baltica, although the rocks themselves may have originated either as Caledonized Precambrian supracrustals and gneisses, or as Lower Palaeozoic cover rocks.

In the single paper in the section on **Palaeontology and Biostratigraphy**, *D. L. Bruton, D. A. T. Harper* and *J. E. Repetski* (Oslo, Galway and Washington) describe the Cambro–Ordovician faunas of the Parautochthon and Lower Allochthon in S Norway and show that, although the Lower Cambrian sediments were variably developed across the region, by Tremadoc times the Dictyonema Shales were uniformly developed across the entire shelf. Sediments and faunas in the thrust sheets, restored to their pre-thrusting positions, allow the diachronous influx of turbiditic facies during the Arenig to be explained in terms of early Caledonian deformation of higher nappes in the west.

The tectonic significance of the several Devonian basins along the west coast of Norway has focused considerable recent debate. The paper by *A. Chauvet* and *M. Séranne* (Montpellier) in the section on **Devonian Geology** adds a new dimension to this debate. They analyse the microstructural evidence associated with the basins and show that a major shallow dipping extensional shear detachment underlies the basins. This detachment was active during early to mid-Devonian basin-fill, producing synsedimentary shear criteria within the basin. In addition the Precambrian, Western Gneiss Region basement was ductilely deformed to a depth of at least 1 km beneath the basins, effectively reducing the thickness of Caledonian crust duplicated during the end Caledonian continental collision.

The final section of the book, covering **East Greenland Caledonian Geology**, contains five papers describing and interpreting the East Greenland late Proterozoic sedimentary record and its Caledonian deformation. The inclusion of this section in a work on the Scandinavian Caledonides is important, as East Greenland represents Caledonian evolution on the northwest margin of Iapetus in a position directly juxtaposed to Scandinavia. In the first paper, *M. J. Hambrey* (Cambridge) outlines the late Proterozoic sedimentary record of East Greenland and introduces the remaining four contributions. *P. M. Herrington* and *I. J. Fairchild* (Birmingham) discuss carbonate shelf and slope environments recorded in the pre-Vendian Upper Limestone–Dolomite ''series'' of the Eleonore Bay Group, and conclude that slope instability may reflect tectonic movements. *I. J. Fairchild* (Birmingham) describes an unusual fine shallowing-upwards sequence between the Varangian Tillites and Early Cambrian sediments that is characterized by well-developed stromatolites and evaporites. The sequence is comparable with a similar (but not coeval) one in the Scottish Dalradian. *A. C. M. Moncrieff* (Cambridge) documents the Varangian tillites in the Central East Greenland fjord region that lie above a 13 km-thick Riphean sequence, and correlates them with isolated tillite occurrences resting directly on early Proterozoic basement to the southwest. Thrusts present within the latter suggest that the Caledonian foreland lay to the west, beneath the Greenland ice cap. Vendian palaeogeography is interpreted as a basin in the fjord region with a large land mass to the southwest. The facies interpretation suggests the development of sub-polar ice sheets near sea level, implying a major glaciation. In the final paper, *G. M. Manby* and *M. J. Hambrey* (London and Cambridge) describe for the first time details of westward-directed thrusting, not only in the fjord region of central East Greenland but in the so-called ''foreland'' windows. No estimates of shortening or translation distances are given, but it is considered that no truly autochthonous foreland sequences have yet been observed west of the central East Greenland fold belt. Evidence of late Caledonian sinistral transcurrence is apparent but producing far less intense and pervasive deformation than the period of main Caledonian westward thrusting.

I would like to express my grateful thanks to all those who have made this book possible; in particular to the authors who kept strictly to a tight publication schedule, not only producing the original articles on time but also responding rapidly to the helpful, constructive criticisms of the referees. I should also like to thank the many unnamed referees who also worked hard within a short time-schedule to raise the standard of science of the articles. Most of all I must thank Graham and Trotman's senior editor, Arthur Valenzuela, who patiently answered many queries and quietly encouraged me to stick to the original deadlines; without his help at all stages of the preparation, this book would not have materialized.

R. A. Gayer
Department of Geology,
University of Wales,
Cardiff, UK
February 1989.

Part I

Finnmarkian Geology

1 U–Pb ages of nepheline syenite pegmatites from the Seiland Magmatic Province, N Norway

R. B. Pedersen*, G. R. Dunning† and B. Robins*

*Geologisk Institutt, avd. A, Allégt. 41, 5007 Bergen – Universitetet, Norway
†Department of Geology, Royal Ontario Museum, 100 Queen's Park, Toronto, Ontario, Canada M5S 2C6

Zircons from two pegmatites belonging to the alkaline magmatic suite in the Seiland Province have yielded concordant U–Pb ages of 531 ± 2 and 523 ± 2 Ma. These ages replace an earlier estimate of 490 ± 27 Ma for the crystallization of the alkaline rocks and provide a new, younger age limit for the Finnmarkian Orogeny. Deductions regarding the environment of emplacement of the alkaline suite suggest that the principal tectonothermal development of the Seiland Province was terminated by Middle Cambrian uplift. Published K–Ar and $^{40}Ar–^{39}Ar$ mineral ages suggest that the Seiland Province was subsequently affected by separate Late Cambrian–Early Ordovician and Silurian tectonothermal cycles. Obduction of the oldest ophiolite complexes known in the Scandinavian Caledonides and the deformation and metamorphism that pre-dated the alkaline magmatism in the Seiland Province cannot have been contemporaneous.

INTRODUCTION

The Seiland Magmatic Province is developed within the uppermost (Sørøy) nappe of the Kalak Nappe Complex, which constitutes part of the Middle Allochthon of the North Norwegian Caledonides (Roberts and Gee, 1985). Its evolution was complex and involved the emplacement of gabbroic plutons, calc-alkaline intrusions, ultramafic complexes, alkaline rocks, carbonatites and swarms of mafic dykes (Robins and Gardner, 1975). The igneous rocks are emplaced in a sequence of metasediments (the "Sørøy Succession") of presumed Late Proterozoic–Middle Cambrian age and their gneiss basement (Ramsay *et al.*, 1985). They are generally believed to have been intruded during polyphasal regional deformation and concurrent Barrovian metamorphism initially termed the Finnmarkian Phase of the Caledonian Orogeny (Sturt *et al.*, 1978; Ramsay *et al.*, 1985), but recently elevated in rank to the Finnmarkian Orogeny (Ramsay and Sturt, 1986). The latter has been loosely constrained by Rb–Sr whole-rock isochrons from the intrusives and their contact-metamorphic aureoles to have occurred between ca. 540 and ca. 490 Ma (Sturt *et al.*, 1978).

Alkaline rocks, carbonatites and mafic dykes are the youngest igneous rocks in the Seiland Province. Field relationships in the Lillebukt Alkaline Complex, Stjernøy (Robins, 1980), and elsewhere, demonstrate unequivocally that nepheline syenite pegmatites, distributed widely on Seiland, Sørøy, Stjernøy and the northwestern part of the Øksfjord Peninsula, represent the very latest phase of the alkaline magmatic activity. They nevertheless appear to be syn-orogenic in that they were intruded into previously deformed rocks and were themselves subsequently deformed.

This contribution presents U–Pb data on zircon crystals and a crystal of sphene collected from two nepheline syenite pegmatite dykes, one from Seiland and the other from the Lillebukt Alkaline Complex, and discusses the implications of their U–Pb ages for the duration of Seiland magmatism and the timing of the Finnmarkian Orogeny.

GEOLOGICAL RELATIONSHIPS OF THE SAMPLED PEGMATITES

One of the zircon crystals analysed was collected from the margin of a member of a swarm of nepheline syenite pegmatite dykes that extend more than 2 km from the eastern shore of Store Kufjord into the interior of the island of Seiland (Fig. 1). The pegmatites are up to 25 m wide, lenticular and commonly organized into *en echelon* strings. They cross-cut deformed and metamorphosed olivine gabbro, gabbronorite, syenodiorite and perthosite within the layered Seiland Syenogabbro of Robins and Gardner (1974) as well as the sheeted contact, Marginal Zone and Central Zone of the younger Melkvann Ultramafic Com-

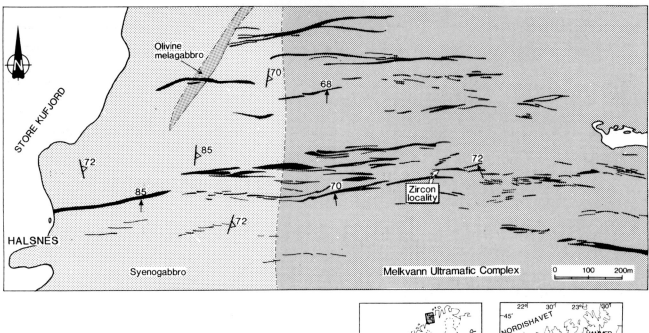

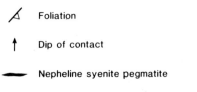

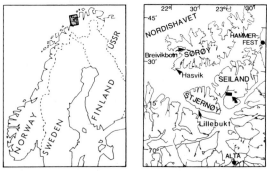

Fig. 1.   Geological map of the Store Kufjord swarm of nepheline syenite pegmatites, Seiland, showing the locality from which one of the analysed zircons was collected. The inset map shows the location of the mapped area as well as that of Lillebukt, Stjernøy, the source of the other analysed zircon.

plex (Bennett *et al.*, 1986). In addition, pegmatites cut across a dyke of olivine melagabbro emplaced during the later stages of evolution of the Melkvann Ultramafic Complex.

The pegmatites generally show little sign of strain but are locally transformed into gneiss and in narrow zones of intense deformation are reduced to mylonite. Foliations in the deformed domains are subparallel to dyke margins. The pegmatites have a simple mineralogy and consist principally of alkali feldspar (antiperthite), nepheline and biotite with minor amounts of Fe–Ti oxides, apatite, corundum and calcite.

Zircon is normally an accessory phase in the pegmatites but is a major mineral at the sampled locality (Fig. 1), where it forms euhedra up to 5 cm long set within a matrix of biotite, Fe–Ti oxides and alkali feldspar. Here, the zircon-rich rocks form a narrow zone immediately within the southern contact of the 120 m long and up to 10 m wide pegmatite with the adjacent, strongly-fenitized olivine gabbro (Robins and Tysseland, 1983).

The other crystal of zircon analysed and a sphene were collected from a nepheline syenite pegmatite within the Lillebukt Alkaline Complex, Stjernøy (Heier, 1961) (for location, see inset in Fig. 1). In this complex, pegmatites were emplaced into hornblende pyroxenite, alkali syenite, nepheline syenite and carbonatite as well as the enveloping metagabbroic rocks and their metasomatic derivatives (Robins, 1980). The pegmatites postdated swarms of mafic dykes that cut the other alkaline rocks and carbonatite but were emplaced before deformation of the complex (Skogen, 1980). Rare, little-metamorphosed mafic dykes

that cut folds and tectonic foliations are the only rocks in the complex that were intruded after the pegmatites.

Zircon and sphene are normally accessory minerals in the Lillebukt pegmatites. The crystals analysed, however, were collected from a short pegmatite dyke where zircon occurs as large, fractured euhedra together with more abundant sphene. This pegmatite is emplaced in inter-banded nepheline syenite and metasomatized gabbroic rocks that are locally foliated and/or folded.

ANALYTICAL TECHNIQUES AND RESULTS

Two fractions of each of the euhedral, 3 cm-long, trans-parent-to-translucent zircons from the Seiland and Stjernøy nepheline syenite pegmatites were analysed. One fraction from each crystal was abraded using the pro-cedure of Krogh (1982). All samples were washed in distilled water, hot 6N HNO_3 and distilled acetone. They were spiked with a mixed $^{205}Pb/^{235}U$ spike (Krogh and Davis, 1975) and processed according to the procedure of Krogh (1973). Lead and uranium were loaded onto a single rhenium filament with H_3PO_4 and silica gel and analysed on VG 354 and VG 30 mass spectrometers housed in the laboratories of the Royal Ontario Museum.

Analytical results are presented in Table I and Fig. 2. Both fractions of the Seiland zircon yield identical $^{207}Pb/$ ^{206}Pb ages of 531 Ma and the abraded fraction was concordant within analytical uncertainties. The best esti-

Table I. U–Pb data for Seiland Magmatic Province

Fractions		Concentrations		Measured[b]			Atomic ratios[c]				Age (Ma)
No* Properties	Wt (mg)[a]	U	Pb rad (ppm)	Total common Pb (pg)	$^{206}Pb/^{204}Pb$	$^{208}Pb/^{206}Pb$	$^{206}Pb/^{238}U$	$^{207}Pb/^{235}U$	$^{207}Pb/^{206}Pb$		$^{207}Pb/^{206}Pb$
1–1 Clear fragments abraded	1.071	60	5.8	5	61247	0.254	0.08550 ± 16	0.6842 ± 14	0.05804 ± 4		531
1–2 One angular shard	2.475	57	5.5	47	15483	0.252	0.08506 ± 29	0.6807 ± 23	0.05804 ± 6		531
2–1 Angular shards abraded	5.742	7.4	1.7	65	3309	2.058	0.08393 ± 22	0.6692 ± 18	0.05782 ± 5		523
2–2 Clear shards	1.653	6.9	1.4	48	1228	1.761	0.08413 ± 29	0.6704 ± 24	0.05780 ± 8		522
2–3 Sphene frags abraded	1.241	15	6.4	2196	62	4.681	0.08400 ± 26	0.6672 ± 77	0.05761 ± 56		515

Notes:
[a] Uncertainty in weight ± 0.003 mg (1 σ).
[b] Corrected for fractionation and spike.
[c] Corrected for fractionation and spike, 5–10 pg Pb blank and initial common lead calculated from the model of Stacey and Kramers (1975) and 2–5 pg U blank.
   2σ uncertainties on the isotopic ratios calculated using the procedure of Ludwig (1980).
*Fractions numbered 1–1 and 1–2 are from the Seiland zircon.
 Fractions numbered 2–1 and 2–2 are from the Stjernøy zircon.
 Fraction 2–3 is from the Stjernøy sphene.

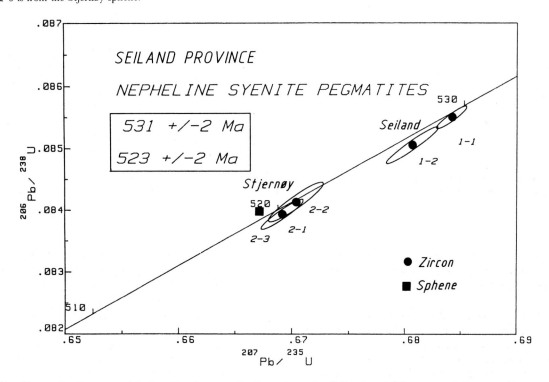

Fig. 2. Concordia diagram with data for zircon and sphene from the Seiland and Stjernøy nepheline syenite pegmatites.

mate of the age of crystallization is therefore 531 ± 2 Ma (2σ).

The two fractions of the Stjernøy zircon have lower uranium contents than the Seiland zircon (Table I). They give $^{207}Pb/^{206}Pb$ ages of 523 and 522 Ma. Both fractions are concordant within analytical error (Fig. 2) and give a crystallization age of 523 ± 2 Ma (2σ).

Sphene from the Stjernøy pegmatite has a high content of common lead (Table I). This results in large uncertainties in $^{207}Pb/^{206}Pb$ and $^{207}Pb/^{235}U$ ratios. The $^{206}Pb/^{238}U$ age of 520 ± 5 Ma (2σ) (Fig. 2) is the only reasonably reliable age that may be extracted from the analytical data for this mineral. This age is indistinguishable from that of the co-existing zircon.

The U–Pb ages for the Seiland and Stjernøy zircons differ by 8 ± 4 Ma (2σ). Since the uncertainties in the U–Pb ages do not overlap even at the 3σ level, it seems reasonable to conclude that there were real differences in the time of emplacement of the Seiland and Stjernøy

pegmatites. The authors are not aware, however, of field relationships that indicate the existence of different generations of pegmatites.

DISCUSSION

Previous geochronological work carried out on rocks from the Seiland Province are summarized and compared with the new U–Pb ages in Fig. 3. With one exception, Rb–Sr whole-rock and Sm–Nd isochron ages of rocks that pre-date nepheline syenite pegmatites in the field are within error of the new ages. The exception is an Rb–Sr whole-rock isochron age of 490 ± 27 Ma (2σ) (Sturt et al., 1978) on a fine-grained syenitic dyke from the Breivikbotn Alkaline Complex, Sørøy (Sturt and Ramsay, 1965) (for location see inset in Fig. 1), which is 33 Ma younger than the youngest zircon U–Pb age reported in this paper. At extreme error limits the two ages differ by only 4 Ma, but

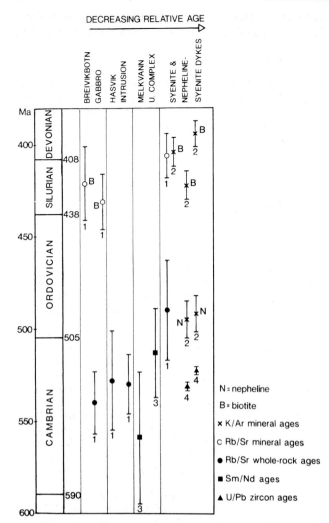

Fig. 3. Summary of published geochronological data from the Seiland Province. Data are taken from: (1) Sturt *et al.* (1978); (2) Sturt *et al.* (1967); (3) Snow *et al.* (1986); (4) this work. Rb–Sr and K–Ar data have been recalculated in accordance with the recommendations of Steiger and Jäger (1977) and Dalrymple (1979). The time-scale is after Harland *et al.* (1982).

since similar syenite dykes in the Lillebukt Alkaline Complex are known to pre-date nepheline syenite pegmatites, the U–Pb and Rb–Sr ages seem to be incompatible. On the basis of the concordance of the U–Pb ratios for the zircons, we favour 531–523 Ma as the best estimate of the true age of the youngest alkaline magmatism within the Seiland Province. The ca. 490 Ma Rb–Sr isochron is regarded as an unreliable estimate of the time of crystallization of the Breivikbotn syenite.

Brueckner (1973) dated a syenite (perthosite) from the Seiland Province by a three-point Rb–Sr whole-rock isochron that yielded an age of 614 ± 135 Ma (recalculated to the decay constant recommended by Steiger and Jäger, 1977) (not shown in Fig. 3). Including data for nearby gabbroic and peridotitic rocks in the regression, the uncertainties were reduced to 612 ± 17 Ma. These rocks have not, however, been demonstrated to be cogenetic with the syenite and their inclusion in an isochron serves only to constrain the initial Sr isotopic ratio to a possible, but not necessarily correct, value. We conclude that there are at present no data that demonstrate unequivocally that the Seiland Province had been magmatically active for a substantial period of time prior to 531 Ma.

The concept of the Finnmarkian Orogeny is closely linked to the structural, metamorphic and magmatic evolution of the Seiland Province (Sturt *et al.*, 1978;

Ramsay and Sturt 1986). According to Sturt and Ramsay (1965) and Sturt *et al.* (1978), the alkaline rocks of the province were both intruded and deformed during the latest stages of D2, the youngest phase of Finnmarkian deformation in the Sørøy Nappe. The 523–531 Ma ages for the alkaline magmatism therefore define a new, younger age limit for the Finnmarkian Orogeny, as envisaged by Sturt *et al.* (1978), close to the Middle–Late Cambrian boundary in the Harland *et al.* (1982) time scale. However, the new U–Pb data preclude the correlation of the youngest magmatic events in the Seiland Province with cleavage formation in the Laksefjord Nappe and Parautochthon suggested by Sturt *et al.* (1978) and Ramsay and Sturt (1986).

Nepheline from nepheline syenite pegmatites has yielded K–Ar ages of ca. 490 Ma (Fig. 3), and these have been interpreted as dating passage through the nepheline "closure temperature" during uplift and erosion of the Finnmarkian Orogen (Sturt *et al.*, 1978, Ramsay and Sturt, 1986). Hornblende from basement samples from the Sørøy Nappe also give $^{40}Ar/^{39}Ar$ plateau ages of 490 ± 5 Ma (Dallmeyer, 1988). However, nepheline and hornblende from the alkaline rocks of the Seiland Province exhibit $^{40}Ar/^{39}Ar$ plateau ages of ca. 431–425 Ma (Dallmeyer, 1988) that are compatible with the K–Ar biotite ages and the Rb–Sr biotite age reported from similar rocks by Sturt *et al.* (1967, 1978). The $^{40}Ar/^{39}Ar$ and K–Ar ages may reflect superimposition of separate pre-Early Ordovician and Silurian tectonothermal cycles (Dallmeyer, 1988).

Krill and Zwaan (1987) questioned earlier interpretations of field relationships on which the concept of the Finnmarkian Orogeny was based and suggested that the igneous rocks of the Seiland Province were pre-orogenic rather than synorogenic. They proposed that the magmatic activity was associated with crustal attenuation and the formation of a rifted margin to the Baltoscandian continent. A similar hypothesis has also been advanced by Andréasson (1987). Krill and Zwaan (1987) regard the regional deformation and metamorphism in the Seiland Province as possibly Silurian in age. This is clearly not the case. Structural relationships in the area shown in Fig. 1 demonstrate unequivocally that the nepheline syenite pegmatite dykes were intruded into rocks that had been both deformed and metamorphosed at an earlier stage in the plutonic development of the province. Furthermore, the olivine gabbro envelope to the Melkvann Ultramafic Complex, cut by the dyke swarm shown in Fig. 1, has yielded Sm–Nd internal isochron ages of 559 ± 36 Ma and 513 ± 24 Ma (Snow *et al.*, 1986), indistinguishable from the U–Pb ages reported here (Fig. 3). Tectonic fabrics in the Breivikbotn (meta) Gabbro, emplaced at 540 ± 17 Ma (Sturt *et al.*, 1978) (Fig. 3), are also cut by nepheline syenite pegmatites (Sturt and Ramsay 1965). These, and similar relationships that exist at many other places in the province, are conclusive evidence that emplacement of the alkaline magmatic suite took place after the principal tectonothermal development of the Seiland Province.

The nature of the relationship between the pre-531 Ma deformation and metamorphism and Early Ordovician cooling of the Seiland Province through the hornblende "closure temperature" is not explicit. Interpretation of the Lillebukt Alkaline Complex as a deformed and metamorphosed ring complex (Skogen, 1980) does, however, suggest that the alkaline suite, in contrast with the older intrusions in the province, was emplaced at a relatively shallow crustal level. Recent work supports this view and leads the authors to the conclusion that the alkaline rocks and carbonatites were intruded after considerable uplift and unroofing, probably in an extensional tectonic regime. This implies that the Finnmarkian Orogeny was terminated in the Middle Cambrian. Deformation of the alkaline suite and their host rocks during upper greenschist facies metamorphism (Sturt and Ramsay 1965; Ramsay and Sturt, 1970) may have taken place in a separate Late Cambrian–Early Ordovician tectothermal cycle, as evidenced by K–Ar and ^{40}Ar/^{39}Ar data, rather than during a continuous strain history as inferred earlier from fold geometries (Sturt and Ramsay, 1965).

In several interpretations of the tectonothermal evolution of the Scandinavian Caledonides, considerable weight has been placed on relatively imprecise Rb–Sr whole-rock isochron ages from the Seiland Province (Sturt *et al.*, 1978; Sturt, 1984; Ryan and Sturt 1985; Ramsay and Sturt, 1986). The D2 of the Kalak Nappe Complex in Finnmark has been attributed to a major compressional deformation that involved the emplacement of ophiolites onto the Baltoscandian continental margin (Sturt *et al.*, 1984; Furnes *et al.*, 1985; Ramsay *et al.*, 1985; Ramsay and Sturt, 1986). The first tectonothermal events affecting Norwegian ophiolite complexes and their cover sequences, such as the Torvastad Group on Karmøy, have therefore been ascribed to the Finnmarkian Orogeny (Sturt, 1984). U–Pb geochronology on several of the ophiolite complexes has shown, however, that the oldest (Karmøy, Gulfjell and Leka) formed at ca. 490 Ma (Dunning and Pedersen, 1988). These ages exclude any connection between the obduction of the ophiolite complexes and the deformation and metamorphism that pre-dated alkaline magmatism in the Seiland Province.

CONCLUSIONS

U–Pb ages of 531 ± 2 Ma and 523 ± 2 Ma on zircons from two nepheline syenite pegmatites, one from a dyke swarm on Seiland and the other from the Lillebukt Alkaline Complex, Stjernøy, show that alkaline rocks in the Seiland Province crystallized at a time close to the Middle–Late Cambrian boundary in the calibration of Harland *et al.* (1982), and not in the Early Ordovician as believed earlier. The new U–Pb data give a new age constraint for "Finnmarkian" deformation and metamorphism, as originally defined by Sturt *et al.* (1978) on the basis of geological relationships in the Sørøy Nappe, but preclude correlation of the youngest magmatic events in the Seiland Province with cleavage formation in the Laksefjord Nappe Complex and the parautochthon as suggested by Sturt *et al.* (1978) and Ramsay and Sturt (1986). Interpretation of the Lillebukt Alkaline Complex as a ring complex (Skogen, 1980) implies emplacement of the alkaline rocks and carbonatites at a shallow crustal level after late "Finnmarkian" uplift and unroofing of the Seiland Province, probably in an extensional tectonic environment. Since the alkaline suite was later exposed to deformation, it is inferred that the tectonic and metamorphic evolution of the Seiland Province was not only polyphasal, as recognized earlier, but was polycyclic. Deformation and metamorphism of the alkaline suite and their host rocks probably took place during a separate and hitherto unrecognized Late Cambrian–Early Ordovician tectonometamorphic cycle as suggested by K–Ar and ^{40}Ar/^{39}Ar ages of ca. 490 Ma on nepheline (Sturt *et al.*, 1967) and hornblende (Dallmeyer, 1988).

Zircon U–Pb ages show that the oldest ophiolites in the Norwegian Caledonides were generated at ca. 490 Ma (Dunning and Pedersen, 1988). Obduction of these ophiolites and the deformation and metamorphism that pre-

dated the emplacement of the alkaline suite in the Seiland Province cannot, therefore, have been contemporaneous.

Acknowledgments

Funding for this project was provided by a grant from the Norwegian Research Council for Science and the Humanities (N.A.V.F.). The authors are grateful to Dr. J. S. Daly for his helpful comments on an earlier version of the manuscript and to E. Irgens who prepared the illustrations. This is Norwegian contribution no. 38 to the International Lithosphere Programme.

REFERENCES

Andréasson, P.-G. (1987). Early evolution of the Late Proterozoic Baltoscandian Margin: inferences from rift magmatism. *Geol. Fören. Stockh. Förh.* **109**, 336–340.

Bennett, M. C., Emblin, S. R., Robins, B. and Yeo, W. J. A. (1986). High-temperature ultramafic complexes in the North Norwegian Caledonides: I—Regional setting and field relationships. *Norges Geol. Unders.* **405**, 1–40.

Brueckner, H. K. (1973). Reconnaissance Rb–Sr investigation of salic, mafic and ultramafic rocks in the Øksfjord area, Seiland Province, Northern Norway. *Norsk Geol. Tidsskr.* **53**, 11–23.

Dallmeyer, R. D. (1988). Polyorogenic ^{40}Ar/^{39}Ar mineral age record within the Kalak Nappe Complex, Northern Scandinavia. *Geol. Soc. Lond. J.*, **145**, 705–716.

Dalrymple, G. B. (1979). Critical tables for conversion of K–Ar ages from old to new constants. *Geology* **7**, 558–560.

Dunning, G. R. and Pedersen, R. B. (1988). U/Pb ages of ophiolites and arc-related plutons of the Norwegian Caledonides: Implications for the development of Iapetus. *Contrib. Mineral. Petrol.* **98**, 13–23.

Furnes, H., Ryan, P. D., Grenne, T., Roberts, D., Sturt, B. A. and Prestvik, T. (1985). Geological and geochemical classification of the ophiolitic fragments in the Scandinavian Caledonides. In: Gee, D. G. and Sturt, B. A. (eds.) *The Caledonian Orogen—Scandinavia and Related Areas*, pp. 657–670. Wiley, Chichester.

Harland, W. B., Cox, A. V., Llewellyn, P. G., Picton, C. A. G., Smith, A. G. and Walters, R. (1982). *A Geological Time-scale*, p. 131. Cambridge University Press, Cambridge.

Heier, K. S. (1961). Layered gabbro, hornblendite, carbonatite and nepheline syenite on Stjernøy, North Norway. *Norsk Geol. Tidsskr.* **41**, 109–155.

Krill, A. G. and Zwaan, B. (1987). Reinterpretation of Finnmarkian deformation on Western Sørøy, northern Norway. *Norsk Geol. Tidsskr.* **67**, 15–24.

Krogh, T. E. (1973). A low contamination method for hydrothermal decomposition of zircon and extraction of U and Pb for isotopic determinations. *Geochim. Cosmochim. Acta* **37**, 485–494.

Krogh, T. E. (1982). Improved accuracy of U–Pb zircon ages by the creation of more concordant systems using an air abrasion technique. *Geochim. Cosmochim. Acta* **46**, 637–649.

Krogh, T. E. and Davis, G. L. (1975). The production and preparation of ^{205}Pb for use as a tracer for isotope dilution analyses. *Carnegie Institute of Washington, Yearbook 74*, pp. 416–417.

Ludwig, K. R. (1980). Calculation of uncertainties of U–Pb isotopic data. *Earth Planet. Sci. Lett.* **46**, 212–220.

Ramsay, D. M. and Sturt, B. A. (1970). The emplacement and metamorphism of a synorogenic dyke swarm from Stjernøy, northwest Norway. *Amer. J. Sci.* **268**, 264–268.

Ramsay, D. M. and Sturt, B. A. (1986). The contribution of the Finnmarkian Orogeny to the framework of the Scandinavian Caledonides. In: Fettes, D. J. and Harris, A. L. (eds.) *Synthesis of the Caledonian Rocks of Britain*, pp. 221–246. Reidel, Dordrecht.

Ramsay, D. M., Sturt, B. A., Zwaan, K. B. and Roberts, D. (1985). Caledonides of northern Norway. In: Gee, D. G. and Sturt, B. A. (eds.) *The Caledonian Orogen—Scandinavia and Related Areas*, pp. 163–184. Wiley, Chichester.

Roberts, D. and Gee, D. G. (1985). An introduction to the structure of the Scandinavian Caledonides. In: Gee, D. G. and Sturt, B. A. (eds.) *The Caledonian Orogen—Scandinavia and Related Areas*, pp. 55–68. Wiley, Chichester.

Robins, B. (1980). The evolution of the Lillebukt alkaline complex, Stjernøy, Norway (abs.). *Lithos* **13**, 219–220.

Robins, B. and Gardner, P. M. (1974). Synorogenic layered basic intrusions in the Seiland Petrographic Province, Finnmark. *Norges Geol. Unders.* **312**, 91–130.

Robins, B. and Gardner, P. M. (1975). The magmatic evolution of the Seiland Province, and Caledonian plate boundaries in northern Norway. *Earth Planet. Sci. Lett.* **26**, 167–178.

Robins, B. and Tysseland, M. (1983). The geology, geochemistry and origin of ultrabasic fenites associated with the Pollen carbonatite (Finnmark, Norway). *Chem. Geol.* **40**, 65–95.

Ryan, P. D. and Sturt, B. A. (1985). Early Caledonian orogenesis in northwestern Europe. In: Gee, D. G. and Sturt, B. A. (eds.) *The Caledonian Orogen—Scandinavia and Related Areas*, pp. 1227–1239. Wiley, Chichester.

Skogen, J. H. (1980). The structural evolution of the Lillebukt carbonatite, Stjernøy, Norway (abs.). *Lithos* **13**, 221.

Stacey, J. S. and Kramers, J. D. (1975). Approximation of terrestrial lead isotope evolution by a two-stage model. *Earth Planet. Sci. Lett.* **26**, 207–221.

Steiger, R. H. and Jäger, E. (1977). Subcommission on geochronology: convention on the use of decay constants in geo- and cosmochronology. *Earth Planet. Sci. Lett.* **36**, 359–362.

Snow, J., Bennett, M. C., Yeo, W. J. A., Basu, A. R. and Tatsumoto, M. (1986). The Melkvann complex, N. Norway: Nd and Sr isotopic evidence for the origin of Alaskan-type ultramafic complexes (abs.). *ICOG IV Symposium abstracts, Terra Cognita*.

Sturt, B. A. (1984). The accretion of ophiolitic terrains in the Scandinavian Caledonides. *Geologie en Mijnbouw* **16**, 201–212.

Sturt, B. A. and Ramsay, D. M. (1965). The alkaline complex of the Breivikbotn area. Sørøy, northern Norway. *Norges Geol. Unders.* **231**, 1–143.

Sturt, B. A., Miller, J. A. and Fitch, F. J. (1967). The age of alkaline rocks from west Finnmark, northern Norway, and their bearing on the dating of the Caledonian orogeny. *Norsk Geol. Tidsskr.* **44**, 255–273.

Sturt, B. A., Pringle, I. R. and Ramsay, D. M. (1978). The Finnmarkian phase of the Caledonian Orogeny. *Geol. Soc. Lond. J.* **135**, 547–610.

Sturt, B. A., Roberts, D. and Furnes, H. (1984). A conspectus of Scandinavian Caledonian ophiolites. In: Gass, I. G., Lippard, S. J. and Shelton, A. W. (eds.) *Ophiolites and Oceanic Lithosphere, Geol. Soc. Lond. Special Publ.* **13**, pp. 381–391.

2 Chronology of Caledonian tectonothermal activity within the Gaissa and Laksefjord Nappe Complexes (Lower Allochthon), Finnmark, Norway: Evidence from K–Ar and ^{40}Ar/^{39}Ar ages

R. D. Dallmeyer*, A. Reuter*, N. Clauer† and N. Liewig†

*Department of Geology, University of Georgia, Athens, GA 30602, USA
†Centre de Sédimentologie et de Géochimie de la Surface, C.N.R.S., 1 rue Blessig, 67084 Strasbourg, Cedex, France

Determination of quartz-normalized illite crystallinity of bulk $<2\,\mu m$ size fractions isolated from nine samples of penetratively cleaved metapelite (slate/phyllite) from the Lower Allochthon of Finnmark suggest that a very low-grade metamorphism is recorded (transitional middle–upper anchizone within the Gaissa Nappe Complex and upper anchizone within the Laksefjord Nappe Complex). White mica-enriched fractions ranging in grain size from $<0.5\,\mu m$ to $4-8\,\mu m$ were separated from three samples. K–Ar apparent ages of these size fractions display a marked positive correlation with grain size, reflecting a detrital influence that diminishes with decreasing grain size and increasing anchizonal metamorphism. The $<0.5\,\mu m$ size fraction isolated from one of the samples of the Gaissa Nappe Complex records an apparent K–Ar age of $440.2 \pm 9.4\,Ma$. This age is suggested to represent a maximum date for cleavage formation and associated very low-grade metamorphism. Microstructural characteristics suggest the cleavage originated largely through pressure-solution-assisted rotation of detrital mica grains without significant neoformation or recrystallization.

^{40}Ar/^{39}Ar age spectra of the nine whole-rock samples display variable internal discordance, recording total-gas ages between $473.8 \pm 8.6\,Ma$ and $550.5 \pm 9.3\,Ma$. Consistent intrasample variations in apparent K/Ca ratios suggest spectra discordance in part reflects experimental gas release from different mineralogical constituents with contrasting diffusive characteristics (chlorite, white mica and detrital feldspar). Relatively larger and constant apparent K/Ca ratios are recorded within most intermediate-temperature gas fractions and suggest experimental gas evolution dominantly from white mica. Apparent ages systematically increase with extraction temperature, perhaps reflecting an episodic, partial rejuvenation of detrital mica during the anchizonal metamorphic overprint. The extent of rejuvenation appears most extensive in the upper anchizone sample from the Laksefjord Nappe. ^{40}Ar/^{39}Ar age spectra of the 1–2 μm size fractions show systematic internal discordance probably owing to extensive recoil redistribution of ^{39}Ar during irradiation.

The K–Ar and ^{40}Ar/^{39}Ar ages suggest that deformation and metamorphism of the Lower Allochthon in Finnmark occurred in the Late Ordovician–Early Silurian. There is no evidence for the earlier (Late Cambrian–Early Ordovician) Caledonian tectonothermal activity recorded within the structurally overlying Kalak Nappe Complex.

INTRODUCTION

The Caledonide orogen in Scandinavia is largely represented by a sequence of nappe complexes that were emplaced onto the Baltoscandian platform during early to middle Paleozoic closure of the Iapetus Ocean (e.g., Stephens and Gee, 1985). The nappes are of varying metamorphic grade and display marked differences in the complexity of internal deformation. As a result, correlation of specific metamorphic, deformational, and thrusting events along the length of the orogen has been uncertain and controversial (e.g., Gee and Roberts, 1983).

Lower tectonic units within central and southern portions of the Scandinavian Caledonides are represented by low-grade, metasedimentary successions of late Proterozoic to early Silurian age. The older successions originated along the Baltoscandian continental margin during separation of Baltica and Laurentia. Younger sequences were deposited during closure of the Iapetus oceanic tract. Significant thicknesses of *westerly* derived clastic material of Middle Ordovician and Early Silurian age in these successions provide evidence of two periods of outboard instability along the Baltoscandian margin (e.g., Gee, 1975b). Early attempts to describe the tectonic evolution within central portions of the Caledonian orogen noted this instability (e.g., Gee, 1975a); however, most of the internal deformation, metamorphism, and nappe translation has been considered to be of Late Silurian through Early Devonian (Scandian) age (e.g., Gee and Roberts, 1983).

In the Finnmark area of northernmost Norway, deformed and variably metamorphosed miogeoclinal sedimentary sequences within one of the higher allochthons (Sørøy Nappe of the Kalak Nappe Complex) are intruded by plutons recording ca. 530–490 Ma Rb–Sr whole-rock dates (all radiometric ages discussed in this study are based on the decay constants and isotopic abundance ratios presented by Steiger and Jäger, 1977) that have been interpreted as reflecting igneous crystallization ages (Sturt et al., 1975, 1978). Because the plutons were considered syntectonic by Sturt et al. (1975, 1978), a major period of Middle Cambrian–Early Ordovician polyphase folding, metamorphism and nappe imbrication was suggested (the Finnmarkian Orogeny of Sturt et al., 1975, 1978).

The Middle Cambrian–Early Ordovician tectonothermal event recorded within metasedimentary sequences of the Kalak Nappe Complex contrasts markedly with unbroken late Proterozoic through early Silurian autochthonous/parautochthonous sequences in the central Scandinavian Caledonides. The suggested diachronism of tectonothermal activity (Finnmarkian to Scandian) *along* the early to middle Paleozoic continental margin of Baltoscandia has been difficult to reconcile with otherwise apparently continuous sedimentological, stratigraphic, paleontologic, and tectonic trends. Autochthonous sedimentary sequences in the Finnmark area only range up into the early Cambrian (Føyn, 1985), providing little chronological control for translation of overlying Caledonian nappes. Baltoscandian miogeoclinal successions within the Lower Allochthon of Finnmark only extend into the Tremadoc (Henningsmoen, 1961) and also offer little control for the timing of Caledonian tectonothermal events. Pringle (1973) reported an Rb–Sr whole-rock "isochron" age of 504 ± 7 Ma for cleaved slates within what has been mapped by Chapman et al. (1985) as the Gaissa Nappe Complex (initially described by Pringle as autochthonous units). An Rb–Sr whole-rock "isochron" age of 503 ± 45 Ma was also reported for cleaved slates of the Laksefjord Nappe Complex by Sturt et al. (1978). Although both "isochrons" were interpreted as dating cleavage formation, earlier work by Clauer (1974) and Gebauer and Grünenfelder (1974) in similar very low-grade metamorphic terranes had clearly documented the difficulty in establishing complete isotopic equilibration within complex detrital and diagenetic mineral assemblages during cleavage formation. Therefore, the geological significance of the Rb–Sr results reported for the Lower Allochthon of Finnmark must be questioned.

In view of the uncertain chronology of folding and metamorphism within the Finnmark Lower Allochthon, a suite of penetratively cleaved slate/phyllite samples was collected within the Gaissa and Laksefjord Nappe Complexes. Microstructural characteristics of these samples have been evaluated and various grain-size fractions have been prepared. These, together with whole-rock samples, have been analysed by conventional K–Ar and incremental-release $^{40}Ar/^{39}Ar$ techniques in an attempt to date the cleavage and concomitant very low-grade metamorphism. These results are presented here and they more clearly define the extent and intensity of Caledonian tectonothermal events in the northern Scandinavian Caledonides.

REGIONAL GEOLOGICAL SETTING

The regional geological setting of the Finnmark Caledonides has been summarized by Roberts (1985). Autochthonous sequences, which are exposed south and east of the Caledonian orogenic front (Fig. 1), include a crystalline Proterozoic basement with an unconformably overlying late Proterozoic through Lower Cambrian clastic sedimentary cover. The cover units record a progressive westward development of Caledonian cleavage and an associated very low-grade regional metamorphism (diagenesis transitional into the lower anchizone: Bevins et al., 1986; Dallmeyer and Reuter, 1988). Attempts to date this cleavage have yielded inconclusive results (Dallmeyer and Reuter, 1988).

The Gaissa Nappe Complex represents the lowest Caledonian structural unit in Finnmark (Fig. 1). It contains an internally imbricated, very low-grade (lower to middle anchizone: Bevins et al., 1986), variably cleaved, Riphaen–Tremadoc metasedimentary succession with lithostratigraphic characteristics similar to those of the autochthonous cover units of eastern Finnmark (e.g., Føyn, 1985). The structurally overlying Laksefjord Nappe Complex contains a slightly higher grade (epizone: Bevins et al., 1986), more penetratively cleaved metasedimentary sequence. Although protolith age has not been constrained (Føyn, 1985), most workers have suggested a Vendian–Cambrian range.

Variably allochthonous sequences are also exposed within several tectonic windows in Finnmark (Fig. 1). These include a crystalline basement interpreted by Chapman et al. (1985) as representing thrust-related duplex structures, and a variably cleaved and metamorphosed (generally very low-grade), unconformably overlying late Proterozoic–early Paleozoic cover succession (Pharaoh, 1985). A Caledonian cleavage within the cover sequences appears to have formed at ca. 420–425 Ma during emplacement of the structurally overlying Kalak Nappe Complex (Dallmeyer et al., 1988), which is represented by at least eight separate thrust sheets (Gayer et al., 1985a). A variety of unique igneous rocks of the Seiland Igneous Suite constitute an important part of the Kalak Nappe Complex (Sturt et al., 1975, 1978; Gayer et al., 1985b). The igneous suite includes variably layered gabbros and associated ultramafic plutons together with extensive alkaline complexes (carbonatites, nepheline syenites, and alkalic pyroxenites). The metasedimentary succession is polydeformed and locally records up to five distinct folding episodes (Gayer et al., 1985a). Gayer et al. (1985a) and Chapman et al. (1985) suggested that all metamorphism and deformation recorded in the Kalak Nappe Complex represents a continuum of tectonother-

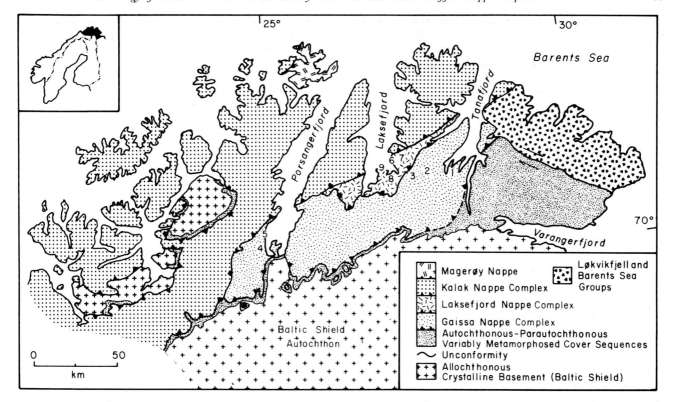

Fig. 1. Generalized geological map of the northern Scandinavian Caledonides, Finnmark, Norway (adapted from Gayer *et al.*, 1985a; Ramsay *et al.*, 1985; Roberts, 1971). Lower allochthon localities sampled in conjunction with the present study are indicated.

mal activity associated with only one orogenic episode, which they interpreted to be Finnmarkian in age. Late Cambrian–Early Ordovician tectonothermal activity in the Kalak Nappe Complex is suggested by ca. 475–480 Ma ^{40}Ar/^{39}Ar hornblende plateau ages reported by Dallmeyer (1988) for some structural units. However, markedly younger hornblende cooling ages (ca. 420–430 Ma) are locally recorded in other structural units, implying a polyorogenic evolution for the Kalak Nappe Complex (involving phases of both Finnmarkian and Scandian tectonothermal activity).

The Magerøy Nappe structurally overlies the Kalak Nappe Complex along a ductile, high-temperature thrust fault. Medium-grade metasedimentary rocks comprise most of the Magerøy Nappe. Fossil-bearing sequences indicate an Early Silurian (Llandovery) and possibly uppermost Ordovician depositional age. The rocks are polydeformed and metamorphosed up to middle amphibolite grade (e.g., Andersen, 1981). Andersen *et al.* (1982) reported Rb–Sr whole-rock isochron crystallization ages of 411 ± 7 Ma for the syn-kinematic Finnvik Granite within the Magerøy Nappe and 410 ± 28 Ma for migmatitic portions of the structurally underlying Kalak Nappe Complex ca. 5 km from the ductile thrust contact. These results suggest that Scandian emplacement of the Magerøy Nappe was generally synchronous with an internally penetrative tectonothermal event within the Kalak Nappe Complex. Widespread maintenance of elevated Scandian metamorphic temperatures within the Kalak Nappe Complex is also indicated by the ca. 420–430 Ma ^{40}Ar/^{39}Ar plateau ages typically recorded by muscovite, biotite, and locally hornblende (Dallmeyer, 1988). Significant Scandian nappe translation is also suggested by the ca. 420–425 Ma ^{40}Ar/^{39}Ar plateau ages recorded by muscovite and whole-rock slate samples adjacent to the basal Kalak thrust exposed around the Komagfjord tectonic window (Dallmeyer *et al.*, 1988).

ANALYTICAL METHODS

Preparation of size fractions

Details of the techniques used during preparation of size fractions from the Finnmark slate/phyllite samples are similar to those listed in Reuter (1985, 1987). After removal of weathered surfaces, the samples were crushed and then ground in a shatter box for 20 s. Constituent grain-size fractions were isolated by differential settling (Atterberg cylinders) and centrifugation. SEM mosaic photomicrographs were prepared for each size fraction, and these were used to evaluate components of the size fractions (rock fragments vs. mineral grains) and to control particle dimensions in the various size separations. In finer fractions, measured grain diameters were systematically slightly larger than the equivalent range, probably owing to an increased settling time for the individual phyllosilicate grains (which are relatively enriched in the finer size fractions). As a result, larger platy grains occur within the finer equivalent fractions. The size fractions are herein referred to their equivalent diameter interval. The edge morphology of constituent grains within the various size fractions was evaluated with TEM photography.

X-ray diffraction

Sample mineralogy was determined by X-ray diffraction analysis of randomly-oriented whole-rock powders and oriented sedimentation slides of the size fractions. Quartz-normalized illite crystallinity of bulk <2 µm size fractions from each sample were determined on oriented sedimentation slides by comparison of the (001) (illite) and (100) (external quartz standard) reflections, following Weber (1972). Cross-calibration of 28 samples (correlation coefficient = 0.97) from Reuter (1985, 1987) suggests the following boundary values (according to Teichmüller *et al.*, 1979) for the operational parameters employed at the University of Georgia: epizone/anchizone, 85; anchizone/diagenesis, 290–425. In the present study, boundaries between the upper anchizone/middle anchizone and middle anchizone/lower anchizone are defined at crystallinity values of 150 and 220 respectively.

Conventional K–Ar dating

Conventional argon analyses were made in Strasbourg following a method similar to that described by Bonhomme *et al.* (1975). The samples were heated to 80° C under vacuum for at least 12 h prior to argon extraction. The spike used for calibration was enriched ^{38}Ar at 99.99% purity. This spike was calibrated in comparison with the GLO glauconite standard with a value of 24.80 × $10^{-6} cm^3$ (STP) radiogenic ^{40}Ar. Potassium concentrations were determined by flame photometry with a precision better than $\pm 2\%$. Ages were calculated using the constants reported by Steiger and Jäger (1977). Independent analysis of different mineral standards allows an error estimate of ca. $\pm 2\%$ for the calculated ages.

$^{40}Ar/^{39}Ar$ analyses

Techniques used during $^{40}Ar/^{39}Ar$ analysis generally followed those described in detail by Dallmeyer and Keppie (1987). The whole-rock shale/slate samples were prepared by crushing and sieving to 80/100 mesh followed by thorough washing. These and aliquots of the 1–2 μm size fractions were wrapped in aluminium-foil packets, encapsulated in sealed quartz vials, and irradiated for 40 h at 1000 kW in the central thimble position of the U.S. Geological Survey reactor in Denver, Colorado. Variations in the flux of neutrons along the length of the irradiation assembly were monitored with the MMhb-1 hornblende standard (519.5 Ma; Alexander *et al.*, 1978). The samples were heated incrementally until fused with an RF generator. Each heating step was maintained for 30 min. Measured isotopic ratios were corrected for the effects of mass discrimination and interfering isotopes produced during irradiation, using factors reported by Dalrymple *et al.* (1981) for the reactor used in the present study. Apparent $^{40}Ar/^{39}Ar$ ages were calculated from the corrected isotopic ratios using the decay constants and isotopic abundance ratios listed by Steiger and Jäger (1977).

Two categories of uncertainties are encountered in $^{40}Ar/^{39}Ar$ incremental-release dating. One group involves intralaboratory uncertainties related to measurement of the isotopic ratios used in the age equation. The other group considers interlaboratory uncertainties in the other parameters used in the age equation (monitor age, J-value determination, etc.), and are the same for each gas increment evolved from a particular sample. Therefore, to evaluate the significance of potential incremental age variations within a sample, only intralaboratory uncertainties should be considered. These are reported here for the incremental gas ages. They have been calculated by statistical propagation of uncertainties associated with measurement of each isotopic ratio (at two standard deviations of the mean) through the age equation. Total-gas ages have been computed for each sample by appropriate weighting of the age and percentage ^{39}Ar released within each temperature increment. For direct comparison with the K–Ar results, 2σ interlaboratory errors are reported for the total-gas ages. Analyses of the MMhb-1 monitor indicate that apparent K/Ca ratios may be calculated for the incremental gas fractions through the relationship $0.518(\pm 0.005) \times (^{39}Ar/^{37}Ar)$ corrected.

SAMPLE CHARACTERISTICS

Whole rock

Following petrographic examination of the entire suite, nine representative, penetratively cleaved slate/phyllite samples were selected for isotopic study. Their locations are indicated in Fig. 1 and are described in the Appendix. The nine samples display generally similar whole-rock mineralogical characteristics (Table I) and are generally composed largely of very fine-grained $2M_1$ white-mica and quartz with subordinate chlorite, detrital plagioclase, and hematite. Sample 3 contains a larger modal percentage of chlorite. Samples 4 and 5 (Gaissa Nappe) also contain subordinate detrital potassium feldspar. The portions of the samples selected for analysis were penetratively cleaved pelite-rich slate/phyllite with only sparse, thinly-laminated bedding horizons. The cleavage is illustrated with representative photomicrographs (Fig. 2) and *in situ* SEM mosaic photographs (Fig. 3). The cleavage is spaced, disjunctive, and appears to be largely defined by the preferred three-dimensional orientation of: (1) elong-

Table I. Mineralogy (determined by X-ray diffraction analysis) of whole-rock slate/phyllite samples and quartz-normalized illite crystallinity of constituent <2 μm size fractions from the Gaissa and Laksefjord Nappe Complexes, Finnmark, Norway.[a]

Sample	Quartz	White mica	Chlorite	Plagioclase	Potassium feldspar	Haematite	Quartz-normalized[b] illite crystallinity
Gaissa Nappe complex							
1	+	+	+	+	−	+	136
2	+	+	+	+	−	+	174
3	+	+	+	−	−	−	142
4	+	+	−	−	(+)	+	168
5	+	+	−	+	+	+	172
Laksefjord Nappe complex							
6	+	+	+	+	−	−	99
7	+	+	+	+	−	−	103
8	+	+	+	+	−	−	113
9	+	+	+	+	−	−	110

[a] + = Present in significant amounts (>5 modal per cent), (+) = Present in insignificant amounts (0–5 modal per cent). − = Not detectable.
[b] Comparison of (001) reflection in illite and (100) reflection in quartz (after Weber, 1972) of <2 μm size fraction.

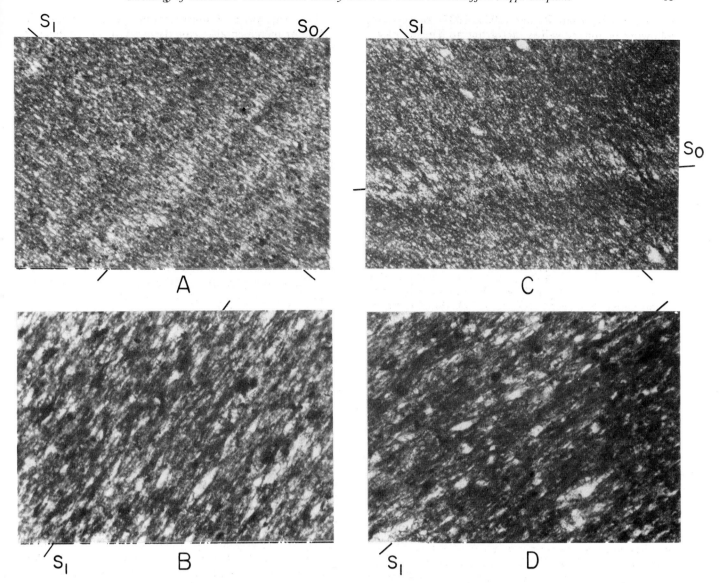

Fig. 2. Representative photomicrographs illustrating the character of cleavage within the Finnmark Lower Allochthon (S_0 = bedding; S_1 = cleavage): *A* and *B* from the Gaissa Nappe Complex (sample 5); *C* and *D* from the Laksefjord Nappe Complex (sample 8). Horizontal field of view is 2.75 mm (*A* and *C*) or 0.85 mm (*B* and *D*).

ate grains of quartz and feldspar whose initial detrital shapes have been modified by pressure-solution processes; (2) seams of very fine-grained opaque material that are probably concentrations of insoluble residue; and (3) relatively larger, presumably detrital phyllosilicate grains. Locally, exceedingly fine-grained phyllosilicate grains appear to have grown within some pressure shadows developed along detrital quartz and feldspar grains (Figs. 2 and 3); however, there is no evidence for any significant neomineralization or recrystallization along the cleavage.

Quartz-normalized illite crystallinity indices for bulk <2 μm size fractions separated from samples of the Gaissa Nappe Complex (1–5) range between 174 and 136 (Table I: Fig. 4), suggesting metamorphic conditions transitional between the middle and upper anchizone, according to the classification of Teichmüller *et al.* (1979). Samples from the Laksefjord Nappe Complex record indices between 113 and 99, indicating an upper anchizone metamorphism. These estimated metamorphic grades appear to be systematically lower than those reported for the same rocks by Bevins et al. (1986); however, these workers based their estimates on the metamorphic subdivisions proposed by Kubler (1967). According to Kubler, minimum illite crystallinity indices are reached *within* the epizone and the anchizone–epizone boundary must therefore be defined by

an arbitrarily assigned illite crystallinity index. On the other hand, Teichmüller *et al.* (1979) have shown that illite crystallinity reaches a minimum, unchanging value in conjunction with the appearance of greenschist facies mineral assemblages. They therefore questioned the geological significance of the epizone, and suggested that the anchizone-greenschist boundary should be established at the appearance of minimum illite crystallinity indices. As a result of the contrasting divisions, rocks suggested to reflect epizonal metamorphism according to Kubler (1967) are classified as upper anchizone according to Teichmüller *et al.* (1979). The Finnmark results obtained here are compared with those of Bevins *et al.* (1986) in Fig. 4. For direct comparison with detailed K–Ar and ^{40}Ar/^{39}Ar studies by the present authors in other very low-grade metamorphic terranes, the metamorphic divisions proposed by Teichmüller *et al.* (1979) will be used in the subsequent discussion.

Size fractions

Size fractions were prepared from two samples (1 and 5) collected within the Gaissa Nappe Complex and one sample (8) from the Laksefjord Nappe Complex. Those fractions comprised of grains <2 μm (<0.5 μm, 0.5–

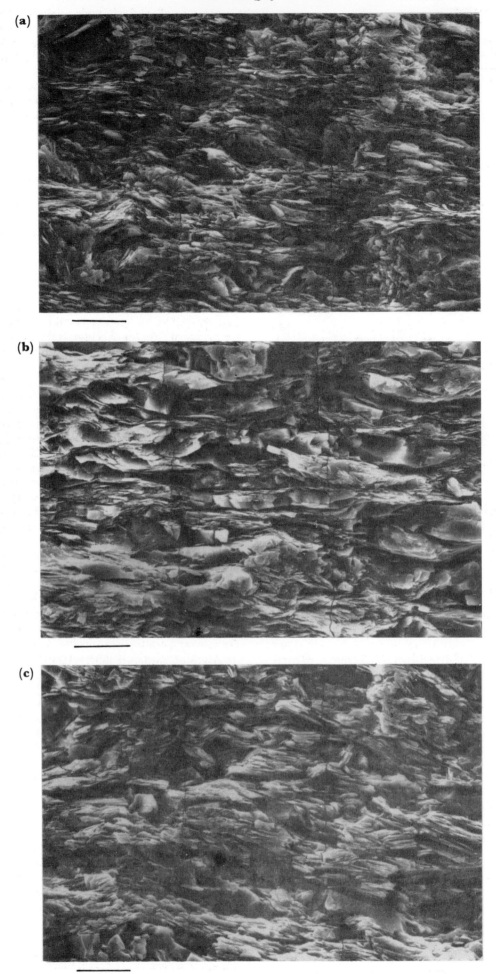

Fig. 3. Representative *in situ* SEM photomicrograph mosaics illustrating the character of cleavage within the Finnmark Lower Allochthon (S_1 oriented horizontally in each photograph: bar represents 10 μm): *A* and *B* from the Gaissa Nappe Complex (samples 1 and 5); *C* from the Laksefjord Nappe Complex (sample 8).

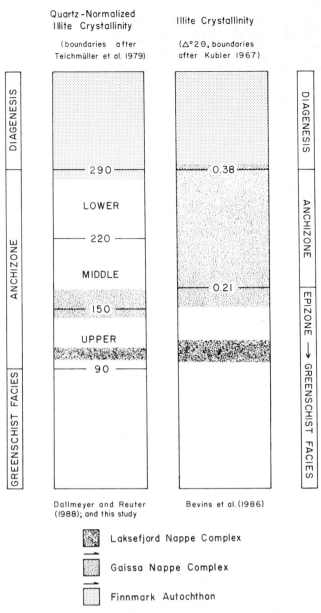

Fig. 4. Comparison of illite crystallinity determinations for various Finnmark tectonic units based upon the contrasting methods of Kubler (1967) (from Bevins *et al.*, 1986) and Teichmüller *et al.* (1979) (from Dallmeyer and Reuter, 1988, and the present study).

1.0 μm, and 1–2 μm) are comprised almost entirely of individual mineral grains (Table II). The larger size fractions (2–4 μm and 4–8 μm) contain significant percentages of rock fragments. The mineralogy of the various size fractions is comparable with that of the whole-rock samples with the exception of a decrease in the abundance of detrital feldspars in those size fractions comprised of grains <2 μm (Table II).

The morphological characteristics of constituent grains within the 1–2 μm size fraction from the three samples are illustrated on the TEM photographs presented in Fig. 5. The three samples display generally similar characteristics with most grains bordered by relatively straight, well-defined edges.

RESULTS

Conventional K–Ar ages

Conventional K–Ar ages have been determined for five size fractions and a representative whole-rock of samples 1, 5 and 8 (for sample 8 insufficient <0.5 μm material was

recovered to allow for isotopic analysis). The analytical data are listed in Table III and are presented graphically in Fig. 6. The K–Ar apparent ages range between 624.4 ± 14.0 Ma (4–8 μm size fraction of sample 1) and 440.2 ± 9.4 Ma (<0.5 μm size fraction of sample 5). For each sample there is a consistent relationship between apparent age and grain-size variation, with the finest fraction recording the youngest apparent age. Sample 1 displays a greater overall range in size fraction apparent ages than do samples 5 and 8.

^{40}Ar/^{39}Ar incremental-release ages

Whole-rock samples

Whole-rock ^{40}Ar/^{39}Ar incremental-release analyses have been carried out on samples 1–5 collected within the Gaissa Nappe Complex. The analytical data are listed in Table IV and are portrayed as apparent age and apparent K/Ca spectra in Fig. 7. The age spectra display variable internal discordancy, recording total-gas ages ranging between 473.8 ± 8.6 Ma (sample 5) and 550.5 ± 9.3 Ma (sample 1). The youngest apparent ages in each spectrum are recorded by gas fractions released at the lowest experimental temperatures. In samples 1–3, apparent ages generally increase systematically throughout intermediate- and high-temperature gas fractions to reach maximum values in fusion increments. In samples 4 and 5, apparent ages recorded by intermediate-temperature gas fractions are less variable, but the apparent ages increase markedly in the high-temperature gas fractions. The apparent K/Ca spectra of the five Gaissa whole-rock samples are also marked by considerable internal discordancy. Apparent K/Ca ratios of gas increments released at low temperatures from samples 1–3 show considerable fluctuations. The ratios increase to slightly higher and less variable values in intermediate-temperature gas fractions. Apparent K/Ca ratios of the high-temperature gas fractions released from samples 1–3 systematically decrease to minimum values in fusion increments. The apparent K/Ca ratios of increments released at lowest experimental temperatures from samples 4 and 5 also display slight fluctuations; however, the ratios systematically decrease throughout the intermediate-temperature increments. Apparent K/Ca ratios increase through high-temperature portions of both experiments and drop markedly in the fusion increments.

Whole-rock, ^{40}Ar/^{39}Ar incremental-release analyses have also been carried out for samples 6–9, collected within the Laksefjord Nappe Complex. The analytical data are listed in Table IV and are portrayed as apparent age and apparent K/Ca spectra in Fig. 8. The age spectra are internally discordant, and record total-gas ages ranging between 484.6 ± 9.2 Ma (sample 8) and 535.4 ± 8.7 Ma (sample 7). The four samples display mutually similar patterns of discordancy, with considerable variation in apparent ages defined by the lower-temperature increments. Apparent ages generally increase systematically through the intermediate-temperature increments to define maximum values in fusion increments. The apparent K/Ca spectra of the four samples are similarily discordant, with K/Ca ratios fluctuating in the lower-temperature gas fractions and maintaining slightly larger and generally constant values throughout intermediate-temperature portions of each analysis. Apparent K/Ca ratios systematically decrease in higher-temperature gas fractions to reach minimum values in the fusion increments.

Size fractions

^{40}Ar/^{39}Ar incremental-release analyses have been carried out on the 1–2 μm size fractions isolated from samples 1,

Table II. Mineralogy (determined by X-ray diffraction analysis) and morphological grain types (observed in SEM photomicrographs) of size fractions isolated from slate/phyllite samples from the Gaissa and Laksefjord Nappe Complexes, Finnmark, Norway.[a]

Size fraction (μm)	Quartz	White mica	Chlorite	Plagioclase	Potassium feldspar	Hematite	Grain types[b]
			Sample 1				
<0.5	+	+	+	+	−	+	M
0.5–1	+	+	+	+	−	+	M
1–2	+	+	+	+	−	+	M
2–4	+	+	+	+	−	+	F(M)
4–8	+	+	+	+	−	+	F
			Sample 5				
<0.5	+	+	(+)	+	−	+	M
0.5–1	+	+	(+)	+	−	+	M
1–2	+	+	(+)	+	−	+	M
2–4	+	+	(+)	+	−	+	F(M)
4–8	+	+	(+)	+	−	+	F(M)
			Sample 8				
0.5–1	+	+	+	+	−	−	M
1–2	+	+	+	+	−	−	M
2–4	+	+	+	+	−	−	M(F)
4–8	+	+	+	+	−	−	M(F)

[a] + = Present in significant amounts (> 5 modal per cent). (+) = Present in insignificant amounts (0–5 modal per cent). − = Not detectable.
[b] F = rock fragments. M = individual mineral grains. F(M) = predominantly rock fragments. M(F) = predominantly individual mineral grains.

5 and 8. The analytical data are listed in Table V and are portrayed as apparent age and apparent K/Ca spectra in Fig. 9. The age spectra are markedly discordant and define total-gas ages of 497.9 ± 9.0 Ma (1), 471.9 ± 9.0 Ma (5) and 475.8 ± 9.3 Ma (8). The patterns of discordancy are different from those displayed in the corresponding whole-rock age spectra. In general, apparent ages are variable in gas fractions released at the lowest experimental temperatures and increase to maximum values in initial intermediate-temperature gas fractions. Apparent ages systematically decrease throughout the remaining intermediate- and high-temperature increments. Slight fluctuations in apparent age occur within the highest-temperature and fusion increments. The apparent K/Ca spectra are also internally discordant. They show characteristics similar to the corresponding whole-rocks with apparent K/Ca ratios fluctuating in the lowest-temperature gas fractions and increasing to slightly higher and more constant values in the intermediate-temperature gas fractions. Apparent K/Ca ratios systematically decrease in higher-temperature gas fractions.

Comparison of K–Ar and ⁴⁰Ar/³⁹Ar results

The ^{40}Ar/^{39}Ar total-gas and K–Ar apparent ages determined for the whole-rock and 1–2 μm size fractions isolated from samples 1, 5, and 8 are compared in Fig. 6. The ^{40}Ar/^{39}Ar total-gas ages are similar within analytical uncertainties to the corresponding K–Ar apparent ages.

INTERPRETATION

Variations in grain size and mineralogy

The mineralogical compositions of samples from the Gaissa and Laksefjord Nappe Complexes are not significantly different. However, there is a slight enrichment of white mica in the finest fractions isolated from sample

8 (Laksefjord Nappe Complex) compared with those from samples 1 and 5 (Gaissa Nappe Complex). In addition, there is a slight increase in grain size within sample 8 compared to samples 1 and 5. This is suggested by a reduction in <0.5 μm material and the fact that the 2–4 μm size fraction is dominated by individual mineral grains whereas this size fraction within samples 1 and 5 is dominated by rock fragments. Illite crystallinity indices suggest that a slightly higher metamorphic grade is recorded within the Laksefjord Nappe Complex (upper anchizone) compared to the Gaissa Nappe Complex (transitional middle–upper anchizone). The variations in grain size probably reflect this difference. Similar grain-size variations between middle and upper anchizone metapelites have been previously documented by Reuter (1985, 1987) and Huon (1985), who related it to crystal growth during metamorphism.

Conventional K–Ar ages

The extent of partial rejuvenation of the K–Ar isotopic system within a population of very fine-grained white mica during very low (anchizonal) metamorphism is likely controlled by: (1) a change of mineralogical characteristics, such as the transition from a 1 Md to a 2M₁ illite polytype (Hunziker *et al.*, 1986); (2) the maximum metamorphic temperatures attained (affecting the diffusion and/or migration of intracrystalline dislocations); and/or (3) the extent of concomitant deformation and associated neomineralization and/or recrystallization of white mica. Reconstruction of a 1 Md to 2 M₁ white mica cannot be responsible for the age relationships observed in the Finnmark samples, because they do not contain detectable 1 Md white mica. Cleavage formation could have involved only rotation of detrital phyllosilicate grains (pressure-solution-assisted grain rotation). If so, the observed grain-size dependence of apparent K–Ar age would reflect variable diffusive loss of radiogenic argon due to contrasting temperatures of anchizonal metamorphism.

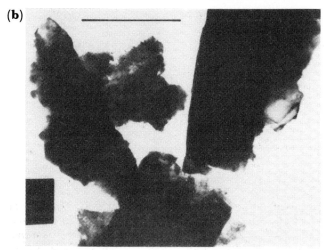

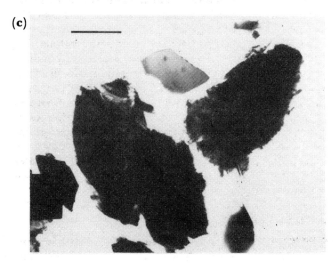

Fig. 5. Representative TEM photomicrographs illustrating the edge morphology of constituent grains within the 1–2 μm size fraction isolated from samples of the Finnmark Lower Allochthon (length of bar is 1.5 μm): *A* and *B* from Gaissa Nappe Complex (samples 1 and 5); *C* from Laksefjord Nappe Complex (sample 8).

On the other hand, the observed grain size-vs.-age relationship could reflect variably completed syn-kinematic recrystallization of very fine-grained white mica during cleavage formation. This would require an enrichment of syn-kinematically recrystallized mica relative to detrital white mica in the fine-vs.-coarse fractions to produce the observed grain-size dependence of K–Ar apparent ages. Microstructural characteristics of the Finnmark samples do not suggest extensive neomineraliz-

ation and/or recrystallization of phyllosilicate minerals along the Caledonian cleavage; therefore, the observed K–Ar age variations are more likely to reflect contrasting anchizonal temperature conditions resulting in variable rejuvenation of argon systems within the detrital mica grains.

The largest size fractions (4–8 μm) isolated from the three samples record older K–Ar apparent ages than the corresponding whole-rock samples. These size fractions are largely comprised of rock fragments that consist of coarse detrital and variably rejuvenated finer mineral grains. The contrast between whole-rock and size fraction K–Ar apparent ages suggests a relatively smaller portion of finer grains in the rock fragments relative to the whole-rock samples.

^{40}Ar/^{39}Ar incremental-release ages

K–Ar vs. ^{40}Ar/^{39}Ar results

The consistent correlation of K–Ar and ^{40}Ar/^{39}Ar total-gas ages suggests that a minimum recoil loss of ^{39}Ar occurred during neutron irradiation. This is not surprising in view of the morphologically distinct edge character of constituent grains, which results in reduced surface/volume ratios (e.g., Reuter and Dallmeyer, 1987a,b, 1988).

Whole-rock analyses

The nine whole-rock samples display variably discordant ^{40}Ar/^{39}Ar apparent age spectra that are difficult to interpret directly. Although these samples generally consist primarily of very fine-grained white mica, systematic intrasample variations in apparent K/Ca ratios suggest that several other phases contributed gas at various stages of the analyses. These appear to have included: (1) a less-retentive phase with a low K/Ca ratio and present in variable modal abundance; and (2) a more refractory phase, also with a low K/Ca ratio, and present in a minor modal abundance. Mineralogical characteristics suggest that these phases are chlorite and plagioclase feldspar, respectively. Fluctuations in the apparent K/Ca ratios determined for high-temperature gas fractions released from samples 4 and 5 suggest the presence of another relatively refractory phase with a large K/Ca ratio. These samples are the only ones containing detectable potassium feldspar, which is considered to be the additional relatively refractory phase. It is proposed that most of the observed spectra discordancy reflects progressively changing mixtures of gas experimentally evolving from these contrasting minerals that had experienced variable rejuvenation during the very low-grade regional metamorphism recorded in the Gaissa and Laksefjord Nappe Complexes.

Within samples 1–3 (Gaissa Nappe Complex) the intermediate-temperature gas fractions display relatively little intrasample variation in apparent K/Ca ratios, suggesting experimental evolution of gas from an individual white mica phase over a considerable temperature interval. Apparent ages recorded by these increments show similar, systematically increasing trends. If not significantly contaminated by mixing of gas released from chlorite (at lower temperatures) or from detrital plagioclase (at higher temperatures), the increasing age trends are similar to those predicted by Turner (1968) to characterize episodic, volume-diffusive loss of argon during a geological overprint. Such patterns of discordance have been documented for pure concentrates of coarser-grained white mica by Snee *et al.* (1988) and Dallmeyer and Lécorché (1987, 1988). Samples 4 and 5 are also from the Gaissa Nappe Complex; however, they are mineralogically more com-

Table III. K–Ar analytical data from whole-rock and constituent size fractions isolated from slate/phyllite samples from the Gaissa and Laksefjord Nappe Complexes, Finnmark, Norway.

Size fraction (μm)	K_2O (%)	$^{40}Ar^a$ (%)	$^{40}Ar^a$ ($10^{-6} cm^3 g$)STP	Apparent age (Ma)[b]
Sample 1—Gaissa Nappe Complex				
<0.5	7.18	97.4	122.3	463.6 ± 9.9
0.5–1	7.05	95.1	123.2	474.2 ± 10.4
1–2	5.20	97.9	95.8	496.6 ± 10.8
2–4	4.39	97.9	102.0	606.6 ± 13.4
4–8	4.08	96.5	98.1	624.3 ± 14.0
WR[c]	4.39	97.6	90.2	546.0 ± 12.1
Sample 5—Gaissa Nappe Complex				
<0.5	8.13	97.6	130.6	440.2 ± 9.4
0.5–1	7.68	97.6	130.3	461.8 ± 9.9
1–2	6.13	97.9	103.9	461.7 ± 9.9
2–4	5.01	97.6	88.0	476.5 ± 10.4
4–8	4.72	97.8	87.0	496.8 ± 10.9
WR[c]	5.10	97.9	92.0	487.6 ± 10.6
Sample 8—Laksefjord Nappe Complex				
0.5–1	4.59	85.3	77.4	459.5 ± 11.4
1–2	4.38	89.5	76.4	473.6 ± 11.3
2–4	4.70	94.1	83.5	481.1 ± 10.9
4–8	5.28	96.5	95.9	490.3 ± 10.8
WR[c]	4.40	96.6	82.5	504.5 ± 11.5

[a]Radiogenic.
[b]Two-sigma analytical uncertainties.
[c]WR = whole-rock.

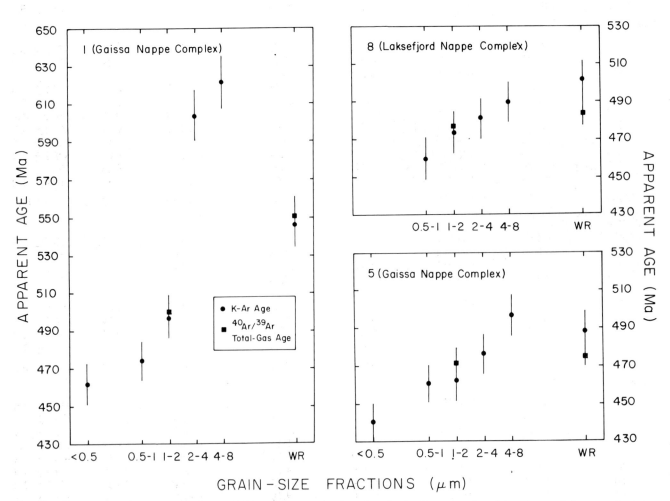

Fig. 6. Comparison of apparent K–Ar age (with a 2σ uncertainty) and grain size from samples of the Finnmark Lower Allochthon (WR = whole-rock). $^{40}Ar/^{39}Ar$ total-gas ages (with a 2σ interlaboratory uncertainty) of the 1–2 μm and whole-rock samples are plotted for comparison.

Table IV. ^{40}Ar/^{39}Ar analytical data for incremental heating experiments on whole-rock slate/phyllite samples from the Gaissa and Laksefjord Nappe Complexes, Norway.

Release temp. (°C)	^{40}Ar/^{39}Ar[a]	^{36}Ar/^{39}Ar[a]	^{37}Ar/^{39}Ar[b]	^{39}Ar (% of total)	^{40}Ar (% non-atmos.)[c]	^{36}Ar/Ca (%)	Apparent age (Ma)[d]
			Gaissa Nappe Complex				
Sample 1: $J = 0.009470$							
425	35.90	0.06492	0.064	0.94	46.56	0.03	265.1 ± 6.6
450	36.14	0.01495	0.073	1.11	87.78	0.13	473.9 ± 3.4
475	31.64	0.00448	0.039	2.36	95.81	0.24	454.9 ± 3.2
500	32.74	0.00664	0.036	3.97	93.16	0.15	457.7 ± 1.7
525	32.15	0.00461	0.037	8.92	95.75	0.22	461.5 ± 1.8
550	34.12	0.00146	0.032	10.01	98.72	0.60	499.5 ± 1.8
575	34.32	0.00103	0.032	13.58	99.10	0.85	503.7 ± 1.8
600	35.21	0.00105	0.042	8.02	99.12	1.08	515.1 ± 1.9
625	36.52	0.00122	0.037	8.50	99.01	0.83	531.2 ± 1.3
675	38.84	0.00151	0.043	8.66	98.84	0.77	559.5 ± 2.1
750	43.55	0.00186	0.040	9.06	98.73	0.58	616.2 ± 2.2
Fusion	47.92	0.00275	0.149	24.86	98.32	1.48	665.5 ± 2.3
Total	38.72	0.00301	0.066	100.00	97.57	0.85	550.5 ± 9.3
Sample 2: $J = 0.009382$							
450	26.19	0.01113	0.038	2.96	87.43	0.09	351.1 ± 2.7
500	26.75	0.00256	0.040	10.77	97.17	0.42	393.7 ± 1.6
525	33.98	0.00142	0.025	23.19	98.75	0.48	493.7 ± 2.7
550	33.91	0.00073	0.026	25.22	99.35	0.96	495.5 ± 1.5
575	36.03	0.00107	0.030	14.96	99.11	0.75	521.3 ± 1.8
600	39.56	0.00174	0.031	7.38	98.70	0.49	563.1 ± 2.0
650	46.22	0.00182	0.035	7.24	98.83	0.52	643.4 ± 2.7
700	56.98	0.00361	0.060	4.73	98.13	0.46	760.8 ± 4.4
750	60.76	0.00584	0.146	2.22	97.17	0.68	795.2 ± 6.5
Fusion	57.88	0.02727	2.192	1.34	86.37	2.19	694.6 ± 4.5
Total	36.56	0.00221	0.062	100.00	98.22	0.65	521.3 ± 9.8
Sample 3: $J = 0.010455$							
425	27.45	0.02003	0.013	3.79	78.42	0.02	366.2 ± 3.6
450	19.41	0.00296	0.015	8.15	95.46	0.14	319.5 ± 1.9
475	28.52	0.00279	0.021	8.60	97.09	0.21	458.6 ± 1.6
500	32.09	0.00107	0.012	26.11	99.07	0.31	517.0 ± 1.9
525	32.88	0.00086	0.013	26.36	99.21	0.40	529.4 ± 2.1
550	33.25	0.00096	0.015	8.48	99.13	0.43	534.2 ± 2.0
575	35.18	0.00167	0.015	7.64	98.58	0.25	558.1 ± 2.2
625	37.51	0.00244	0.014	4.70	98.06	0.15	587.1 ± 4.0
650	47.33	0.00722	0.016	1.44	95.48	0.06	698.1 ± 5.4
725	47.38	0.00783	0.022	3.06	95.11	0.08	696.4 ± 3.6
Fusion	53.64	0.02427	0.127	1.66	86.64	0.14	714.4 ± 4.2
Total	33.04	0.00282	0.016	100.00	97.40	0.28	521.6 ± 11.8
Sample 4: $J = 0.009232$							
425	19.03	0.00997	0.014	2.74	84.50	0.04	249.7 ± 5.2
450	26.48	0.00172	0.017	4.68	98.06	0.27	387.6 ± 3.2
500	32.42	0.00061	0.057	25.24	99.45	2.61	470.1 ± 1.4
525	33.05	0.00041	0.065	26.05	99.63	4.32	478.9 ± 1.2
550	32.54	0.00017	0.075	18.81	99.85	12.17	473.3 ± 1.2
575	33.33	0.00050	0.125	11.85	99.57	6.84	482.1 ± 1.9
625	40.48	0.00227	0.099	4.06	98.35	1.19	564.7 ± 2.4
700	51.76	0.00428	0.052	2.50	97.55	0.33	690.3 ± 2.9
Fusion	52.21	0.01069	0.191	4.06	93.97	0.49	674.1 ± 2.1
Total	33.68	0.00133	0.075	100.00	98.80	4.97	482.1 ± 8.6
Sample 5: $J = 0.009313$							
425	21.21	0.01910	0.041	2.09	73.38	0.06	244.2 ± 4.6
450	25.81	0.00385	0.042	3.14	95.58	0.29	373.1 ± 3.1
500	30.49	0.00115	0.052	16.30	98.88	1.23	446.4 ± 2.4
525	32.13	0.00059	0.103	25.43	99.47	4.77	470.1 ± 2.2
550	31.78	0.00075	0.229	16.81	99.34	8.33	465.0 ± 1.2
575	31.53	0.00115	0.576	14.24	99.05	13.68	460.8 ± 1.3
600	32.79	0.00061	0.234	7.28	99.49	10.50	478.7 ± 1.6
650	37.86	0.00278	0.106	4.18	97.84	1.04	534.6 ± 1.8
725	43.13	0.00255	0.064	4.21	98.25	0.68	600.1 ± 2.2
Fusion	44.36	0.00451	0.137[b]	6.31	97.01[c]	0.82	608.1 ± 2.0[d]
Total	32.82	0.00170	0.190	100.00	98.35	5.66	473.8 ± 8.6

Table IV.—*Contd.*

Release temp. (°C)	$^{40}Ar/^{39}Ar^a$	$^{36}Ar/^{39}Ar^a$	$^{37}Ar/^{39}Ar^b$	^{39}Ar (% of total)	^{40}Ar (% non-atmos.)^c	$^{36}Ar/Ca$ (%)	Apparent age (Ma)^d
			Laksefjord Nappe Complex				
Sample 6: $J = 0.008551$							
400	154.26	0.37362	0.032	0.41	28.43	0.00	574.5 ± 15.9
450	58.64	0.08151	0.068	5.22	58.92	0.02	467.0 ± 4.7
475	38.89	0.00544	0.040	3.63	95.86	0.20	499.1 ± 1.7
500	36.91	0.00244	0.031	4.64	98.04	0.35	486.2 ± 1.4
525	38.92	0.01131	0.033	0.93	91.41	0.08	479.1 ± 4.9
550	37.71	0.00617	0.028	2.14	95.15	0.13	482.7 ± 2.3
600	35.89	0.00209	0.024	5.37	98.27	0.32	475.5 ± 1.6
675	34.64	0.00064	0.027	22.93	99.44	1.13	465.6 ± 1.5
725	35.33	0.00164	0.022	11.70	98.62	0.37	470.4 ± 1.2
775	37.16	0.00271	0.027	11.59	97.83	0.27	490.3 ± 1.6
825	37.93	0.00113	0.038	14.52	99.11	0.44	502.8 ± 1.2
Fusion	51.53	0.00425	0.398	16.91	97.61	2.55	645.4 ± 2.2
Total	40.52	0.00799	0.091	100.00	95.97	0.88	508.6 ± 10.2
Sample 7: $J = 0.008675$							
425	99.18	0.18968	0.188	0.71	43.49	0.03	573.5 ± 12.9
475	35.61	0.00599	0.088	2.58	95.03	0.40	464.4 ± 2.2
500	35.88	0.00311	0.066	6.03	97.43	0.58	477.9 ± 1.9
575	36.15	0.00128	0.060	13.26	98.95	1.27	487.6 ± 1.3
650	36.01	0.00043	0.047	24.27	99.64	2.99	488.9 ± 1.0
700	38.28	0.00106	0.061	14.78	99.25	10.00	514.0 ± 1.2
750	40.94	0.00176	0.070	10.14	98.73	1.08	542.3 ± 1.4
825	43.99	0.00114	0.106	25.42	99.24	2.53	579.4 ± 1.8
Fusion	92.39	0.01351	2.850	2.82	95.92	5.74	1030.3 ± 10.8
Total	40.91	0.00296	0.199	100.00	98.54	3.33	535.4 ± 8.7
Sample 8: $J = 0.008783$							
425	221.24	0.52592	0.037	0.29	29.75	0.00	823.1 ± 23.3
450	73.46	0.04403	0.065	0.38	82.29	0.04	768.2 ± 16.2
475	25.66	0.01045	0.023	2.90	87.96	0.06	326.3 ± 3.3
500	36.29	0.00414	0.012	5.39	96.61	0.08	484.3 ± 2.1
550	34.31	0.00135	0.008	15.84	98.82	0.15	470.3 ± 2.3
600	33.63	0.00102	0.009	20.51	99.02	0.23	463.1 ± 2.3
650	35.53	0.00124	0.010	17.62	98.95	0.23	485.5 ± 1.6
700	36.36	0.00085	0.011	25.15	99.29	0.26	496.8 ± 2.2
750	37.06	0.00156	0.024	10.12	98.75	0.42	502.7 ± 1.7
Fusion	58.86	0.01157	0.446	1.81	94.24	1.05	716.2 ± 7.3
Total	36.16	0.00342	0.019	100.00	98.23	0.25	484.6 ± 9.2
Sample 9: $J = 0.008766$							
425	57.31	0.08171	0.055	1.20	57.86	0.02	460.3 ± 4.5
500	33.12	0.00151	0.026	9.03	98.64	0.48	454.3 ± 2.3
575	33.40	0.00067	0.025	20.55	99.39	1.01	460.8 ± 1.2
650	34.66	0.00047	0.028	26.82	99.59	1.62	476.8 ± 1.2
700	36.95	0.00047	0.029	26.89	99.61	1.68	504.4 ± 1.4
750	40.88	0.00103	0.099	12.73	99.26	2.62	549.1 ± 1.3
Fusion	46.20	0.00273	0.625	2.78	98.35	6.23	605.0 ± 2.7
Total	36.26	0.00172	0.054	100.00	98.89	1.65	491.5 ± 8.8

^{a} Measured.

^{b} Corrected for post-irradiation decay of ^{37}Ar (35.1-day half-life).

^{c} $[^{40}Ar_{tot} - (^{36}Ar_{atmos})(295.5)]/^{40}Ar_{tot}$.

^{d} Calculated using correction factors of Dalrymple *et al.* (1981); 2σ intralaboratory errors reported for incremental ages; 2σ interlaboratory errors reported for total-gas ages.

plex, containing microcline as an additional constituent. Apparent K/Ca ratios recorded by gas fractions evolved over intermediate experimental temperatures systematically decrease. This suggests that no portions of either analysis are related to gas evolved from an individual, chemically distinct mineral phase, but reflect mixed ages of uncertain significance.

The apparent K/Ca ratios of gas fractions evolved at intermediate experimental temperatures from samples 6–9 (Laksefjord Nappe Complex) also display generally minor intrasample variations. The apparent ages recorded by these gas fractions show systematically increasing age trends similar to those of samples 1–3; however, the overall intrasample age range is markedly reduced. If these trends are related to experimental evolution of gas from populations of episodically rejuvenated detrital white-mica grains, the results clearly suggest more extensive argon loss in samples 6–9 relative to those from the Gaissa Nappe Complex.

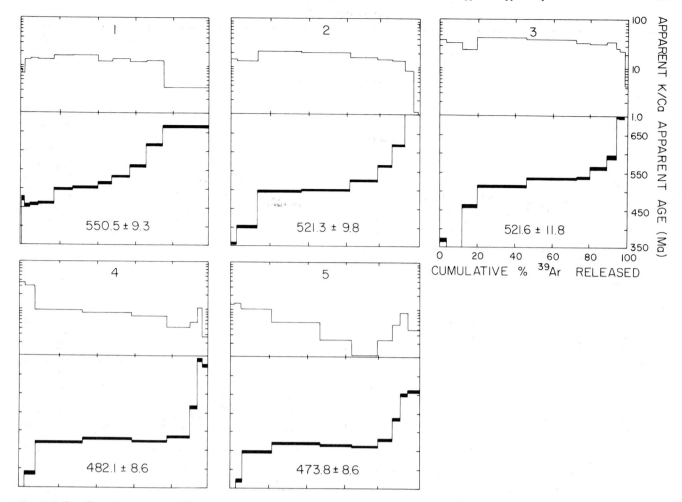

Fig. 7. $^{40}Ar/^{39}Ar$ age and apparent K/Ca spectra of whole-rock slate/phyllite samples from the Gaissa Nappe Complex of the Finnmark Lower Allochthon. All spectra have the coordinates shown at upper right. Uncertainties in incremental ages (2σ intralaboratory uncertainty) are indicated by vertical width of bars. Experimental temperatures increase from left to right. Total-gas ages are listed on each spectrum with a 2σ interlaboratory uncertainty. Sample locations are indicated on Fig. 1 and are described in the Appendix.

1–2 μm size fractions

The 1–2 μm size fractions separated from samples 1, 5 and 8 display similarly discordant $^{40}Ar/^{39}Ar$ apparent age and apparent K/Ca spectra. Intrasample variations in apparent K/Ca ratios are generally similar to those displayed by the nine whole-rock samples, suggesting experimental evolution of gas dominantly from chlorite (low temperatures), white mica (intermediate temperatures) and plagioclase (high temperatures). The patterns of age discordancy are markedly different from those that characterize the whole-rock samples. The 1–2 μm size fractions generally show considerable variation of apparent ages in the low-temperature portions of each analysis. Maximum apparent ages are typically defined in initial intermediate-temperature increments. Ages systematically decrease throughout the remainder of each experiment. The variations in apparent age combined with mineralogical characteristics and the observed variations in apparent K/Ca ratios suggest that significant recoil-redistribution of ^{39}Ar occurred within the 1–2 μm size fractions during irradiation. This appears to have involved recoil loss of ^{39}Ar from white mica with plagioclase feldspar representing the dominant receptor.

GEOLOGICAL SIGNIFICANCE

The consistent variation of K–Ar apparent age with size fraction observed within the Finnmark samples is identical to that described by Reuter (1985, 1987) for penetratively cleaved slate/phyllite within an anchizone terrane in the Rheinisches Schiefergebirge (Federal Republic of Germany), where the age of anchizonal metamorphism (and associated cleavage formation) is closely bracketed by the depositional age of the metamorphosed protoliths and an unconformably overlying, non-metamorphic succession. Reuter reported that the K–Ar apparent ages recorded by all size fractions isolated from metapelite samples in the middle anchizone are older than the geological constraints. Reuter interpreted this to indicate that a detrital memory was retained within all size fractions. Within metapelite samples from the upper anchizone, only the K–Ar apparent ages recorded by the very smallest size fraction (<0.63 μm) approached the geological constraints. The Finnmark metapelite samples display a cleavage of similar character, have the same mineralogy and are from a comparable metamorphic grade to that of the Rheinisches Schiefergebirge. Therefore, the K–Ar apparent age variations are interpreted as reflecting a similar incomplete metamorphic rejuvenation of detrital phyllosillicate grains. On the basis of the conclusions drawn by Reuter (1985, 1987) in her study of samples from the Rheinisches Schiefergebirge, the closest approximation to the actual age of cleavage formation in the Finnmark Lower Allochthon would be provided by the 440.2 ± 9.4 Ma age recorded by the <0.5 μm size fraction from sample 5 collected within the Gaissa Nappe Complex. However, this is considered only to reflect a maximum date.

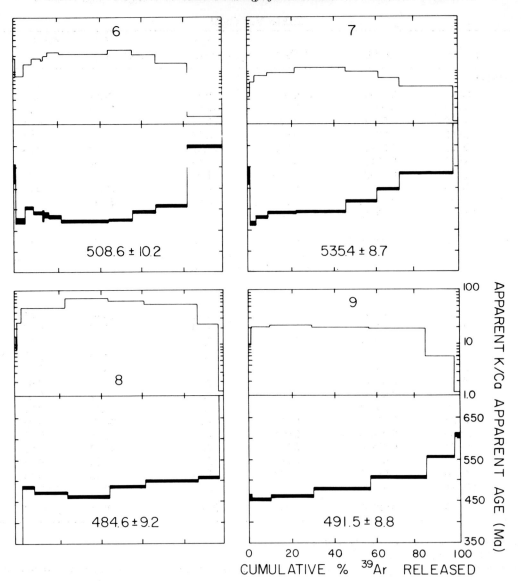

Fig. 8. $^{40}Ar/^{39}Ar$ age and apparent K/Ca spectra of whole-rock slate/phyllite samples from the Laksefjord Nappe Complex of the Finnmark Lower Allochthon. All spectra have the coordinates shown at lower right. Data are plotted as in Fig. 7.

Table V. $^{40}Ar/^{39}Ar$ analytical data for incremental heating experiments on 1–2 µm size fractions isolated from slate/phyllite samples from the Gaissa and Laksefjord Nappe Complexes, Norway.

Release temp. (°C)	$^{40}Ar/^{39}Ar^a$	$^{36}Ar/^{39}Ar^a$	$^{37}Ar/^{39}Ar^b$	^{39}Ar (% of total)	^{40}Ar (% non-atmos.)^c	$^{36}Ar/Ca$ (%)	Apparent age (Ma)^d
			Gaissa Nappe Complex				
Sample 1: J = 0.0080605							
375	38.96	0.03596	0.019	2.11	72.71	0.01	371.2 ± 4.9
400	43.27	0.01131	0.030	7.55	92.27	0.07	503.5 ± 1.6
420	33.62	0.00614	0.032	7.74	94.59	0.14	411.9 ± 2.1
435	43.68	0.00387	0.013	7.45	97.37	0.09	532.0 ± 0.8
445	44.43	0.00233	0.012	9.62	98.44	0.28	545.0 ± 1.0
455	43.37	0.00203	0.011	10.68	98.60	0.15	534.6 ± 2.4
465	42.45	0.00197	0.010	9.05	98.61	0.14	524.8 ± 1.8
475	41.06	0.00149	0.013	11.82	98.92	0.23	511.2 ± 1.0
485	39.97	0.00104	0.012	15.26	99.22	0.31	500.6 ± 0.9
495	39.32	0.00124	0.010	10.40	99.06	0.23	492.8 ± 1.0
505	38.87	0.00278	0.021	4.63	97.87	0.21	482.7 ± 2.6
515	38.44	0.00667	0.048	1.59	94.87	0.20	465.1 ± 8.5
Fusion	114.68	0.32066	0.337	2.09	17.39	0.03	269.2 ± 10.2
Total	42.55	0.01039	0.023	100.00	95.55	0.18	497.9 ± 9.0
Sample 5: J = 0.008380							
375	23.37	0.02636	0.063	1.70	66.67	0.06	221.4 ± 12.4
400	33.54	0.01450	0.026	3.35	87.21	0.05	395.5 ± 3.3
420	34.97	0.00776	0.022	3.27	93.43	0.08	436.6 ± 3.6

Table V.—*Contd.*

Release temp. (°C)	$^{40}Ar/^{39}Ar^a$	$^{36}Ar/^{39}Ar^a$	$^{37}Ar/^{39}Ar^b$	^{39}Ar (% of total)	^{40}Ar (% non-atmos.)^c	$^{36}Ar/Ca$ (%)	Apparent age (Ma)^d
Sample 5: J=0.008380							
440	38.42	0.00334	0.008	10.46	97.42	0.06	492.2 ± 1.7
455	38.19	0.00187	0.015	10.82	98.54	0.22	494.5 ± 1.1
470	38.62	0.00163	0.030	15.54	98.74	0.51	500.2 ± 1.2
485	37.96	0.00114	0.031	18.08	99.11	1.26	494.4 ± 0.9
495	36.68	0.00148	0.032	21.82	98.80	0.59	478.4 ± 1.3
510	33.85	0.00136	0.022	11.43	98.80	0.44	445.7 ± 0.9
525	27.23	0.00624	0.070	1.92	93.23	0.31	348.0 ± 11.1
540	30.30	0.01113	0.087	0.70	89.15	0.21	368.1 ± 14.5
565	43.21	0.04464	0.278	0.22	69.51	0.17	405.0 ± 51.0
Fusion	177.44	0.49149	0.207	0.70	18.16	0.01	431.2 ± 24.1
Total	37.62	0.00640	0.033	100.00	96.76	0.53	471.9 ± 9.0

Laksefjord Nappe Complex

Release temp. (°C)	$^{40}Ar/^{39}Ar^a$	$^{36}Ar/^{39}Ar^a$	$^{37}Ar/^{39}Ar^b$	^{39}Ar (% of total)	^{40}Ar (% non-atmos.)^c	$^{36}Ar/Ca$ (%)	Apparent age (Ma)^d
Sample 8: J=0.008465							
375	79.56	0.19947	0.038	2.21	25.91	0.01	290.2 ± 12.4
400	50.66	0.03779	0.066	8.00	77.96	0.05	520.3 ± 2.1
420	47.86	0.01992	0.052	5.97	87.69	0.07	548.4 ± 3.4
435	40.48	0.00937	0.051	14.77	93.16	0.15	499.8 ± 2.8
445	40.15	0.00560	0.027	15.80	95.87	0.13	508.8 ± 1.2
455	39.27	0.00572	0.008	9.18	95.68	0.04	498.2 ± 2.0
465	38.61	0.00683	0.005	5.89	94.77	0.20	486.8 ± 5.3
475	37.99	0.00624	0.006	4.85	95.13	0.01	481.5 ± 2.7
485	40.08	0.01601	0.055	1.81	88.19	0.09	472.2 ± 6.8
495	37.84	0.00813	0.034	2.57	93.64	0.11	473.2 ± 8.0
510	34.23	0.00466	0.036	12.96	95.97	0.21	442.6 ± 2.8
525	31.11	0.00649	0.035	10.14	93.83	0.15	398.4 ± 1.9
540	32.50	0.01300	0.070	3.13	88.18	0.15	391.8 ± 9.4
Fusion	149.21	0.39126	0.591	2.74	22.54	0.04	452.1 ± 10.0
Total	43.10	0.02499	0.053	100.00	89.25	0.12	475.8 ± 9.3

^{a} measured.
^{b} Corrected for post-irradiation decay of ^{37}Ar (35.1-day half-life).
^{c} $[^{40}Ar_{tot} - (^{36}Ar_{atmos})(295.5)]/^{40}Ar_{tot}$.
^{d} Calculated using correction factors of Dalrymple *et al.* (1981); 2σ intralaboratory errors reported for incremental ages; 2σ interlaboratory errors reported for total-gas ages.

TECTONIC IMPLICATIONS

The 440.2 ± 9.4 Ma K–Ar apparent age recorded by the < 0.5 μm size fraction isolated from metapelite within the Gaissa Nappe Complex is suggested to represent a maximum date for metamorphism and deformation of the Finnmark Lower Allochthon. This is consistent with the 420–425 Ma age bracket proposed by Dallmeyer *et al.* (1988) for emplacement of the Kalak Nappe Complex and associated imbrication of underlying parautochthonous sequences exposed within the Finnmark structural windows. The 440.2 ± 9.4 Ma age maximum is also consistent with the ca. 425 Ma thermal overprint recorded by all muscovite and some hornblende argon systems within the Kalak Nappe Complex (Dallmeyer, 1988).

Interpretation of the overall regional significance of the Finnmark results depends upon calibration of the Ordovician and Silurian time-scales (e.g., Palmer, 1983). Snelling (1985) and Kunk *et al.* (1985) propose that the Ordovician–Silurian boundary (base of the Llandovery) is ca. 435–440 Ma. This, together with a 420 Ma calibration of the Ludlow (Wybourn *et al.*, 1982), indicates that the deformation and metamorphism recorded within the Finnmark Lower Allochthon developed in the Late Ordovician to Early Silurian. This tectonothermal chronology is consistent with that of the Scandian orogenesis recorded within late Proterozoic through Early Silurian autochthonous/parautochthonous metasedi-

mentary sequences exposed in the central and southern Scandinavian Caledonides.

Retrograded basement and imbricated miogeoclinal metasedimentary cover sequences within the Kalak Nappe Complex clearly record the affects of Late Cambrian–Early Ordovician tectonothermal activity (the Finnmarkian Orogeny of Sturt *et al.*, 1975, 1978). However, K–Ar and $^{40}Ar/^{39}Ar$ ages reported for proximal Baltic miogeoclinal sequences exposed both in eastern Finnmark (present study) and the Finnmark windows (Dallmeyer *et al.*, 1988) show no record of this early Caledonian tectonothermal event, and thereby question whether Kalak metasedimentary units originated within the early Paleozoic miogeocline of continent Baltica. The uncertainty of this relationship is further highlighted by ca. 800 Ma crystallization ages (both Rb–Sr whole-rock and U–Pb zircon) reported by Daly *et al.* (1987) for *syntectonic* granitic plutons within Kalak metasedimentary sequences. These characteristics, together with the occurrence of basic-alkalic plutons of the Seiland Igneous Province (unique within the entire Scandinavian Caledonides), combine to suggest that the Kalak Nappe Complex was probably not derived from the Baltic miogeocline. It should be considered an exotic tectonostratigraphic terrane with an uncertain palinspastic derivation within the early Paleozoic Iapetus Oceanic tract.

Although early Caledonian (Middle Cambrian–Early

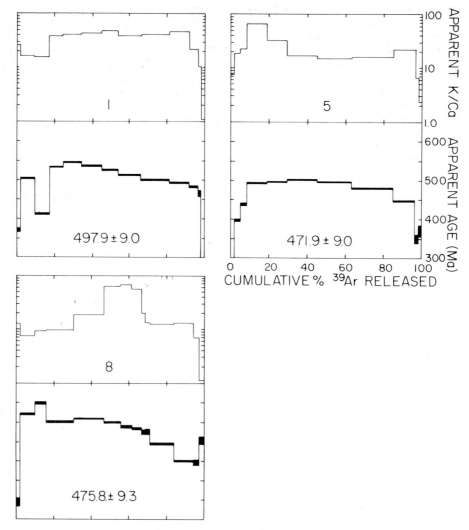

Fig. 9. $^{40}Ar/^{39}Ar$ age and apparent K/Ca spectra of the 1–2 µm size fractions isolated from representative samples of the Gaissa and Laksefjord Nappe Complexes of the Finnmark Lower Allochthon. All spectra have the coordinates shown at upper right. Data are plotted as in Fig. 7.

Ordovician) tectonothermal activity has been described from several locations within lithostratigraphic sequences believed to have originated in distal, outboard (westernmost) portions of the Baltic miogeocline (Seve units of the upper allochthon and associated imbricated basement units: e.g., Dallmeyer *et al.*, 1985, 1987; Dallmeyer and Gee, 1988). There is no clearly defined record of early Caledonian metamorphism or deformation within inboard (easternmost) portions of the Baltic miogeocline.

Acknowledgements

Analytical aspects of this work were supported in part by a grant from the U.S. National Science Foundation, Crustal Structure and Tectonics Program (EAR-8407027). We also acknowledge a postdoctoral research grant to A. Reuter from the Deutsche Forschungsgemeinschaft. Dr. A. Andresen collaborated during sample collection. The original manuscript benefitted from critical reviews by R. A. Gayer, A. H. N. Rice, and C. Townsend.

REFERENCES

Alexander, E. C. Jr., Michelson, G. M. and Lanphere, M. A. (1978). A new $^{40}Ar/^{39}Ar$ dating standard. In: Zartman, R. E. (ed.) *Short Papers of the Fourth International Conference on Geochronology, Cosmochronology and Isotope Geology*: U.S. Geological Survey Open-File Report, vol. 78–701, pp. 6–8.

Andersen, T. B. (1981). The structure of the Magerøy Nappe, Finnmark, North Norway. *Norges Geol. Unders.* **363**, 1–23.

Andersen, T. B., Austrheim, H., Sturt, B. A., Pedersen, S. and Kjaersrud, K. (1982). Rb–Sr whole-rock ages from Magerøy, north Norwegian Caledonides. *Norsk Geol. Tidsskr.* **62**, 79–85.

Bevins, R. E., Robinson, D., Gayer, R. A. and Allman, S. (1986). Low-grade metamorphism and its relationship to thrust tectonics in the Caledonides of Finnmark, North Norway. *Norges Geol. Unders. Bull.* **404**, 31–42.

Bonhomme, M., Thuizat, R., Pinault, Y., Clauer, N., Wendling, A. and Winkler, R. (1975). Méthode de datation potassium-argon. *Appareillage et téchnique; Note téchnique de l'Institut de Géologie, Univ. L. Pasteur, Strasbourg*, vol. 3.

Chapman, T. J., Gayer, R. A. and Williams, G. D. (1985). Structural cross-sections through the Finnmark Caledonides and timing of the Finnmarkian event. In: Gee, D. G. and Sturt, B. A. (eds.) *The Caledonian Orogen—Scandinavia and Related Areas*, pp. 593–610. Wiley, Chichester.

Clauer, N. (1974). Utilisation de la méthode rubidium-strontium pour la datation d'une schistosité de sédiments peu metamorphiques: Application au Précambrian II de la boutonnière de Bou Azzer-El Graara (Anti-Atlas). *Earth Planet. Sci. Lett.* **22**, 404–412.

Dallmeyer, R. D. (1988). Polyorogenic $^{40}Ar/^{39}Ar$ mineral age record within the Kalak Nappe Complex, Northern Scandinavian Caledonides. *J. Geol. Soc. Lond.*, **145**, 707–716.

Dallmeyer, R. D. and Gee, D. G. (1986). $^{40}Ar/^{39}Ar$ mineral dates from retrogressed eclogites within the Baltoscandian miogeocline: Implications for a polyphase Caledonian orogenic evolution. *Bull. Geol. Soc. Amer.* **97**, 26–34.

Dallmeyer, R. D. and Gee, D. G. (1988). Polyorogenic $^{40}Ar/^{39}Ar$ mineral age record in the Seve and Köli nappes of the Gäddede area, northwestern Jämtland, central Scandinavian Caledonides. *J. Geol.* **96**, 181–198.

Dallmeyer, R. D. and Keppie, J. D. (1987). Late Paleozoic tectonothermal evolution of the southwestern Meguma Terrane, Nova Scotia: *Can. J. Earth Sci.* **24**, 1242–1254.

Dallmeyer, R. D. and Lécorché, J. P. (1987). $^{40}Ar/^{39}Ar$ polyorogenic mineral age record within the central Mauritanides. In: R. D. Dallmeyer (ed.) *Geotraverse Excursion Across the Central Mauritanide Orogen*, pp. III-1–III-68, I.G.C.P. Project 233 Field Guide Series, University of Georgia.

Dallmeyer, R. D. and Lécorché, J. P. (1988). $^{40}Ar/^{39}Ar$ polyorogenic mineral age record within the central Mauritanide Orogen. *Bull. Geol. Soc. Amer.*, in press.

Dallmeyer, R. D. and Reuter, A., (1988). $^{40}Ar/^{39}Ar$ whole-rock dating and the age of cleavage in the Finnmark authochthon, northernmost Scandinavian Caledonides. *Lithos*, in press.

Dallmeyer, R. D., Gee, D. G. and Beckholmen, M. (1985). $^{40}Ar/^{39}Ar$ mineral age record of early Caledonian tectonothermal activity in the Baltoscandian miogeocline. *Amer. J. Sci.* **285**, 532–568.

Dallmeyer, R. D., Clark, A., Cumbest, R. J., Hames, W. E. and McKinney, J. (1987). Polyphase Caledonian tectonothermal evolution of the western gneiss terrane, Senja, Troms, North Norway. In: Gayer, R. A. and Townsend, C. (eds.) *The Caledonian and Related Geology of Scandinavia*. Abstracts p. 12. University College, Cardiff.

Dallmeyer, R. D., Mitchell, J. G., Pharaoh, T. C., Reuter, A. and Andersen, A. (1988). K–Ar and $^{40}Ar/^{39}Ar$ whole-rock ages of slate/phyllite from autochthonous-parautochthonous sequences in the Komagfjord and Alta-Kvaenangen tectonic windows, northern Scandinavian Caledonides: Evaluating the extent and timing of Caledonian tectonothermal activity. *Bull. Geol. Soc. Amer.*, **100**, 1493–1501.

Dalrymple, G. B., Alexander, E. C., Lanphere, M. A. and Kraker, G. P. (1981). Irradiation of samples for $^{40}Ar/^{39}Ar$ dating using the Geological Survey TRIGA reactor. *U.S. Geological Survey Professional paper 1176*.

Daly, J. S., Cliff, R. A., Gayer, R. A. and Rice, A. H. N. (1987). A new Precambrian terrane in the Caledonides of Finnmark, Arctic Norway. In: Gayer, R. A. and Townsend, C. (eds.) *The Caledonian and Related Geology of Scandinavia*, Abstracts p. 15. University College, Cardiff.

Føyn, S. (1985). The Late Precambrian in northern Scandinavia. In: Gee, D. G. and Sturt, B. A. (eds.) *The Caledonian Orogen—Scandinavia and Related Areas*, pp. 233–245. Wiley, Chichester.

Gayer, R. A., Hayes, S. J. and Rice, A. H. N. (1985a). The structural development of the Kalak Nappe Complex of eastern and central Porsangerhalvoya, Finnmark, Norway. *Norges Geol. Unders. Bull.* **400**, 67–87.

Gayer, R. A., Humphreys, R. J., Binns, R. E. and Chapman, T. J. (1985b). Tectonic modelling of the Finnmark and Troms Caledonides based on high level igneous rock geochemistry. In: Gee, D. G. and Sturt, B. A. (eds.) *The Caledonide Orogen—Scandinavia and Related Areas*, pp. 931–951. Wiley, Chichester.

Gayer, R. A., Rice, A. H. N., Roberts, D., Townsend, C. and Welbon, A. (1987). Restoration of the Caledonian Baltoscandian margin from balanced cross-sections: the problem of excess continental crust. *Trans. R. Soc. Edinburgh*, 197–217.

Gebauer, D. and Grünenfelder, M. (1974). Rb–Sr whole-rock dating of late diagenetic to anchimetamorphic, Paleozoic sediments in southern France (Montagne Noire). *Contrib. Mineral. Petrol.* **47**, 113–130.

Gee, D. G. (1975a). A geotraverse through the Scandinavian Caledonides—Östersund to Trondheim. *Sver. Geol. Unders. ser. C*, **717**.

Gee, D. G. (1975b). A tectonic model for the central part of the Scandinavian Caledonides. *Amer. J. Sci.* **275A**, 468–515.

Gee, D. G. and Roberts, D. (1983). Timing of deformation in the Scandinavian Caledonides. In: P. E. Schenk (ed.) *Regional Trends In The Geology Of The Appalachian–Caledonian–Hercynian–Mauritanide Orogen*, pp. 279–292. Riedel, New York.

Henningsmoen, G. (1961). Cambro-Silurian fossils in Finnmark, North Norway. *Norges Geol. Unders.* **213**, 289–314.

Huon, S. (1985). Clivage ardoisier et réhomogénéisation isotopique K–Ar dans les schistes paléozoiques du Maroc: Thèse 3ème cycle, Université Louis Pasteur, Strasbourg, 124pp.

Hunziker, J. C., Frey, M., Clauer, N., Dallmeyer, R. D., Friedrichsen, H., Flehmig, W., Hochstrasser, K., Roggwiler, P. and Schwander, H. (1986). The evolution of illite to muscovite: mineralogical and isotopic data from the Glarus Alps, Switzerland. *Contrib. Mineral. Petrol.* **92**, 157–180.

Kubler, B. (1967). La cristallinité de l'illite et les zones tout à fait supérieures du métamorphisme. Colloque sur les "Etages Tectoniques", 18–21 avril 1966, Festschrift, p. 105–122.

Kunk, M. J., Sutter, J. F., Obradivitch, J. and Lanphere, M. A. (1985). Age of biostratigraphic horizons within the Ordoivician and Silurian systems. In: Snelling, N. J. (ed.) *The Chronology of the Geological Record. Geol. Soc. Lond.*, Memoir 10, pp. 89–92.

Palmer, A. R. (1983). The Decade of North American Geology 1983 Geologic Time Scale. *Geology* **11**, 503–504.

Pharaoh, T. C. (1985). The stratigraphy and sedimentology of autochthonous metasediments in the Repparfjord–Komagfjord tectonic window, west Finnmark. In: Gee, D. G. and Sturt, B. (eds.) *The Caledonian Orogen—Scandinavia and Related Areas*, pp. 347–357. Wiley, Chichester.

Pringle, I. R. (1973). Rb–Sr age determinations on shales associated with the Varanger Ice Age. *Geol. Mag.* **109**, 465–472.

Ramsay, D. M., Sturt, B. A., Zwaan, K. B. and Roberts, D. (1985). Caledonides of northern Norway. In: Gee, D. G. and Sturt, B. A. (eds.) *The Caledonide Orogen—Scandinavia and Related Areas*, pp. 163–184. Wiley, Chichester.

Reuter, A. (1985). Korngrössenabhängigkeit von K–Ar Datierungen und Illitkristallinität anchizonaler Metapelite und assoziierter Metatuffe aus dem östlichen Rheinischen Schiefergebirge. *Göttinger Arb. zur Geol. Paläontol.* **27**, 91.

Reuter, A. (1987). Implications of K–Ar ages of whole-rock and grain-size fractions of metapelites and intercalated metatuffs within an anchizonal terrane. *Contrib. Mineral. Petrol.* **97**, 105–115.

Reuter, A. and Dallmeyer, R. D. (1987a). Significance of $^{40}Ar/^{39}Ar$ age spectra of whole-rock and constituent grain-size fractions from anchizonal slates. *Chem. Geol. (Isotope Geosci. Sec.)* **66**, 73–88.

Reuter, A. and Dallmeyer, R. D. (1987b). $^{40}Ar/^{39}Ar$ dating of cleavage formation in tuffs during anchizonal metamorphism. *Contrib. Mineral. Petrol.* **97**, 352–360.

Reuter, A. and Dallmeyer, R. D. (1988). K–Ar and $^{40}Ar/^{39}Ar$ dating of cleavage formed during very low-grade metamorphism. A review. In: S. Daly (ed.) *Evolution of Metamorphic Belts*, in press.

Roberts, D. (1971). Timing of Caledonian orogenic activity in the Scandinavian Caledonides. *Nature (Phys. Sci.)* **232**, 22–23.

Roberts, D. (1985). The Caledonian fold belt in Finnmark: a synopsis. *Norges Geol. Unders.* **403**, 161–178.

Snee, L. W., Sutter, J. F. and Kelly, E. C. (1988). Thermochronology of economic mineral deposits: Dating stages of mineralization at Panasqueira, Portugal, by high-precision $^{40}Ar/^{39}Ar$ age-spectrum techniques on muscovite. *Economic Geol.*, **83**, 335–354.

Snelling, N. J. (1985). *The Chronology of the Geological Record. Geological Society of London*, Memoir 10.

Steiger, R. H. and Jäger, E. (1977). Subcommission on geochronology convention on the use of decay constants in geo- and cosmochronology. *Earth Planet. Sci. Lett.* **36**, 359–362.

Stephens, M. B. and Gee, D. G. (1985). A tectonic model for the evolution of the eugeoclinal terranes in the central Scandinavian Caledonides. In: Gee, D. G. and Sturt, B. A. (eds.) *The Caledonide Orogen—Scandinavia and Related Areas*, pp. 953–978. Wiley, Chichester.

Sturt, B. A., Pringle, I. R. and Roberts, D. (1975). Caledonian nappe sequence of Finnmark, northern Norway, and the timing of orogenic deformation and metamorphism. *Bull. Geol. Soc. Amer.* **86**, 710–718.

Sturt, B. A., Pringle, I. R. and Ramsay, D. M. (1978). The Finnmarkian phase of the Caledonian orogeny. *J. Geol. Soc. Lond.* **135**, 597–610.

Teichmüller, M., Teichmüller, R. and Weber, K. (1979). Inkohlung und Illitkristallinität—Vergleichende Untersuchungen im Mesozoikum und Palaozoikum von Westfalen. *Fortschr. Geol. Rheinland u. Westfalen* **27**, 201–276.

Turner, G. (1968). The distribution of potassium and argon in chondrites. In: Ahrens, L. H. (ed.) *Origin and Distribution of the Elements*, vol. 30, pp. 387–398. Pergamon Press, Oxford.

Weber, K. (1972). Notes on determination of illite crystallinity. *Neues Jahrbuch Mineralogie, Monatshefte*, **6**, 267–276.

Wybourn, D., Owen, M., Compston, W. and McDougall, I. (1982). The Ludlow volcanics: A Late Silurian point on the geological time-scale. *Earth Planet. Sci. Lett.*, **59**, 90–100.

APPENDIX

Sample localities

Sample 1.   Roadcut along E-6 at Sjursjork.
Sample 2.   Roadcut along E-6, 7.4 km west of Sjursjork.
Sample 3.   Roadcut along E-6, 7.5 km east of Ifjord.
Sample 4.   Roadcut along E-6 at Iggeldas bridge.
Sample 5.   Roadcut along E-6, 1.2 km north of Iggeldas.
Sample 6.   Roadcut, 6.8 km south of Bukta; eastern shore of Ifjorden.
Sample 7.   Roadcut, 5.6 km north of Ifjord.
Sample 8.   Friar Fjorden slate quarry.
Sample 9.   Roadcut along E-6 at Tarnvik.

3   Regional correlations in NE Troms–W Finnmark: the demise of the "Finnmarkian" orogeny?

Richard E. Binns

Strindveien 64, N–7015 Trondheim, Norway

The age discrepancy between the Upper Ashgill fossil-bearing strata in NE Troms and the previously correlated assumed mid-Cambrian and older Sørøy strata in Finnmark is examined in its full breadth for the first time. Previous proposals unreservedly supporting radiometric evidence for the Sørøy strata age and seeking means of reinterpreting other data to avoid the correlation, are shown to be largely unacceptable. The grounds for the original correlation are examined and new points in its favour are noted, including indications that intrusions from the Seiland Igneous Province (the source of most adduced radiometric evidence) extend further southwest than usually assumed, to intrude the fossil-bearing segment. Thrusts are traced through the region. The tectonothermal alteration of strata on either side of the synorogenic Skibotn/Helligskogen Nappe Complex boundary seems everywhere to result from a single orogenic cycle, assumed on radiometric evidence to be Finnmarkian in Finnmark, but proved by the fossils in the overriding nappe to be Scandian in Troms. Hence, despite some apparently irrefutable radiometric data indicating a late Cambrian–early Ordovician orogenic event, abundant field evidence added to the fossils supports a late-Silurian age for the main orogeny affecting all these strata. Although final acceptance of this view should await more data (not least additional fossils), the radiometric data and consequently the Finnmarkian orogenic phase as defined in this area should be regarded as highly suspect.

INTRODUCTION

During the present decade it has become common practice to make a two-fold division of the Caledonian nappe pile of Finnmark and Troms, distinguishing Finnmarkian from overlying Scandian Nappes. By definition (Sturt *et al.*, 1978), the former consist of rocks whose primary age pre-dates an orogenic phase that Ramsay and Sturt (1976) had named the Finnmarkian (recently raised to the Finnmarkian Orogeny by Ramsay and Sturt (1986)). Published radiometric data (chiefly Rb–Sr whole-rock dates on Seiland Igneous Province (SIP) intrusions) have suggested that this took place about 540–490 Ma ago (e.g., Sturt *et al.*, 1967, 1978; Pringle and Sturt 1969), but new data (Pedersen *et al.*, this volume) indicate that intrusion ceased earlier (530–520 Ma ago). Most or all polyphasal tectonothermal alteration of these rocks is generally attributed to this Finnmarkian event, and in the case of some older units, to earlier (Precambrian) event(s) (e.g. Daly *et al.*, 1987). Since the second group of nappes contains fossil-bearing Ordovician–Silurian strata (e.g., Henningsmoen, 1961; Binns and Gayer, 1980), the approximately 90–130 Ma younger Scandian orogenic phase is judged responsible for the polyphasal deformation, metamorphism and magmatism to which at least most of the bedrock forming those nappes was subjected.

Ductile southeastward thrusting of the apparently older set of nappes in Finnmark is believed to have occurred synorogenically with respect to the Finnmarkian; later east–southeastward brittle movement on lower thrusts is also judged to be a Finnmarkian event (e.g., Sturt *et al.*, 1975, 1978; Gayer *et al.*, 1985; 1987; Rice, 1987a; Townsend, has been found by these workers, although Sturt *et al.* (1975, 1978) mention evidence in its favour. These views, however, conflict with the fossil and tectonometamorphic evidence from Troms (and areas further south) that indicates that the Scandian phase is responsible for most or all synorogenic thrust transport and the final piggy-back emplacement of the entire nappe pile, effected by brittle movement on the regional sole thrust, as previously argued in respect of both Troms and the surrounding region (Binns, 1978). Zwaan and Roberts (1978) and Zwaan and Gautier (1980) suggested that the final emplacement may have taken place in Scandian time.

In Finnmark, the largest tectonic unit attributed to the Finnmarkian phase has been named the Kalak Nappe Complex (Roberts, 1974), its uppermost portion now being called the Sørøy Nappe. The overlying Magerøy Nappe (Ramsay and Sturt, 1976) on Magerøy (Fig. 1) is the only unit recognized as being emplaced by Scandian thrusting.

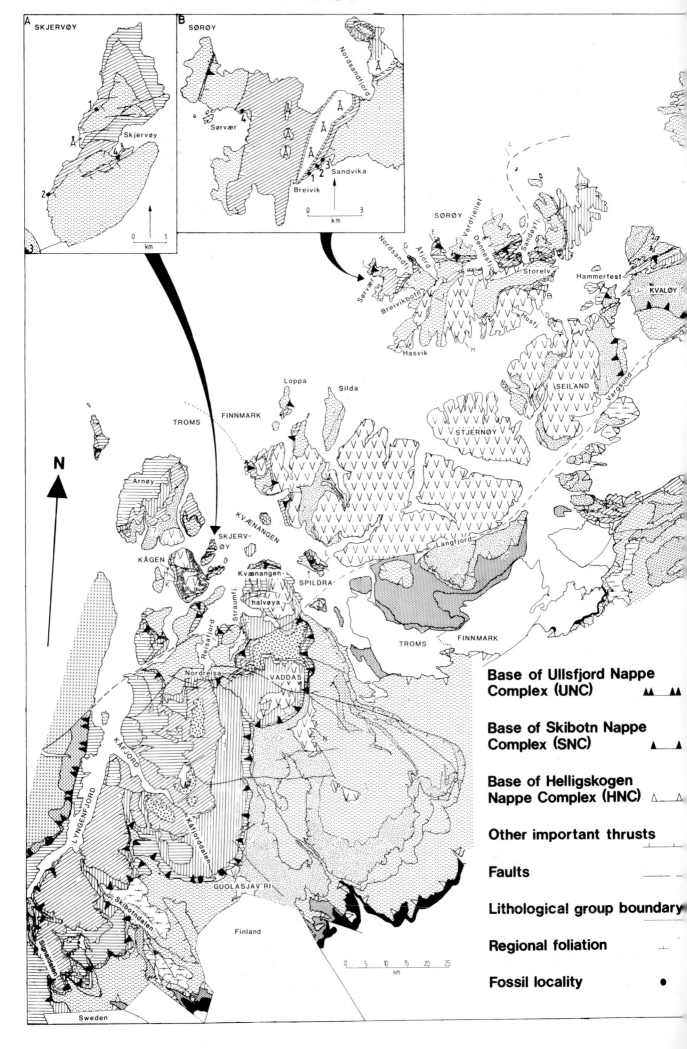

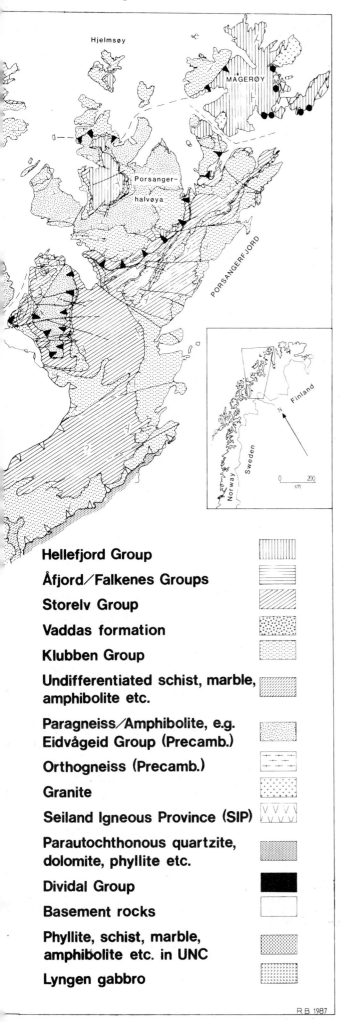

Hellefjord Group

Åfjord/Falkenes Groups

Storelv Group

Vaddas formation

Klubben Group

Undifferentiated schist, marble, amphibolite etc.

Paragneiss/Amphibolite, e.g. Eidvågeid Group (Precamb.)

Orthogneiss (Precamb.)

Granite

Seiland Igneous Province (SIP)

Parautochthonous quartzite, dolomite, phyllite etc.

Dividal Group

Basement rocks

Phyllite, schist, marble, amphibolite etc. in UNC

Lyngen gabbro

Fig. 1. Simplified map of the bedrock geology of NE Troms and W Finnmark showing the main lithological group divisions and nappe boundaries of the Caledonian rocks. (Based on many sources, principally Roberts (1974, 1985a), Sigmond *et al.* (1984), Gayer *et al.* (1985), Roberts and Andersen (1985), Zwaan (1988) and own work.) Lithological boundaries on inset A are based on Ash (1968) and those on inset B on Ramsay (1971b); nappe boundaries originate from own observations.

In NE Troms, strata believed to pass laterally south-westwards from the Kalak Nappe Complex (KNC) were divided into the Helligskogen and overlying Skibotn Nappe Complexes (HNC and SNC) (Binns, 1978), the latter being equated with the Sørøy Nappe (then called the Seiland Nappe). At that time this was a generally accepted (e.g., Hooper and Gronow, 1970; Armitage *et al.*, 1971; Gayer, 1973; Roberts, 1974; Sturt *et al.*, 1978; Zwaan and Roberts, 1978) tectono- and lithostratigraphic correlation involving metasediments supposed (e.g., Ramsay and Sturt, 1963; Roberts, 1968a; Holland and Sturt, 1970; Ramsay, 1971a) to be of late Riphean or Vendian to about mid-Cambrian age in the Sørøy district (Fig. 1). However, doubt about the age and/or correlation was voiced by Binns (1978) because the assumed youngest unit in both areas (particularly as developed in NE Troms) resembled strata (the Furulund Group and assumed correlates) further south in northern Norway and northern Sweden proved by fossils to be of Ordovician–Silurian age. Shortly afterwards, fossils of mid–late Ordovician to Silurian age were unexpectedly found preserved at Guolasjav'ri in NE Troms (Fig. 1) in a distinctive, widely traceable, mid-amphibolite facies, carbonate-bearing formation (Binns and Gayer, 1980) stratigraphically underlying the queried unit. These fossils proved that the Scandian phase was responsible for all the tectonometamorphic alteration of at least the bulk of the SNC.

This clearly called for re-evaluation of the basis for the age of SIP and the Sørøy region strata, the litho- and tectonostratigraphic correlations across the Finnmark–Troms border (Fig. 1), the distribution of Finnmark-ian–Scandian orogenic belt rocks in this region, and even the existence of the Finnmarkian here.

A number of proposals have been put forward to solve the discrepancy. Ramsay *et al.* (1981, 1985a) and Ramsay and Sturt (1986) claimed the existence of a major unconformity in the Kvaenangen–Kåfjord district of NE Troms (Fig. 1) underlying the fossil-bearing strata and transecting a major overfold involving beds correlated with all but the youngest member of the Sørøy-district sequence. This situation was said to be found in the basal part of the lowest SNC nappe unit, thus implying that the HNC/SNC boundary was thought to cut the Sørøy succession, since that was also shown occupying the uppermost nappe unit (Sørøy Nappe) in KNC (= HNC) (Ramsay *et al.*, 1985a, Fig. 1). The revised 1:1 million bedrock map of Norway (Sigmond *et al.*, 1984) also places the Sørøy Nappe uppermost in the HNC horizon in NE Troms and depicts the Finnmarkian/Scandian boundary coinciding with the HNC/SNC boundary there. Other-wise, it shows some notable differences in lithological correlations in this general district, some of which contra-dict the unconformity theory and imply a different overfold from that mentioned above. Finally, Krill and Zwaan (1987) presented evidence from Sørøy that they interpreted as showing that the radiometrically-dated intrusive activity (SIP) was entirely pre- rather than broadly synorogenic, and was rift-related (see also Rice, 1987b). They also concluded that the intrusions are confined to the basal (arkosic) unit of the Sørøy sequence and the underlying Precambrian gneiss and implied that an unconformity (presumably broadly equivalent to that reported by Ramsay *et al.* (1985a) and Ramsay and Sturt (1986)) separates the arkosic unit from the overlying part of the Sørøy sequence.

In my opinion, too much emphasis has so far generally been placed on unreserved reliance on radiometric data for constraining the age of the bedrock and the tectono-thermal activity (Finnmarkian). Such data should not be

uncritically preferred to field, faunal and other evidence. As it is, the primary evidence for the original regional correlations has been disregarded without being disproved or significantly discussed.

This paper aims to examine the problem more broadly and, in addition to considering the recent proposals, will summarize the substantial evidence from regional litho-stratigraphic development, relations between igneous intrusions and their country rock, tectonostratigraphic relationships and tectonothermal history, which, in view of the Guolasjav'ri fossil evidence, does not support the conclusions based on the radiometric data.

AGE EVIDENCE

The discovery in 1979 at Guolasjav'ri in Kåfjorddalen (Fig. 1) of a variety of fossils including tabulate corals (Binns and Gayer, 1980), proved that most of the bedrock sequence in NE Troms which for more than a decade had been correlated with the "type succession" on Sørøy, was Ordovician–Silurian in age, at least 100 Ma younger than the supposed age of the Sørøy bedrock. The age estimate of the fossil-bearing strata at Guolasjav'ri has now been improved by the identification of the Upper Ashgill brachiopod *Holorhynchus* (Bassett and Cherns, personal communication, 1987).

A Riphean to mid-Cambrian age for the correlated Klubben–Hellefjord Group succession in the Sørøy region rests entirely on radiometric age determinations (e.g., Sturt *et al.* 1967, 1978; Pedersen *et al.*, this volume) mostly dating SIP bodies that intrude the metasediments. No fossils have been found, Archaeocyatha claimed by Holland and Sturt (1970) being disproved by Debrenne (1984).

Prior to the first age determinations (Sturt *et al.*, 1967) it had been natural to assume that the tectonometamorphic and magmatic activity affecting the rocks of this region was late Silurian in age. This agreed with the presence of fossil-bearing Ordovician–Silurian metasediments on Magerøy (Fig. 1) (e.g., Henningsmoen, 1961), and was in keeping with most of the Scandinavian Caledonides even though there were scattered signs of a late Cambrian-early Ordovician event of ill-defined nature.

Most radiometric age data (e.g., Sturt *et al.*, 1967, 1978: see Pedersen *et al.* (this volume) for a review of these and some new data) indicate broadly syntectonic SIP intrusive activity ca. 540–490 Ma ago. However, partial lack of published analytical data and information on sampling procedures, along with some imprecision, help to put their interpretation in doubt, especially in view of the conflicting faunal and correlation evidence. The appar-ently more reliable zircon ages of 531 Ma and 523 Ma now published by Pedersen *et al.* (this volume) give a new age estimate for the late-synorogenic intrusion of alkaline magma previously dated at ca. 490 Ma. Those authors take this to show that "Finnmarkian" deformation and metamorphism ended before 531 to 523 Ma ago, and point out that this precludes the correlation made by Sturt *et al.* (1978) and Ramsay and Sturt (1986) between the youngest magmatic events in SIP, which date the D2 event of the Sørøy region, and the age of cleavage at lower levels in East Finnmark, dated imprecisely at around 493–504 Ma (Pringle, 1973).

Apparent absence of tillitic material (Sturt *et al.*, 1975) has usually been taken to indicate that Klubben Group deposition post-dated the ca. 650 Ma old Varangian tillites, although Rice (1987b) has recently inferred that all Klubben–Hellefjord Group deposition was completed before that time. Radiometric evidence is indeed indicating an older history for the Klubben Group. Daly *et al.* (1987), for example, reported preliminary, imprecise ages for granite intrusions into both HNC and SNC Klubben Group portions of KNC (see below) in Porsanger. This led them to suggest that the D2 deformation of this group predated 800 Ma and may have resulted from the Sveconorwegian or Morarian orogenies.

STRATIGRAPHY AND STRATIGRAPHIC CORRELATION

Tables I and II detail the type lithologies and probable stratigraphies of the two areas under discussion, and Table III and Fig. 2 show how they seem to correlate. Figure 1 shows the simplified geology of the region based on group divisions established on Sørøy and in NE Troms. For convenience, Sørøy terminology is used throughout, even though revision is called for.

The stratigraphy for NE Troms shown in Table I resembles that worked out by Pearson (1970, 1971) in the area just southwest of Kvaenangen, which formed the main basis for the standard correlation to Sørøy in the 1970s. However, field work carried out by the author there and elsewhere in the region has shown a need for certain revisions, the basis for these being given below and in a forthcoming publication. Way-up information is sparse; some is located in and close to the area mapped by Pearson (Pearson, 1970; Lindahl, 1974), and some near Guolasjav'ri (Padget, 1955; Binns and Gayer 1980; Binns, unpublished observations). The present scheme differs in certain respects from that proposed by Zwaan (1988), because of disagreement about some structures and lithological correlations.

Before the Guolasjav'ri fossils were found in 1979, there was broad agreement (e.g., Hooper and Gronow, 1970; Armitage *et al.*, 1971; Gayer, 1973; Roberts, 1974; Binns,

Table I. Proposed lithostratigraphy of the Signaldalen–Kåfjorddalen area, based on own work and that of Padget (1955). The double line indicates the HNC/SNC boundary. This compilation is based on information obtained from the whole area, derived from several nappe units and major overfolds none of which have the full sequence preserved.

Stratigraphic divisions	Important lithologies and assemblages	Characteristic features
Ankerlia formation	qz-rich am-(cl)zois-pl sch bi-pl-qz-(cl)zois-(am) sch am-rich qz-(cl)zois-pl-(gar) sch diop-am-pl-(cl)zois sch hbl sch	interbanded (cm-dm) giving lt. to dk. grey & green + white (± reddish) flaggy/platy rock; often pitted weath, and w/am garben; turbiditic dk, green to black, fine-grained, lava
Ak'kejav'ri formation	calc & qz-rich bi-am-(cl)zois sch calc-bi-pl-qz-am-(cl)zois sch calc-bi-am-gar-(cl)zois-qz sch polymict cglm	calcite mainly ankeritic, giving orange- brown & pitted weathering; lithols, interbanded w/qtzitic layers dominant very local, at base
Guolasjav'ri formation	hbl sch/lava impure mbl, w/am phl ep pl etc. marble, sometimes gr-rich various calc silicate sch qz-mu-gar-am sch qz-mica-gar-(al.sil.) sch	pillows, etc. locally preserved sil. mins. esp. in mm-cm bands grey, banded, relat. coarse compositional colour-banded lt. grey, nonaligned dk. am blades violet-grey; rusty weathered
Cizzenvar'ri formation	qz-pl-mica sch qtzite feld-bearing qtzite qz-pl-mica-gar-(gr)-(al.sil.) sch	grey; rusty to red-brown weath. white to light buff, platy, light rust brown to reddish weath. grey, w/rusty to red-brown weath.
Markusriep'pi formation	qz-gar-(al.sil.)-mica sch marble w/diop pl (vesuv fluor) etc. skarn w/ep gross vesuv etc. calc sil. sch./qz-mica-gar-(am) sch	relat. coarse; grey; rusty weath. coarse, light col.; grey or buff weath. fine gr.; mm-cm green and grey bands
Markusfjellet formation	qz-rich pl-bi-hbl-ep-gar sch qz-pl-mica-ep-gar sch qz-pl-mu-ep-gar sch qztitic (± feld) sch qz-pl-kfeld-mica-(al.sil.) sch	grey, w/green ep/hbl/diop/calc lenses light grey, med. grained, often fairly massive layered qtzitic/subarkosic rock w/mica-rich laminae/layers light col., rusty weath., qz/feld-rich
Addjet formation	mica-qz-pl-gar-al.sil.sch mica-qz-pl-gar sch	relat. coarse, grey; rusty weath., relat. mica-rich, but many qtzitic ribs
Goatteras'sa formation	qtzite feld-bearing qtzite qz-gar-mica sch qz-pl-kfeld-si-bi-gar sch	massive to blocky, lt. grey to buff, nr. glassy, dk. mica laminae; ± gar-hbl interlayered w/qtzite knots/flecks of bi & si, v. local
	partly/wholly recryst. myl./blasto- myl.meta-arkose w/thrust sheets of granitoid gn., hbl gn., etc. often w/migm. gn./arkose boundary zones; also bi-hbl sch./dol.mbl. unit	arkose is banded, grey or green, often flaggy; migm. zones often med/coarse grained w/abundant red & white feld

Table II. Lithostratigraphy of the Sørøy region, based mainly on Roberts (1968a, 1974) and Ramsay (1971a). The double line indicates the HNC/SNC and other thrust boundaries.

Stratigraphic divisions	Important lithologies and assemblages	Characteristic features
Hellefjord Group	qz-feld-act-(gar)-(bi) sch bi-gar-qz-feld-(act) sch calc-sil.-qz-feld-am sch	flaggy, med/fine grained greenish-grey sch. & fine grained grey/purplish-grey phyllitic sch. (distal turbid.) lt. green/white ribs (± 1 cm) in grey sch., esp. in W
Åfjord Group	ky-mica-(gr)-(gar) sch/phyll qtzite gr-ky-(st)-mica-(gar) sch/phyll	rusty weath., grey, partly coarse grained rusty weath., nr. white to lt. buff, granular rusty weath., dk. grey w/calc sil. bands (ca. 15 cm)
Falkenes Group	gar-mica-al.sil. sch calc-sil. sch mbl w/calc-sil. ribs/laminae gr-ky-mica-(gar)-(si) sch calc-sil.(act $\pm$ diop) sch ky-gr-mica sch. $\pm$ qtzite mbl tr/act-diop-phl sch/phyll ep sch	rusty weath., garnet-rich blue-grey, rel. massive, coarse, brown weath., rusty weath., grey to violet lt. col. to greenish and purplish rusty weath.; qtzite lt. buff, granular, local in E & W bands of thin, pure to impure, lt. grey to blue-grey mbls & various green, nr. white and med. grey calc-sil. schs
Storelv Group	mu-bi-(gar) sch. $\pm$ calc-sil qtzitic & mu sch gar-mu-(gr)-(al.sil.) sch	fairly massive, grey weath., grey to violet grey weath., lt. col. to grey, often clearly layered rusty weath., grey, micaceous w/qtzitic ribs
Klubben Group	qtzite feld qtzite meta-arkose qz-rich mica-(gar) sch mica-(si)-(gar) sch granitic gneiss	distinctly layered, light/med grey to lt. buff mainly meta-arkosic rock, w/darker micaceous layers and laminae, X-bedding, etc. common in places; pockets and lenses of carb.-rich meta-sst in places; flaggy when affected by ductile strain; arkose beds mainly 5–20 cm thick; extensive belts of migmatized arkose

Table III. Proposed stratigraphical correlations through NE Troms–W Finnmark. (Based on various sources, chiefly: column 1, own work; column 2, Pearson (1970), own work; column 3, Ash (1968), own work; column 4, Hooper and Gronow (1970), own work on Spildra (Sp. 1–5); column 5, e.g., Ramsay (1971a), Roberts (1974).)

Signal./Kåfj.	Straumfjord	Skjervøy	Kvaenangen	Sørøy
Ankerlia fm.				
	Oksfjord fm.		Sp.5	Hellefjord Gp.
Ak'kejav'ri fm.				
	local rusty qtzite/sch. lenses	Bratteidet fm.	Sp. 4	Åfjord Gp.
Guolasjav'ri fm.	Straumfjord fm.	Engenes fm.		
Cizzenvar'ri fm.	Fossvann fm.		Loppa gp./Sp. 3	Falkenes Gp.
Markusriep'pi fm.	Fosselv fm.	Prestberget fm.		
Markusfjellet fm.		Lysthus fm.		
	Heindalstind fm.		Mevaer gp.Sp. 2	Storelv Gp.
Addjet fm.		Vågen fm.		
Goåtteras'sa fm.	Vaddas fm.	?Trollfjellet fm.	Sp. 1	
		Thrust separated		
Stordalen fm.	Raesvar'ri fm.	Skattefjellet fm.	Brynilen fm.	Klubben Gp.

1978; Sturt *et al.*, 1975, 1978; Zwaan and Roberts, 1978) that the lithostratigraphic sequence of the Sørøy district strongly resembled and could readily be correlated with that in SNC along with the meta-arkose of HNC, although a few minor (Binns, 1978) and more major (Zwaan and Roberts, 1978) differences of opinion were apparent. That there was an overall similarity in the sedimentary environment throughout the region, also above the meta-arkose horizon, was thus accepted, even though it was already clear from work by, for example, Padget (1955) and Lindahl (1974) that there was a distinct increase in volcanic horizons in the southern area (Tables I and II, Fig. 2).

After the discovery of the Guolasjav'ri fossils, SNC in NE Troms has been stated to have "markedly different lithostratigraphic character" (Ramsay *et al.*, 1985a) from the Sørøy sequence (see also Ramsay and Sturt, 1986). The fossil-bearing Ordovician?–Silurian sediments of the Magerøy Nappe, on the other hand, have now been correlated with SNC strata (Ramsay *et al.*, 1985a), even though they also lack volcanics. No new information was offered to support these reversals of opinion.

Figure 2 and Tables I and II demonstrate the strong similarity between the Sørøy and NE Troms sequences, on group, formation and assemblage levels. The overall consistency in lithofacies development supports the initial view that these rocks originated in a single sedimentary basin. Even though this apparent consistency could be

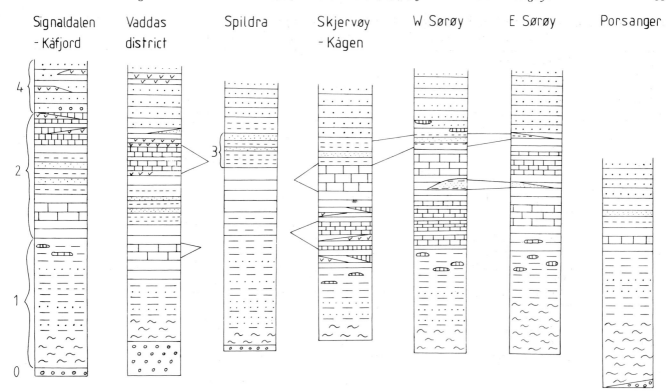

| Signaldalen – Kåfjord | Vaddas district | Spildra | Skjervøy – Kågen | W Sørøy | E Sørøy | Porsanger |

Fig. 2. Synthesis of the Caledonian lithostratigraphical sequence (in SNC) in NE Troms–W Finnmark, showing the strong consistency in development through the region and emphasizing some facies variations. The compilation attempts to take account of variations at different tectonic levels. Individual columns also largely represent specific parts of the basin, i.e. the two left-hand columns chiefly show the westernmost, probably most predominantly distal, facies development, the two central columns more central developments, the two Sørøy columns are again relatively distal, and the left-hand column probably represents the most proximal (eastern) facies development. 0 = Vaddas formation, 1 = Storelv Gp., 2 = Falkenes Gp., 3 = Åfjord Gp., 4 = Hellefjord Gp. Brick-like ornament indicates marbles/calc–silicate rock—most massive when bricks are large; v indicates lava etc.; dots indicate psammitic horizons/breds; lines and waves indicate dominantly pelitic schists.

coincidental, resulting from repeated development of similar facies, no locations for breaks can be found that adequately satisfy limitations placed by structural features and intrusions apparently related to SIP (see below).

Three significant points of similarity may be emphasized.

(1) The consistent appearance of calcareous lenses in the upper part of the Storelv Group on Sørøy (Roberts, 1968a; Ramsay, 1971a), in the Loppa area (Hooper and Gronow, 1970), on Skjervøy (Ash, 1968) and widely south of Kvaenangen (own observations).

(2) Graphitic (and often aluminium silicate-rich) schist and white quartzite, both showing characteristic rusty weathering, are found at two horizons related to the Falkenes Group. Firstly, above the uppermost marble (Ramsay's (1971a) Åfjord Group on Sørøy); they are also found on Skjervøy, Kågen and Spildra (own observations), and perhaps as fragmentary slices in the Kvaenangen–Nordreisa district (reinterpretation of Pearson (1970) and Armitage (1972) and further south (own observations). Secondly, between the marble formations where, as the Cizzenvar'ri formation (Tables I and III), it is widely and locally thickly developed in Troms (e.g., Padget's (1955) Nitsim Varre series, and Pearson's (1970) Fossvann quartzite—see Table III), but also appears on western Sørøy (own observations west of Dønnesfjord) and eastern Sørøy (Roberts 1968a, Fig. 10).

(3) My recent field work has strengthened the impression gained from the literature that some Ordovician?– Silurian strata on Magerøy, notably the Lower Silurian Sardnes and Juldagnes Formations, where they occur in an

equivalent amphibolite facies state in northern Magerøy, strongly resemble the common development of the Hellefjord Group on Sørøy and its suggested equivalent in NE Troms. Absence of volcanics on Magerøy is the chief difference, but is a predictable facies development from Troms, especially in view of their scarcity on Sørøy (Roberts, 1974). The apparent Troms–Magerøy link is not surprising considering that the relevant Troms strata immediately overlie the Upper Ashgill fossil-bearing marble/schist formation at Guolasjav'ri. However, Ramsay and Sturt (1976) and others with broad knowledge of Sørøy geology (e.g., Roberts, 1985b; Roberts and Andersen, 1985) have not referred to the similarity with the Hellefjord Group, perhaps because of their opinions that the two groups are of different age.

Notwithstanding dominant lithostratigraphic similarity, there are some differences. Apart from local and some more far-reaching repetitions or excisions attributable to tectonics, these seem readily accountable for by facies variations within one depositional basin, especially since they are most marked in those parts of the sequence apparently deposited in a shelf-sea environment (Storelv–Åfjord Groups). Most variations occur *within both* parts of the region, rather than *between* them, e.g. psammite/pelite proportions in the Storelv Group, aluminium silicate content of Storelv, Falkenes and Åfjord Group pelites, and changes in the character of marbles in the Falkenes Group. Figure 2 attempts to bring out the first and last mentioned variations. Concerning the last one, field observations and literature study show that the character of the two main marble horizons varies, as regards purity of individual beds, proportion of silicates in them and proportion of calc silicate/pelite beds between marble beds. One change occurs crossing Sørøy (cf. Roberts, 1968a, 1974; and Ramsay, 1971a), the other between Skjervøy and Straumfjord (Fig. 2), but here in different nappe units. Several lensoid marble members may be present.

The only marked lithological difference that is almost confined to only one part of the region is the occurrence of extrusions, mostly as basic lava (with locally preserved pillow structures). This becomes an increasingly important element of the Falkenes Group southwestwards, seen first on Skjervøy (Fig. 2). Basic volcanic material is also more widespread in the youngest group in the southwest, but tuff is found in the Hellefjord Group of NE Sørøy (Roberts, 1974). Hence, extrusive volcanic activity was probably largely confined to western and southern parts of the basin. Indeed the eastern part seems to have contracted substantially after deposition of the Storelv Group. This is indicated by the thin and local development of the Falkenes/Åfjord Groups (Gayer *et al.*, 1985; Ramsay *et al.*, 1985b; Roberts and Andersen, 1985), the proximal character of the Hellefjord Group on Hjelmsøy (Roberts and Andersen, 1985) and the tendency for proximal and flysch-type turbidites (Andersen, 1984) in the possibly equivalent strata on Magerøy, in contrast to the distal turbidites of the Hellefjord Group on Sørøy (Roberts, 1968b).

Figure 2 and Tables I–III show two main differences from previously published correlations.

(1) *The "Vaddas quartzite" (Vaddas formation) correlation with the Klubben Group of Sørøy (and equivalents) is rejected.*

The correlation by Pearson (1970, 1971) and Armitage *et al.* (1971) between the "Vaddas quartzite" and a meta-arkosic schist west of Straumfjord (Fig. 1), which itself was correctly correlated with Klubben Group meta-arkose on Skjervøy, across Kvaenangen and on to Sørøy, was rejected by Binns (1978) and Zwaan and Roberts (1978). However, Ramsay *et al.* (1985a) and Ramsay and Sturt (1986) have again correlated the "Vaddas quartzite" with the Klubben Group. Zwaan and Roberts (1978) and Zwaan (1988) linked the former to the Falkenes/Åfjord Groups of the Sørøy sequence. Sigmond *et al.* (1984) and Zwaan (1988) claim correlation with the Cizzenvar'ri Formation (cf. Table I). In Zwaan's case, the two-way correlation implies assumption of the Sørøy–NE Troms (SNC) link and therefore a common Scandian age for all the strata above the Klubben Group, as postulated by Krill and Zwaan (1987).

The possibility that the Vaddas and Cizzenvar'ri Formations are equivalent has been a recurring problem during years of mapping in this region, owing to local similarities in appearance and a tendency, for structural reasons, for the units to occur between two marbles. Comparisons are further hindered by the apparent correlative of the normal Storelv Group being only thinly and discontinuously present in lower and middle SNC levels in most of NE Troms. Structural and stratigraphic considerations in at least most of NE Troms seem best served by distinguishing two main quartzite-bearing formations, as in Table I and as explained in more detail elsewhere (below, and Binns, in prep.). These are additional to the Klubben Group.

The typical banded character of the Klubben Group over large parts of Sørøy, Kvaløy and Porsangerhalvøya (Table II, Fig. 1) is often, there and particularly southwestwards into Troms, substantially changed by strong ductile layer-parallel strain leading to flagginess, varying degrees of (blasto-)mylonitization, and a variably grey or green colour depending on mineral content and degree of

mylonitization and subsequent recrystallization (Binns, 1978). Extensive belts are migmatized. Only very locally does it resemble typical Vaddas formation (Table I), mainly in thin beds, and insofar as gneiss or migmatite occurs in some Vaddas formation fold cores (e.g. Fig. 4). On Spildra (see below), typical Vaddas formation quartzite seems to have been placed uppermost in the Klubben Group during the original mapping; this may not be a unique case. The formation is not known northeast of Spildra except perhaps in Porsanger, where Vaddas-type quartzite has recently been noted (own observation) lowermost in one of the Storelv Group occurrences of Gayer *et al.* (1985) (Figs. 1 and 2). The unit is tentatively interpreted as a basal formation of the Storelv Group deposited locally in proximal (mainly southeastern) parts of the sedimentary basin.

(2) *The Klubben Group (and equivalents) is no longer held to be necessarily part of a single stratigraphic sequence.*

Reasons will be given later suggesting that it may be an older unit or be derived from a more external part of the basin.

The Hammerfest map-sheet (Roberts, 1974) shows that Spildra in Kvaenangen (Fig. 1), and neighbouring islands to the northwest, occupy a key position in any discussion concerning correlations across the region, since they provide "stepping stones" at a relatively narrow part of the fjord. In contrast to the Hammerfest Sheet (based, as regards, Spildra, on mapping by Hahn (1968) and Johnson (1968)), the recent Bedrock map of Norway (Sigmond *et al.*, 1984) shows allochthonous Precambrian gneiss instead of Hellefjord Group intruded by SIP gabbro. Recent work indicates that the original interpretation was correct, at least in the south where the full Caledonian stratigraphy seems to be present (Table III, Fig. 2, and Fig. 3 K–L), containing variations from both Sørøy and NE Troms developments.

On Spildra, quartz schists (Table III, Sp. 5) resembling W Sørøy-type Hellefjord Group, intruded, migmatized and hornfelsed by SIP gabbro, are underlain by strata typical of the Åfjord Group (Sp. 4). These are followed by a semipelitic lithology (Sp. 3) like one of those found in the marble-bearing group in the Kåfjorddalen–Signaldalen area. This grades up from thinly (cm to dm) interlayered schists and quartzites (Sp. 2) identical with the Storelv Group in part of western Sørøy. The lower micaceous schist of this unit is conformably underlain by typical Vaddas formation quartzite (Sp. 1). All these strata dip moderately northeastwards. They are underlain after a break in exposure by only gently dipping and flexured meta-arkose of typical Klubben Group aspect, containing lenses of amphibole- or feldspar-rich pegmatitic gneiss.

More work is called for on and near Spildra, but a thrust is likely to occupy the exposure gap above the Klubben Group, and the island appears to link the two areas that seem correlatable. No geological or geophysical evidence (Chroston, 1974) is known for a major fault along the fjord, which is the site of open, late-stage NW–SE trending folds widely recognized in NE Troms (Fig. 3 K–L).

This suggested link across the fjord agrees with the view of Hooper and Gronow (1970), the former having co-ordinated the Swansea-group mapping astride Kvaenangen in the 1960s and early 1970s. This apparent absence of marble beds on Spildra contrasts with their great thickness in the cliffs on Kvaenangenhalvøya, but such swelling and wedging out of marble horizons is a typical feature of the group in NE Troms.

The (tectono)stratigraphic and correlation interpretations given above conflict with the claim made by Ramsay

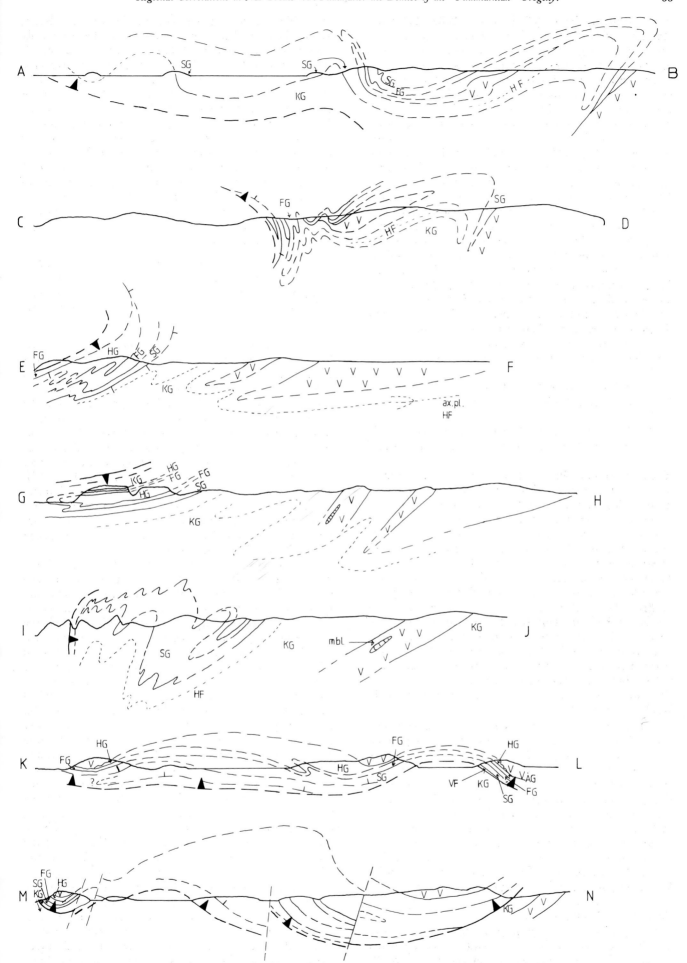

Fig. 3. Sketch profiles (not to scale) across Fig. 1 (see Fig. 1 for positions of profile lines). (Based on various sources, chiefly: A–B, C–D, Ramsay and Sturt (1963), Roberts (1974), Speedyman (1983); E–F, Roberts (1974); G–H, I–J, Ramsay (1971b), Roberts (1974); K–L, M–N, Ash (1968), Pearson (1970), Armitage (1972), own work.)

et al. (1981, 1985a) and Ramsay and Sturt (1986, e.g. Fig. 3) that a primary unconformity southwest of Kvaenangen separates the fossil-bearing Ordovician–Silurian strata from a Finnmarkian basement forming the basal part of SNC and composed of the "Sørøy Succession". However, results of detailed mapping by, for example, the Swansea group referred to above, and observable (tectono)stratigraphical relationships do not support such an unconformity.

For example, although it is true that conglomerate locally fills depressions (Fig. 4) in what may be an eroded surface of "Vaddas quartzite" in the Vaddas area (Fig. 1), lenses of identical conglomerate and of marble can be found at least 7 m below the stratigraphical and structural top of the quartzite near Vaddas and on Kvaenangsfjellet some 8 km farther north (Fig. 4). Conglomerate containing similar, mainly quartzitic clasts (resembling Vaddas formation) and vein quartz, along with schist and graphitic marble clasts, occurs locally (in places thickly) in impure marble overlying the quartzite in the Vaddas area (also reported by Lindahl, 1974). Most conglomerate pockets may represent channels, and since they do not occur at a single level they are thought to mark a period of variable shallow-water deposition, with possible breaks but not on the scale of a major unconformity between Scandian and Finnmarkian strata.

Secondly, although units are locally missing or added owing to tectonic discontinuities and primary facies variations (e.g., lenses of lava, conglomerate and marble), study of a wider area indicates that the sequence above and below the quartzite is very similar and located in a huge isoclinal, subrecumbent fold (Fig. 4), proposed here by Pearson (1970, 1971) and Armitage (1972) (see also Armitage *et al.*, 1971; Hooper, 1971a; Binns, in prep.). This fold is not identical to that proposed by Ramsay *et al.* (1981, 1985a) and Ramsay and Sturt (1986), but seems to be implied by Sigmond *et al.* (1984). It is difficult to see how the fold described by Ramsay *et al.* (1985a) and Ramsay and Sturt (1986, Fig. 3) fits into the map picture revealed by the work of Pearson (1970), Lindahl (1974), Zwaan (1988) and myself.

Finally, the concept of a major unconformity is still less acceptable when the westward continuation, as shown by Ramsay *et al.* (1985a, Fig. 1), past the north side of Guolasjav'ri (Fig. 1) is considered. Apparently eager to couple conglomerates to major unconformities to explain the correlation/age problem raised by the fossil discovery, these authors have overlooked that the conglomerate pocket recorded by Padget (1955) just north of Guolasjav'ri is immediately underlain by Upper Ashgill fossil-bearing marbles; way-up evidence (Binns and Gayer, 1980) confirm that it is younger than the latter.

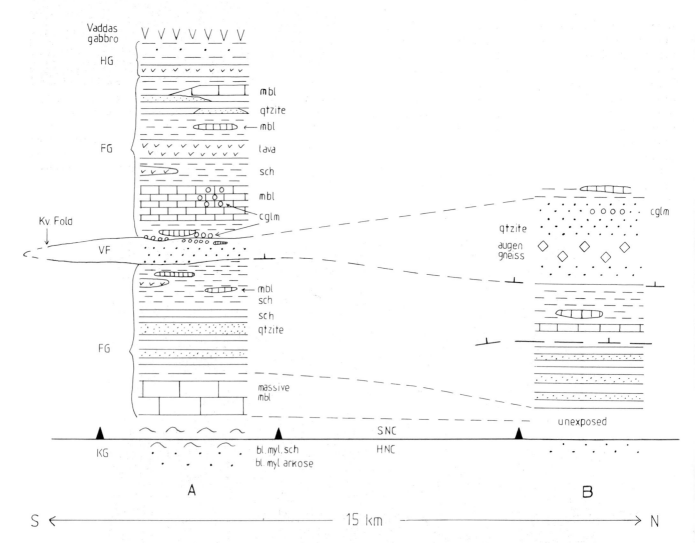

Fig. 4. Simplified sketch of the lower portion of SNC along the hillside from Rieppeelv in the south past Vaddas towards Kvaenangen, NE Troms, to give an impression of the lithostratigraphy through the probable Kvaenangen Fold. This is an alternative interpretation of the tectonostratigraphic relationships from those proposed by Lindahl (1974), Ramsay *et al.* (1985a) and Ramsay and Sturt (1986). Not to scale. Profile A represents a ca. 400 m broad profile from Frokosthaugen southwards, and B one of similar breadth near Jiednajav'ri. Heights of profiles are ca. 600 m and 250 m, respectively.

IGNEOUS ACTIVITY

Krill and Zwann (1987) put forward evidence from western Sørøy which they claimed showed that all igneous activity related to SIP was pre-orogenic and rift-related. As they pointed out, this would better explain the nature of the intrusive association, which fits an extensional environment better than a synorogenic one. They further postulated that all the intrusive activity pre-dated deposition of the Storelv–Hellefjord Groups and that the orogenic event that affected all the rocks was the Scandian phase. Unfortunately, no field evidence was given to support this implied return to the pre-1979 correlation of the Storelv–Hellefjord Groups between Sørøy and NE Troms (but now ascribed an Ordovician–Silurian age). Nor was any attempt made to justify the first assertion in the light of many accounts giving evidence for the extensive, broadly synorogenic age span of SIP and its intrusion into the entire Klubben–Hellefjord Group sequence (e.g., Sturt and Ramsay, 1965; Sturt, 1970; Hooper, 1971b; Sturt and Taylor, 1972; Bennett, 1974; Roberts, 1974; Robins and Gardner, 1974, 1975; Speedyman, 1983).

Although some examples of field relationships justify Krill and Zwaan's interpretation that dyke-intruded folds have a flow rather than tectonic origin, observations along the same road-cuts, and elsewhere, indicate that most intrusive sheets cut at least one foliation and some tectonically-formed isoclinal folds (Fig. 5). Krill and Zwaan also overlooked basic sheets in the Falkenes and Storelv Groups (e.g., at 414335 and 387365, respectively), and cutting the tectonic Klubben–Storelv Group boundary (at 383369), in the western part of the area they studied—Breivik to Sørvaer (Fig. 1). There is also evidence for post-thrusting intrusion of basic sheets in western Dønnesfjord (Sørøy) and on Skjervøy (Fig. 1), although other sheets pre-date an important episode of ductile strain (Fig. 6).

The literature and my work show that most major intrusions post-date deposition of all the stratigraphic units. For example, Falkenes and probably Storelv Group

Fig. 6. Central part of the thrust zone on Kågen (loc. 3, inset A, Fig. 1) showing quartzofeldspathic augen and lenses in the Storelv Group to the left and fine-grained, mylonite-banded Klubben Group migmatitic arkose to the right. The car key lies on a dark amphibolite lens at the group boundary, resulting from layer-parallel ductile strain of a normally up to ca. 1 m thick basic sheet.

Fig. 5. Two generations of basic dykes presumably related to SIP cutting already foliated and tectonically folded hornfelsed and partially migmatized Klubben Group in the aureole of the Breivikbotn gabbro at 463277 on the Hasvik–Breivikbotn road.

rafts and inclusions are reported from several large intrusions (e.g., Holland and Sturt (1970) and S. Bjerkenes (personal communication, 1972) from the Storelv gabbro, and by Ball *et al.* (1963) and Hooper (1971b) from the area northeast of Kvaenangen (Fig. 1, Fig. 3 G–H, I–J)). Speedyman (1983), in addition to noting such rafts, described the Husfjord gabbro in eastern Sørøy to intrude overfolded Klubben, Storelv, Falkenes and Hellefjord Groups (Fig. 1, Fig. 3 A–B, C–D). The bodies are often reported to form a ghost stratigraphy relatable to the gabbro envelope, thus suggesting that they do not represent exotic lithologies. The Spildra gabbro, probably a continuation of Hooper's (1971b) Olderfjord layered norite, intrudes apparent Hellefjord Group, inclusions of which have isoclinal D1 minor folds; its margin is weakly foliated by the S₂ regional foliation in its envelope, indicating a post-D1 to early-D2 age for its intrusion. However, the recent work on Sørøy, Skjervøy and Kågen, and evidence reported from Porsangerhalvøya (e.g., Gayer *et al.*, 1978), suggests that some minor basic sheets in at least the Klubben Group may be syn- or pre-D1 in age (see also Zwaan, 1987).

The link across Kvaenangen implied by the gabbro-intruded sequence on Spildra led to the suggestion (Binns, 1978) that SIP extends much further west and southwest than has otherwise been conceded. The strong resemblance that much of the Vaddas gabbro (as described by Lindahl, 1974) bears to gabbro north of Langfjord (Fig. 1) (own data, confirmed by B. Zwaan, personal communication 1987) supports this. Gabbros have, moreover, once been far more extensive southwest of Kvaenangen than their present outcrop suggests. Hooper (1971b) attributed hornfelsing of schists (presumed Hellefjord Group equivalent) on isolated peaks north of Vaddas (Fig. 1) to now-eroded gabbro. A large mass of hornfels is cut off by the thrust at the top of the Hellefjord? Group horizon between Nordreisa and Kåfjorddalen (Fig. 1) (own observations, Zwaan, 1988). Hence, there are grounds for proposing that one or more intrusions emanated from the main SIP mass to intrude SNC in the

Kågen-Arnøy area in the west, and at least as far south as Kåfjorddalen, in much the same way as the sheets crossing Sørøy (Figs. 1,3). Together with gabbro bodies in HNC south of Kvaenangen (some are shown on Fig. 1), which may also belong to SIP (Zwaan and Roberts, 1978), this almost doubles the extent of SIP.

Little is known about the age of these gabbros relative to the tectonometamorphic history. The sheared amphibolitized state of those on Kågen and Kvaenangenhalvøya (e.g., Hooper, 1971b) suggests that they were intruded pre- or syn-D2. The Vaddas gabbro on the other hand, being said by Lindahl (1974) to be little deformed and largely non-metamorphosed, may be younger, but scant evidence is given. In all these cases, this must be Scandian deformation and metamorphism, since they intrude strata stratigraphically overlying the Guolasjav'ri fossil-bearing beds.

Finally, the conjectured correlation between the Hellefjord Group and strata forming the Magerøy Nappe implies that the igneous complex intruding the Magerøy Nappe (e.g., Ramsay and Sturt, 1976) may bear some relation to SIP.

DEFORMATION HISTORY

Table IV summarizes the tectonometamorphic history recorded in the SNC in the Signaldalen–Kåfjorddalen area. As pointed out by Binns (1978) (see also Zwaan and Roberts, 1978), the table shows that overthrusting of HNC by SNC was synorogenic, culminating late in the D2 event recorded in SNC. D3 was common to both complexes (Binns, 1978; Zwaan and Roberts, 1978), F3 and younger folds deforming the thrust zone (Binns, in prep.). In HNC, the main ductile mylonitization, tectonic interleaving of gneisses, meta-arkose, etc., and rarely preserved early isoclinal folds are attributed to a D1 phase. Partial recrystallization of porphyroclasts, abundant folding (F2) with development of axial planar regional foliation is attributed to D2. The highly non-cylindrical F2a folds

Table IV. Summary of the tectonometamorphic development of the SNC rocks in the Signaldalen–Kåfjorddalen area. This is entirely a product of the Scandian orogenic phase.

Deformation episode	Fold phase	Axial trend	Style	Distribution	Foliation
D1	F1	?	isoclinal, recumbent-intrafolial	seldom	Sl preserved as Si in porphs.
D2	F2	NW–SE to WNW–ESE, rarely NE–SW	tight to isoc., (sub)-recumbent, part intra-fol., mostly cylind.,	common, spec. in qtzites, carbs. & semipelites.	S2 reg. fol. penetrative, parallel layering except in hinges
	F2a	c. NE–SW	tight to isocl., sub-recumb., non-cylind.	drag folds in main thrust zones	
**Culmination of synorogenic thrusting and nappe formation**					
	F3	mainly c. NW–SE	conjug., varied, mainly monocl. to overturned with open, upright parasites, often non-cylind.	widespread	S3 strain slip cleav., mainly in fold hinges, also in micaceous laminae/layers
D3	F4	NW–SE	upright, open to tight, mainly slightly asymmetrical	widespread, frequent in carb./sch. lithols.	
	F5	mainly NE–SW, also N–S, E–W	open, nearly symmetrical flexures	widespread	
	F6	ENE–WSW	open monocl. flexures	a few large folds	
**Brittle thrusting mainly on regional sole thrust**					

deform this and may be shear zone structures related to the inter-nappe complex thrusting that culminated in late-D2 when HNC bedrock (except for exotic higher-grade tectonic sheets) was also at its peak metamorphic grade—but only in upper greenschist to low amphibolite facies.

No evidence has yet been found suggestive of a break within this sequence of events. The only possibility would seem to be between D1 and D2 in HNC. In NE Troms there is a more obvious retrograde phase associated with thrust zones within SNC than there is in its sole thrust zone. Moreover, the undoubted single Scandian deformation cycle of the strata in the overlying Ullsfjord Nappe Complex registers a temporary, general retrograde phase following nappe formation that is much more distinct than diaphthoresis seen in HNC rocks near or in its roof thrust zone.

Since SNC in this area is formed of an essentially continuous (except for relatively small tectonic inclusions) lithostratigraphical sequence that contains Upper Ashgill fossils, the tectonometamorphism affecting it cannot be older than Scandian. Hence, at any rate, the D2 and D3 events affecting HNC here are also Scandian. The final piggy-back emplacement of the nappe pile (both here and elsewhere in the region, unless improbably substantial diachroneity existed) was effected by largely brittle movement on the regional sole thrust in late-D3 time (Binns, 1978; Zwaan and Roberts, 1978; Zwaan and Gautier, 1980).

Work on Porsangerhalvøya by, for example, Rice (1984, 1987b), Gayer *et al.* (1985, 1987) and Townsend (1987), has also failed to reveal a break in the tectonometamorphic history of the strata correlated with the "Sørøy Succession" (Klubben–Hellefjord Groups). However, despite acknowledging (Gayer *et al.*, 1985) the similarity to the Scandian deformation sequence found in HNC and SNC in NE Troms, and despite full agreement on the stratigraphic correlation at the Klubben Group level throughout the Porsanger–Sørøy–NE Troms region, these workers maintain that the tectonometamorphic history, including all the thrusting, on Porsangerhalvøya was completed during Finnmarkian time.

Workers in the Sørøy region have distinguished two main deformation episodes, referred to nowadays as D1 and D2. Both are described as polyphasal, shown principally for D2 by, for example, Roberts (1968a, 1974), Ramsay (1971b), Sturt and Taylor (1972) and Ramsay and Sturt (1973). Scarcity of detailed published maps and descriptions (with the principal exception of Roberts (1968a)), and a few apparent discrepancies from one area to another, hinder assessment of how the published tectonometamorphic history relates to that recorded elsewhere in the region, but comparison with the data from NE Troms and Porsangerhalvøya implies significant differences. Study of the literature and recent field work suggest, however, that at least some aspects can be reinterpreted, thereby bringing the histories more into line. Two are dealt with here, and others in the next two main sections.

Migmatization

Ramsay & Sturt (1976) made a valuable contribution to the Finnmarkian?–Scandian relationship question when they recognised a major thrust crossing Magerøy separating previously migmatized and twice-folded arkoses (assumed Klubben Group) from overlying Silurian strata forming the Magerøy Nappe. This nappe was shown to have been emplaced late in the D1 episode of the Scandian

phase recorded in the Silurian rocks; the earlier history in the migmatites was assumed to belong to the Finnmarkian phase. However, despite referring to the extensive tracts of such migmatized strata further west, Ramsay and Sturt did not attempt to trace the thrust boundary in that direction, and the possible implications of the discoveries on Magerøy for interpretations of Sørøy geology have never been mentioned.

One migmatite belt traverses the northern peninsulas (e.g. Roberts, 1974). Its southern boundary has recently been examined. At Vardfjellet, northwest of Dønnesfjord (Fig. 1), massive, coarse migmatite is sheared and blastomylonitized through a few metres marginal to non-migmatized Storelv Group schists. The first decimetre or two of the schists carry quartzofeldspathic augen and lenses enveloped by the regional foliation, which seems to be coeval with the shearing of the migmatite. When traced westwards to the Sørvaer area (Fig. 1), this horizon is found to form a discontinuity transecting group boundaries in the younger sequence on the upper flank of Ramsay's Sørvaer Fold, as shown by Ramsay (1971b). Ramsay explained this excision by sliding *during his F1 episode*. But relating it to F1 (see proposed revision of labelling below) conflicts with the opinion held by himself and other workers (e.g., Ramsay and Sturt, 1963; Roberts, 1968a, 1974; Hooper, 1971a; Sturt *et al.*, 1978; Roberts and Andersen, 1985) that greenschist facies conditions prevailed until the post-D1 to syn-D2 metamorphic peak, although in contrast to Roberts and Andersen (1985), Ramsay and Sturt (1976) believed that migmatization on Magerøy occurred in late-D1.

Speedyman (1983) attributed migmatization near Husfjord to the period following intrusion of his late-D1 Husfjord gabbro on the grounds that Klubben Group rafts and xenoliths, and immediately enveloping gabbro, are migmatized along strike from migmatized Klubben Group beyond the gabbro. Since the gabbro is also shown to intrude Storelv to Hellefjord Group strata (Fig. 1, Fig. 3 A–B, C–D) this implies a post-Hellefjord Group age for the migmatization. However, no explanation is offered for why large-scale migmatization of metasediments within and beyond the gabbro is restricted to the Klubben Group, why the gabbro is not more extensively migmatized in the area concerned, or why there is a relict early regional kyanite–sillimanite fabric in Klubben Group xenoliths that should not have exceeded greenschist facies prior to gabbro intrusion if there was only a single tectonothermal event. It seems at least as reasonable to interpret the local migmatization of gabbro near these migmatized Klubben Group inclusions by invoking anatectic remobilization of *already feldspar-rich migmatite inclusions*.

The widespread occurrence of such migmatite belts in the Klubben Group, also far from gabbros and both on outer coastal islands and the mainland east and south of Kvaenangen, indicates that at least some of this large-scale migmatization is unrelated to gabbro intrusion. It may be appreciably older than both this and the deposition of the Storelv–Hellefjord Groups, even though those were locally affected by (?Scandian) migmatization (e.g., Roberts, 1968a). Alternatively, the partially migmatized Klubben Group may belong to tectonic units derived from a separate portion of the same orogenic belt as the younger groups.

Folding

Observations in roadcuts in Storelv Group schists near Sandvika in western Sørøy (Fig. 1, inset B, loc. 1) show that the younger strata also have a more complex history than hitherto described from this area. According to Ramsay (1971b), they are in the upright lower limb of a major F1 fold (Fig. 1, inset B). Figure 7, however, shows examples of isoclinal folds deformed by isoclinal and overturned folds of, for example Ramsay's, F1 and F2 (=D1 and D2) generations. Two more examples were found in nearby roadcuts (427342, 424340). This suggests that these F1 and F2 folds, but perhaps not those in the Klubben Group (e.g., Roberts (1968a) F1 Hønseby Fold

(Fig. 3 A–B, C–D, E–F, G–H)), should be relabelled F2 and F3. The new F1 phase probably only gave rise to rarely preserved small-scale folds and a related foliation, this apparently being the case in NE Troms (Table IV, Olesen, 1971).

If the deformation scheme is to be brought more into line with that found in the apparently correlatable strata in NE Troms (Table IV), the new F1 and F2 folds should be assigned to D1 and D2, respectively, and the F3 folds to D3. Roberts (1968a) may be alone in having identified the

(a)

(b)

Fig. 7. Two examples of small-scale isoclinal folds predating structures previously referred to as F1 and F2 by Ramsay (1971b) and others. In road-cuts through thinly interbanded pelitic/psammitic middle/upper Storelv Group in the lower limb of Ramsay's (1971b) "F1" Breivik Fold. SW of Sandvika, western Sørøy: (*a*) at 422337 (loc. 1 on inset B, Fig. 1); (*b*) at 427342 (loc. 2 on inset B, Fig. 1). The dotted lines accentuate the axial traces of the earliest (new F1) folds. These are folded by tight to isoclinal F2 (ex-F1) folds. F3 folds are less distinct monoclinal structures.

same F1 and F2 phases, mainly on fabric evidence, but data from other parts of Sørøy are inadequate for deciding how the history and individual major structures he described fit into those described elsewhere.

THE HNC/SNC BOUNDARY

The HNC/SNC boundary in NE Troms south of Kvaenangen (Fig. 1) is located at a distinct tectonometamorphic discontinuity described (Binns, 1978) from this sector of Roberts' (1974) Kalak Nappe Complex. SNC was correlated (Binns, 1978) with the uppermost Kalak Nappe (Sørøy Nappe) in line with then generally agreed views on the Klubben Group/Vaddas quartzite correlation, though Gayer (1973) had correlated all meta-arkosic strata above the Kolvik Thrust in Porsanger (base of HNC in Fig. 1) with the Klubben Group.

Probably with this latter correlation in mind, but chiefly to try to solve the age discrepancy between the Guolasjav'ri fossils and the supposed 100–200 Ma older age of the Sørøy region strata, the Sørøy Nappe has recently been placed uppermost in HNC (e.g., Ramsay *et al.*, 1981, 1985a; Sigmond *et al.*, 1984; Ramsay and Sturt, 1986; Zwaan, 1988), entailing rapid wedging out of the nappe south of Kvaenangen.

Recent re-examination of specific parts of Sørøy, Kvaløy, Magerøy and the Kvaenangen area, and review of relevant literature, suggest another path for the HNC/SNC boundary east of Kvaenangen (Fig. 1) that is more in keeping with structural, fossil and stratigraphic correlation evidence. However, some of this path is more tentative than the map is able to show and alternative lines are possible.

In NE Troms the boundary coincides with the upper contact of the Klubben Group except in the Kvaenangen area (Fig. 1, Fig. 3 K–L, M–N) where what seem to be south-southeastward wedging-out tectonic slices containing Klubben and apparent Storelv–Hellefjord Groups conceivably extending from western Sørøy, are thought to form the lower portion of SNC (see below). Allochthonous Precambrian para- and orthogneiss bodies (including the Eidvågeid Group) underlie it or form imbricate slices in the thrust zone. The HNC/SNC boundary probably lies below sea level in the Skjervøy district; it is not clear whether the thrust boundary on Loppa, Spildra and the adjacent Kvaenangen coasts represents this (as on Fig. 1) or another of the thrusts.

The distinction seems less clear in Finnmark. Based mainly on the work of Gayer *et al.* (1985, 1987), the best choice for the boundary on Porsangerhalvøya seems to be the marked discontinuity at the base of the Havvatnet Imbricate Stack and Seivika Nappe (Fig. 1); the allocation of the Gardevar'ri Imbricate Stack is doubtful. Falkenes–Hellefjord Group occurrences overlie these units, but so do significant bodies of Klubben Group and strata correlated with the Eidvågeid Group; moreover, bodies of apparent Storelv Group are found below as well as above it. Thrust tectonics seem capable of explaining the former differences, which also continue further west in Finnmark, but the presence of Storelv Group (and perhaps equivalent undifferentiated schist, etc., shown on Fig. 1) below the boundary may require another explanation. Some of these occurrences are distinguished by Gayer *et al.* (1985) as the Kokelv Schist Group, a mainly biotite schist unit separating normal Klubben and Storelv Groups, and instead of the suggested correlation with the Storelv Group, tentatively shown on Fig. 1, these bodies (and also the "undifferentiated schists") may represent such a stratigraphic

unit related to the Klubben Group and deposited in southern parts of the sedimentary basin.

Following criteria used in NE Troms (Binns, 1978, and Table IV) and the descriptions of Ramsay and Sturt (1976) and Roberts and Andersen (1985), it seems natural to interpret the Magerøy Nappe sole thrust as the HNC/SNC boundary. However, although the subjacent migmatites can be traced westwards to the thrust-limited belt crossing northern Sørøy (see above) via the northwest tip of Porsangerhalvøya, the outer islands (Hjelmsøy, etc.) are problematic, since descriptions of these (Ramsay *et al.*, 1979; Roberts and Andersen, 1985) give no grounds for placing a thrust at the Klubben/Storelv boundary, which those authors interpret as a sedimentary transition. The Storelv/Hellefjord boundary on the other hand is said to be unexposed (Roberts and Andersen, 1985) and, especially since the Falkenes Group seems to be lacking, this may be thought to mark a thrust. The HNC/SNC boundary crossing Sørøy is considered quite well founded (see above).

On its passage from Magerøy to the area south of Kvaenangen, the HNC/SNC boundary makes several major turns, and is also temporarily inverted at least in northwestern Sørøy (Fig. 3 E–F, G–H, I–J, and Fig. 8) where this causes the westward opening of Ramsay's (1971b) Sørvaer Fold; Ramsay and Sturt (1973) give another explanation for this. The boundary follows the arcuate fold axial trend that most Sørøy workers (e.g., Roberts, 1968a; Ramsay, 1971b; Ramsay and Sturt, 1973) have explained as being brought about during the F1 (or

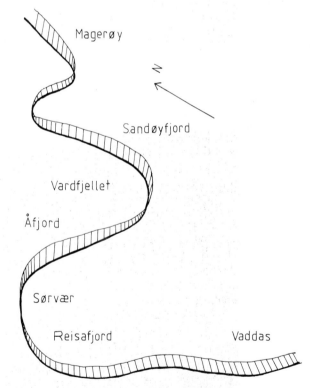

Fig. 8. The HNC/SNC boundary between Magerøy and NE Troms, illustrating its arcuate path and changing dip attitude, roughly parallel to the trend of the earlier-formed major folds and foliation. The changes in trend and in the attitude of structural planes, with overturning on at least one stretch, are interpreted as being due to refolding in D3—mainly F3 folds in the Sørøy area and F4 or younger in NE Troms. From Vaddas the boundary can be envisaged running towards ca. ENE to Porsangerhalvøya, with a dominantly gentle dip to ca. NW–N, resulting in a highly asymmetrical, trough-like thrust surface in profile, transverse to the belt.

perhaps F2 (Roberts, 1968a)) deformations. Despite their appealing arguments for this, it seems that since the late-D2 nappe complex boundary and other thrusts, which must have been deformed simultaneously, also follow this trend, the arcuate changes are chiefly caused by roughly NW–SE trending deformation of the earlier structures during F3 (early D3, cf. Table IV), though primary non-cylindrism probably also had some influence. Speedyman (1983) also seems to hold this view. F4 and younger phases have also deformed the structures, at least in Troms. Similar, but generally less severe, F3 (Scandian) and younger deformation of nappe complex boundaries has been recorded widely in Troms (e.g., Landmark, 1973; Binns, 1978; Binns and Matthews, 1981; Zwaan and Roberts, 1978) and also elsewhere in Finnmark (e.g., Gayer *et al.*, 1987 e.g. Fig. 3). The extra severe deformation in the Sørøy region may be related to the major SIP intrusions disturbing the compressive stresses, as Roberts (1968a) suggested in respect of his F2 deformation.

Figure 1 shows the HNC/SNC boundary crossing northern Kvaløy where the Klubben Group beneath to the north is largely migmatized and/or fairly intensely blastomylonitized, resembling its common state in HNC in NE Troms. In the hills east of Hammerfest, this borders sharply up to a non-migmatized, tightly synformally folded, condensed pile of strata containing the entire younger sequence (simplified in Fig. 1), which probably has a tectonic boundary to the south against the Eidvågeid Group (Ramsay and Sturt, 1977; Sturt *et al.*, 1978; own observations).

ADDITIONAL THRUSTS

The literature contains passing references to sliding and shearing at a number of horizons on Sørøy, causing, for example, excision of groups and marginal deformation of gabbros (e.g., Sturt and Ramsay, 1965; Ramsay, 1971b;

Roberts, 1974). Their importance generally seems to have been underestimated (but see Gayer *et al.*, 1987). In addition to the HNC/SNC boundary proposed above, two other significant tectonic discontinuities were recognized during the recent field work.

East of Sørvaer, where the Klubben/Storelv boundary is deformed by tight F3 overfolds and generally disguised by small- to medium-scale F3 folding, one road-cut (Fig. 1, inset B, loc. 4) reveals evidence of dislocation at the boundary. The actual boundary between arkose and overlying schist is sharp, slightly discordant and clearly tectonic. Lenses of calc silicate rock, skarn and marble, perhaps derived from a Falkenes Group horizon caught up in the thrusting, are found here and ca. 100 m below. The pattern of thin, small-folded quartzitic ribs in the Storelv schist implies that a larger F2 fold is disrupted by the boundary. The boundary both here and on the nearby shore is cut by basic sheets that have been deformed by F3 folds. Figures 1 and 3 suggest that this thrust can be traced across northern Sørøy.

At Sandvika in western Sørøy (Fig. 1, inset B, loc. 3) a road-cut exposes a thrust at the Klubben/Storelv boundary in the lower limb of Ramsay's (1971b) "F1" Breivik Fold (now looked upon as probable F2, see above). The upper portion of the banded meta-arkose is flaggy, but contains occasional overturned folds with their middle limbs more or less cut out by shears. These are probably the F2 folds of Ramsay (1971b) and are thought to be representatives of F2 folds in the Klubben Group according to the scheme used here, i.e. they may not be strictly, if at all, equivalent to F2 folds in the younger sequence. Figure 9 shows that the uppermost of these folds is cut off by a more low-angled thrust plane at the base of ca. 2 m of micaceous schist of Storelv Group affinity containing tectonic lenses of meta-arkose. This is overlain by ca. 9 m of ordinary (Storelv) quartz-mica schist, and then ca. 6 m of coarse, phyllonitic schist with boudins of quartz-rich material, elongate lenses of meta-arkose and quartzite and

Fig. 9. Thrust in the lower part of the thrust zone separating underlying Klubben Group from Storelv Group SW of Sandvika, western Sørøy (432345—loc.3, inset B, Fig. 1). The thrust plane cuts off a probable F2 fold in flaggy meta-arkose of the Klubben Group. The relationship of this and nearby folds to the fold scheme in the Storelv and younger groups is unclear, but they have the sense of overturning as if they were parasites to the F2 Breivik Fold.

a few isoclinal fold remnants. Above that is ordinary Storelv schist with thin layers and ribs of quartzite.

Flaggy arkose directly underlies phyllonitic schist at several localities along strike to Nordsandfjord (Fig. 1) where a ca. 15° angular discordance can be seen in the hillside south of the fjord; only slight discordance is visible east of the fjord. The arkoses are strongly deformed by two generations of isoclinal to tight subrecumbent folds that do not seem to affect the thrust zone or overlying strata, and in the cliffs south of the beach such folds are seen to be cut by a number of minor thrusts. Throughout this area folds are rare or absent in the upper 15–40 m of the arkose, which is always flaggy to nearly fissile in contrast to its more massive layering lower down where right-way-up (allowing for folding) cross bedding is common. Similar evidence of tectonization, along with shear-derived alternations of the two groups, is found along this horizon to Dønnesfjord.

Both these thrusts probably culminated in late-D2 (Scandian) time, being essentially developed as slides replacing limbs of F2 isoclinal folds such as the Breivik Fold. The HNC/SNC boundary seems to have been active latest and to cut off some of the other thrusts and the rock bodies they enclose, but it also culminated in D2, since it is deformed by F3 folds.

There seems little doubt about the inherent validity of lithostratigraphic correlations across outer Kvaenangen (Hooper and Gronow, 1970; Hooper, 1971b). In clear weather, the Klubben Group can be seen passing across the sea from Sørøy to Silda and Loppa in one direction and from Kågen and Skjervøy to the northeast coast of Kvaenangen in the other direction (Fig. 1).

Ash (1968) did not recognize a break between the Klubben Group and the younger sequence on Skjervøy, but thought it occupied the core of a major subrecumbent overfold. However, at several localities here, on neighbouring Kågen and on the adjacent mainland, there is evidence for thrusting at the group boundaries. Four examples will be described briefly.

At the rifle range northwest of Skjervøy town (Fig. 1, inset A, loc. 1) the Klubben Group overlies Storelv Group. Typically, no exposure of the actual boundary could be found, but the lower portion of the group is flaggy and has conspicuous mylonitic banding lowermost; the uppermost schist is crenulated phyllonite.

The Klubben Group also overlies Storelv Group schists near Tennskjaerneset (Fig. 1, inset A, loc. 2). The boundary zone on the hillside and shore section is marked by narrow shear zones, including (on the hillside) impersistent highly micaceous–garnetiferous bands, forming a transition between the two groups. Basic sheets cut the zone subconcordantly, and these and the tectonic fabric are folded by overturned folds, apparently of F2 age. These have an opposite vergence from more monoclinal second-generation folds about 15 m up in the Klubben Group, implying that the two fold sets belong to different tectonic units if they really belong to the same generation.

On Kågen, west of the bridge to Skjervøy (Fig. 1, inset A, loc. 3), the upper boundary of the Klubben Group underlies typical lower Storelv Group schist. Here (in contrast to the boundary on the easterly peninsula on Kågen) the thrust zone has cut down to migmatized Klubben Group. It is marked by a ca. 12 m broad zone of shearing and augen development spanning both groups nearly equally. Basic sheets, normally around 0.4–1 m thick in these groups, are reduced to slivers and wispy lenses (Fig. 6).

A road-cut on the hill leading out of Skjervøy town reveals a tectonic, sheared boundary between the Klubben

and Storelv Groups at the lower limit of the schist wedge (Fig. 1, inset A, loc. 4). Mylonitic banding is developed through about 2.5 m, especially in the arkose, and a ca. 35 cm thick basic sheet cuts this slightly obliquely and is itself weakly foliated.

If the basic sheets at these localities belong to the same generation, they provide evidence for movement on the thrusts at localities 2 and 4 being completed earlier than that at locality 3. Comparison with Sørøy suggests that recognition of thrusts at arkose boundaries does not necessarily rule out the recumbent overfold concept of Ash (1968) and Hooper *et al.* (1968).

The tectonic sheets bounded by these thrusts (which may be the same as those at Sandvika and east of Sørvaer on Sørøy) can be traced southeastwards on the mainland and seem to be transected by the underlying HNC/SNC boundary near Vaddas where the upper thrust forms the lower boundary of the Vaddas formation (Figs. 1, 3, 4).

DISCUSSION AND CONCLUSIONS

Following the finding of Upper Ordovician fossils in strata in Troms that had been confidently correlated with assumed Cambrian beds in Finnmark, there has been a need for a broadly-based examination of this age discrepancy. It has become common practice to accept without reservation the radiometric age evidence on which the Cambrian assumption was largely based. The grounds for the original correlation itself have been disregarded, or unsatisfactory attempts have been made to circumvent them. Nevertheless, although some of the Sørøy region radiometric data are of dubious quality, no clear reasons can be seen in present geochronological theory or local geological setting for rejecting all the isochrons, not least the most recent ones. Nor can alternative interpretations of them be proposed that are more amenable to the new fossil evidence. On the other hand, they should be regarded with suspicion, since radiometric data should not be uncritically accepted when they conflict with such primary evidence as fossil-based age determinations and other field observations. It is not enough for them to appear to fit field observations on a local scale (e.g., relations between intrusions and tectonic structures), they should also harmonize on a broader scale before being unreservedly accepted.

In this case, the fossil evidence is geographically somewhat removed from the localities for the radiometric data. But the foregoing account has shown that previously well-established field-based arguments, and new ones, strengthen the main aspects of the formerly accepted correlation that linked these localities. This correlation was largely founded on the striking similarity in lithofacies development and stratigraphy, together with the natural continuity of the strata, throughout the region. An apparent link provided by the SIP intrusions had also been pointed out. The present review of all these aspects has brought out additional points in their favour rather than weakening them, even though there is always the risk of still-unrecognized complications when there are significant stretches of sea that may, for example, conceal major faults.

The recent field observations on Sørøy, Kvaløy and in the Kvaenangen area, even though mainly of a reconnaissance nature and requiring confirmation through more detailed field and petrographic study, indicate a stronger similarity in deformational history between the two contested areas than has hitherto been apparent. If correct, this too strengthens the original correlation. The most important aspects here are:

1. the evidence indicating an *apparently* separate tectonometamorphic event, including regional migmatization and two episodes of folding, in the Klubben Group relative to that found in the Storelv–Hellefjord Groups;
2. the presence of tectonic boundaries between these parts of the succession;
3. the interpretation that these thrusts are deformed by F3 and younger structures.

However, neither in NE Troms (Binns, 1978; Binns, in prep.) nor in Porsangerhalvøya (e.g., Rice 1984) where comparable fold deformed thrusts have also been recognized, nor in the Sørøy region, has evidence yet been found for a clear break in the metamorphic or deformational history fitting the event mentioned in point 1. Ramsay and Sturt (1976) infer such a break at the corresponding horizon on Magerøy, but their view only seems to be based on an assumption that the tectonometamorphic history of the migmatites is Finnmarkian and therefore distinct from that of the overthrust Magerøy Nappe. The thrusting is synmetamorphic with respect to D2 in both affected parts of the rock pile, at least in NE Troms (e.g., Binns, 1978; Zwaan and Roberts, 1978) and Porsangerhalvøya (e.g., Gayer *et al.*, 1985; Rice, 1987a). Hence, rather than being a truly time-separated event, if there is a partially different history it seems likely to reflect derivation of the HNC segment of the rock pile (and similarly migmatized belts of Klubben Group in assumed SNC parts of, for example, Sørøy) from a different part of the same orogenic belt as the other segment. However, the preliminary radiometric data from Porsangerhalvøya reported by Daly *et al.* (1987) suggest that a completely separate and substantially earlier event, perhaps Morarian or even Sveconorwegian, may be viable.

In view of the fossil control on the age of the NE Troms sector and the continuity of the nappe pile at the HNC level between NE Troms and Porsangerhalvøya, it follows that if this history really is continuous, and if improbably substantial lithological and/or tectonometamorphic diachroneity is not present, it can scarcely represent other than Scandian activity or perhaps a very protracted "Finnmarkian"–Scandian event. The apparent intrusion of SIP into SNC, including the horizon stratigraphically overlying the Upper Ashgill beds, as well as HNC, supports this view. However, since there may still be room for so far indeterminable breaks in the lithostratigraphic, tectonostratigraphic, tectonometamorphic and magmatic continuity through the region concerned, final acceptance of the validity of this hypothesis must perhaps await such decisive progress as the finding of mid-Ordovician–Silurian fossils in, for example, the Falkenes Group marbles on Sørøy.

In the meantime, there seem to be good grounds for reconsidering current interpretations of the Sørøy region radiometric data and little virtue in retaining the concept of a "Finnmarkian" phase or orogeny, as it has been defined, in this region.

Acknowledgements

This paper has benefited greatly from helpful comments on the initially submitted version received from David Gee, Geoff Milnes and Bouke Zwaan. I am also grateful to Kirsten Vik for invaluable assistance in redraughting the figures. This is publication no. 6 of the project "The age of the Caledonian orogeny in northern Norway".

REFERENCES

Andersen, T. B. (1984). The stratigraphy of the Magerøy Supergroup. Finnmark, North Norway. *Norges Geol. Unders.* **395**, 25–37.

Armitage, A. H. (1972). The stratigraphy, structure and metamorphism of western Nordreisa, Troms, North Norway. Unpubl. Ph.D. thesis. University College, Swansea.

Armitage, A. H., Hooper, P. R., Lewis, D. and Pearson, D. E. (1971). Stratigraphic correlation in the Caledonian rocks of S. W. Finnmark and North Troms. *Norges Geol. Unders.* **269**, 318–322.

Ash, R. P. (1968). The geology of Skjervøy, North Troms, Norway. *Norges Geol. Unders.* **255**, 37–54.

Ball, T. K., Gunn, C. B., Hooper, P. R. and Lewis, D. (1963). Preliminary geological survey of the Loppen district, West Finnmark. *Norsk Geol. Tidsskr.* **43**, 215–246.

Bennett, M. C. (1974). The emplacement of a high temperature peridotite in the Seiland province of the Norwegian Caledonides. *J. Geol. Soc. Lond.* **130**, 205–228.

Binns, R. E. (1978). Caledonian nappe correlation and orogenic history in Scandinavia north of lat 67° N. *Bull. Geol. Soc. Amer.* **89**, 1475–1490.

Binns, R. E. and Gayer, R. A. (1980). Silurian or Upper Ordovician fossils at Guolasjav'ri, Troms, Norway. *Nature* **284**, 53–55.

Binns, R. E. and Matthews, D. W. (1981). Stratigraphy and structure of the Ordovician–Silurian Balsfjord Supergroup, Troms, North Norway. *Norges Geol. Unders.* **365**, 39–54.

Chroston, P. N. (1974). Geological interpretation of gravity data between Tromsø and Øksfjord (Finnmark) *Norges Geol. Unders.* **312**, 59–90.

Daly, J. S., Cliff, R. A., Gayer, R. A. and Rice, A. H. N. (1987). A new Precambrian terrane in the Caledonides of Finnmark, Arctic Norway. *The Caledonian and Related Geology of Scandinavia.* Abstracts, p. 15. University College, Cardiff.

Debrenne, F. (1984). Archaeocyatha from the Caledonian rocks of Sørøy, North Norway—a doubtful record. *Norsk Geol. Tidsskr.* **64**, 153–154.

Gayer, R. A. (1973). Caledonian geology of Arctic Norway. In: Pitcher, M. G. (ed.) *Arctic Geology*, pp. 453–468. Assoc. Petrol. Geol. Amer. Memoir 19.

Gayer, R. A., Powell, D. B. and Rhodes, S. (1978). Deformation against metadoleritic dykes in the Caledonides of Finnmark, Norway. *Tectonophysics* **46**, 99–115.

Gayer, R. A., Hayes, S. J. and Rice, A. H. N. (1985). The structural development of the Kalak Nappe Complex of eastern and central Porsangerhalvøya, Finnmark, Norway. *Norges Geol. Unders.* **400**, 67–87.

Gayer, R. A., Rice, A. H. N., Roberts, D., Townsend, C. and Welbon, A. (1987). Restoration of the Caledonian Baltoscandian margin from balanced cross-sections: the problem of excess continental crust. *Trans. R. Soc. Edinburgh, Earth Sci.* **78**, 197–217.

Hahn, U. F. (1968). The geology of Spildra, Norway,. Unpubl. B.Sc. thesis, University College, Swansea.

Henningsmoen, G. (1961). Cambro-Silurian fossils in Finnmark, northern Norway. *Norges Geol. Unders.* **213**, 93–95.

Holland, C. H. and Sturt, B. A. (1970). On the occurrence of archaeocyathids in the Caledonian metamorphic rocks of Sørøy and their stratigraphical significance. *Norsk Geol. Tidsskr.* **50**, 341–355.

Hooper, P. R. (1971a). A review of the tectonic history of S. W. Finnmark and North Troms. *Norges Geol. Unders.* **269**, 11–14.

Hooper, P. R. (1971b). The mafic and ultramafic intrusions of S. W. Finnmark and North Troms. *Norges Geol. Unders.* **269**, 147–158.

Hooper, P. R. and Gronow, C. W. (1970). The regional significance of the Caledonian structures of the Sandland peninsula, West Finnmark, northern Norway. *Q. J. Geol. Soc. London.* **125**, 193–217.

Hooper, P. R., Pearson, D. E. and Lewis, D. (1968). Recent observations on the southern coast of Skjervøy and the opposite coast of Kågen. In: Ash, R. P., *The geology of Skjervøy, North Troms, Norway* (Appendix, pp. 53–54). *Norges Geol. Unders.* **255**, 37–54.

Johnson, D. K. (1968). The geology of Spildra, Norway. Unpubl. B.Sc. thesis, University College, Swansea.

Krill, A. G. and Zwaan, K. B. (1987). Reinterpretation of Finnmarkian deformation on western Sørøy, northern Norway. *Norsk Geol. Tidsskr.* **67**, 15–24.

Landmark, K. (1973). Beskrivelse til de geologiske kart "Tromsø" og "Målselv", II, Kaledonske bergarter. *Tromsø Museums Skr.* **15**, 1–263.

Lindahl, I. (1974). Økonomisk geologi og prospektering i Vaddas-Rieppe feltet, Nord-Troms. Unpubl. Lic. thesis, University of Trondheim.

Olesen, N. Ø. (1971). The relative chronology of fold phases, metamorphism and thrust movements in the Caledonides of Troms, North Norway. *Norsk Geol. Tidsskr.* **51**, 355–377.

Padget, P. (1955). The geology of the Caledonides of the Birtavarre region, Troms, northern Norway. *Norges Geol. Unders.* **192**, 1–107.

Pearson, D. E. (1970). The structure and metamorphic history of Straumfjord, North Troms, Norway. Unpubl. Ph.D. thesis, University College, Swansea.

Pearson, D. E. (1971). Problems concerning the Kvaenangen nappe fold. *Norges Geol. Unders.* **269**, 85–88.

Pedersen, R. B., Dunning, G. R. and Robins, B. (1988). U–Pb ages of nepheline syenite pegmatites from the Seiland Magmatic Province, North Norway, *this volume.*

Pringle, I. R. (1973). Rb–Sr age determinations on shales associated with the Varanger ice age. *Geol. Mag.* **109**, 465–472.

Pringle, I. R. and Sturt, B. A. (1969). The age of the peak of the Caledonian orogeny in west Finnmark, northern Norway. *Norsk Geol. Tidsskr.* **49**, 435–436.

Ramsay, D. M. (1971a). Stratigraphy of Sørøy. *Norges Geol. Unders.* **269**, 314–317.

Ramsay, D. M. (1971b). The structure of north-west Sørøy. *Norges Geol. Unders.* **269**, 15–20.

Ramsay, D. M. and Sturt, B. A. (1963). A study of fold styles, their associations and symmetry relationships, from Sørøy, North Norway. *Norsk Geol. Tidsskr.* **43**, 441–430.

Ramsay, D. M. and Sturt, B. A. (1973). An analysis of noncylindrical and incongruous fold pattern from the Eo-Cambrian rocks of Sørøy, northern Norway. I & II. *Tectonophysics* **18**, 81–121.

Ramsay, D. M. and Sturt, B. A. (1976). The syn-metamorphic emplacement of the Magerøy Nappe. *Norges Geol. Tidsskr.* **56**, 291–307.

Ramsay, D. M. and Sturt, B. A. (1977). A sub-Caledonian unconformity within the Finnmarkian nappe sequence and its regional significance. *Norges Geol. Unders.* **334**, 107–116.

Ramsay, D. M. and Sturt, B. A. (1986). The contribution of the Finnmarkian orogeny to the framework of the Scandinavian Caledonides. In: Fettes, D. J. and Harris, A. L. (eds.) *Synthesis of the Caledonian Rocks of Britain*, pp. 221–246. Reidel, Dordrecht.

Ramsay, D. M., Sturt, B. A. and Andersen, T. B. (1979). The sub-Caledonian unconformity on Hjelmsøy—new evidence of primary basement/cover relations in the Finnmarkian nappe sequence. *Norges Geol. Unders.* **351**, 1–12.

Ramsay, D. M., Sturt, B. A., Zwaan, K. B. and Roberts, D. (1981). Caledonides of northernmost Norway. *Terra Cognita* **1**, 67–68.

Ramsay, D. M., Sturt, B. A., Zwaan, K. B. and Roberts, D. (1985a). Caledonides of northern Norway. In: Gee, D. G. and Sturt, B. A. (eds.) *The Caledonide Orogen—Scandinavia and Related Areas.* Vol. 1, pp. 163–184. Wiley, New York.

Ramsay, D. M., Sturt, B. A., Jansen, Ø., Andersen, T. B. and Sinha-Roy, S. (1985b). The tectonostratigraphy of western Porsangerhalvøya, Finnmark, north Norway. In: Gee, D. G. and Sturt, B. A. (eds.) *The Caledonide Orogen—Scandinavia and Related Areas.* Vol. 1, pp. 611–619. Wiley, New York.

Rice, A. H. N. (1984). Metamorphic and structural diachroneity in the Finnmarkian nappes of north Norway. *J. Metam. Geol.* **2**, 219–236.

Rice, A. H. N. (1987a). Continuous out-of-sequence ductile thrusting in the Norwegian Caledonides. *Geol. Mag.* **124**, 249–260.

Rice, A. H. N. (1987b). A tectonic model for the evolution of the Finnmarkian Caledonides of North Norway. *Can. J. Earth Sci.* **24**, 602–616.

Roberts, D. (1968a). The structural and metamorphic history of the Langstrand–Finfjord area, Sørøy, northern Norway. *Norges Geol. Unders.* **253**, 1–160.

Roberts, D. (1968b). Hellefjord Schist Group—a probable turbidite formation from the Cambrian of Sørøy, West Finnmark. *Norsk Geol. Tidsskr.* **48**, 231–244.

Roberts, D. (1974). Hammerfest. Beskrivelse til det 1:250 000 berggrunnsgeologiske kart. *Norges Geol. Unders.* **301**, 1–66.

Roberts, D. (1985a). Geologisk kart over Norge, berggrunns geologisk kart Honningsvåg—M 1:250 000, foreløpig utgave. *Norges Geol. Unders.*

Roberts, D. (1985b). The Caledonian fold belt in Finnmark: a synopsis. *Norges Geol. Unders. Bull.* **403**, 161–177.

Roberts, D. and Andersen, T. B. (1985). Nordkapp. Beskrivelse til det bergrunnsgeologiske kartbladet M 1:250 000. *Norges Geol. Unders. Skr.* **61**, 1–49.

Robins, B. and Gardner, P. M. (1974). Synorogenic basic intrusions in the Seiland petrographic province, northern Norway—a descriptive account. *Norges Geol. Unders.* **312**, 91–130.

Robins, B. and Gardner, P. M. (1975). The magmatic evolution of the Seiland province, and Caledonian plate boundaries in northern Norway. *Earth Planet. Sci. Lett.* **26**, 167–178.

Sigmond, E. M. O., Gustavson, M. and Roberts, D. (1984). Berggrunns kart over Norge—M 1:1 million. *Norges Geol. Unders.*

Speedyman, D. L. (1983). The Husfjord plutonic complex, Sørøy, northern Norway. *Norges Geol. Unders.* **378**, 1–48.

Sturt, B. A. (1970). Exsolution during metamorphism with particular reference to feldspar solid solution. *Mineral. Mag.* **37**, 815–832.

Sturt, B. A. and Ramsay, D. M. (1965). The alkaline complex of the Breivikbotn area, Sørøy, northern Norway. *Norges Geol. Unders.* **231**, 1–164.

Sturt, B. A. and Taylor, J. (1972). The timing and environment of emplacement of the Storelv gabbro. *Norges Geol. Unders.* **272**, 1–34.

Sturt, B. A., Miller, J. A. and Fitch, F. J. (1967). The age of alkaline rocks from West Finnmark, northern Norway, and their bearing on the dating of the Caledonian orogeny. *Norsk Geol. Tidsskr.* **47**, 255–273.

Sturt, B. A., Pringle, I. R. and Roberts, D. (1975). Caledonian nappe sequence of Finnmark, northern Norway, and the timing of orogenic deformation and metamorphism. *Bull. Geol. Soc. Amer.* **86**, 710–718.

Sturt, B. A., Pringle, I. R. and Ramsay, D. M. (1978). The Finnmarkian phase of the Caledonian orogeny. *J. Geol. Soc. Lond.* **135**, 597–610.

Townsend, C. (1987). Thrust transport directions and thrust sheet restoration in the Caledonides of Finnmark, North Norway. *J. Struct. Geol.* **9**, 345–352.

Zwaan, K. B. (1987). The intrusion mechanism of a rift-related diabase swarm, as deduced from field relations preserved in the Corrojav'ri mega lens, North Norway. *The Caledonian and Related Geology of Scandinavia.* Abstracts, p. 38. University College, Cardiff.

Zwaan, K. B. (1988). Geologiske kart over Norge, berggrunns geologisk kart Nordreisa—M 1:250 000. Norges Geol. Unders. , in press.

Zwaan, K. B. and Gautier, A. M. (1980). Alta og Gargia: Beskrivelse til de berggrunnsgeologiske kart 1834 I og 1934 IV-M 1:50 000. *Norges Geol. Unders.* **357**, 1–47.

Zwaan, K. B. and Roberts, D. (1978). Tectonostratigraphic succession of the Finnmarkian nappe sequence, North Norway. *Norges Geol. Unders.* **343**, 53–71.

4 Evidence of intracratonic Finnmarkian orogeny in central Norway

Daniel Tietzsch-Tyler

Department of Geology, University College, Belfield, Dublin, Ireland.

Two Caledonian orogenic events are recognized in the Steinkjer–Snåsavatnet area on stratigraphic and structural evidence. Deformation of a discrete volcanosedimentary, intracratonic rift basin during an early (Finnmarkian) orthotectonic event gave rise to rooted nappe structures. Following uplift and erosion, and a period of island arc-related volcanosedimentary accumulation, Scandian thin-skinned paratectonic deformation occurred. A model is presented from which several possible implications arise. First, the single broad miogeoclinal model for early Caledonian sedimentation and magmatism on the Baltoscandian margin of Iapetus might be replaced by one invoking a broad zone of crustal extension characterized by a series of basins in which oceanic lithosphere was sometimes generated. Secondly, nappe structures may be confined to these older rocks. They may be rooted locally beneath younger Lower Palaeozoic successions and they may be largely parautochthonous rather than highly allochthonous. Thirdly, unconformable or disconformable Middle Ordovician and younger successions might then be largely autochthonous or parautochthonous and have had a Baltoscandian continental basement.

INTRODUCTION

Work in recent years has seen the accumulation of evidence in favour of the idea that the Scandinavian arm of the Caledonides is the product of multiple orogeny. In northernmost Norway an early Caledonian tectonothermal event, dated to ca. 530–490 Ma (Sturt *et al.*, 1978; Gee and Roberts, 1983), has been termed the Finnmarkian (Ramsay and Sturt, 1976). A similar event, dated to ca. 535–468 Ma (Bryhni and Sturt, 1985; Furnes *et al.*, 1985), has also been identified in SW Norway. Until comparatively recently, however, the greater part of the Scandinavian Caledonides were considered to have been deformed during the post-Llandovery Scandian (Gee, 1975) tectonothermal event (e.g., Gee, 1975, 1978; Roberts and Sturt, 1980). This was despite the recognition by Gale and Roberts (1974) of a major ophiolite emplacement event that predates the unconformable deposition of upper Middle Arenig sediments (Ryan *et al.*, 1980). Ophiolites are now widely recognized in Scandinavia and while some represent back-arc marginal basin activity and are Middle Ordovician in age, many are earlier and were probably emplaced during latest Cambrian and earliest Ordovician times (Furnes *et al.*, 1980, 1985; Sturt 1984; Sturt *et al.*, 1984). More recently, numerous radiometric age determinations have confirmed the existence of early Caledonian tectonothermal activity.

Syn- and post-tectonic intrusive ages of ca. 526–503 Ma (Nissen, 1986) and ca. 478–447 Ma (Klingspor and Gee, 1981) respectively, and mylonite ages of ca. 485 Ma (Claesson, 1980) all support a Cambro–earliest Ordovician event. So too do numerous and widespread mineral age determinations, particularly on hornblendes, providing evidence of ca. 510–455 Ma post-metamorphic uplift and cooling throughout the central Scandinavian Caledonides (e.g., Welin *et al.*, 1983; Dallmeyer *et al.* 1985; Dallmeyer and Gee, 1986). On the basis of this accumulated evidence, Dallmeyer and Gee (1986) postulated the development of an early Ordovician accretionary prism at the edge of the Baltoscandian platform above a subduction zone that dipped away from Baltoscandia. Middle Ordovician island arc-related volcanicity and sedimentation then took place on the eroded remnants of this subduction complex. This model replaced earlier ones that invoked an uninterrupted evolution from a late Precambrian–early Ordovician mio-eugeocline to a Middle Ordovician–early Silurian marginal basin.

If a widespread late Cambrian to earliest Ordovician orogenic event can be demonstrated along the entire length of the Scandinavian Caledonides there must also be a major unconformable relationship throughout the belt that remains largely undetected (Sturt, 1984). It is the contention of this paper that in part of the central Norwegian Caledonides there exists stratigraphic and

structural evidence for this early Caledonian tectonothermal event. The event appears to be confined within the
framework of the Baltoscandian margin of Iapetus and
thus might reasonably be included in the Finnmarkian
orogeny. The post-Finnmarkian unconformity is also
recognized and regional correlations suggest that uplift
preceding it was of Arenig age. A tectonic model will be
presented to account for the two-part Caledonian evolution of the area in question and some observations on its
application to the whole of central Scandinavia will be
made. In order to put the model into its context, the
geology of the Steinkjer–Snåsavatnet area has first to be
summarized. A more complete description of the area is
currently in preparation for publication elsewhere.

THE STEINKJER–SNÅSAVATNET AREA

The work presented here relates to the Steinkjer–Snåsavatnet area (Fig. 1), which covers part of the Caledonian
allochthon and its parauthochthonous basement at the
southwestern end of the Snåsa Synform in North Trøndelag, central Norway. The succession in the area has
previously been subdivided and correlated with most of the
regionally recognized nappes within the Lower, Middle
and Upper Allochthons (Gee *et al.*, 1985) in central
Scandinavia. The most complete succession is found in the
southern limb of the Snåsa Synform, where representative
parts of the basement and its cover, and of the Offerdal,
Särv, Seve, Köli, Gula and Støren Nappes have been
variously recognized (Gee, 1977, 1978, 1980; Andreasson,
et al. 1978, 1979; Dyrelius *et al.* 1980; Tietzsch-Tyler, 1983;
Tietzsch-Tyler and Roberts, 1985) (Table III, Fig. 1
inset). Only the Seve Nappe has been recognized in the
northern limb of the synform in the area of Fig. 1, where
it is in direct contact with granitic basement. The Gula
and Särv Nappes have, however, been identified in the
northern limb of the synform further to the northeast
(Andreasson and Johansson 1983). The two limbs are
separated by the Mollelva Fault Zone (MFZ, Fig. 1), and
no synformal hinge is present (Tietzsch-Tyler, 1983).

Stratigraphy

South of the MFZ two tectonostratigraphic divisions are
recognized (Table I), and the internal stratigraphy of each
is summarized in Table II. Inferred ages and likely
correlations are also included in Table II (columns 4 and
5). In the remainder of this section several significant
points concerning the stratigraphy of the area will be
highlighted with reference to Table II. The stratigraphic
nomenclature used here is largely new and as yet remains
informal (Tietzsch-Tyler and Roberts, 1985). Its relationship to previous nomenclature is shown in Table III.

The lower tectonostratigraphic division has been shortened in at least three largely concordant ductile deformation zones. These now separate the Parautochthonous
basement, the Kvernabekk Nappe, the Leksdals Nappe
and the Skjøtingen Nappe (the latter two terms having
been redefined from the earlier usage of Wolff (1976, 1979)
and Roberts and Wolff (1981), (Table III). Between these
tectonic units it is possible to reassemble a succession that
commences with up to 750 m of locally derived arkosic
sandstones. In the highest, most northwesterly derived
Skjøtingen Nappe, the arkosic sandstones are intruded by
the doleritic Ramstad Dyke Suite (RDS) that locally has
given rise to up to 33% extension. They pass up into
a 200 m thick succession that is dominated by two

essentially mafic volcanic sequences that are separated by
a metasedimentary formation.

Geochemically, the metavolcanics of the Reinsvatn
Group have almost unequivocal tholeiitic affinites, while
the RDS has a more transitional character with both
tholeiitic and alkaline affinities (Andréasson *et al.*, 1979;
Tietzsch-Tyler, 1983). Metagabbroic, probably intrusive,
amphibolites, in the Reinsvatn Formation appear to be
representative of the RDS, based on only a limited number
of analyses. In terms of palaeomagmatic environments
both the RDS and the Reinsvatn Group metavolcanics
share a character that is transitional between within plate
basalts and ocean floor or mid-ocean ridge basalts (Fig.
2(*a*), (*b*)). Such a suite of rocks is usually taken to
correspond to an environment of continental extension
and rifting such as precedes ocean opening (Roberts, 1975;
Solyom *et al.*, 1979a,b; Pearce, 1980; Stephens 1980). The
alternative of an ocean ridge and ocean island combination (Pearce, 1980) is precluded by the overall geological
framework of the lower division. Thus, the greater part of
the Reinvsvatn Group comprises predominantly basaltic
volcanics, fed by the RDS during a period of continental
extension. Even the more intermediate metavolcanics
classify as icelandites rather than as orogenic andesites
(Middlemost, 1985) and thus are typical of a rifting
environment. The metasedimentary Byaelv Formation
between the two metavolcanic formations implies either
episodic extension or a hiatus in the magmatism associated
with that extension.

From Table II it can be seen that the Leksdalsvatn
Group and the Reinsvatn Formation have been correlated
with the successions within the Särv and Seve Nappes
respectively. The intrusive Ottfjället dolerites in the Särv
Nappe and the amphibolites of the Seve Nappe have
themselves been related to continental rifting (Solyom *et al.*,
1979b; Hill, 1980) during the late Precambrian. The
upper two formations of the Reinsvatn Group can be
correlated with the lowermost Seima Formation of the
Köli Supergroup (Table II, column 4, and Tietzsch-Tyler,
1983). Furthermore, the graphitic phyllites and schists of
the Byaelv Formation may be correlated with the Middle
Cambrian to Tremadocian Alum Shale Formation in the
autochthon and Lower Allochthon of central Sweden. The
Alum Shale Formation is typically rich in organic material
(up to 20%), and includes several limestone members at
lower levels. Lithological correlations, mostly on the basis
of organic carbon, uranium and vanadium contents have
already been made with several parts of the Middle and
Upper Allochthons (Bergström and Gee, 1985). Although
uranium and vanadium contents have not been determined for the Byaelv Formation, their highly carbonaceous nature supports such a correlation. The Gula Nappe,
which has been identified further northwest in the
southern limb of the Snåsa Synform by Andreasson and
Johansson (1983) is a direct extension of the Byaelv
Formation. Though it may have a tectonic contact with
the rocks ascribed to the Seve Nappe in that area, there is
a gradational relationship between the two in the area
described here (Tietzsch-Tyler, 1983) and separation
into the Seve and Gula Nappes seems unnecessary.

The two groups that define the upper tectonostratigraphic division south of the MFZ both commence with
thick polymict metaconglomerates but are now separated
by a thrust. Although an element of diachroneity is likely
to exist between them, the basal metaconglomerates are
interpreted as one linking unit. Above the metaconglomerates the succession exhibits an evolution that varies
across the area from southeast to northwest. In the
southeast, the thickest part of the Snåsavatn Group

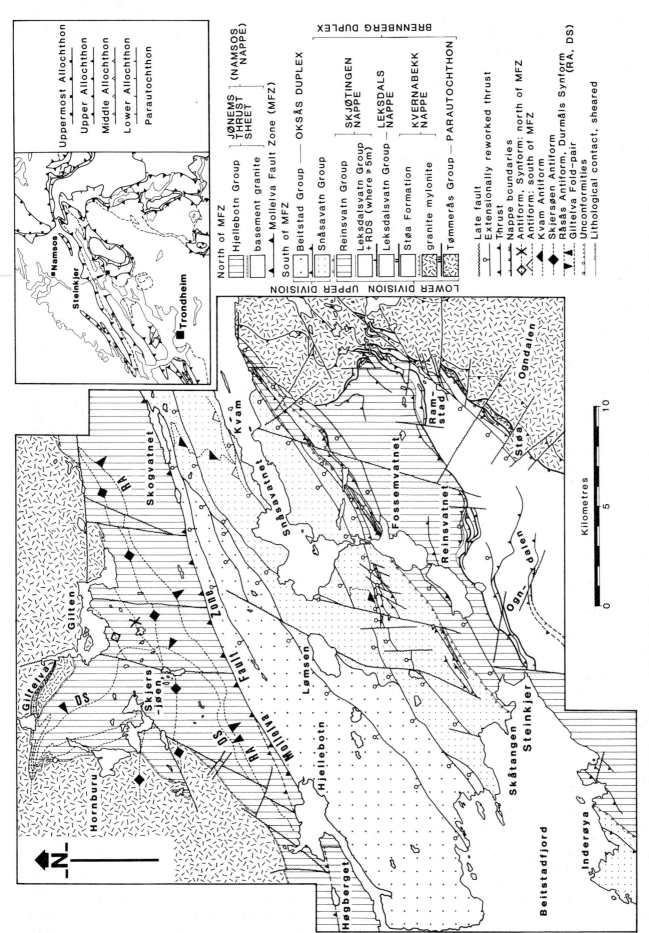

Fig. 1. Geological map of the Steinkjer-Snåsavatnet area. Inset location map shows the major tectonic subdivisions of the central Scandinavian Caledonides (from Gee *et al.*, 1985).

Table I. Tectonostratigraphy of the Steinkjer–Snåsavatnet area.

UPPER DIVISION: Scandian Units		JØNEMS THRUST SHEET	OKSÅS DUPLEX	BRENNBERG DUPLEX			
		?	Beitstad Group	Snåsavatn Group			
	LOWER DIVISION: Finnmarkian Units	NAMSOS NAPPE		SKJØTINGEN NAPPE	LEKSDALS NAPPE	KVERNABEKK NAPPE	PARAUTOCH-THON
		Hjellebotn Group	?	Reinsvatn Group	?		(Bjørndalen Formation)
				Leksdalsvatn Group (RDS)	Leksdalsvatn Group	Støa Formation	
		basement granite	basement granite	?	?	basement granite	Tømmerås Group

The Bjørndalen Formation is not exposed in the mapped area but occurs immediately outside. (RDS, Ramstad Dyke Suite)

represents a period of volcanosedimentary accumulation in a progressively deepening basin, ending with an extended interval of relative stability represented by the marbles of the Snåsa Formation. Further northeast, the marbles are succeeded by another volcanic formation (Roberts, 1967a, 1982). In the centre of the area only an incomplete succession is developed between the basal metaconglomerates and the marbles, suggesting the existence of an elevated basement block. The coarse clastics of the Skåtang and Rungstadvatn Formations in the southeast of the area represent a littoral facies indicative of continued erosion of active faults at the margins of the elevated block. Northwest of that block, the thick clastic sediments of the Beitstad Group reflect the development of a second basin, though lacking the magmatic activity of its neighbour. The suite of sedimentary structures preserved in the Beitstad Group is thought to point to a fluvial/shallow marine depositional environment (D. Roberts, personal communication, 1984). The basal metaconglomerates of the Steinkjer Formation contain a wide variety of clasts that could all have been derived by erosion of the lower division, including its parautochthonous basement (Table II, and Tietzsch-Tyler, 1983). Towards the northwest, on the raised block, granite clasts become common. The metaconglomerates at the base of the Beitstad Group are dominated by granitic clasts but do retain a variety of other lithologies that are typical of the lower division. The increase in granitic clasts in these basal metaconglomerates reflects variations in the basement beneath them across the Steinkjer–Snåsavatnet area. In the parautochthonous Tømmerås Group, the higher and more westerly thrust sheets contain more granitic than prophyritic lithologies, and the basement of the overlying Kvernabekk Nappe is entirely granitic (Tietzsch-Tyler, 1983). North of the MFZ, the basement is predominantly granitic and east of the immediate area of Fig. 1, on the north side of Beitstadfjord, the Beitstad Group rests directly on granitic basement (Carstens, 1956, 1960: Fig. 2, Tietzsch-Tyler, 1983). The granitic rocks beneath the Beitstad Group and north of the MFZ may of course not be the same age as those within the Tømmerås Group (cf. Table II).

The metavolcanic Lauvås and Skrattås Formations, mafic and felsic respectively, make up as much as 30% of the Snåsavatn Group. They comprise basalts and basaltic andesites, and less commonly andesites, quartz keratophyres and rhyolites together with a variety of tuffs and agglomerates. Taken together, these have a mixed tholeiitic and calc-alkaline chemistry (Tietzsch-Tyler, 1983). In terms of palaeomagmatic environments, the metavolcanics have a predominantly island arc character with subordinate ocean floor and within-plate affinities (Fig. 2(c), (d)). Consequently, an island arc setting, probably transitional to a back-arc basin (Pearce, 1980; Stephens, 1980) is favoured for the Snåsavatn Group. The coexistence of the metavolcanics with coarse, continentally derived metasediments suggests that the arc was on or closely adjacent to continental crust.

The two groups of the upper division may be correlated with both the central Trøndelag succession of the Støren Nappe and the type succession of the Köli Supergroup in central Sweden. In the former case the correlation is with the Ordovician Hovin Groups, extending up to the Upper Ordovician Tautra Limestone, which has been tentatively correlated with the Snåsa Formation (Spjeldnaes, 1985). Alternatively, almost the whole of the Snåsavatn Group may be correlated with the Middle–Upper Ordovician Gilliks Formation of the Köli Supergroup, while the capping Snåsa Formation may be correlated with the Ashgillian Slätdal Limestone of the Bellovare Formation (Table II; Tietzsch-Tyler, 1983).

North of the MFZ, basement granites pass directly into the Hjellebotn Group. Like the Reinsvatn Group in the lower division south of the MFZ, the Hjellebotn Group comprises two predominantly mafic metavolcanic formations that are separated by a metasedimentary formation (Table II). The intermediate metasediments are usually calcareous psammitic schists and thus compare well with much of the Byaelv Formation in the Reinsvatn Group. However, no graphitic sediments are evident. It would appear that the Gula Nappe, which has been identified on the north side of the Snåsa Synform by Andreasson and Johansson (1983: Figs. 1 and 2, F–F) is an extension of the Følinghei Formation. As with the Byaelv Formation in the Reinsvatn Group, south of the MFZ, there is no apparent necessity to define the formation as a separate nappe. Generally, the few Hjellebotn Group amphibolites that were analysed compare well with those of the Reinsvatn Group (Fig. 2(a), (b)). The geochemical and gross stratigraphical similarity between the Hjellebotn and Reinsvatn Groups, taken together with their structural position and close proximity, supports their correlation as advocated in Table II. The absence of a correlative of the Leksdalsvatn Group in the area of Fig. 1 and its only limited occurrence further to the northwest (the Särv Nappe of Andreasson and Johansson 1983, Fig. 1) suggests that the main area of its deposition may have been restricted to south of the MFZ.

Table II. Summary of lithostratigraphic units in the Steinkjer–Snåsavatnet area: unit thicknesses (column 2), lithological distributions and variations (column 3), and suggested regional correlations (column 4) and stratigraphic ages (column 5) are summarized by group and formation (column 1).

Division	Group, FORMATION	Maximum thickness exposed	lithological summary	Correlations	Probable age
South of M.F.Z. — UPPER DIVISION	**Beitstad Group**	4500m	Grey, grey-green metagreywackes; locally interbedded with 2.5-8m dark blue-grey, green or black slates. Sedimentary structures, including X-lamination, convolution, slump folds, intraformational breccias & conglomerates, indicate fluviatile to, shallow marine deposition (DR). 200m basal formation of unsorted polymict metaconglomerates. Occasional greywacke members. Commonest clasts are granite; grey, pink felsic volcanics. Also amphibolite; greenstone; gabbro; quartzite; metasandstone.	Basal metaconglomerates correlated with Steinkjer Fm (below).	Caradoc?
	Snåsavatn Group SNÅSA FM	1000m	Dark blue-grey, creamy white marbles; occasional beds, members up to 400m thick of grey, grey-green calcareous metapelites or metasandstones. Rarely, more diverse lithologies.	Tautra Limestone, Trøndelag (Spjeldnaes 1985): Slåtdal Limestone, Västerbotten (though of European faunal affinity).	"North American" fauna of recrystallized gastropods, cephalopods, brachiopods, bivalves & corals; considered late Middle to Upper Ordovician (Spjeldnaes 1985): Caradoc–Ashgill?
	RUNGSTADVATN FM	500m	Grey metasandstones, metagreywackes. Thick polymict metaconglomerate toward base in NE with granite, felsic volcanic, greenstone, epidote & quartzite clasts. Basal member of dark blue-grey, grey-white marbles. Subordinate black, grey calcareous phyllites.	Upper Hovin Gp, Trøndelag; Gilliks Fm, Västerbotten.	Caradoc?
	SKRATTÅS FM	500m	Fine, medium grained felsic volcanics, lithic & crystal tuffs, volcaniclastics. Often sheared to quartz sericite, talc schists. Infrequent marbles, calcareous phyllites, greenstones, tuffitic or volcaniclastic greenschists. Skrattås massive sulphide deposit; dispersed pyrite & magnetite throughout. 150m basal member of pale grey marbles with dark grey, black calcareous phyllites.	Upper Hovin Gp, Trøndelag; Gilliks Fm, Västerbotten.	Caradoc?
	LAUVÅS FM	800m	Chloritic, actinolitic greenstones; coarse volcaniclastic breccias; banded tuffitic & volcaniclastic greenschists. Several 1.5-9m white felsic volcanic members. Dispersed pyrite & magnetite throughout. Occasional thin blue-grey quartzites, one thin marble.	Upper Hovin Gp, Trøndelag; Gilliks Fm, Västerbotten.	Caradoc?
	SKÅTANG FM	1500m	Grey-green psammites, often differentiated into 2-25mm felsic, pelitic bands. Locally calcareous, pebbly. Occasional graphitic phyllites.	Upper Hovin Gp, Trøndelag; Gilliks Fm, Västerbotten.	Caradoc?
	STEINKJER FM	250m	Ill-sorted polymict metaconglomerate with grey-green phyllitic, psammitic matrix: diamictite. Common clasts of pink felsic volcanic; epidote; dark blue quartzite; amphibolite; chloritic greenstones; tuffitic greenschists. Also granite; grey quartz arenite; metasiltstone; marble clasts. Occasional felsic volcanics in NE.	Upper Hovin Gp, Trøndelag; Gilliks Fm, Västerbotten.	Caradoc?
South of M.F.Z. — LOWER DIVISION	**Reinsvatn Group** BYHALL FM	600m	Fine-grained chloritic, actinolitic greenstones & greenschists. Locally tuffitic, volcaniclastic. Occasional 0.3-2.6m white, pink felsic volcanic members, some with spessartine. Marble, calcareous phyllite member in SW. In NE, dark amphibolites, locally garnetiferous, cf Reinsvatn Fm.	Støren Gp, Trøndelag; Seima Fm, Västerbotten.	Upper Cambrian–Arenig (cf Byaelf Fm, below)?
	BYAELV FM	650m	Pale, dark grey phyllites; commonly pelitic, locally psammitic, quartzitic. Frequently graphitic, locally a distinctive dark blue-grey quartz mylonite with up to 23% graphite; elsewhere variably calcareous. Towards base, 2m member of blue-grey marble & calcareous phyllite. Discontinuous quartz conglomerate, greenschist toward NE. In NE, coarse psammitic schists; variably graphitic pale, dark blue-grey mica schists; quartzites; amphibolites.	Gula Group, Trøndelag; Seima Fm, Västerbotten; Alum Shale, Jämtland.	Upper Cambrian?
	REINSVATN FM	750m	Massive fine, medium grained amphibolites, variably feldspathic, locally garnetiferous. Towards top, amphibole content reduced & epidote, chlorite increased. Large lenticular bodies, lenses, augen of coarse, variably feldspathic metagabbroic amphibolite retaining random fabric; locally rhythmically layered. Locally volcaniclastic amphibolites; tuffitic greenstones; amphibolitic 'metaconglomerates'. Discontinuous coarse & garnetiferous, pelitic & psammitic schists; sometimes with thin amphibolite bands. Up to 2m marble beds occur infrequently.	Amphibolites & schists of Upper & Lower Seve Nappe, Jämtland & Västerbotten.	Latest Precambrian – Upper Cambrian?
	Leksdalsvatn Group	750m	Usually thick-bedded, pale grey feldspathic sandstones; often texturally greywackes but chemically arkosic; with X-lamination load & flame structures, slump folds. Subordinate dark siltstones. In footwall of Skjøtingen Nappe, intruded by RDS dolerite (up to 33%); locally sheared to interbanded amphibolites & psammitic schists. Up to 250m basal zone of tectonic schists; increasingly recrystallized downwards into coarse-grained garnet mica schists.	Sediments: Tossåsfjället & Risback Gps, Jämtland; RDS: Ottfjället Gp, Hedmark; RDS: Ottfjället Dyke Suite, Jämtland.	Upper Proterozoic.
	STØA FM	80m	Cap on granites of Kvernabekk Nappe. Sheared & tightly folded feldspathic psammitic schists.		
	Tømmerås Group	>500m	Fine grained, usually porphyritic, grey & pink dacites to rhyodacites. Occasional volcanic breccia. Adamellitic & monzonitic microgranites in highest parautochthonous thrust sheet; mylonitized adamellite in Kvernabekk Nappe. Intruded by suite of gabbro, ophitic dolerite & metadolerite sheets, lenses & lopoliths up to 300m thick.	Volcanics & granites: Dala volcanics & granites, Sub-Jotnian Complex (pre-caledonian basement): intrusive suite: (CSDG) or Ottfjället Dyke Suite.	Lower Proterozoic (c.1650Ma)
North of M.F.Z.	**Hjellebotn Group** VESTERDAL FM	>900m	Fine, medium grained amphibolites & greenstones, often garnetiferous. Occasional white, pale green quartzofeldspathic rocks, grey pink toward top of formation. Greenstone: white, pale green metamorphic schists, often garnetiferous: pale grey marbles & calcareous schists: dark blue quartzites: thinly interbanded greenstone, quartzfeldspathic rock. Fine-grained mafic & felsic mylonites in MFZ.	Byhall Fm (above).	Upper Cambrian–Arenig?
	FØLINGHEI FM	850m	Blue-green & grey, pelitic to psammitic schists, usually calcareous, sometimes garnetiferous: occasional white marble.	Byaelv Fm (above).	Upper Cambrian?
	RØRDALSBUKT FM	450m	Variably feldspathic garnetiferous amphibolites; white, grey & green garnetiferous quartzofeldspathic rocks, locally with amphibole bands, garnetiferous interbanded amphibolite, quartzofeldspathic rock: coarse metagabbroic amphibolite. Green-grey psammitic schists, usually with amphibolite bands, sometimes coarse quartz-feldspar-amphibole-biotite schist; locally massive white marbles. Zone of mylonitic semipelitic & psammitic schists, amphibolites, quartzofeldspathic rocks at base toward NE.	Reinsvatn Fm (above).	Latest Precambrian–Upper Cambrian.
	Basement granite		Variably foliated coarse red granite, locally grey or white; often with concordant pegmatites.	Tømmerås Gp (above).	Lower Proterozoic?

DR, personal communication by D. Roberts; MFZ, Mollelva Fault Zone; RDS, Ramstad Dyke Suite; CSDG Central Scandinavian Dolerite Group (Gorbatschev *et al.*, 1979).

Table III. Stratigraphic correlations and interpretations for the Steinkjer–Snåsavatnet area. The Bjørndalen Formation sedimentary cap to the Tømmerås Group has been omitted for clarity. RDS Ramstad Dyke Suite (used to define the Särv Nappe in column 4).

Stratigraphy (this paper)			Carstens 1960, Springer–Peacey 1964, Roberts 1975	Wolff 1976, 1979, Roberts & Wolff 1981	Andreasson et al. 1978, 1979, Gee 1980, Dyrelius et al. 1980
Beitstad Group			Upper Hovin Group	*Trondheim Nappe Complex*	
Snåsavatn Group	Snåsa Fm		Lower Hovin Group	Støren or Köli Nappe	Trondheim or Köli & higher nappes
	Rungstadvatn Fm				
	Skrattås Fm				
	Lauvås Fm				
	Skåtang Fm				
	Steinkjer Fm				
Reinsvatn Group	Byhall Fm		Støren Group		
	Byaelv Fm				
	Reinsvatn Fm		Gula Group	Skjøtingen, Gula or Seve Nappe	Seve Nappe
(RDS)			Leksdalsvatn Group	Leksdal Nappe	Särv Nappe
Leksdalsvatn Group (Støa Fm)					Offerdal Nappe
Tømmerås Group			Tømmerås Group	basement	basement

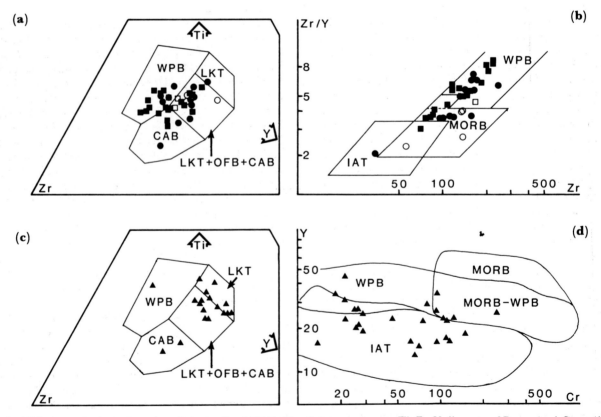

Fig. 2. Palaeomagmatic environment diagrams for the Steinkjer–Snåsavatnet area: Ti–Zr–Y diagram of Pearce and Cann (1973), Zr/Y–Zr and Y–Cr diagrams of Pearce (1980). WPB, within plate basalts; LKT, low-potassium tholeiites; CAB, calc-alkaline basalts; MORB, mid-ocean ridge basalts; IAT, island arc tholeiites. ■ Ramstad Dyke Suite; □ Tømmerås Group metadolerites; ● Reinsvatn Group amphibolites & greenschists; ○ Hjellebotn Group amphibolites; ▲ upper division greenschists.

Structure

A complex and variable polyphase structural and metamorphic history is evident in the rocks of the Steinkjer–Snåsavatnet area. The specific elements that define this history are summarized in some detail in Table IV. As with the previous section, this section will highlight only the more significant aspects of the structure and metamorphism of the area with reference to Table IV.

Several lines of evidence indicate that the lower tectonostratigraphic division to the south of the MFZ was subject to a distinct tectonometamorphic history that is absent from the upper division. Subsequently both divisions were imbricated within the Brennberg Duplex (D_S1d, Table IV). Pre-imbrication deformation of the upper division is restricted to the ductile fold and cleavage development at low metamorphic grades and is entirely consistent with an initial stage of D_S1 shortening leading up to imbrication. Imbrication of the lower division was, however, superimposed in an out-of-sequence manner on an already assembled nappe-pile that had been subject to polyphase deformation and metamorphism (Table IV, and Tietzsch-Tyler, 1983). The basal metaconglomerates of the upper division rest with apparent unconformity across both of the upper formations of the Reinsvatn Group. They also cut across the metamorphic isograds which obliquely transect the same two formations towards the northeast of the area. Between Inderøya and immediately northeast of Fossemvatnet the metaconglomerates rest on low-grade greenschists and micaceous phyllites of the Byhall and Byaelv Formations respectively. On the southern shore and immediately southwest of Snåsavatnet, however, the low-grade metaconglomerates rest directly on garnetiferous amphibolites and schists of the same two formations (Tietzsch-Tyler, 1983, Fig. 107).

The range of clast compositions in the basal metaconglomerates of the upper division (Table II) demonstrates that a complete cross-section of the lower division must have been available for sampling by erosion at the time of their deposition. Looking progressively northwest, in the core of the Kvam Antiform and at the base of the Beitstad Group, a greater proportion of granitic clasts is evident—indicative of the direct contact between the Beitstad Group and granitic basement nearby (above). Considerable uplift and a variable degree of erosion of the lower division must have taken place prior to and during the deposition of the upper division. The time interval for this uplift and erosion could have spanned as much as 30 Ma (Ross *et al.*, 1982; Gale, 1985; McKerrow *et al.*, 1985), between late Arenig and Caradocian times (Table II, column 5).

The polyphase deformation and polymetamorphism that accompanied nappe development in the lower division is unevenly distributed. The characteristic intense shearing and occasional folding, accompanied by up to two episodes of amphibolite facies recrystallization (Andreasson and Lagerblad, 1980; Tietzsch-Tyler, 1983), was mostly confined to the two deformation zones that broadly define the boundaries between the nappes. The upper zone is essentially confined to the Reinsvatn Formation, the RDS-intruded Leksdalsvatn Group and the upper 20 m of the dyke-free Leksdalsvatn Group, and is up to 950 m wide. The upper boundary of the zone climbs across the Byaelv and Byhall Formations towards the southern shore of Snåsavatnet. Within the zone, sizeable low-strain tectonic lenses have preserved primary textures in metagabbroic bodies within the Reinsvatn Formation and primary textures and relationships in the RDS-intruded Leksdalsvatn Group. The latter have also

escaped appreciable metamorphism in the otherwise amphibolite facies zone. A second, lower zone of deformation and recrystallization gives rise to the 250 m Basal Tectonic Schist Zone (BTSZ) and the deformation in the thin, underlying Støa Formation. The mylonitization of basement granites in the Kvernabekk Nappe and the deformation and metamorphism in the tectonic schist zone (TSZ) at the base of the parautochthon define two lower, less well-developed zones. Quite high metamorphic grades were again reached in the TSZ, seen in particular where a thin sedimentary cover on the parautochthon (the Bjørndalen Formation, which is not exposed in the area of Fig. 1) has taken up much of the deformation (Andreasson and Lagerblad, 1980). Deformation between these discrete zones and in most of the basement was minimal and was accompanied by significantly lower grades of metamorphism—usually low-greenschist facies. Consequently, it is suggested that the nappes were developed by translation across broad, low-angle ductile shear zones rather than on discrete thrusts, but still at a relatively high crustal level to account for the regional low grade of metamorphism. Emplacement of a higher, dominant nappe or nappe-complex must have been responsible for the ductile imbrication of the lower division.

The polyphase deformation and polymetamorphic history of the Hjellebotn Group closely parallels that observed in the Skjøtingen Nappe (Table IV), but, instead of being displaced on a major shear zone, the Hjellebotn Group was twice isoclinally folded together with its basement (D_F1, D_F2). However, the sheared basement-cover contact in the inverted limb of the Durmåls Synform with its locally preserved blastomylonite zone (Table II) may indicate that much of this early deformation north of the MFZ was also taken up in shear zones. The Hjellebotn Group and its basement must represent the dominant nappe or nappe-complex (The Namsos Nappe, Table II) referred to above.

The deformation that followed deposition of the upper division centres on its imbrication together with the lower division (D_S1, Table IV). This deformation took place within two duplexes and was foreland-propagating. Within each duplex, deformation was initiated by the ductile shortening that is evident in an overturned macrosyncline in the Oksås Duplex and, within the Brennberg Duplex, by the isoclinal macrofolding of the thinner Snåsavatn Group succession towards the northwest. This latter folding probably reflects the thinner succession being scraped off its elevated basement beneath the overriding Oksås Duplex. Only a weak cleavage developed during initial folding within the Oksås Duplex, but a more complex composite fabric was developed in the Snåsavatn Group involving some metamorphic differentiation prior to the formation of a phyllitic cleavage. WNW–ESE stretching is related to these early structures, which are at least partly controlled by facies and thickness variations. This suggests that the syn-depositional raised block had a NNE–SSW trend.

Imbrication was on relatively few thrusts in the lithologically homogeneous Oksås Duplex, while a much more complex thrust system was responsible for imbrication within the Brennberg Duplex (Fig. 3). The evidence of direct contact between the Beitstad Group and basement granite west of the area shows that the floor-thrust of the Oksås Duplex climbed out of the basement towards the southeast and into its commonly observed position within the basal metaconglomerates of the Beitstad Group (Fig. 3). Progressive imbrication of the Brennberg Duplex was accompanied by meso- and macrofolding within

Table IV. Structural and metamorphic history of the Steinkjer–Snåsavatnet area, summarized in a sequential synopsis of structural and metamorphic events and correlated across the area within the tectonostratigraphic framework defined in this paper.

	JØNEMS THRUST SHEET	OKSÅS DUPLEX	BRENNBERG DUPLEX				parautochthon
			Snåsavatn Gp	Skjøtingen Nappe	Leksdals Nappe	Kvernabekk Nappe	
D_S3?	Late faults; cataclasis	Late faults	Late faults.	Late faults.	Late faults	Late faults	Late faults; cataclasis
D_S2b	Recument open to close mesofolds in steep belts: down-dip vergence, shallow NE, SW plunge.	Recumbent open macro- & meso-folds in steep belts: down-dip vergence, shallow WSW plunge.	Ubiquitous recumbent open to close, assymetric, disharmonic & non-cylindrical mesofolds: down-dip vergence, shallow WSW plunge. Frequent coaxial crenulation, occasional axial planar crenulation cleavage.	Recumbent open to close mesofolds in steep belts: down-dip vergence, shallow WSW plunge.	Rare recumbent mesofolds, down-dip vergence, crenulation, in steep belts.		
D_S2a	Extensional reworking of earlier thrusts; down-dip shearing in former thrust zones.	Extensional reworking of earlier thrusts; down-dip shearing in former thrust zones.	Extensional reworking of earlier thrusts; brecciation, cataclasis, minor down-dip directed "thrusts", recumbent folds & spaced, crenulation cleavage in fault zones. Rotation of early down-dip verging folds into steep NNW plunging extension direction.	Extensional reworking of earlier thrusts.	Extensional reworking of earlier thrusts; associated open drag folds.		Extensional reworking of earlier thrusts; down-dip verging folds in foliated hanging wall metadolerites.
D_S1d M_S1d	Assymetric macrofolds with steep SSE dipping axial planes; open meso-folds; rare kink- & box-folds: shallow NNE plunge.	Imbrication, duplex formation: flat & 19° ramp geometry causes open folds with ENE–WSW trend. Pebble deformation increases towards floor–thrust in basal metaconglomerates. Brecciation, assymetric & irregular folds with shallow WSW plunge, steeply axial planes in thrust zones; cleavage rare.	Imbrication, duplex formation, antiformal stacking of thrust sheets; flat & 18–33° ramp geometries with meso-HWA, FWS ramp geometries. brecciation, mylonitization in thrust zones. Rotation of deformed pebbles into NNW plunging stretching direction. Close to tight, assymetric to overturned mesofolds with sheared interlimbs associated with thrusts; SSE vergence, shallow W–WSW plunge. Infrequent steep NNW dipping axial planar cleavage; Bedding/cleavage intersection coaxial with mesofolds. Upright open to close macrofolds, 150–200m wavelengths, steep NNW dipping axial planes, occasional axial planar cleavage: shallow W–WSW plunge. Progressive steepening of planar structures (24° to 42–54°) & tightening of macrofolds toward NNW, into antiformal stack. Late metamorphic pulse: layer–//, axial planar chl, bi.	Imbrication, duplex formation; general 33° dipping flat & 19° ramp geometry causes open folds with NE–SW trend. Brecciation, mylonitization, isoclinal intrafolial meso- & microfolds in thrust zones. Shearing & recrystallization of RDS dyke margins; shallow NNW plunging stretching lineation defined by hbl alignment. Limited rotation of earlier hbl folds into NNW direction. Late metamorphic pulse: porphyroblastic chl, bi at upper levels.	Imbrication, duplex formation: 9° dipping flat & 19° ramp geometry causes open folds with NE–SW trend. Stretching lineation, defined by qz elongation. shallow NNW plunge. Close to tight mesofolds, locally with axial planar cleavage. Thin concordant mylonites in USZ. Some reworking of BTSZ?	Stretching lineation with shallow NNW plunge?	Imbrication, duplex formation: 9° dipping flat & 17° ramp geometry causes open folds with ENE–WSW trend. Limited reworking of TSZ, which forms floor–thrust, in thrust zones. Shearing & recrystallization in metadolerites. Stretching lineation, defined by qz elongation, hbl alignment: shallow NNW plunge.
D_S1c M_S1c			Phyllitic cleavage, weak metasandstone cleavage. Axial planar to rare recumbent isoclinal folds, notably in high strain zone in Skåtang Fm; shallow, moderate WSW, occasionally NNW plunge. Overturned anticlinal macrofold, NNW dipping axial plane, refolds Kvam Antiform. Generally prolate strain accompanied by volume loss in metaconglomerates; prolate strain in area of mesofold. Stretching lineation, defined by pebble long axes, qz elongation: shallow WNW plunge. Greenschist facies metamorphism.				
D_S1b M_S1b			Compositional banding formed at low angle to bedding by cataclasis and metamorphic differentiation in high strain zone in Skåtang Fm & associated with recumbent southward closing isoclinal macrofold, the Kvam Antiform (~5000m interlimb). Greenschist facies metamorphism.				
D_S1a M_S1a	Upright Skjersjøen Antiform with steep NNW dipping axial plane; open to close mesofolds, steeply inclined axial planes: shallow ENE plunge.	Overturned synclinal macrofold (Snåsa Syncline?), axis faulted out. Weak metagreywacke cleavage, slaty cleavage. Axial planar to close to tight mesofolds: bedding/cleavage intersection plunges WSW. Stretching lineation, defined by qz elongation, pebble long axes: shallow WNW plunge. Low greenschist facies metamorphism.					

	?	Deposition	Erosion
M_F4	Retrogression: gt to chl? Mylonites in MFZ?	Retrogression: gt to chl?	Weak foliation at low angle to primary lamination. Thin tectonic schist zone (TSZ) at 500m into parautochthon: progressive recrystallization from phyllites to coarse blastomylonitic schists. Axial planar to rare SE verging assymetric mesofolds. Low greenschist facies metamorphism.
M_F3	Post-schistosity pegmatites? Late metamorphic pulse: skeletal garnets overgrow post-schistosity folds.		
D_F3	Close to isoclinal intrafolial, post-schistosity mesofolds; local crenulation: shallow ENE plunge. Layer-// shearing: foliation flattened around, augening porphyroblasts; fracturing & extension of same.	Assymetric open to recumbent tight, post-schistosity mesofolds, often with sheared interlimbs: SSE vergence, shallow WSW plunge. Axial planar crenulation cleavage inclined at moderate, low angle to schistosity. Layer-// shear: foliation flattened around, augening porphyroblasts; fracturing & extension of same.	Layer-// shear: foliation flattened around, augening garnets in USZ, BTSZ.
M_F2	Metamorphic peak. Grade varies from upper greenschist to mid-amphibolite facies (Vesterdal to Rørdalsbukt Fm). Gt, hbl porphyroblasts overgrow schistosity.	2nd metamorphic peak. Grade varies from upper greenschist to mid-amphibolite facies (Byhall to Reinsvatn Fm). Gt, hbl porphyroblasts overgrow schistosity, P-T estimated at 450-500°C, 4-5.5kb.	Metamorphic peak. Grade reaches low to mid-amphibolite facies in USZ, BTSZ. T estimated at c.425°C in BTSZ. Low greenschist facies in bulk of nappe.
D_F2	Schistosity, defined by chl, bi, mu, hbl: essentially // compositional layering. Axial planar to isoclinal macrofolds - notably to 15km interlimbs - notably tight to isoclinal mesoform; Durmåls Synform, Råsås Antiform: shallow ENE plunge. Stretching lineation, defined by qz elongation, hbl alignment: shallow ENE plunge.	Schistosity, defined by chl, bi, mu, hbl; strong L-fabric in amphibolites. Often blastomylonitic in schists. Axial planar to rare recumbent isoclinal folds: plunge varies ENE-WSW to NNW. Stretching lineation, defined by hbl alignment, qz elongation: shallow WNW plunge. Metagabbros in Reinsvatn Fm, core of RDS-intruded Leksdalsvatn Fm largely escaped deformation: marginal blastomylonitization at metagabbro margins: limited rotation, pinch & swell of dykes in less deformed RDS.	Støa Fm: isoclinal folds, refold earlier folds. Axial planar fabric. Basement granite: blastomylonitization. Stretching lineation: shallow WNW plunge. Upper greenschist facies metamorphism. 20m Upper Shear Zone (USZ): blastomylonitic fabric: rare assymetric tight to isoclinal mesofolds with fabric axial planar. 250m Basal Tectonic Schist Zone (BTSZ): progressive recrystallization from phyllites to coarse blastomylonitic schists toward base of nappe. Isoclinal folds.
D_F1 M_F1	Isoclinal macrofolds of basement & cover with up to 4.5km interlimbs - notably Giltelva Fold-pair. Shearing of basement-cover contact. Formation of compositional layering, origin of inclusion trails oblique to schistosity in garnets: early metamorphic event?	Pre-schistosity mylonitic fabric. 1st metamorphic peak, only preserved locally: upper amphibolite facies. P-T estimated at 550-600° C, 4-6kb (ky, stau grade).	

RDS, Ramstad Dyke Suite; D_F1-D_F3, M_F1-M_F4 Finnmarkian events; D_S1-D_S3, M_S1 Scandian events; bi, biotite; chl, chlorite; gt, garnet; hbl, hornblende; mu, muscovite; qz, quartz; ky, kyanite; st, staurolite. Thrust-ramp angles for D_S1d deformation are relative to the associated thrust-flat, and metamorphic P-T data come from Tietzsch-Tyler (1983).

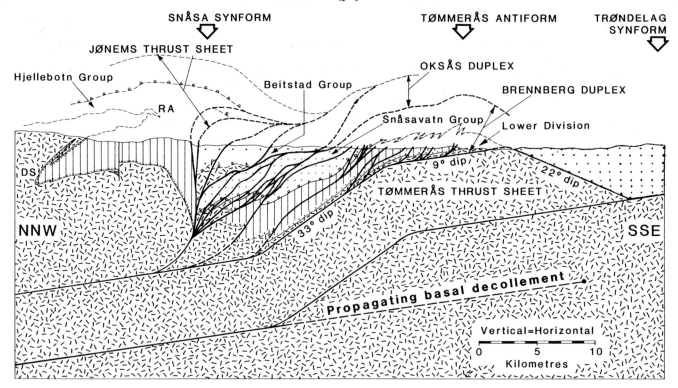

Fig. 3. Structural profile across the Steinkjer–Snåsavatnet area illustrating the geometry of the Tømmerås Thrust Sheet. DS, Durmåls Synform; RA, Råsås Antiform. Shading as for Fig. 1, but with the Leksdalsvatn Group included with the Reinsvatn Group (vertical hatching).

individual thrust sheets and by antiformal stacking. In the lower division, structures related to imbrication are usually confined to geometrically required hangingwall anticlines (HWA) and footwall synclines (FWS). Poly-phase mylonite and microfold development in the basal TSZ of the earlier parautochthon probably reflects the reactivation of its basal shear zone as the floor-thrust of the Brennberg Duplex.

The D_S1a Skjersjøen Antiform on the north side of the MFZ (Fig. 1) is interpreted as a large-scale HWA within a dominant Jønems Thrust Sheet (Table II), the emplace-ment of which was responsible for the development of the Oksås and then the Brennberg Duplexes beneath and in front of it. The macrofolds south of Gilten (Fig. 1) that have SSE dipping axial planes (D_S1d, Table IV) were probably formed as antithetic backfolds in response to the antiformal stacking of the Oksås and Brennberg Duplexes.

The final widespread deformation in the area is recog-nized in extensional reworking of thrusts within the two duplexes (D_S2a, Table IV) and by the development of ubiquitous down-dip verging assymmetric mesofolds in steeply dipping strata (D_S2b). Such structures are recognized throughout central Scandinavia as a late-stage gravitational collapse of the orogen (Roberts, 1967b, 1979; Trouw, 1973; Stephens, 1977; Sjøstrand, 1978). In the present area, these structures can be directly related to extension associated with ramping during the develop-ment of the Tømmerås Antiform as a basement thrust culmination (Fig. 3). In its simplest form, a 5 km thrust sheet is required geometrically to produce the observed structure as a major HWA, though the sheet could also take the form of a duplex involving an initially thinner slice of basement.

Before presenting the model, the variation in stretching directions (Table IV) requires some comment. In the Hjellebotn Group a uniform ENE plunging stretching lineation is developed, while an equally uniform WNW plunging stretching lineation characterizes early events on

the south side of the MFZ. One explanation for this might be that the WSW–ENE trending MFZ, the principal basement structure throughout deformation of the area, had a dextral transpressional sense of displacement at that time, and thrust the Hjellebotn Group and its basement obliquely towards the south-southeast. The close proximity of the preserved Hjellebotn Group to the MFZ might account for its ubiquitous fault-parallel stretching lineation. The equivalent zone on the south side of the fault is no longer visible. At a later stage post-D_F3, pre-D_S1 sinistral transtension on the MFZ could have given rise to NNE–SSW trending sinistral oblique-slip subfaults capable of producing the elevated block between the Beitstad and Snåsavatn Group basins. These faults subsequently controlled the initial ESE-directed stages of imbrication that was once again oblique to the regional strike. Further imbrication was by dip-slip thrusting toward the south-southeast evident in ramp orientation and a second stretching direction in the lower division. A similar orientation for ramping of the basement Tøm-merås Thrust Sheet is indicated by the regional strike and by association with D_S2a NNW–SSE extension. A similar duality of stretching lineations is observed throughout the Scandinavian Caledonides (Hossack and Cooper, 1986), also usually indicating a more east–west early translation followed by a more north–south translation closer toward the foreland.

DISCUSSION

In recent years the tectonic evolution of the central Scandinavian Caledonides has been modelled in some detail, particularly by Gee (1975, 1978) and more recently by Dallmeyer and Gee (1986). These authors have argued that the late Precambrian to Silurian Jämtland Supergroup of the autochthon, parautochthon and lower allocthon represent a single westward-thickening miogeoclinal wedge that extended from the

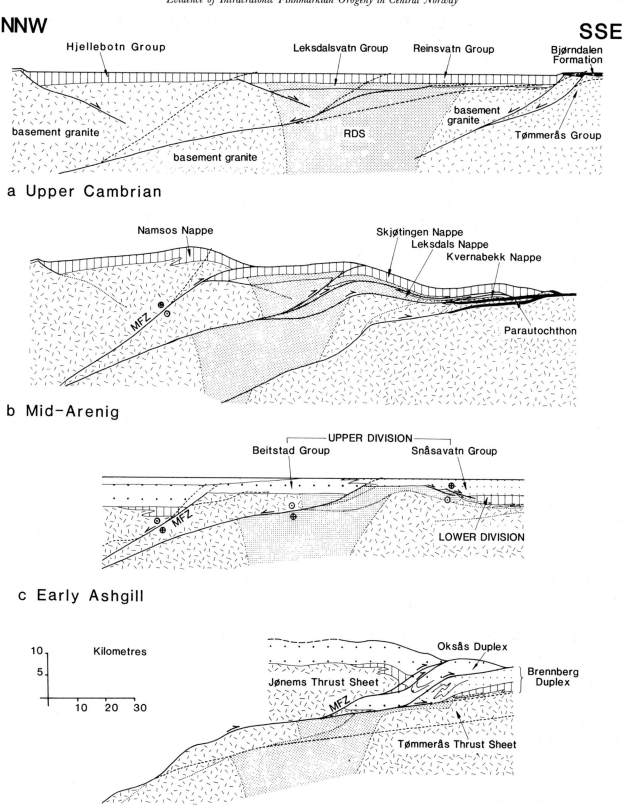

Fig. 4. A tectonic model for the evolution of the Steinkjer–Snåsavatnet area. RDS, Ramstad Dyke Suite; MFZ, Mollelva Fault Zone. Shading as for Fig. 1, but with the Leksdalsvatn Group included with the Reinsvatn Group (vertical hatching) in (*c*) and (*d*). For discussion, see accompanying text.

Caledonian front to include the Bjørndalen Formation (Stephens and Gee, 1985: Fig. 2). At the margin of Baltsocandia, these sediments are modelled as passing into thicker late-Precambrian rift-related sediments (the Offerdal and Särv Nappes), which, towards the west, were

intruded by a major dyke swarm (the Ottfjället dykes) that fed a thick Precambrian to Cambrian rift-related volcanic sequence (the Seve Nappe). In the earlier models this miogeocline was succeeded directly by the subduction-related sedimentation and volcanism of the Köli Supergroup. In the light of radiometric evidence for an early Ordovician tectonothermal event in the Seve Nappe, these models have been modified so that the outer miogeocline was deformed into an early Ordovician accretionary prism above a westward-dipping Benioff Zone (Dallmeyer and Gee, 1986). Since the Bjørndalen Formation is considered to be a part of the inner miogeocline and the outer miogeocline to have originated largely beyond the west coast of Norway (Gee, 1975), it has been necessary to envisage translation of the Scandinavian nappes, including the remnants of the accretionary prism, 500 km or more across central Scandinavia (Gee, 1978). In that the various subdivisions recognized in the Steinkjer–Snåsavatnet area have been directly correlated with the major nappes (Table III, column 4), the area has already been integrated with the model of a single broad miogeocline.

In mapping the Steinkjer–Snåsavatnet area it has become apparent that between the Hjellebotn Group in the north (which generally lacks a correlative for the Leksdalsvatn Group), the RDS-intruded Leksdalsvatn Group and succeeding Reinsvatn Group, the RDS-free Leksdalsvatn Group, the Støa Formation on its granitic basement in the Kvernabekk Nappe, and finally the Tømmerås Group with its thin Bjørndalen Formation cover in the southeast, the area could represent a former discrete basin in itself. The primary significance of the rift-related sedimentation, intrusion and volcanism would remain, but would be confined to a largely self-contained basin within the Baltoscandian margin—presumably one of several such basins, now represented by other comparable successions. As would be expected, since they represent the same basic pattern of events, each succession is broadly similar—so much so that in existing models they have been directly correlated. Differences do exist, however, as between the largely amphibolite-free Gula Group in central Norway and the amphibolite-dominated Seve Nappe. Such a situation would compare more closely with the models of a braided system of rift centres (giving rise to many future ophiolites!) that have been advocated for the early Tethyan evolution of the Eastern Mediterranean by Sengor and Yilmaz (1981) and Robertson and Dixon (1985), rather than the current Scandinavian model of a single broad miogeocline. This realization has led to the generation of the following tectonic model, which applies most immediately to the Steinkjer–Snåsavatnet area but could have wider implications for the whole of the central Scandinavian Caledonides. This model is not a total contradiction of those of Gee (1975, 1978) and Dallmeyer and Gee (1986), but represents a modification of them. It may not prove to be entirely correct in the long run, but in the interim it may provide a useful focus for discussion of the existing model.

A tectonic model

On the basis of the geological evidence outlined above, the following four-stage model for the tectonic evolution of the Steinkjer–Snåsavatnet area is suggested (Fig. 4).

(1) *Late Precambrian to Upper Cambrian*

Extension of the Baltoscandian continental margin on a series of low-angle faults leading to the development of a discrete ENE–trending basin. Initial arkosic Leksdalsvatn Group sedimentation was followed by the lengthy period of Reinsvatn Group rift-related volcanicity that was interrupted during the Upper Cambrian by an interval of quiet sedimentation in organic-rich waters. During the period of volcanicity, the original sedimentary basin was

extended at least 80 km NNW, beyond the area of Leksdalsvatn Group sedimentation. Quite possibly it covered much of the Baltoscandian margin of Iapetus. At its SE margin the basinal sediments and volcanics passed laterally into a thin sedimentary cover on the basement, the Bjørndalen Formation (Andreasson and Lagerblad, 1980; formerly the Bjørntjern Formation of Gee, 1974). On the basis of lithological correlation with the Caledonian foreland succession (Gee, 1977) the quartzites, graphitic shales and limestones of the Bjørndalen Formation probably range from the latest Precambrian into the lowest Ordovician (Gee *et al.*, 1974). The consistency of this succession across the Baltoscandian margin suggests that the whole margin was part of a marine shelf during earliest Palaeozoic times, and that thick Leksdalsvatn-like sediments and Reinsvatn-like volcanics represent localized centres of active rifting and basin deepening. These rift-centres may have been interconnected in a braided system. It would then be possible that the direct identification of the Byaelv and Følinghei Formations with the Gula Group by Andreasson and Johansson (1983) would be correct, though between one basin and another. That second basin would now be represented by the Gula Group and the various ophiolite fragments in the Trondheim area, the latter suggesting the local development of oceanic lithosphere in this basin. During volcanicity, the arkosic sediments and their immediate basement at the centre of the Leksdalsvatn basin were intruded by the RDS, the likely feeder system for the surface magmatism. There appears to be no reason for advocating more than a relatively restricted intrusive belt along the axis of the basin.

(2) *Latest Cambrian to earliest Ordovician*

Inversion of the basin occurred on a number of broad, low-angle ductile shear zones, probably in part reusing the existing extensional fault system. These shear zones now define the boundaries of a sequence of locally rooted nappes that were stacked at this time. The deformation occurred under a dextral transpressive regime centred on the MFZ that thrust the Namsos Nappe to the east-southeast. This oblique thrusting across the centre of the basin caused the oblique ductile imbrication of the lower division. At the lowest level, the deformation involved parautochthonous imbrication of the uppermost basement and its thin veneer of Bjørndalen Formation sediments along the SSE margin of the basin. Except where ophiolite fragments indicate otherwise, the deformation is one of crustal imbrication or 'A'-type subduction. Elsewhere in the Scandinavian Caledonides, more extreme subduction of Baltoscandian crust and its rift-related cover might have given rise to the eclogitic metamorphic conditions developed locally (Dallmeyer and Gee, 1986), as is currently advocated for Scandian eclogites in SW Norway (Cuthbert *et al.*, 1983; Griffin *et al.*, 1985). This would serve as an alternative to the accretionary prism model suggested by Dallmeyer and Gee (1986) to explain the early Ordovician eclogites. As an accretionary prism the Särv and Seve nappes are unconvincing when compared with modern accretionary prisms (Taira *et al.*, 1982; Byrne, 1982; Bachman, 1982) and with other proposed Lower Palaeozoic analogues (e.g., Leggett *et al.*, 1979), which are dominated by pelite-rich distal turbidites that have suffered chaotic polyphase synsedimentary deformation.

(3) *Early Ordovician to Lower Silurian*

Following the uplift associated with inversion a lengthy, perhaps 30 Ma period of prolonged, locally deep erosion accompanied by differential subsidence ended in a period of further sedimentation and volcanicity during Caradoc times (the upper division). This deposition and volcanicity took place within and between a pair of basins separated by a raised, fault-controlled NNE–SSW-trending basement block. Subsidence occurred as part of a sinistral transtensional system centred once again on the MFZ, with subsidiary NNE–SSW sinistral transtensional faults controlling basin topography. A hiatus in detrital input, probably Ashgillian (Table IV, column 5; Spjeldnaes, 1985), is reflected in the pure fossiliferous limestones of the Snåsa Formation. In the north-east of the Snåsa Synform, volcanism resumed at a later stage (Roberts, 1967a, 1982).

(4) *Mid to end Silurian*

Though there is no record of post-Ashgill sedimentation in the immediate area, it is clear from regional considerations that sedimentation was completed by Lower Silurian times and a second period of basin inversion was initiated in response to Scandian shortening. The second inversion once again centred on the MFZ, now flooring the dominant Jønems Thrust Sheet. Beneath this, the upper division succession and its remnant early Ordovician orogenic basement were deformed in two major duplexes. Initial imbrication may once again have been dextral transpressional resulting in oblique ESE shortening, but later shortening was orthogonal to the belt—perhaps in response to a fundamental change in plate motions and plate boundary configurations. Major deformation in the immediate area concluded with ramping of a 5 km thick basement thrust sheet, giving rise to the Tømmerås Antiform. Assuming that propagation of the Scandian thrust belt was towards the foreland (Gee, 1978: Fig. 7), repeated basement imbrication presumably gave rise to the Sylarna and Mullfjället Antiforms farther east. It should be noted that as a result of early Ordovician and then Scandian shortening, the RDS root-zone in the basement could have been completely concealed beneath the overthrust Jønems Thrust Sheet (Fig. 4(d)), which originated beyond the root-zone. Similar crustal imbrication in the Western Gneiss Region (WGR) in SW Norway, a lateral continuation of the Jønems Thrust Sheet, led to extreme crustal thickening and eclogite facies metamorphism (Cuthbert *et al.*, 1983; Griffin *et al.*, 1985).

CONCLUSIONS

The model presented here, if applied more widely to central Scandinavia, could lead to a significant modification of current ideas on the evolution of the region.

In the first place, it would suggest that all units equivalent to the Hjellebotn Group and the lower division of the Steinkjer–Snåsavatnet area, such as the Gula and Støren Groups, the Särv and Seve Nappes and perhaps the Seima Formation at the base of the Köli Supergroup (Table IV, column 5), were involved in the early Caledonian orogenic event to some extent. From this it would follow that a disconformable, if not unconformable relationship must exist between younger parts of the Støren and Köli Supergroups and those parts that were involved in the early orogeny, as suggested by Sturt in 1984.

It would appear that the early orogenic event in central Scandinavia might be related to the development of discrete basins in continental crust, infilled by locally derived sediments and rift-related volcanics, and the subsequent inversion of these basins. Early Ordovician and then Scandian imbrication in the Steinkjer–Snåsavatnet

area could have combined to conceal the root-zone of the RDS that fed the local volcanics.

Much of the argument for derivation of the Offerdal, Särv and Seve Nappes from progressively farther west of the Snåsa Synform is based on consideration of the Ottfjället dykes and all possible correlatives (including the RDS) as a single dyke swarm, and on the absence of a suitable root-zone anywhere in the central Scandinavian basement (Gee, 1975, 1978). If a relatively localized root-zone for the Ottfjället dykes could have been concealed by Baltoscandian crustal imbrication, as is suggested for the RDS, the question of a root-zone in western Norway becomes largely irrelevant. In a further argument, eclogites in the Seve Nappe were correlated with those in the basement of the WGR and thus it was suggested that the Seve Nappe might have originated along strike of the WGR, beyond the Norwegian coast (Gee, 1978). More recent radiometric evidence indicates that such eclogites were likely to be an integral part of an early Caledonian succession (Dallmeyer and Gee, 1986) and as such they could have been generated much more locally during 'A'-type subduction as suggested in the revised model. Finally it has been suggested that basement rocks in the Offerdal Nappe can most easily be correlated with the Tømmerås Group and correlatives in western Norway, and thus that the nappe is likely to have originated in that region (Gee, 1978). The most likely correlative of the Tømmerås Group with the basement of the Swedish autochthon is the sub-Jotnian Complex (Tietzsch-Tyler, 1983), which chiefly comprises acid to intermediate porphyries and granites (Lundqvist, 1968). Only a minimal translation for the Offerdal Nappe would take its source into the likely northward extension of the sub-Jotnian Complex beneath the Caledonian allochthon (Sveriges Geologiska Undersökning, 1958). Thus, it would appear possible for the Särv and Seve Nappes also to have originated in a discrete basin, though larger than that of the Leksdalsvatn and Reinsvatn Groups, and for that basin to have been within Baltoscandia. One possibility would be that their source area is now concealed beneath the Upper Köli Nappes and the Gula Nappe of the synformal Trondheim Nappe Complex. Similar apparently rift-related successions on the basement in Vestranden, west of the Snåsa Synform, could be taken to reflect other rift-centres much closer to the main spreading centre of Iapetus.

Uniformity of certain characteristics of sedimentation from the foreland to the Steinkjer–Snåsavatnet area, most notably the organic Upper Cambrian shales, suggests that all of the early Caledonian basins envisaged here were merely centres of enhanced subsidence, sediment accumulation and rift-related magmatic activity within a broad Baltoscandian marginal shelf sea. The elevated basement areas that must have separated individual basins from each other are now represented in thrust culminations, like that of Tømmerås, along the Sylarna and Mullfjället axes.

Placing of the early Ordovician orogenic event entirely within Baltoscandia suggests that it may be correlated directly with the Finnmarkian orogenic event as currently defined in north Norway, and makes the definition of a new orogenic event for this area unnecessary.

Ophiolite emplacement at the time of this event shows that at least some of these Finnmarkian basins developed an oceanic lithosphere. Others, such as that modelled here, may perhaps be included in the stillborn ocean concept of Ramsay (1973). Mantle upwelling beneath these basins is indicated by the intrusion of quantities of peridotite in the Rørdalsbukt and Reinsvatn Formations, throughout the Seve Nappe and perhaps in the Lower Køli Nappes.

If Finnmarkian orogenesis was entirely confined within Baltoscandia, then all of the younger Ordovician and Silurian sedimentation and volcanicity in the Upper Allochthon must have taken place on or immediately adjacent to Baltoscandia too. Certainly in Steinkjer–Snåsavatnet area, the clast content of conglomerates has a continental source. This lends credence to the arguments in favour of Ordovician island arc and back-arc activity taking place on the Baltoscandian margin of Iapetus above an eastward-dipping Benioff Zone, as suggested by basin polarity and spatial variation of magmatic geochemistry (Ryan *et al.*, 1980; Roberts, 1980; Roberts *et al.*, 1984, 1985), and puts in doubt the validity of models envisaging Laurentian margin subduction (Gee, 1975; Stephens and Gee, 1985) based on simple models of faunal affinity (Sturt, 1984). Palaeontologists actively working with these faunas have themselves stressed their limited value in establishing the geographic position of island arcs (Bruton and Harper, 1981).

Since the model for Finnmarkian orogeny presented here obviates the necessity for large-scale nappe translations, and indeed may confine nappe development to pre-Scandian deformation, it is likely that the Middle Ordovician to Silurian successions of the Upper Allochthon remain largely autochthonous or parautochthonous on their Finnmarkian basement. Ramping of basement thrust sheets to form the Tømmerås, Sylarna and Mullfjället Antiforms and their thrusts climbing through the Finnmarkian and post-Finnmarkian successions could then be responsible for much of the repetition that is evident in the successions—the western and eastern belts of the Seve Nappe, and the Lower, Middle and Upper Köli Nappes.

Finally, no attempt should be made to define Caledonian terranes in Scandinavia (Stephens and Gee, 1985) until the regional extent and the exact nature of the Finnmarkian orogeny has been fully realized. Only then can the relationship between post-Finnmarkian successions and their Finnmarkian basement be defined, and a time framework established within which strictly age-equivalent tectonic units may be evaluated in terms of terrane analysis. The model put forward in this paper makes it possible that the definition of terranes will remain largely unnecessary, the exception being the Uppermost Allochthon, which is probably part of the Laurentian marginal miogeocline and its basement (Stephens and Gee, 1985).

Acknowledgments

The tenure of a N.E.R.C. research studentship, held at University College Cardiff, and the field-assistance of A/S Sulfidmalm, Oslo, during the original mapping of the Steinkjer-Snåsavatnet area is greatly acknowledged. The final manuscript has benefitted from the comments and criticism of an initial draft by Dr P. G. Andreasson and Dr R. A. Gayer, both of whom are thanked.

REFERENCES

Andreasson, P. G. and Johansson, L. (1983). The Snåsa Megalens, west-central Scandinavian Caledonides. *Geol. Fören. Stockh. Förh.* **104**, 305–326.

Andreasson, P. G. and Lagerblad, B. (1980). Occurrence and significance of inverted metamorphic gradients in the western Scandinavian Caledonides. *J. Geol. Soc. Lond.* **137**, 219–230.

Andreasson, P. G., Gee, D. G. and Kumplainen, R. (1978). Some remarks on the Steinkjer Mega-boudin. *Norsk Geol. Tidsskr.* **58**, 305–307.

Andreasson, P. G., Solyom, Z. and Roberts, D. (1979). Petrochemistry and tectonic significance of basic and alkaline-ultrabasic dykes in the Leksdal Nappe, northern Trondheim region, Norway. *Norges Geol. Unders.* **348**, 47–72.

Bachman, S. B. (1982). The Coastal Belt of the Franciscan: youngest phase of northern California subduction. *Geol. Soc. Lond. Spec. Publ.* **10**, 401–417.

Bergström, J. and Gee, D. G. (1985). The Cambrian in Scandinavia. In: D. G. Gee and B. A. Sturt (eds.) *The Caledonide Orogen—Scandinavia and Related Areas*, pp. 247–271. Wiley, Chichester.

Bruton, D. L. and Harper, D. A. T. (1981). Brachiopods and trilobites of the early Ordovician serpentine Otta Conglomerate, south central Norway. *Norsk Geol. Tidsskr.* **61**, 153–181.

Bryhni, I. and Sturt, B. A. (1985). Caledonides of southwestern Norway. In: D. G. Gee and B. A. Sturt (eds.) *The Caledonide Orogen—Scandinavia and Related Areas*, pp. 89–107. Wiley, Chichester. 89–107.

Byrne, T. (1982). Structural evolution of coherent terranes in the Ghost Rocks Formation, Kodiak Island, Alaska. *Geol. Soc. Lond. Spec. Publ.* **10**, 229–242.

Carstens, H. (1956). Geology. In: *Fosdalens Bergverk 1906–1956*, pp. 149–158. Fosdalens Bergverk-Aktieselskab.

Carstens, H. (1960). Stratigraphy and volcanism in the Trondheimsfjord area—Guide to excursions no. A1 and C1. *Int. Geol. Congr. 21*, Session Norden, 23p.

Claesson, S. (1980). A Rb-Sr isotope study of granitoids and related mylonites in the Tännäs Augen Gneiss Nappe, southern Swedish Caledonides. *Geol. Fören. Stockh. Förh.* **102**, 403–420.

Cuthbert, S. J., Harvey, M. A. and Carswell, D. A. (1983). A tectonic model for the metamorphic evolution of the Basal Gneiss Complex, Western South Norway. *J. Metam. Geol.* **1**, 63–90.

Dallmeyer, R. D. and Gee, D. G. (1986). $^{40}Ar/^{39}Ar$ mineral data from retrogressed eclogites within the Baltoscandian miogeocline: implications for a polyphase Caledonian orogenic evolution. *Bull. Geol. Soc. Amer.* **97**, 26–34.

Dallmeyer, R. D., Gee, D. G. and Beckholmen, M. (1985). $^{40}Ar/^{39}Ar$ mineral age record of early Caledonian tectonothermal activity in the Baltoscandian miogeocline. *Amer. J. Sci.* **285**, 532–568.

Dyrelius, D., Gee, D. G., Gorbatschev, R., Ramberg, H. and Zachrisson, E. (1980). A profile through the central Scandinavian Caledonides. *Tectonophysics* **69**, 247–284.

Furnes, H., Roberts, D., Sturt, B. A., Thon, A. and Gale, G. H. (1980). Ophiolite fragments in the Scandinavian Caledonides. In: *Proc. Int. Ophiolite Symp., Nicosia, 1979*, pp. 582–599.

Furnes, H., Ryan, P. D., Grenne, T., Roberts, D., Sturt, B. A. and Prestvik, T. (1985). Geological and geochemical classification of the ophiolitic fragments in the Scandinavian Caledonides. In: Gee, D. G. and Sturt, B. A. (eds.), *The Caledonide Orogen—Scandinavia and Related Areas*, pp. 657–670. Wiley, Chichester.

Gale, G. H. and Roberts, D. (1974). Trace element geochemistry of Norwegian Lower Palaeozoic basic volcanics and its tectonic implications. *Earth Planet. Sci. Lett.* **22**, 380–390.

Gale, G. H. (1985). Numerical calibration of the Palaeozoic time-scale; Ordovician, Silurian and Devonian Periods. In: N. J. Snelling (ed.) *The Chronology of the Geological Record*, pp. 81–88. Geol. Soc. Lond. Memoir No. 10. Blackwell, London.

Gee, D. G. (1974). Comments on the allochthon in northern Trøndelag, central Scandinavian Caledonides. *Norsk Geol. Tidsskr.* **54**, 435–440.

Gee, D. G. (1975). A tectonic model for the central part of the Scandinavian Caledonides. *Amer. J. Sci.* **275-A**, 468–515.

Gee, D. G. (1977). Extension of the Offerdal and Särv Nappes and the Seve Supergroup into northern Trøndelag. *Norsk Geol. Tidsskr.* **57**, 163–170.

Gee, D. G. (1978). Nappe displacement in the Scandinavian Caledonides. *Tectonophysics* **47**, 393–419.

Gee, D. G. (1980). Basement-cover relationships in the central Scandinavian Caledonides. *Geol. Fören. Stockh. Förh.* **102**, 455–474.

Gee, D. G. and Roberts, D. (1983) Timing of deformation in the Scandinavian Caledonides. In: P. E. Schenk (ed.), *Regional trends in the geology of the Appalachian-Caledonian-Hercynian-Mauritanide Orogen*, pp. 279–292. Reidel, Dordrecht.

Gee, D. G., Karis, L., Kumplainen, R. and Thelander, T. (1974). A summary of Caledonian front stratigraphy, northern Jämtland/southern Västerbotten, central Swedish Caledonides. *Geol. Fören. Stockh. Förh.* **96**, 389–397.

Gee, D. G., Guezou, J. C., Roberts, D. and Wolff, F. C. (1985). The central-southern part of the Scandinavian Caledonides. In: D. G. Gee and B. A. Sturt (eds.) *The Caledonide Orogen—Scandinavia and Related Areas*, pp. 109–133. Wiley, Chichester.

Gorbatschev, R., Solyom, Z. and Johansson, I. (1979). The Central Scandinavian Dolerite Group in Jämtland, central Sweden. *Geol Fören. Stockh. Förh.* **101**, 177–190.

Griffin, W. L. *et al.* (1985). High-pressure metamorphism in the Scandinavian Caledonides. In: D. G. Gee and B. A. Sturt (eds.) *The Caledonide Orogen—Scandinavia and Related Area*, pp. 783–801. Wiley, Chichester.

Hill, T. (1980). Geochemistry of the greenschists in relation to the Cu–Fe deposit in the Ramundberget area, central Swedish Caledonides. *Norges Geol. Unders.* **360**, 195–210.

Hossack, J. R. and Cooper, M. A. (1986). Collision tectonics in the Scandinavian Caledonides. *Geol. Soc. Lond. Spec. Publ.* **19**, 287–304.

Klingspor, I. and Gee, D. G. (1981). Isotopic age-determination studies of the Trøndelag trondhjemites. Abstract. *Terra Cognita* **1**, 55.

Leggett, J. K., McKerrow, W. S. and Eales, M. H. (1979). The Southern Uplands of Scotland: A Lower Palaeozoic accretionary prism. *J. Geol. Soc. Lond.* **136**, 755–770.

Lundqvist, T. (1968). Precambrian geology of the Los-Hamra region, central Sweden. *Sver. Geol. Unders.* **Ba23**, 255pp.

McKerrow, W. S. *et al.* (1985). The Ordovician, Silurian and Devonian Periods. In: N. J. Snelling (ed.), *The Chronology of the Geological Record*, pp. 73–80. Geol. Soc. Lond. Memoir No. 10. Blackwell, London.

Middlemost, E. A. K. (1985). *Magmas and Magmatic Rocks*. Longman, London.

Nissen, A. L. (1986). Rb–Sr age determination of intrusive rocks in the southeastern part of the Bindal Massif, Nord Trøndelag, Norway. *Norges Geol. Unders.* **406**, 83–92.

Pearce, J. A. (1980). Geochemical evidence for the genesis and eruptive setting of lavas from the Tethyan ophiolites. In: *Proc. Int. Ophiolite Symp., Nicosia, 1979*, pp. 261–272.

Pearce, J. A. and Cann, J. R. (1973). Tectonic setting of basic volcanic rocks determined using trace element analysis. *Earth Planet. Sci. Lett.* **19**, 290–300.

Ramsay, D. M. (1973). Possible existence of a stillborn marginal ocean in the Caledonian orogenic belt of northwest Norway. *Nature (Phys. Sci.)* **245**, 107–109.

Ramsay, D. M. and Sturt, B. A. (1976). The syn-metamorphic emplacement of the Magerøy Nappe. *Norsk Geol. Tidsskr.* **56**, 291–308.

Roberts, D. (1967a). Geological investigations in the Snåsa-Lurudal Area, Nord Trøndelag. *Norges Geol. Unders.* **247**, 18–38.

Roberts, D. (1967b). Structural observations from the Kopperå–Riksgrens area and discussion of the tectonics of Stjørdalen and the NE Trondheim region. *Norges Geol. Unders.* **245**, 64–122.

Roberts, D. (1975). Geochemistry of dolerite and metadolerite dykes from Varanger Peninsula, Finnmark, North Norway. *Norges Geol. Unders.* **322**, 55–72.

Roberts, D. (1979). Structural sequence in the Limingen–Tunnsjøen area of the Grong district, Nord Trøndelag. *Norges Geol. Unders.* **345**, 101–114.

Roberts, D. (1980). Petrochemistry and palaeogeographic setting of the Ordovician volcanic rocks of Smøla, central Norway. *Norges Geol. Unders.* **359**, 43–60.

Roberts, D. (1982). Disparate geochemical patterns from the Snåsavatn Greenstone, Nord Trøndelag, central Norway. *Norges Geol. Unders.* **373**, 63–73.

Roberts, D. and Sturt, B. A. (1980). Caledonian deformation in Norway. *J. Geol. Soc. Lond.* **137**, 241–250.

Roberts, D. and Wolff, F. C. (1981). Tectonostratigraphic development of the Trondheim region Caledonides, central Norway. *J. Struct. Geol.* **3**, 487–494.

Roberts, D., Grenne, T. and Ryan, P. D. (1984). Ordovician marginal basin development in the central Norwegian Caledonides. *Geol. Soc. Lond. Spec. Publ.* **16**, 233–245.

Roberts, D., Sturt, B. A. and Furnes, H. (1985). Volcanite assemblages and environments in the Scandinavian Caledonides and the sequential development history of the mountain belt. In: D. G. Gee and B. A. Sturt (eds.) *The Caledonide Orogen—Scandinavia and Related Areas*, pp. 919–930. Wiley, Chichester.

Robertson, A. H. F. and Dixon, J. E. (1985). Introduction: aspects of the geological evolution of the Eastern Mediterranean. *Geol. Soc. Lond. Spec. Publ.* **17**, 319–335.

Ross, R. J. *et al.* (1982). Fission-track dating of British Ordovician and Silurian Stratotypes. *Geol. Mag.* **119**, 135–153.

Ryan, P. D., Skevington, D. and Williams, D. M. (1980). A revised interpretation of the Ordovician stratigraphy of Sør Trøndelag and its implications for the evolution of the Scandinavian Caledonides. In: D. R. Wones (ed.) *The Caledonides in the USA*, pp. 99–103. Virginia Polytech Inst. and State Univ., Dept. Geol. Sci. Mem. 2.

Sengor, A. M. C. and Yilmaz, Y. (1981). Tethyan evolution of Turkey: a plate tectonic approach. *Tectonophysics*, **75**, 181–241.

Sjöstrand, T. (1978). Caledonian geology and the Kvarnbergsvattnet area, northern Jämtland, central Sweden. *Sver. Geol. Unders.* **C735**, 107p.

Solyom, Z., Gorbatschev, R. and Johansson, I. (1979a). The Ottfjället Dolerites: geochemistry of a dyke swarm in relation to the geodynamics of the Caledonide orogen of central Scandinavia. *Sver. Geol. Unders.* **C756**, 1–38.

Solyom, Z., Andreasson, P. G. and Johansson, I. (1979b). Geochemistry of amphibolites from Mount Sylarna, central Scandinavian Caledonides. *Geol. Fören. Stockh. Förh.* **101**, 17–25.

Spjeldnaes, N. (1985). Biostratigraphy of the Scandinavian Caledonides. In: D. G. Gee and B. A. Sturt (eds.) *The Caledonide Orogen—Scandinavia and Related Areas*, pp. 317–329. Wiley, Chichester.

Springer-Peacey, J. (1964). Reconnaissance of the Tømmerås Anticline. *Norges Geol. Unders.* **227**, 13–84.

Stephens, M. B. (1977). Stratigraphy and relationship between folding, metamorphism and thrusting in the Tärna–Björkvatnet area northern Swedish Caledonides. *Sver. Geol. Unders.* **C726**, 146p.

Stephens, M. B. (1980). Spilitization, element release and the formation of massive sulphides in the Stekkenjokk volcanites, central Swedish Caledonides. *Norges Geol. Unders.* **360**, 159–193.

Stephens, M. B. and Gee, D. G. (1985). A tectonic model for the evolution of the eugeoclinal terranes in the central Scandinavian Caledonides. In: D. G. Gee and B. A. Sturt (eds.) *The Caledonide Orogen—Scandinavia and Related Areas*, pp. 953–978. Wiley, Chichester.

Sturt, B. A. (1984). The accretion of ophiolitic terranes in the Scandinavian Caledonides. *Geologie en Mijnbouw* **63**, 201–212.

Sturt, B. A., Pringle, I. R. and Ramsay, D. M. (1978). The Finnmarkian phase of the Caledonian orogeny. *J. Geol. Soc. Lond.* **135**, 597–610.

Sturt, B. A., Roberts, D. and Furnes, H. (1984). A conspectus of Scandinavian Caledonian ophiolites. *Geol. Soc. Lond. Spec. Publ.* **13**, 381–391.

Sveriges Geologiska Undersökning (1958). Karta över Sveriges berggrund i tre blad, skala 1:1 000 000. *Sver. Geol. Unders.*, **Ba16**.

Taira, A., Okada, H., Whitaker, J. H. McD. and Smith, A. J. (1982). The Shimanto Belt of Japan: Cretaceous–lower Miocene active margin sedimentation. *Geol. Soc. Lond. Spec. Publ.* **10**, 5–26.

Tietzsch-Tyler, D. (1983). The Caledonian geology of the southwestern part of the Snåsa Synform in the central Norwegian Caledonides and its regional significance. Unpublished Ph.D. thesis, University of Wales, Cardiff, 417p.

Tietzsch-Tyler, D. and Roberts, D. (1985). Steinkjer, berggrunnsgeologisk kart 1723 3–1:50 000, foreløpig utgave. *Norges Geol. Unders.*

Trouw, R. A. J. (1973). Structural geology of the Marsfjällen area Caledonides of Västerbotten, Sweden. *Sver. Geol. Unders.*, **C689**, 115p.

Welin, E., Claesson, S. and Kähr, A. M. (1983). K–Ar ages from the Lower Seve Nappe, Mt. Åreskutan, Sweden. *Geol. Fören. Stockh. Förh.* **105**, 275–277.

Wolff, F. C. (1976). Geologisk kart over Norge, berggrunnskart Trondheim 1:250,000. *Norges Geol. Unders.*

Wolff, F. C. (1979). Beskrivelse til de berggrunnsgeologisk kart Trondheim og Østersund 1:250 000. *Norges Geol. Unders.* **353**, 1–76.

5 The timing of orogenesis in northern Norway: did the Finnmarkian Orogeny occur?

C. Townsend[1] and R. A. Gayer[2]

[1]Cambridge Arctic Shelf Programme, University of Cambridge, West Building, Gravel Hill, Huntingdon Road, Cambridge, UK
[2]Department of Geology, University of Wales, Cardiff, P.O. Box 914, Cardiff, CF1 3YE, UK

During the Scandinavian Caledonides conference held at University College, Cardiff in September 1987 a session was devoted to discussion of the status of the Finnmarkian Orogeny. This session was chaired by R. D. Dallmeyer (Univ. of Georgia, USA) and was freely contributed to by several other participants, in particular: B. Robins (Bergen), S. Daly (U.C. Dublin), K. B. Zwaan (N.G.U. Trondheim), T. B. Andersen (Oslo), R. E. Binns (Trondheim), A. J. Barker (Southampton), and Susan Aitcheson (U.C. Dublin).

A pre-Silurian orogenic event within the Scandinavian Caledonides was first recognized by the initial dating of the Seiland Igneous Province, which was interpreted to have been synorogenically emplaced into the Klubben Group psammites (Stumpfl and Sturt, 1965; Sturt and Ramsay, 1965). Radiometric dates of 540–490 Ma established the Mid-Cambrian–Early Ordovician age of this Finnmarkian Orogeny (Sturt *et al.*, 1967, 1978; Pringle and Sturt, 1969). Tectonic models for the evolution of the Scandinavian Caledonides, in particular the northern part, have since incorporated this earlier event along with the later Silurian-aged Scandian Orogeny (e.g., Sturt and Roberts, 1978; Roberts and Sturt, 1980; Ramsay *et al.*, 1985).

Recently, the evidence for the earlier Finnmarkian Orogeny has been contested. Cambrian Archaeocyathids found in rafts within a gabbro within the Seiland Igneous Province (Holland and Sturt, 1970) have since been discredited (Debrenne, 1984), allowing correlations between the Sørøy sequence and the Silurian age sequences in Nordreissa (Binns, this volume). Furthermore, intrusive relationships between folds and dykes that lie along their axial planes have been questioned by Krill and Zwaan (1987). Sturt and Ramsay (1965) interpreted the dykes to be post-folding, whereas Krill and Zwaan (1987) reinterpreted them to be pre-folding, after the models of Gayer *et al.* (1978) and Rice (1986). This reinterpretation led Krill and Zwaan (1987) to suggest that the folding around the Seiland Igneous Province on Sørøy occurred after magmatic activity and was therefore probably due to the Scandian Orogeny.

Recent radiometric dating by R. D. Dallmeyer, S. Daly and B. Robins along with their co-workers has shed further light on the status of the Finnmarkian Orogeny. These dates were presented and discussed at the Cardiff conference, but many have yet to be published. In addition, several participants questioned the validity of the earlier published dates from the Seiland Igneous Province, owing to their non-reproducibility, MSWD were not published and several points appear to plot at an unacceptable distance from the suggested isochrons. It was suggested that both an investigation of the relationship between deformation and intrusion, and a new isotopic dating programme, should be carried out on Sørøy to resolve these uncertainties.

As a result of this recent controversy, a discussion was organized to consider the present status of pre-Silurian orogenies in the Scandinavian Caledonides, with particular emphasis on Finnmark. The following account results from the free discussion between many of the participants at the Cardiff meeting and outlines what is known, possible hypotheses that need testing, and what is unknown. Individual contributions are not acknowledged, but all those who took part are thanked for their input of information and ideas. As authors of this article, we lay claim to few of these ideas but have attempted to express the main feeling of the discussion, in the hope of clarifying the position.

The information, ideas and problems are laid out as a time sequence. The problems and areas for further work are highlighted again in a concluding paragraph.

(1) Precambrian deformation is known to have affected the cover metasediments of the Kalak Nappe Complex on Porsangerhalvøya, as granites dated at around 800 Ma cut tectonic folds, interpreted at D_2 structures. These cover metasediments must also be older than this age.

(2) Dolerite dyke emplacement occurred throughout large areas of the Kalak and Laksefjord Nappe Complexes (Middle and Lower Allochthons) in Finnmark and also in the Upper Allochthon (Seve Nappe) further south in Scandinavia. These intrusions, which in places are of sheeted dyke character, suggest large-scale crustal extension and are of Precambrian age. No radiometric dates for these dykes have yet been published from the main area of Finnmark, although similar dykes cutting the Barents Sea Group give an age of 640 Ma (Beckinsale *et al.*, 1975) and further south in Scandinavia the Ottfjället dyke swarm has been dated at 665 ± 10 Ma (Claesson and Roddick, 1983). Crustal extension on such a large scale must have caused considerable deformation, which has so far gone unrecognized. There was some disagreement as to whether the dykes cut a pre-existing fabric.

(3) The Seiland Igneous Province is a very complex series of intrusions and was formed at deep crustal levels during crustal extension. Regional metamorphism, deformation and gabbro intrusion occured at 6–8 kb as evidenced by assemblages in the gabbro aureoles. Some doubt about the validity of these pressures was expressed, as a similar situation occurs around the Sulitjelma Gabbro, where subsequent work has shown these high pressures to be in error and that the true pressures were nearer to 3 kb. These metamorphosed and deformed rocks are cut by a suite of alkaline, nephelene-syenite dykes dated at 530 ± 2 Ma. Further evidence exists for this pre-530 Ma event, where some xenoliths from within intrusions exhibit earlier tectonic fabrics. Therefore, deformation on Sørøy, at least in part, is older than 530 Ma. However the alkaline suite is itself deformed, so there must also be a younger phase of deformation. The older deformation may have occurred shortly before the intrusion of the nephelene-syenite dykes, although a significantly older age cannot be ruled out. The question must be raised as to how the pre-530 Ma deformation relates to the pre-800 Ma event (1) and the Precambrian crustal extension in the middle and upper allochthons (2).

(4) The alkaline suite is thought to have been intruded at a high crustal level. Therefore, approximately 20 km of unroofing must have occurred after the gabbro intrusion and prior to 530 Ma.

(5) The Sørøy sequence (Klubben, Storelv, Falkenes, Åfjord and Hellefjord Groups) is intruded by the Hussfjord Igneous Complex. If this complex is part of the Seiland Igneous Province, as has been previously suggested, then the Sørøy sequence must be older than mid-Cambrian and not of Ordovician to Silurian age as postulated earlier.

(6) ^{40}Ar/^{39}Ar step heating suggests that hornblende retention of argon during cooling through 500°C occurred at 480–490 Ma (Early Ordovician). No field evidence for a tectonic event was put forward. However, it was argued that since the Mid-Cambrian alkaline suite was intruded into the Kalak Nappe Complex at a high crustal level, the Early Ordovician cooling ages must reflect a subsequent burial event. This was possibly related to an Early Caledonian collision (?arc/continent or microcontinent/

continent) that may have caused the deformation of the alkaline suite.

(7) The Silurian-aged Scandian Orogeny (430–410 Ma) deformed the Laksefjord, Gaissa and Magerøy Nappes and partly deformed and reactivated the Kalak Nappe Complex. This event is equated with the major foreland-propagating linked thrust system in Finnmark.

The above summary suggests that the Finnmarkian Orogeny *sensu stricto* (540–490 Ma) did not occur. However, there is evidence for probably two and possibly three pre-Silurian phases of orogenesis. We need now to examine the evidence from this new perspective for these earlier orogenies and thereby increase our knowledge of the pre-Silurian events. Particular points which need to be closely studied are: the relationship between the pre-800 Ma and the pre-530 Ma event; the relationship between the pre-530 Ma event and deformation related to Precambrian crustal extension during dyke emplacement, possibly at 600–650 Ma; field evidence for the possible collision prior to the 490–480 Ma cooling; and how the Scandian Orogeny reactivated earlier structures.

Some of these problems may prove to be intractable owing to uncertainties in correlating geological events, especially where such tools as poly-phase fold events are used. However, some of these questions may be answered as tectonic models are refined and developed. A deeper understanding of the Caledonian evolution of northern Norway is likely to be developed with the recent advances made in extensional tectonics and its significance on the deformation of the crust.

The discussion highlighted the dangers of building simplistic models to explain the complex relationships that must have developed during the collisional tectonic evolution of the Scandinavian Caledonides. Such an orogenic belt will involve many discrete events separated in both time and space that preclude orogen-wide correlations. Inherent in the principle of terrane analysis is that each terrane will have its own history independent of that of neighbouring terranes.

REFERENCES

Beckinsale, R. D., Reading, H. G. and Rex, D. C. (1976). Potassium-argon ages for basic dykes from East Finnmark: stratigraphic implications. *Scott. J. Geol.* **12**, 51–65.

Binns, R. E. (1988) Finnmarkian–Scandian relationships in NE Troms–W. Finnmark, *this volume.*

Claesson, S. and Roddick, J. C. (1983). ^{40}Ar/^{39}Ar data on the age and metamorphism of the Ottfjället dolerites, Särv Nappe, Swedish Caledonides. *Lithos* **16**, 61–73.

Debrenne, F. (1984). Archaeocyatha from the Caledonian rocks of Sørøy, North Norway—a doubtful record. *Norsk Geol. Tidsskr.* **64**, 153–4.

Gayer, R. A., Powell, D. B. and Rhodes, S. (1978). Deformation against metadolerite dykes in the Caledonides of Finnmark, Norway. *Tectonophysics* **46**, 99–115.

Holland, C. H. and Sturt, B. A. (1970). On the occurrence of archaeocyathids in the Caledonian metamorphic rocks of Sørøy, and their stratigraphical significance. *Norsk Geol. Tidsskr.* **50**, 34–55.

Krill, A. and Zwaan, K. B. (1987). Reinterpretation of Finnmarkian deformation on western Sørøy, northern Norway. *Norsk Geol. Tidsskr.* **67**, 15–24.

Pringle, I. R. and Sturt, B. A. (1969). The age of the peak of the Caledonian Orogeny in west Finnmark, northern Norway. *Norsk Geol. Tidsskr.* **49**, 435–436.

Ramsay, D. M., Sturt, B. A., Zwaan, K. B. and Roberts, D. (1985). Caledonides of northern Norway. In: Gee, D. G. and

Sturt, B. A. (eds.) *The Caledonide Orogen—Scandinavia and Related Areas*, pp. 163–184. Wiley, Chichester.

Rice, A. H. N. (1986). Structures associated with superimposed inhomogeneous shearing of basic dykes from Finnmark, Norway. *Tectonophysics* **128**, 61–75.

Roberts, D. and Sturt, B. A. (1980). Caledonian deformation in Norway. *J. Geol. Soc. Lond.* **137**, 241–250.

Stumpfl, E. F. and Sturt, B. A. (1965). A preliminary account of the geochemistry and ore mineral parageneses of some Caledonian basic igneous rocks from Sørøy, northern Norway. *Norges Geol. Unders.* **234**, 196–230.

Sturt, B. A., Pringle, I. R. and Ramsay, D. M. (1978). The Finnmarkian phase of the Caledonian Orogeny. *J. Geol. Soc. Lond.* **135**, 597–610.

Sturt, B. A. and Ramsay, D. M. (1965). The alkaline complex of the Breivikbotn area, Sørøy, northern Norway. *Norges Geol. Unders.* **231**, 1–164.

Sturt, B. A. and Roberts, D. 1978. Caledonides of northernmost Norway. In Schenk, P. (ed.), *Caledonian–Appalachian orogen of the North-Atlantic Region. Geol. Surv. Canada, Pap.*, **78–13**, 17–24.

Part II

Regional Geology

6 The Middle Allochthon in Västerbotten, northern Sweden: tectonostratigraphy and tectonic evolution

R. O. Greiling

Department of Geology, University of Wales, Cardiff, P.O. Box 914, Cardiff CF1 3YE, UK. (Present address: Geologisch-Paläontologisches Institut, Ruprecht-Karls-Universität, INF 234, D-6900 Heidelberg, FRG.)

The Middle Allochthon (MA) in Västerbotten forms an almost continuous belt along the eastern margin of the Caledonides. There, the MA overlies the Lower Allochthon and is itself covered by the Seve unit of the Upper Allochthon. Farther west, the MA reappears in isolated windows, of which the Bångonåive, Fjällfjäll and Børgefjell windows are treated here. Two major units build up the MA at the eastern Caledonian margin, the Stalon Nappe Complex (SNC) and the Särv unit. The SNC forms an imbricate stack of alternating slices of "cover" metasediments (arkoses, conglomerates, subordinate shales) and crystalline "basement" rocks (acidic to intermediate gneisses, schists, metagabbros, and mafic dykes of Baltic Shield, "Jotnian" affinity). The Särv unit, overlying the SNC, is composed of Late Proterozoic metapsammites with metabasite dykes, and is restricted to the southern part of the Västerbotten Caledonian margin. It may reappear in the Bångonåive window, where similar metabasites are found, however, in "basement" gneisses of the "Njereutjakke unit". This unit overlies basement/cover imbricates of the "Gukkesjauretje unit", a possible SNC equivalent. In the Børgefjell window, the "Rainesklumpen unit" of alternating basement/cover slices is comparable to the SNC. It is overlain by the Fjällfjäll unit, composed exclusively of arkosic, schistose metapsammites. Basement/cover imbricates similar to and including the SNC generally occupy the lower part of the MA, whereas Särv and Fjällfjäll units form an upper part.

Caledonian-age metamorphism of greenschist grade and relatively intense deformation distinguish the MA from overlying (higher-grade) and underlying (lower-grade) units. The peak metamorphic conditions (garnet + biotite + muscovite) in the MA are synchronous with an early, penetrative foliation (D1). After further deformation (D2), thrusting generated mylonites with lower greenschist-grade mineral assemblages and the imbricate tectonostratigraphy of the SNC and related units. At the Caledonian margin in the east, metamorphism waned immediately after nappe transport (D3), in the west, however, mylonites associated with D3 are overprinted metamorphically. In both areas the MA was subsequently folded.

It is speculated that during its early evolution the MA was "subducted" as part of the Baltic continental margin underthrusting a western plate. Underplating onto the western plate led to upward motion of the MA and mylonitization, stacking of different units and, finally, to regional nappe transport. Overlying units blanketed the MA in the west, thus allowing further metamorphic overprint, in contrast to the eastern, marginal orogenic domains.

INTRODUCTION

The Middle Allochthon (MA) in Västerbotten County, northern Sweden, is best known from the eastern margin of the Caledonides, where it forms an almost continuous belt along the strike of the mountain range (Fig. 1). There, the MA overlies the Lower Allochthon and is itself covered by the Seve unit of the Upper Allochthon (e.g., Gee and Zachrisson, 1979; Gee *et al.*, 1985; Table I). Farther west, the Ma reappears in four isolated windows beneath the Upper Allochthon, from north to south the Ammarnäs, Bångonåive, Fjällfjäll and Børgefjell windows (Kulling, 1955), of which the three latter ones will be treated here. This work describes for the first time within the Bångonåive and Børgefjell windows the lithologies characteristic of the MA, represented at the eastern Caledonian margin by the Stalon Nappe Complex (SNC —Kulling, 1942; Greiling, 1985) and the Särv unit (Gee *et al.*, 1985).

In the first part of the paper, therefore, an account of the lithology and tectonostratigraphy of the SNC and the Särv unit is given and a comparison is made with the MA units

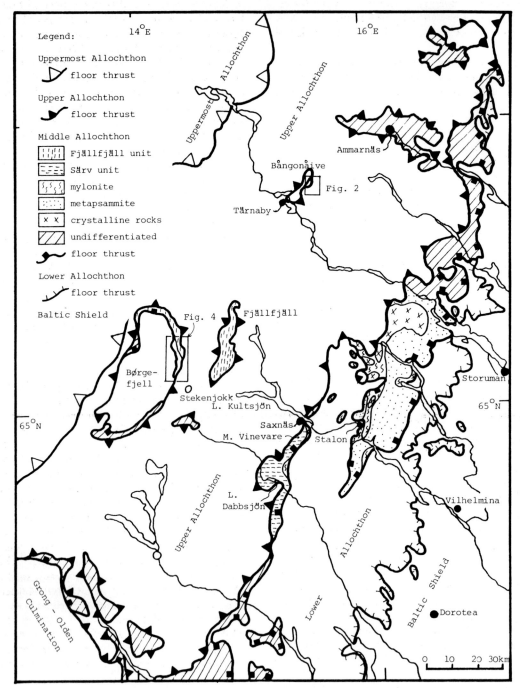

Fig. 1. Sketch map of the north-central Scandinavian Caledonides, showing the distribution of the Middle Allochthon (MA) in Västerbotten (northern Sweden) and adjacent areas and the location of detailed maps, Figs. 2 and 4. The parts of the Middle Allochthon not treated here are shown as "undifferentiated". (Revised from Gee *et al.*, 1985.)

Table I. Tectonostratigraphy of the Middle Allochthon in Västerbotten, northern Sweden

Major Caledonian tectonic units	Western domain, tectonic windows			Eastern domain, Caledonian margin
	Børgefjell	*Fjällfjäll*	*Bångonåive*	
Upper Allochthon	Köli unit	Seve and Köli units	Seve and Köli units	Seve unit
Middle Allochthon	Fjällfjäll unit (metapsammites, "Fjällfjäll arkose")	Fjällfjäll unit ?	Njereutjakke unit (gneisses and metabasites)	Särv unit (metapsammites with metabasite dykes)
	Rainesklumpen unit (mylonitic basement/ cover horses)	Not exposed	Gukkesjauretje unit (mylonitic basement/ cover horses)	Stalon Nappe Complex (basement/cover/ mylonite horses)
Lower Allochthon/ ? Autochthon	Basement/cover imbricates	Not exposed	Basement/cover imbricates	Blaik Nappe and equivalents

of the Bångonåive and Børgefjell windows. In the latter, the SNC equivalents underly the Fjällfjäll unit of the MA (Zachrisson, 1964, 1969), which is, thus, a higher unit of the MA.

The second part of the paper summarizes the structural and metamorphic information on the MA in Västerbotten, leading, finally, to a preliminary model for the tectonic evolution of the MA.

TECTONOSTRATIGRAPHY

The major tectonostratigraphic units of the Middle Allochthon (MA) in Västerbotten are shown in Table I. They are briefly described here, starting from the eastern Caledonian margin and then treating the MA in tectonic windows farther west.

The tectonostratigraphy of the Stalon Nappe Complex (SNC) has been described in detail by Greiling (1985) and is only briefly summarized here: the major components of the SNC are pre-Caledonian, crystalline rocks, conglomerates, feldspathic sandstones with subordinate pelites, and distinct units of mylonites (Fig. 1). The crystalline rocks comprise predominantly coarse-grained gabbroic varieties and minor granitoid gneisses and some mica-rich gneisses. All of these rock types are cut by mafic dykes, which have geochemical affinities with those of "Jotnian" (about 1200 Ma) age in the adjacent Baltic Shield. The crystalline rocks are now disrupted into numerous tectonic lenses embedded in mylonites. The latter are, according to their geochemical composition, derived mostly from the crystalline rocks and not from the psammitic, quartz-rich metasediments of the SNC.

Metasediments of the SNC of probable Late Proterozoic age (Kulling, 1942, 1972) apparently start with conglomerates of two particular types. One of these types is characterized by a biotite-rich, schistose matrix, in which coarse, mainly crystalline fragments "float". The second type also contains biotite-rich layers, though rarely, but is mostly composed of clast-supported coarse and fine conglomerate beds. Clasts are mainly composed of granitoid, crystalline rocks and of quartzitic sediments. Upwards, this latter type of conglomerate passes into relatively finer-grained psammites with rare conglomeratic interbeds. Pelite layers within the psammites are very rare and thin (millimetres), although, in places, centimetre to decimetre size mudflakes are common in conglomeratic interlayers. Primary textures, such as cross bedding, are frequently preserved in the metapsammites. Both crystalline and sedimentary rocks and mylonitic units are stacked together so that they now form a duplex of alternating basement and cover horses separated by individual mylonitic horses so that no primary basement/cover relationships are preserved.

Towards the north, the SNC and its equivalents are continuous into northern Västerbotten and probably even into Norrbotten (Kulling, 1942, 1982). However, towards the south and southwest the SNC pinches out almost completely. Only the characteristic mylonitic units of the SNC can be traced somewhat farther south and southwest beneath the overlying Seve unit of the Upper Allochthon or the Särv unit of the MA, for example south of Lake Kultsjön (Fig. 1). It is thus evident that the Särv unit lies structurally above the SNC. On a regional scale, the SNC and Särv unit apparently substitute each other at the eastern Caledonian margin in Västerbotten since the Särv unit appears to pinch out towards north and is not observed north of Lake Kultsjön.

The Särv unit, most completely developed and studied in detail at its type locality in southern Jämtland (see, e.g., Gee and Zachrisson, 1979), is composed mainly of Late Proterozoic psammitic sediments of Jämtland Supergroup equivalent (Kumpulainen and Nystuen, 1985), which are cut by mafic dyke swarms of predominantly tholeiitic composition (Solyom *et al.*, 1979). Similar psammites and dyke swarms are observed south of Lake Kultsjön and at Lake Dabbsjön (Fig. 1). The geochemical composition of the mafic dykes implies their correlation with the mafic dykes of the Särv unit farther south (Greiling *et al.*, 1984a). South of Lake Kultsjön, at Mount Vinevare, massif metabasites with psammites subordinate or lacking are exposed (Vinevare diabas of Kulling (1942)). They form isolated lenses of kilometre size (e.g. $X = 1$ km, $Y = 500$ m, $Z = 100$ m), similar to the lenses of the SNC.

The only other area of the Västerbotten Caledonides where possible Särv unit equivalents are exposed is the Bångonåive window northeast of Tärnaby (Fig. 1). There, a metabasite in strongly sheared gneisses revealed a geochemical composition characteristic of Särv-type dykes. However, owing to strong deformation, the metabasite margin is now parallel with the banding in the surrounding rocks and it is not clear whether the metabasite was primarily a discordant dyke or part of the gneisses. Furthermore, crystalline rocks are rather an exception in the Särv unit at its type locality (Gee and Zachrisson, 1979). The gneisses build up the upper, Njereutjakke unit of the MA in the Bångonåive window (earlier mapped as Seve equivalent by Kieft (1952)), which overlies the lower, Gukkesjauretje unit (Figs. 2 and 3). The Gukkesjauretje unit is a duplex formed of alternating psammitic and mica rich gneissic horses (earlier mapped as psammite, equivalent to the sedimentary cover of the Bångonåive window, Kieft (1952)). The psammites are relatively pure quartzites, with little feldspar and white mica and no primary features preserved. As elsewhere in the MA, the quartzites are disrupted into decimetre to kilometre size lenses. Gneisses are composed of coarse mica flakes, and "augen" of feldspar and quartz. The Gukkesjauretje duplex probably forms an antiformal stack, overlain to the east by the Seve unit of the Upper Allochthon and to the west by the Njereutjakke unit, which forms either a thrust slice of its own or the uppermost horse of the Gukkesjauretje duplex (Fig. 3). The Njereutjakke gneisses are partly composed of strongly deformed mica-rich gneisses of intermediate and acidic composition. Inter-

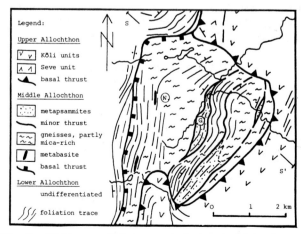

Fig. 2. Field-controlled photogeological map of the major part of the MA at the Bångonåive window. (The northern "end" of the MA is probably a lateral thrust ramp parallel to nappe transport?) MA rocks are strongly sheared and partly mylonitic. N—(Mount) Njereutjakke; G—(Lake) Gukkesjauretje; S–S'— line of section in Fig. 3; for location, see Fig. 1.

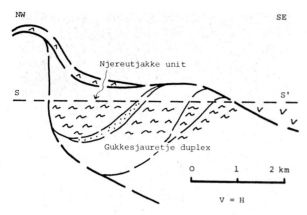

Fig. 3. Section through the MA at the (eastern part of the) Bångonåive window. The MA probably forms a duplex. The uppermost horse, the Njereutjakke unit, may be related to the Särv unit; the Gukkesjauretje duplex compares well with the SNC at the eastern Caledonian margin. For location, see Fig. 2.

calated in these gneisses are fine-grained metabasites forming metre to ten-metre size lenses or sheets. Although only lithological correlations are possible, the MA succession of the Bångonåive window compares well with the tectonostratigraphy at the eastern Caledonian margin, where the basement/cover horses of the SNC, comparable to the Gukkesjauretje duplex, are overlain by the Särv unit, of which the Njereutjakke unit is a possible equivalent.

In contrast, the MA in the Fjällfjäll and Børgefjell windows farther west appears to represent a somewhat different tectonostratigraphy. The Fjällfjäll window is made up exclusively of the metapsammitic Fjällfjäll unit beneath the Seve and Köli units of the Upper Allochthon, although its base is not exposed (Zachrisson, 1964, 1969). However, the northeastern (Västerbotten) part of the Børgefjell window reveals a more complete tectonostratigraphic section of the MA (Fig. 4). There, the MA is built up of two distinct units, an upper, metapsammitic, Fjällfjäll unit and an underlying basement/cover duplex, the Rainesklumpen unit. The lowermost horse of the Rainesklumpen unit is composed of "cover" rocks represented by greenish-grey arkoses and feldspathic quartzites with pelitic interlayers. These Rainesklumpen psammites underly fine-grained, feldspar-rich mica schists that contain imbricate slices of greenish metapsammites similar to those described above. South of Saxfjället black, graphite-bearing schists are also exposed, together with a thin band of highly deformed carbonate rocks. These schists are intimately interbanded with greenish and sometimes yellowish quartzites and may be part of the cover sequence, but owing to the intense deformation and imbrication they cannot, at least at map scale, be distinguished from the mica-rich gneisses of the basement slices.

Although the Rainesklumpen unit occupies only a few tens of square kilometres and is generally only a few hundred metres thick, it is distinguished here as a separate unit, since it is an obvious equivalent of the much larger Dearka unit, which dominates the MA at the southern margin of the Børgefjell window (Greiling *et al.*, 1984b).

Basement cover imbricates of the Lower Allochthon in the footwall of the Rainesklumpen unit are distinct from the latter both lithologically (granitic gneisses, light quartzites) and by their considerably less-deformed state.

The Rainesklumpen unit is overlain towards the northeast by the Fjällfjäll unit, described in detail by Zachrisson (1964, 1969). The Fjällfjäll unit consists exclusively of

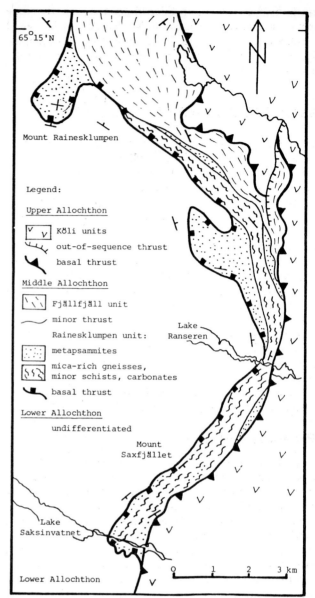

Fig. 4. Geological sketch map of the MA at the northeastern part of the Børgefjell window. The different units of the MA, invariably separated by minor thrusts, dip generally towards east and northeast, forming a simple stack. The lens-shape of psammitic units is probably related mainly to late, gravitational deformation. The latter also led to the uplift of the Børgefjell window and a relatively steep, partly subvertical attitude of the MA roof and floor thrusts east of Mount Saxfjället. The out-of-sequence thrust affecting Fjällfjäll and Köli units, recorded by Zachrisson (1964) may also be related to such a late phase of deformation. For location, see Fig. 1.

schistose, mica-rich arkoses and feldspathic quartzites, banded on centimetre to decimetre scale.

The MA as a whole wedges out southwards so that first the Fjällfjäll unit disappears southwards and the Upper Allochthon footwall rests directly on the Rainesklumpen unit, which, in turn, is cut out, leaving the Upper Allochthon floor thrust resting directly on the Lower Allochthon units. Towards the northwest, the Fjällfjäll unit at the northern margin of the Børgefjell window is at least several hundred metres thick. Similarly, farther east, where the Fjällfjäll unit reappears beneath the Upper Allochthon in the Fjällfjäll window, its type locality (Zachrisson, 1964), the succession is at least 500 m thick, although exposure does not extend down to its base.

The MA tectonostratigraphy of southern Västerbotten, shown in Table I, suggests possible relationships between the individual structural units of the MA. No Särv-type metabasites have been found in the Fjällfjäll and Børgefjell windows and it is thus not clear, whether, for example, the Fjällfjäll unit is equivalent to the Särv unit or structurally above or below.

STRUCTURE AND METAMORPHISM

Prior to nappe transport, the structural and metamorphic evolution in the MA of Västerbotten, as far as its structures and mineral assemblages are preserved, appear to be rather uniform throughout the area. However, post-thrusting evolution differs between the eastern, marginal or external Caledonian domains and the western MA domains exposed in tectonic windows and in a more internal position relative to the Caledonian orogen. Metamorphism in the eastern parts ceased during thrusting and nappe transport, leaving the associated mylonites relatively well preserved. In contrast, metamorphism in the western parts outlasted the nappe transport and overprinted associated mylonites.

The following account of the structural and metamorphic evolution is divided accordingly. The first and second parts, which cover the pre- and syn-thrust history respectively, deal with the MA as a whole. Subsequently, two sections on the post-thrust evolution separately treat the eastern and western parts of the MA respectively.

Pre-thrust evolution

The earliest structural features preserved in the SNC are a penetrative foliation, associated locally with recumbent, tight to isoclinal folds (D1). They are overprinted by mainly cylindrical folds associated with an axial surface crenulation cleavage (D2). Synchronous with D1, garnet, biotite and muscovite were formed. Garnet is rotated during D1, but the latest, marginal phase of garnet growth is not deformed by D1 and thus outlasted (D1) deformation (Greiling 1984, 1985). According to associated mineral assemblages, host-rock composition and its colour, garnet is assumed to be almandine, which forms at pressures not less than 4 kb (Winkler, 1979).

Garnet growth ceased, however, before D2 deformation, which is associated with muscovite and biotite growth parallel to D2 axial surfaces (Greiling, 1985; compare Fig. 9). This early evolution is best observed at the eastern Caledonian margin, where subsequent overprint is relatively weak. The metamorphic evolution of the Särv unit in southern Västerbotten is comparable to that in the SNC. An early peak of metamorphism is characterized by coexisting garnet, biotite (Fig. 5) and muscovite in metapsammite, or garnet, hornblende and biotite in metabasite. However, these metamorphic assemblages developed only in domains of appropriate composition, mostly along dyke margins but neither in pure metabasites nor in pure metapsammites. The metamorphic mineral assemblages are synchronous with an early deformation, with garnet sometimes growing along a foliation and biotite developed in its pressure shadows (Fig. 5). Owing to relatively high competence of both metabasites and metapsammites, folds are rare and mostly irregularly developed, but some isoclinal folds are also observed.

Farther west, in the Bångonåive window, a similar pre-thrusting history can be deduced from the MA rocks. However, subsequent deformation was considerably more intense than in the east (see below) and earlier features are only preserved in the interior, low-strain parts of structural

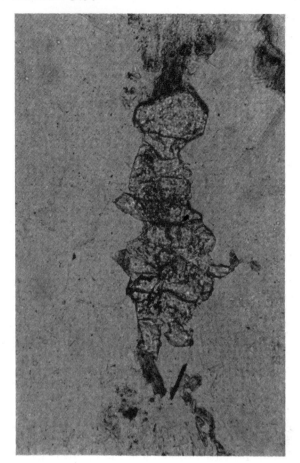

Fig. 5. Micrograph of garnet (centre) growing along a (D1) foliation (vertical), coexisting with biotite in its pressure shadows (above and below garnet). Surrounding quartz (light grey) is randomly recrystallized after D1. Original size 0.47 × 0.70 mm, one polarizer, Särv unit, northern slope of Mount Vinevare (see Fig. 1), sample G80:85.

units. There, garnet, biotite, muscovite assemblages of an early metamorphic stage can still be recognized. In contrast, no garnet is preserved in the MA at the northeastern Børgefjell window, where only rarely are pseudomorphs of biotite and chlorite after garnet found. As is clear from the subsequent history of the Børgefjell MA, overprinting was more intense than in the other areas and, therefore, all of the earlier mineral assemblages are affected. However, evidence of an early garnet growth is taken as an indication that also in the Børgefjell MA the pre-thrusting peak of metamorphism was similar to that in the other MA domains.

This is also assumed for the Fjällfjäll unit, which was only studied in the Børgefjell window.

All of the earlier features were intensely overprinted by the subsequent regional thrusting event (D3) as described below.

Thrusting and regional nappe transport

The most spectacular effects of thrusting on the MA are strong shearing and the formation of huge volumes of mylonites, for which the MA is famous and which led to the recognition of the MA as the highly deformed base of the major Caledonian nappes (e.g., Kulling, 1972; Gee and Zachrisson, 1979; Greiling, 1985). Thrusting also led to the disruption of the MA rock units and the mylonites into tectonic lenses, and their stacking together into major duplexes, for example the formation of the Gukkesjauretje duplex (Figs. 2 and 3; see also Greiling (1985) for further examples).

Fig. 6. Micrograph of mylonitic foliation (top left to bottom right) defined by muscovite (dark grey) and quartz, which is subsequently recrystallized (and, thus, apparently unstrained). The foliation is cut by a crenulation cleavage (top right to bottom left), which leads to an extension of the muscovite grains into a lens or "fish" shape, similar to that in S–C mylonites. Original size 3.0 × 4.5 mm, polarizers × 50°, base of Njereutjakke unit, Bångonåive window (south of Fig. 2), sample G80:153A.

Fig. 7. Micrograph showing disrupted grain of plagioclase with kinked twin lamellae. Between the two parts of the grain is a "shear zone" (top right to bottom centre) with plagioclase and quartz, partly retaining a high-strain state (undulose extinction, bottom centre) but mostly randomly recrystallized (top right, bottom left). Original size 1.2 × 1.8 mm, polarizers × 90°, base of Rainesklumpen unit, Børgefjell window, south of Mount Saxfjället (see Fig. 4), sample G80:159.

Small-scale structures related to thrusting are mylonitic foliations, passing from an almost planar, penetrative foliation within and close to shear zones to an extensional crenulation cleavage (S–C mylonite? Fig. 6) in less-deformed domains. The crenulation cleavage is associated with microfolds and (stretching) lineations, oriented NW–SE (310–330°). Whereas stretching lineations at the eastern Caledonian margin are found mostly close (within a few metres) to shear zones, they tend to be more widely distributed in the western domains, being most intense in shear zones and less intense towards the interior of tectonic units. Similarly, sheath folds in decimetre to metre scale, disrupted within the mylonites, are rare in the east but are more frequent in the Børgefjell window. There also sheath folds in kilometre scale have been observed, covering the southernmost part of the MA shown in Fig. 4, east of Lake Saksinvattnet.

On a microscopic scale, both in western and eastern domains, pre-existing mineral grains are finely sheared, and relatively competent minerals, for example feldspars, are, in places, tectonically rounded and/or cracked and their twin lamellae are kinked (Fig. 7). Mica minerals are torn apart into lens or "fish" shaped grains aligned along the mylonitic foliation (Fig. 6). In between, quartz shows undulose extinction and subgrain formation. Margins of larger grains are thus conspicuously "serrated" (Fig. 8). Penetrative shearing/deformation during thrusting is obviously related to retrogression of earlier mineral as-

semblages. For example, garnet and (brown) biotite are retrogressed to green biotite or chlorite (Greiling, 1985). Retrogression is less intense, where shearing is not penetrative, for example in lenses of lower strain, which now make up individual horses of the MA.

The retrogression is the last metamorphic change observed in the mineral assemblages of the MA at the eastern Caledonian margin (except for some local illite and chlorite, see below). There, the MA was transported to a tectonic position of relatively low sub-greenschist P and T conditions, close to the Earth's surface.

In contrast, in the western areas metamorphic mineral changes continued after the thrusting of the MA into its present structural position (compare Fig. 9). Whereas the MA is well developed at the eastern Caledonian margin, it pinches out dramatically beneath the overlying units of the Upper Allochthon, both along the margin and towards west. This feature may be related to the nappe transport (e.g., horse and duplex formation, Figs. 2 and 3) although a contribution by large-scale boudinage cannot be completely excluded (see below).

Post-thrust evolution at the eastern Caledonian margin

After duplex stacking and thrusting into their present position, the MA units are folded into open, regional folds, which are associated locally with minor folds and

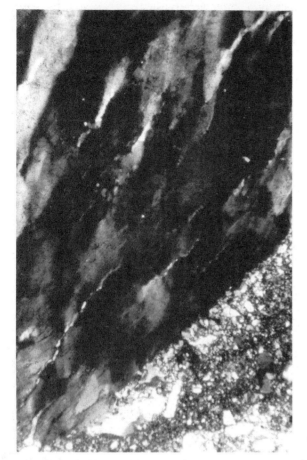

Fig. 8. Micrograph of highly strained quartz grain with a "serrated" margin towards a finely crushed matrix (bottom right), showing undulose extinction and disruption into subgrains by subparallel "serrated" fractures (top right to bottom left), along which tiny new grains are forming (mainly small white spots along fractures, particularly in bottom left quadrant). Original size 1.2 × 1.8 mm, polarizers × 90°, top of MA at southeastern margin of Børgefjell window (see Fig. 1), sample G80:149B.

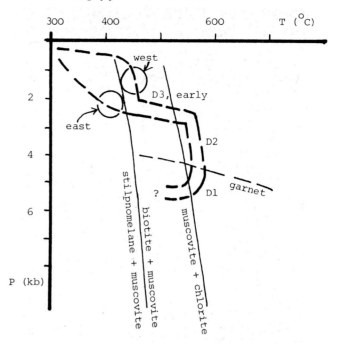

Fig. 9. Sketch diagram to illustrate the metamorphic evolution in the Middle Allochthon as outlined in the text. Mineral assemblage stability fields and corresponding *P–T* data from Winkler (1979). Mineral assemblages related to early burial (D1) comprise garnet + biotite + muscovite (+ hornblende), D2-related assemblages (biotite + muscovite) of relatively lower greenschist grade probably reflect subsequent upwards transport. Early D3, syn-thrust assemblages (biotite + muscovite + chlorite) are overprinted by further greenschist grade assemblages (biotite + muscovite + chlorite) in western domains (west) and by very low grade assemblages (illite, chlorite) or not at all at the eastern Caledonian margin (east). Compare Fig. 10 for tectonic interpretation.

related axial surface crenulation or fracture cleavages. At least three different sets of folds with N–S, NE–SW and E–W respective axial orientations (in that time-sequence) are observed (Greiling, 1985). Except for some illite and, possibly, chlorite, no metamorphic mineral growth is associated with that deformation. It is possible that these folds are due to passive roof deformation during the advance of the underlying Lower Allochthon. However, since the latest, E–W oriented folds also deform the floor thrust of the Lower Allochthon, there is at least one phase of deformation shortening the nappe pile as a whole, including basement windows (unless there is another thrust sheet beneath the Lower Allochthon).

Microscopic evidence of extension veins, cutting *inter alia* the thrust-related mylonites, with fibre growth essentially horizontal points to a post-thrust lateral extension. This extension could relate to large-scale boudinage as envisaged, for example by Gee (1978) or Ramberg (1981), but the geometry of the MA does not show unequivocal large-scale boudins.

Post-thrust evolution of the western MA domains

Folding of the MA in the Bångonåive window is, macroscopically, comparable to that at the eastern Caledonian margin. Microscopically, however, it is evident that the deformation took place under metamorphic conditions,

since, for example, muscovite, biotite and chlorite are growing parallel to the axial surface of folds, which also deform the mylonitic banding related to regional thrusting. The thrust-related mylonites are recrystallized, and quartz in particular shows an apparently undeformed fabric. However, mica minerals, for example, still retain their distinctive lens shape and define shear band margins (Fig. 6). Similarly, rounded and cracked feldspar grains now "float" within almost randomly recrystallized quartz. The metamorphic overprint annealed the mylonites, which give macroscopically, at least on a first sight, the appearance of relatively undeformed metapelites or metapsammites. This effect has been particularly important during early investigations in the Børgefjell and Fjällfjäll windows. There, it led to the earlier neglect of the MA roof (Zachrisson, 1964) and floor thrusts (Greiling, 1974). Regional considerations (discovery of a Seve slice between the Köli nappes of the Upper Allochthon and the Fjällfjäll arkose) led Zachrisson (1969) to separate the Fjällfjäll unit from the Upper Allochthon. Similarly, the Rainesklumpen unit was earlier combined with the basement/cover imbricates of the underlying Lower Allochthon/Parautochthon units of the Børgefjell window (Greiling, 1974). Detailed studies in the MA at the southwestern margin of the Børgefjell window (Greiling *et al.*, 1984b) and in the SNC (Greiling, 1985) allowed regional comparisons and correlation of the Rainesklumpen unit with the SNC. Similar to the evolution at the Caledonian margin, gravitational collapse of the nappe pile after thrusting is documented, for example, by late folds with subhorizontal axial surfaces and "down-dip" vergence (e.g., Zachrisson, 1969).

Post Caledonian collapse yes?

DISCUSSION: A TECTONIC MODEL FOR THE MIDDLE ALLOCHTHON

Early metamorphic conditions observed in the MA prior to thrusting with *P* conditions of probably up to 4 kb imply a burial beneath a rock pile of at least 12 km. As has been argued earlier (Greiling, 1985), this thickness of overburden exceeds the cumulative thickness of the nappe pile overlying the MA (or having overlain the MA prior to erosion), as can be deduced from the regional tectonic context (e.g., Zachrisson, 1969; Stephens *et al.*, 1985; Hossack, 1983, 1985). Consequently, burial of the MA in Västerbotten cannot be related to an early phase of regional-scale stacking of nappe units prior to their final transport to the external parts of the orogen. It is, therefore, suggested that the MA acquired its early metamorphic imprint during an early phase of orogenic shortening, when it still formed part of the subducting margin of Baltica (Fig. 10, D1). Whereas, for example, the Seve unit of the Upper Allochthon, palaeogeographically situated close to and west of the MA, was subducted sufficiently deeply to reach granulite facies conditions (e.g., van Roermund, 1985, and this volume), the MA reached only upper greenschist facies conditions. It is likely that the MA was scraped off from the subducting margin of Baltica and was attached to the overriding plate as can be observed and deduced from present subduction environments (e.g., Leggett, 1987; Fig. 10, D2). Then, as part of the overriding plate, the MA was transported upwards again and some of the earlier metamorphic mineral assemblages were retrogressed (Fig. 10, early D3). Finally, the MA was transported close to the surface again and onto what is now the Baltic Shield, into its present tectonic position at the margin of the Caledonide orogen. There, metamorphism waned immediately after nappe transport, leaving mylonites essentially unchanged (Fig. 10, east). Farther west, however, metamorphic reactions (e.g., in mylonites) could take place well after nappe transport (Fig. 10, west) and mylonites were, in places, annealed. It

is, therefore, speculated that the nappe pile covering the MA was thicker and/or "hotter" in the west than in the east, at the Caledonian margin.

This latter part of the model is supported by recent work (Lindqvist, 1988) in the Nasafjäll window (north of Fig. 1, position comparable to the Børgefjell window), where the grade of post-thrust metamorphism increases towards the west from greenschist facies at the window's eastern margin to amphibolite facies at its western margin.

Regional differences of the nappe pile overlying the MA between the Caledonian margin and the area farther west may have already influenced the mechanical behaviour of the MA earlier, during later stages of the nappe transport. In the east, for example, stretching seems to be restricted to shear zones, whereas in the west the rocks as a whole were stretched and deformed into sheath folds, implying a more ductile state during deformation of the rocks in the west.

Acknowledgements

Field work was carried out on various occasions during a major study on the Stalon Nappe Complex at the Caledonian margin in order to assess the regional extent and lateral variation of the Middle Allochthon. The author thanks the German Research Foundation (D.F.G.) for partly supporting the field work. He also wishes to thank the Swedish Geological Survey (S.G.U.) for providing base maps and some aerial photographs, in particular to E. Zachrisson (Uppsala) for his continued interest and stimulating discussions, especially on field excursions. The paper was prepared when visiting the Department of Geology, University of Wales, Cardiff, as a Heisenberg research fellow of the D.F.G. The author thanks R. A. Gayer (Cardiff) for his patient help and fruitful discussions and, not least, for organizing the stimulating Scandinavian Caledonide meeting, at which this paper was presented. R. A. Gayer, P.-G. Andreasson (Lund) and D. G. Gee (Lund) suggested substantial improvements to an early draft of the manuscript.

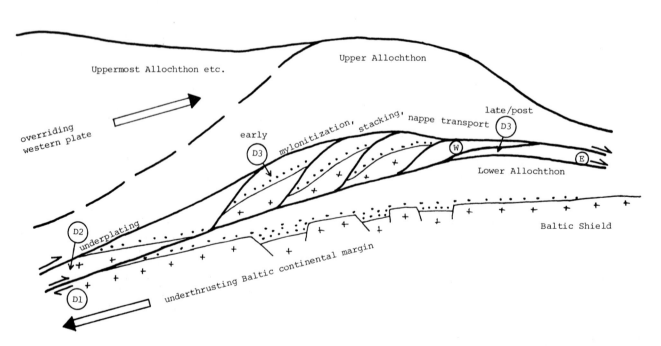

Fig. 10. Speculative synoptic view of tectonic situations reflected in the early evolution of the MA. As part of the underthrusting Baltic continental margin the MA is "subducted", deformed and metamorphosed (D1). Subsequent underplating onto the overriding western plate leads to uplift of the MA and D2 deformation and metamorphism of lower grade relative to D1. Further transport of the MA onto Baltica leads to partial mylonitization of the MA rocks, stacking of horses (early D3) and the regional nappe transport. Relatively thick or hot overburden in the west allows further metamorphic reactions to take place (late/post-D3, 'W'). In the east, at the Caledonian margin, there is no or very weak overprint, probably owing to relatively quick cooling (late/post-D3, 'E').

REFERENCES

Gee, D. G. (1978). Nappe displacement in the Scandinavian Caledonides. *Tectonophysics* **47**, 393–419.

Gee, D. G. and Zachrisson, E. (1979). The Caledonides in Sweden. *Sver. Geol. Unders. C*, **769**, 1–48.

Gee, D. G., Kumpulainen, R., Roberts, D., Stephens, M. B., Thon, A. and Zachrisson, E. (1985). Scandinavian Caledonides—Tectonostratigraphic map. *Sver. Geol. Unders.* **Ba 35**.

Greiling, R. (1974). Das kristalline, präkambrische Grundgebirge im Östlichsten Teil des Børgefjell-Fensters (Zentrale Kaledoniden Skandinaviens). *Geol. Fören. Stockh. Förh.* **96**, 247–251.

Greiling, R. (1984). Structure and metamorphism at the base of major, far-travelled nappes: an example from the Middle Allochthon of the central Scandinavian Caledonides. *Sci. Géol. Bull.* **37**, 13–22.

Greiling, R. (1985). Strukturelle und metamorphe Entwicklung an der Basis grosser, weittransportierter Deckeneinheiten am Beispiel des Mittleren Allochthons in den zentralen Skandinavischen Kaledoniden (Stalon-Deckenkomplex in Västerbotten, Schweden). *Geotekt. Forsch.* **69**, 1–129.

Greiling, R., Gorbatschev, R., Johansson, L. and Eberz, G. (1984a). Chemical variation and tectonic significance of some Middle and Late Proterozoic dyke swarms in the central Scandinavian Caledonides and their foreland. *Terra Cognita* **4**, 88–89.

Greiling, R., Kaus, A. and Leipziger, K. (1984b). Tectonic setting and geochemistry of Seve metabasites at the northern margin of the Grong district (Caledonides of central Norway). *Meddel. Stockh. Univ. Geol. Inst.*, **255**, 76.

Hossack, J. R. (1983). A cross-section through the Scandinavian Caledonides constructed with the aid of branch-line maps. *J. Struct. Geol.* **5**, 103–111.

Hossack, J. R. (1985). The role of thrusting in the Scandinavian Caledonides. In: Gayer, R. A. (ed.) *The Tectonic Evolution of the Caledonide–Appalachian Orogen*, pp. 97–116. Earth Evolution Sciences (Vieweg), Braunschweig.

Kieft, C. (1952). Geology and petrology of the Tärna region, southern Swedish Lapland. Thesis, University of Amsterdam, 98pp.

Kulling, O. (1942). Grunddragen av fjällkedjerandens bergbyggnad inom Västerbottens län. *Sver. Geol. Unders. C*, **445**, 1–320.

Kulling, O. (1955). Den kaledoniska fjällkedjans berggrund inom Västerbottens län. *Sver. Geol. Unders. Ca*, **37**, 101–296.

Kulling, O. (1972). The Swedish Caledonides. In: Strand, T. and Kulling, O. (eds.) *Scandinavian Caledonides*, pp. 149–285. Wiley Interscience, London.

Kulling, O. (1982). Översikt över södra Norrbottensfjällens kaledonberggrund. *Sver. Geol. Unders. Ba*, **26**, 1–295.

Kumpulainen, R. and Nystuen, J. P. (1985). Late Proterozoic basin evolution and sedimentation in the westernmost part of Baltoscandia. In Gee, D. G. and Sturt, B. A. (eds.), *The Caledonide Orogen—Scandinavia and Related Areas*, pp. 213–232. Wiley, Chichester.

Leggett, J. K. (1987). The Southern Uplands as an accretionary prism: the importance of analogues in reconstructing palaeogeography. *J. Geol. Soc. Lond.* **144**, 737–752.

Lindqvist, J.-E. (1988). Metamorphism in basement rocks and their implications for the tectonic evolution, Nasafjäll Window, Scandinavian Caledonides. *Geol. Rdsch.* **76**, 837–850.

Ramberg, H. (1981). The Role of Gravity in Orogenic Belts. *Geol. Soc. London Spec. Publ.* **9**, 125–140.

Roermund, H. van (1985). Eclogites of the Seve Nappes, central Scandinavian Caledonides. In: Gee, D. G. and Sturt, B. A. (eds.) *The Caledonide Orogen—Scandinavia and Related Areas*, pp. 873–886. Wiley, Chichester.

Roermund, H. van (1988). High pressure ultramafic rocks from the allochthonous nappes of the Swedish Caledonides, *this volume*.

Solyom, Z., Gorbatschev, R. and Johansson, I. (1979). The Ottfjäll dolerites: geochemistry of the dyke swarm in relation to the geodynamics of the Caledonide orogen of central Scandinavia. *Sver. Geol. Unders. C*, **756**, 1–38.

Stephens, M. B., Gustavson, M., Ramberg, I. B. and Zachrisson, E. (1985). The Caledonides of central-north Scandinavia—a tectonostratigraphic overview. In: Gee, D. G. and Sturt, B. A. (eds.) *The Caledonide Orogen—Scandinavia and Related Areas*, pp. 135–162. Wiley, Chichester.

Winkler, H. G. F. (1979). *Petrogenesis of Metamorphic Rocks.* Springer, New York.

Zachrisson, E. (1964). The Remdalen syncline. *Sver. Geol. Unders. C*, **596**, 1–53.

Zachrisson, E. (1969). Caledonian geology of northern Jämtland—southern Västerbotten. *Sver. Geol. Unders. C*, **644**, 1–33.

7 The Middle Allochthon of the Scandinavian Caledonides at Kvikkjokk, northern Sweden: Sedimentology and Tectonics

R. O. Greiling[1] *and R. Kumpulainen*[2]

[1]Department of Geology, University of Wales, Cardiff, P.O. Box 914, Cardiff CF1 3YE, UK (Present address: Geologisch-Paläontologisches Institut, Ruprecht-Karls-Universität, INF 234, D-6900 Heidelberg, FRG.)
[2]Geologiska Institutionen, Stockholms Universitet, S-106 91 Stockholm, Sweden

The Middle Allochthon around Kvikkjokk and towards the south comprises two distinct units: a lower Kabla–Stuor Tata Unit (KSTU) and an overlying Kvikkjokk Complex (KC). The KSTU northeast of Kvikkjokk contains a sedimentary sequence, named the Stuortjåkkå Formation. It is little deformed and metamorphosed, at least several hundred metres thick and composed of graded, turbiditic sandstones and shales.

In the upper part of the sequence, intercalated beds of unsorted conglomerates (Kabla conglomerate) contain pebbles of granite and rhyolite. This sequence is foliated and locally stretched, parallel to the nappe transport direction (120–140°). Subsequent folds, trending east–west, are of local importance.

A mylonite zone dipping SSW separates the KSTU from the overlying KC. It is approximately parallel to the nappe transport direction and interpreted as a lateral ramp. The KC, earlier assumed to be composed of mylonitic schists and acidic igneous rocks, is built up of quartz-rich metasediments with alternating bands of quartzitic/arkosic and quartzphyllitic composition. Locally, the metasediments are intruded by metabasite dykes or sheets. An early, penetrative foliation, synchronous with the peak of metamorphism (greenschist grade) overprinted nearly all of the primary textures. It is succeeded by a second foliation, associated locally with intense crenulation cleavage under waning metamorphic conditions. Subsequent small folds are oriented NW–SE, parallel to and probably related to the lateral ramp at the KC/KSTU margin. In places, a transitional zone from the KC to the Seve unit is developed, with garnet growth in the uppermost KC apparently induced by heating from the overriding Seve unit. Deformation after nappe transport is restricted to local, conjugate kink folds and the Middle Allochthon floor thrust is undeformed on a large scale.

Metabasite dykes or sheets cutting KC metasediments imply a correlation with the Särv unit of the Middle Allochthon. The KC is separated from overlying igneous rocks (*inter alia* Ruoutevare anorthosite) by micaschists, which are here related to those of the Seve unit in the Upper Allochthon.

In contrast to the Middle Allochthon at the Caledonian margin towards the north and south, where basement and basement-cover units are frequent, the KSTU and the KC are almost exclusively built up of metasediments. Sedimentological evidence implies that the KSTU sediments were deposited in a restricted marine basin at the margin of the Baltoscandian continent. We speculate that the relatively thick sedimentary sequences around Kvikkjokk facilitated a decollement during thrusting and that the Middle Allochthon floor thrust did not cut down into the basement rocks.

INTRODUCTION

In his review on the southern Norrbotten Caledonides, Kulling (1982) recognized two distinct units in the Middle Allochthon of the Kvikkjokk area. The lower one, to the northeast of Kvikkjokk, he called Kabla–Stuor Tata sparagmitfält (sparagmite area); the upper, larger one, distributed mainly to the north, east and south of Kvikkjokk, he named Kvikkjokk complex (KC; Fig. 1). The Kabla–Stuor Tata "sparagmites" are well preserved and their sedimentary character was easily, and early, recognized (e.g., Hamberg, 1910). In contrast, the KC is strongly deformed and primary textures are mainly absent. This led to the (erroneous) classification of all the KC rocks as crystalline schists and acidic igneous rocks (Hamberg, 1910; Gavelin, 1915; Kulling, 1982). As

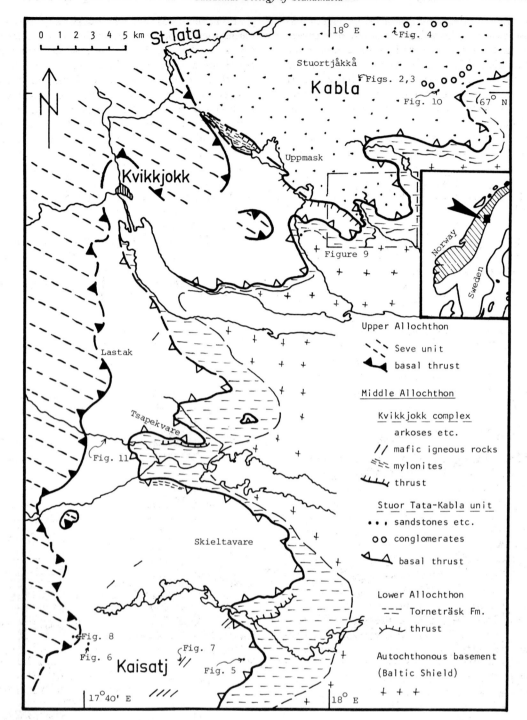

Fig. 1. Geological sketch map of the Middle Allochthon around Kvikkjokk, at the eastern Caledonian margin in northern Sweden. The location is shown in the inset map of part of Scandinavia (Caledonides hatched). (Revised from Kulling (1982) and Gee *et al.* (1985).) The locations of Figs. 2–11 are shown by dots or a square (Fig. 9).

a consequence, the igneous suite, which includes the Ruoutevare iron ore bearing anorthosites overlying the KC, was included in the Middle Allochthon (Kulling, 1982). However, subsequent work (Gee and Zachrisson, 1979; Zachrisson and Stephens, 1984; Gee *et al.*, 1985; and our data) has shown the KC, the lower and major part of the Middle Allochthon, to be clearly distinct from the upper part of Kulling's (1982) Middle Allochthon, since it consists of mainly psammitic metasediments with only subordinate, mafic igneous rocks. Furthermore, our results indicate that the Middle Allochthon does not directly rest on autochthonous sequences but is underlain by a Lower Allochthon, which has not previously been recognized. These new results have important consequences on the tectonostratigraphic interpretation of the Kvikkjokk area

and the Middle Allochthon/Upper Allochthon boundary in general. Therefore, we present here an account of the lithology and sedimentology of the Middle Allochthon of the Kvikkjokk area and discuss the regional tectonic and tectonostratigraphic consequences. The newly discovered Lower Allochthon has not yet been systematically studied and is mentioned only briefly.

LOWER ALLOCHTHON

Although the work was mainly directed towards the Middle Allochthon, our mapping revealed some tectonic imbrications and repetitions in the unmetamorphic sedimentary sequence beneath the Middle Allochthon floor thrust. For example, on the southwestern slope of Tsapek-vare (Fig. 1) at least two allochthonous units are exposed. The lowermost unit is thrusted over black alum shale in the foot-wall and consists of dark grey quartzite, siltstone and black alum shale (from bottom to top), which is cut by a thrust that forms the floor of a slice of quartzite. This quartzite lies in the foot-wall of the basal thrust of the Middle Allochthon.

Further repetitions of a similar sedimentary sequence have been observed on the south slopes of mount Njannja (Fig. 9) and to the east of mount Kaisatj (Fig. 1). Imbrication and stacking within the Lower Allochthon apparently did not affect the Middle Allochthon floor (or Lower Allochthon roof) thrust.

MIDDLE ALLOCHTHON: LITHOLOGY AND SEDIMENTOLOGY

The two major units of the Middle Allochthon in the Kvikkjokk area display important lithological differences and are therefore treated separately, starting with the structurally lower Kabla–Stuor Tata unit and then treating the upper KC and its relation to overlying units. Descriptive terms for the sedimentary rocks are from Dzulynski and Walton (1965), Pettijohn *et al.* (1972) and Reading (1985).

The Kabla–Stuor Tata Unit

The name Kabla–Stuor Tata Unit for the structurally lowermost unit of the Middle Allochthon is adopted from Kulling (1982: "Kabla–Stuor Tata sparagmitfält"). The Kabla–Stuor Tata unit (KSTU) is underlain by the Lower Allochthon and overlain by the Kvikkjokk Complex of the Middle Allochthon. It occupies an area of some 250 km², about 10–15 km north and northeast of Kvikkjokk (Fig. 1), of which we studied the southeastern half. The KSTU is composed of a sedimentary sequence dipping gently towards the west to northwest, and represents a total stratigraphic thickness of several hundred metres. These sedimentary rocks are apparently unfossiliferous and isotopic age data are lacking. On the basis of compositional and tectonostratigraphic similarities, Kulling (1982) correlated these rocks with the Upper Proterozoic, subtillite successions of the Jämtland Supergroup, exposed farther south along the Caledonian front (Kulling, 1942; Kumpulainen and Nystuen, 1985). This correlation is adopted here.

The sequence starts with feldspathic sandstones. In the central-eastern part of the area, these sandstones are exposed on the eastern slope of Stuortjåkkå, the highest peak of the Kabla group of hills, where two sections were studied (see below). These sandstones dip beneath a conglomerate, which is the Kabla conglomerate of Kulling (1982). The contact between these two rock types is not exposed. We propose here the name *Stuortjåkkå Formation* for the whole sedimentary sequence in the KSTU. The Kabla conglomerate (Kulling, 1982) is the uppermost member in the Stuortjåkkå Formation. The base and the top of the formation are cut by Caledonian thrust surfaces. Kulling (1982) reported repetitions of the Kabla conglomerate within the "sparagmites". We assume that these repetitions are also related to Caledonian thrusting.

The sandstones of the Stuortjåkkå Formation are grey, medium- to coarse-grained, and rich in K-feldspar, some of which is altered to white mica. Additionally, granules and pebbles of granitic debris occur in some of the sandy beds. Part of the K-feldspar occurs as scattered, larger

grains in the sandstone. These feldspar grains with smooth embayments appear to be similar to the feldspar phenocrysts in the porphyry clasts of the overlying Kabla conglomerate and thus of volcanic origin. The sandstones are thin- to medium-bedded, being laterally even to uneven in outcrop scale. The beds display consistent normal grading (Figs. 2 and 3), showing that the sequence is right-way-up. The individual beds start with medium- to coarse-grained sand, grading up to fine-grained sand and then to mud. The lower surfaces of the sandy beds are straight and planar and no bottom structures (such as flutes or grooves) other than local load-casting have been observed. Nevertheless, the regularly graded sedimentation units, although mostly lacking the c-division of the Bouma cycle, suggest that these sediments were deposited from turbidity currents, probably on a submarine fan.

Some of the beds have no obvious grading, being composed of granules and small pebbles of granitic debris and also mudflakes, scattered in a homogeneous matrix of

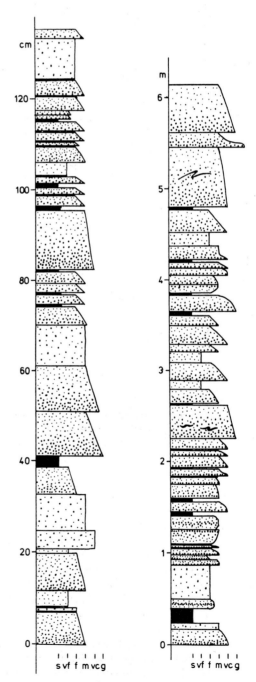

Fig. 2. Two sections from the Stuortjåkkå sandstones with minor units of mudstones. Eastern slope of Stuortjåkkå.

Fig. 3. Graded sandstones, i.e. ca. 10 cm thick sandy units with very thin mud units. Compass for scale; eastern slope of Stuortjåkkå.

medium to coarse sand. These units may have been deposited from high-density turbidity currents and homogenized by late-depositional fluidized flow. Homogenization may have occurred also through post-depositional local liquefaction (Reading, 1985).

The Kabla conglomerate, lying above the sandstones, is composed of conglomerate beds, about 0.5–1.0 m in thickness, and interbedded sandstones similar to those described above. The conglomerate beds are dominated by pebbles and cobbles, their diameter averaging 5–10 cm but grading up to about 20 cm, composed of granite and quartz porphyry. Minor amounts of quartzite and siltstone to fine sandstone occur also as pebbles. The clasts are generally well rounded and have a high sphericity.

These are the characteristics common to all of the conglomerate beds in the KSTU. According to their internal sedimentary features, the conglomerates fall into two different categories.

One type of the conglomerate is unsorted and matrix-supported (Fig. 4). The matrix is apparently massive, medium to coarse sand similar to the matrix in the other conglomerate beds. This type of conglomerate is interpreted as a debris-flow unit and may represent proximal deposits on a submarine fan, in a channel or an area adjacent to it. The other type of conglomerate is stratified, with preferred clast orientation parallel to bedding. This type was probably deposited from traction currents along the channel floor.

The amount of data from the Stuortjåkkå Formation is limited and does not allow a detailed discussion about the original fan type (see Stow *et al.*, 1985). However, the apparent coarsening-upwards sequence from turbiditic sandstones to gravelly units suggests a prograding stage of a fan.

The Kvikkjokk Complex

Towards the southwest the KSTU is overlain structurally by the Kvikkjokk Complex (KC; Kulling, 1982), which occupies about 500 km² to the east and, mainly, to the south of Kvikkjokk. Except for the extreme southern margin, our survey covered most of this area. In contrast to the generally mild deformation of the KSTU (see below), the rocks of the KC are "for a large part mylonitized" (Kulling, 1982). Our investigations confirmed the strong deformation of the KC rocks, but we found true mylonites to be restricted to the base of the KC. Usually, the mylonites are only a few centimetres to a few metres thick. It is only at the margin towards the KSTU that they are more widely developed, being several tens of metres thick. In particular, southeast of Lake Uppmask, the mylonites contain metre size lenses of mica-rich and feldspar-rich rocks, possibly derived from a crystalline protolith.

In general, the KC rocks are well banded in millimetre to decimetre scale with alternating quartz-rich and more micaceous "layers" that are thought to be of sedimentary origin (Fig. 5, see below). As is argued below, overlying micaschists are more closely related to the Seve Unit of the Upper Allochthon and hence excluded from the KC.

The redefined KC is about 300 m thick. The layering dips overall gently towards the west and northwest.

Fig. 4. The Kabla conglomerate, matrix-supported with clasts up to ca. 15 cm in diameter scattered in unstratified coarse sand. Hammer shaft 0.3 m; northeast of Stuortjåkkå.

Fig. 5. Kvikkjokk Complex metasediments with light quartz-rich and dark mica-rich bands. Northeast of Kaisatj, watch (55 mm diameter) for scale.

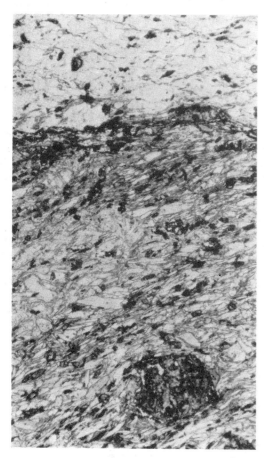

Fig. 6. Micrograph showing part of a quartz-rich (top quarter) and a mica-rich band in the KC. In the mica-rich band the metamorphic minerals muscovite (light grey), biotite (darker grey, in particular towards the quartz-rich band), garnet (bottom right, dark grey) and chlorite (e.g., around and within garnet, slightly lighter than garnet and formed at the expense of garnet) are developed (dark grey and black: epidote, zoisite, opaque minerals). Note an early foliation subparallel to banding (D1) and a later crenulation cleavage (D2, top right to bottom left). "Curving" of the mica minerals around the garnet porphyroblast is due to this late deformation, whereas the early fabric is contiguous throughout the porphyroblast. Original size 3.0 × 4.5 mm, one polarizer, BFRG 84012, WNW Kaisatj (Fig. 1).

Quartz-rich bands/layers consist generally of about 80–90% quartz with minor plagioclase, muscovite, biotite and subordinate K-feldspar, epidote, zoisite, and chlorite. Micaceous layers are distinct in having higher muscovite and lower quartz contents. Biotite is subordinate to muscovite even in the micaceous layers (Fig. 6). There, grain size is distinctly smaller relative to the quartz-rich layers and may reflect a primary difference, although no primary grain shapes are preserved.

In contrast to earlier views (summarized by Kulling, 1982), predominance of quartz and muscovite precludes a syenitic protolith and supports the field impression that the KC rocks in general are of sedimentary origin. However, some thin (centimetre to decimetre) bands of metabasite are exposed in the southern part of the KC, mainly around the hill Kaisatj. In the field, these metabasites appear as finely sheared, green bands with rare white patches, resembling plagioclase phenocrysts. Their margin with the metasediments is either sharp, cutting the banding at a low angle, or transitional, possibly owing to shearing and "tectonic mixing", mainly on the southern slope of mount Kaisatj. Microscopically, the observed "plagioclase phenocrysts" are seen to be replaced by calcite, the mafic minerals are completely altered to chlorite. The only primary texture recognized is that within a small low-strain domain in a metabasite. Plagioclase appears to be essentially unchanged, whereas pseudomorphs of chlorite are all that remain of the mafic minerals hornblende or pyroxene (Fig. 7). The composition of mainly plagioclase and hornblende/pyroxene with quartz content of almost zero implies a basaltic/andesitic protolith for these sheets or dykes, which cut the KC metasediments.

The metasedimentary sequence appears to be dominated by thin- to medium-bedded sandstones ranging from 60 to 90% of the total rock. The interbedded sediments, altered to quartz–muscovite phyllites, represent muddy intercalations in the sedimentary sequence. Both the sandy and shaly beds are laterally persistent to impersistent in the outcrop scale. Some of the impersistency is obviously caused by tectonic deformation (shearing). The composition of sandy beds, mainly quartz, feldspar, white mica and biotite, suggests that the original composition was

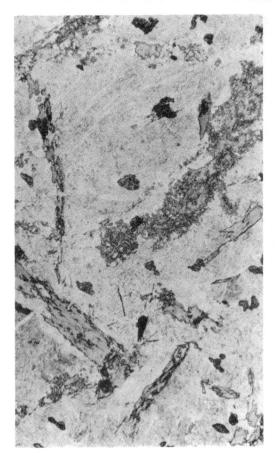

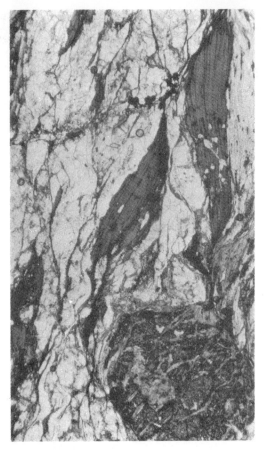

Fig. 7. Micrograph showing plagioclase (light) and chlorite (grey; dark grey and black spots are opaque minerals and bubbles) with a relic primary texture. Note well-developed pseudomorphs of chlorite after hornblende or pyroxene in lower half of figure. Original size 3.0 × 4.5 mm, one polarizer, BFRG 84024, Kaisatj (Fig. 1).

Fig. 8. Micrograph showing Seve micaschist with muscovite and quartz (light), biotite porphyroblasts (dark grey, "striated") and garnet (dark, lower right) with some chlorite along cracks. Note intensive foliation (vertical) and larger grain size relative to KC in Fig. 6, which is at the same scale. Original size 3.0 × 4.5 mm, one polarizer, BFRG 84010, Tjuonak, WNW Kaisatj (Fig. 1).

impure, probably that of a feldspathic to lithic sandstone. Candidates for the depositional environment would be rivers or a submarine fan, but because the internal structures are not preserved, further reasoning is difficult.

Higher units

Upwards, the KC metasediments pass through a transitional zone into garnetiferous muscovite–biotite schists. For example, to the west of mount Kaisatj (Fig. 1) this zone is up to a few hundred metres wide. These mica schists overlying the KC metasediments are composed mainly of coarse (millimetre to centimetre) muscovite and biotite flakes (Fig. 8) with subordinate quartz and epidote, zoisite and chlorite as accessories. Garnet porphyroblasts, with an average size of several millimetres, up to 1 cm, are frequent. Composition, texture and structural evolution (see below) distinguish these micaschists clearly from the KC metasediments. They are very similar to micaschists of the Seve Unit in the Upper Allochthon and are therefore included in the Upper Allochthon (e.g., Gee and Zachrisson, 1979).

The muscovite–biotite schists of Seve affinity underlie the Ruoutevare anorthosites and the related igneous suite and separate them from the KC as defined above. Consequently, the Ruoutevare anorthosites and associated rocks can no longer be regarded as part of the Middle Allochthon but have to be accepted as part of the Seve Unit in the Upper Allochthon. The Middle Allochthon of the Kvikkjokk area, therefore, is reduced to only two

major, essentially metasedimentary units, which are the Kabla–Stuor Tata Unit and the Kvikkjokk Complex (Table I).

MIDDLE ALLOCHTHON: STRUCTURE AND METAMORPHISM

The mineralogical and lithological differences between the KSTU and the KC are due not only to primary compositional differences but also to a different structural and metamorphic evolution. Therefore, the KSTU and the KC are again treated separately. Subsequently, some observations from the overlying units are presented and their bearing on the distinction between the Middle Allochthon and the Upper Allochthon is discussed.

Kabla–Stuor Tata Unit

The sedimentary textures in the Stuortjåkkå Formation are overprinted by an intense foliation, which relates to the axial surface of rare, tight, recumbent folds, with axial trends of about 140°. Foliation is most intense in mica-rich, incompetent rocks and less pronounced in quartz-rich lithologies. Similarly, competent pebbles of the Kabla conglomerate may appear almost undeformed, whereas incompetent ones are strongly compressed. The XY plane of the pebbles is oriented parallel to the foliation, the X axes are mainly pointing towards southwest, at about 120°. The pebbles therefore indicate a NW–SE directed

Table I. Tectonostratigraphy of the Kvikkjokk area

	Kulling (1982)	This paper	
Upper Allochthon	Higher units Seve Nappe	Higher units Major (higher) part	
Middle Allochthon	Syenite–anorthosite complex in the Sarek–Pårte–Kvikkjokk mountains, including the Ruoutevare iron ore-bearing anorthosite Mica schists	Seve Nappe	Ruoutevare anorthosite and related igneous rocks Garnet-mica schists
	Kvikkjokk Complex Kabla–Stuor Tata Sparagmite, including Kabla conglomerate (lowermost nappe unit)	Kvikkjokk Complex Kabla–Stuor Tata Unit	Stuortjåkkå Formation with Kabla conglomerate
Lower Allochthon		Imbricates at Njannja and Tsapekvare	
Autochthon (Baltic Shield)	Torneträsk Formation Crystalline basement	Torneträsk Formation Crystalline basement	

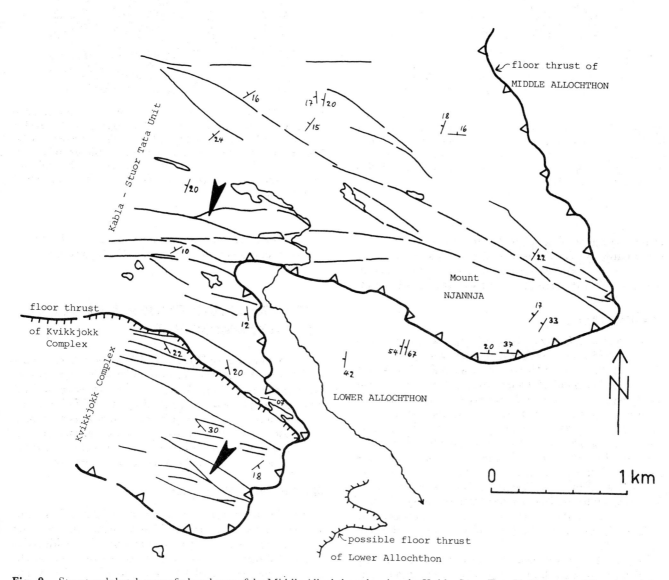

Fig. 9. Structural sketch map of a basal part of the Middle Allochthon showing the Kabla–Stuor Tata Unit at mount Njannja and the overlying Kvikkjokk Complex towards southwest. Based on a field-controlled aerial photointerpretation. Middle Allochthon and Kvikkjokk Complex basal thrust as in Fig. 1. Thin lines represent gently inclined minor shear zones at low angles to bedding (attitude of bedding shown by symbols). These shear zones disrupt the Middle Allochthon metasediments into metre to kilometre scale lenses. Two examples are marked by black arrows. Bedding in the Middle Allochthon, although only gently dipping, is substantially more steeply inclined than the Middle Allochthon floor thrust, which dips at an angle of ca. 2–5° towards WNW, as is evident from the map pattern. Note the relatively steep dip of bedding in the Lower Allochthon, probably due to imbrication.

stretching, which is approximately parallel to the axes of early folds, and the direction of the general nappe transport in the Scandinavian Caledonides (see, e.g., Hossack, 1985). Other lineations related to deformation or nappe transport are rarely observed, similar to conditions in the Middle Allochthon of other areas (e.g., Greiling, 1985).

The orientation of these D1 structures implies a relationship to nappe transport and we presume that the D1 deformation took place either prior to or during nappe transport of the Middle Allochthon.

Strain is apparently inhomogeneous and concentrated at fold hinges and in discrete shear zones, which are oriented at low angles to bedding. These shear zones dissect the KSTU marginally into metre to kilometre scale tectonic lenses, an example of which is shown in Fig. 9.

Subsequent deformation (D2) is restricted to local small folds, mainly of kink type, oriented about E–W (87°) and locally associated with a steeply dipping crenulation cleavage. Although nappe transport is obviously related to the early (D1) deformation, no deformation of the Middle Allochthon floor thrust related to D2 is observed.

Metamorphism in the KSTU is restricted to an early, fine-grained quartz recrystallization at subgrain boundaries in high-strain domains, fine-grained white mica formation and rare growth of green biotite and/or chlorite (Fig. 10). Since these metamorphic minerals are aligned parallel to the early foliation, metamorphism took place prior to or during the early deformation. No mineral changes are observed in relation to subsequent deformation.

Kvikkjokk Complex

The oldest feature preserved in the KC metasediments is a distinct banding in millimetre to decimetre scale with alternating bands of quartzitic/arkosic and quartz-phyllitic composition (Figs. 5, 6). This banding resembles a primary layering, but no primary textures are observed. At the microscopic scale, the early layering shows up as a distinct compositional variation between quartz-rich and muscovite rich bands. Muscovite growth is parallel to this banding and therefore this banding includes an early, metamorphic foliation (D1, Fig. 6). Biotite is formed synchronously with muscovite formation during D1. Biotite porphyroblasts in general are distinctly smaller than muscovite grains (Figs. 6 and 11), except for domains close to mafic rocks, where the bulk composition appears to be more favourable for biotite formation. Garnet growth is restricted to the upper, western parts of the KC, marking a transition towards the Seve micaschists. Occasionally, garnet is seen to overgrow an earlier fabric or, rarely, shows a weak zonation. Generally, however, garnet porphyroblasts appear as simple, idioblastic grains without any internal texture (Fig. 6). The critical mineral assemblage is (garnet +) muscovite + biotite + albite + quartz and indicates the peak metamorphic conditions within the (western part of the) KC to be the higher part of greenschist facies.

Competent layers are seen to be cut by a second foliation

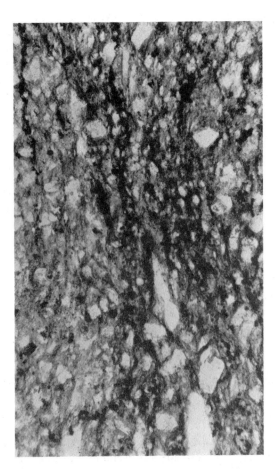

Fig. 10. Micrograph showing pelitic part of KSTU with metamorphic minerals chlorite and illite developed along an early cleavage (vertical, D1). Original size 1.2 × 1.8 mm, one polarizer, BFRG 85006, SE slope Kabla (Fig. 1).

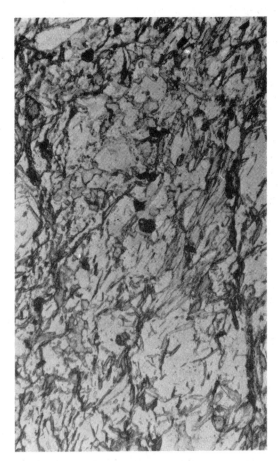

Fig. 11. Micrograph showing KC metasediment with quartz, muscovite (light) and biotite (grey). These minerals are aligned parallel to an early cleavage (D1, top right to bottom left), which is deformed by a second, crenulation cleavage (vertical) along which chlorite (medium to dark grey) is developed. Random dark grey and black dots are epidote and opaque minerals. Original size 1.2 × 1.8 mm, one polarizer, BFRG 84031, Tsielejåkkå, SW Tsapekvare (Fig. 1).

(D2) at a low angle to layering. In incompetent layers this foliation is virtually parallel to the layering and cannot readily be distinguished from the D1 foliation in the field. South of Lastak, an isolated fold hinge is exposed, where the D2 foliation is related to the axial surface. The axial surface foliation is continuous into the foliation outside the fold hinge and it would appear that progressive D2 shortening led to folding and subsequent disruption of the earlier formed D2 folds. Microscopically, the D2 foliation occurs as a penetrative foliation in the mica-rich layers but as a crenulation cleavage in relatively competent domains. There, muscovite and biotite are folded and sometimes a second generation of muscovite is seen to grow parallel to the axial surfaces (Fig. 11). Biotite (and garnet) are retrogressed to green biotite and chlorite along the D2 axial surfaces or minor shear zones, which develop from these axial surfaces. The minor shear zones are also related to larger-scale shear zones observed in the field and on aerial photographs (Fig. 9). These shear zones also occasionally produce "tectonic mixing" at the margins of mafic intrusions and the large-scale disruption of the KC metasedimentary succession into centimetre to kilometre size "tectonic lenses" (Fig. 9). Since no later mineral changes are observed, the D2 event marks the waning of metamorphism in the KC from lower greenschist grade (muscovite + chlorite) at the beginning of D2 (e.g., Fig. 11) to subsequent very low-grade conditions. Later deformation produced small folds in WNW–ESE directions (110–125°), verging mainly towards northeast but also towards southwest. The folds are about parallel to the stretching direction in the KSTU, its nappe transport direction and, probably more importantly, the margin between the KSTU and the KC. There, the KSTU/KC marginal shear zone dips about 20–30° towards southwest with the KC on top of the KSTU. This shear zone, parallel to nappe transport direction, may form a lateral ramp that gave rise to the folds parallel to its strike direction in the overlying KC. Subsequently, these small folds are overprinted by rare, NE–SW-trending open folds (30–40°) with relatively weakly inclined axial surfaces. Finally, a few E–W-trending kink folds with steeply dipping axial surfaces, similar to the latest folds in the KSTU, have been observed, but their relation to other structures could not be established.

Higher units

Besides the lithological criteria put forward above to distinguish the KC metasediments from the overlying Seve micaschists, there is also some structural and metamorphic evidence to support this distinction.

Generally, the Seve schists are completely, and coarsely, recrystallized, whereas the KC metasediments are distinctly finer-grained. The KC grain size varies with the compositional layering, the mica-rich bands being finer-grained and probably still reflecting the primary grain size variations between coarser-grained psammitic and finer-grained pelite-rich bands.

Garnet porphyroblasts are common throughout the Seve schists and distinctly larger (up to 1 cm) than those observed in the uppermost parts of the KC. Furthermore, the Seve garnets show several growth stages, sometimes with an inclusion-free or poor core and an inclusion-rich margin (Fig. 8). Many of these porphyroblasts contain an internal fabric that is curved or rotated and that is generally no longer connected to the external fabric, defined, for example, by mica minerals. In contrast, the small, more uniformly built garnet porphyroblasts in the upper part of the KC are simply overgrowing a (D1) fabric

with internal and external fabric still contiguous (Fig. 6). The garnet porphyroblasts in the Seve schists therefore indicate a much more complex metamorphic and structural evolution in the higher units than that recorded in the KC metasediments of the Middle Allochthon. Absence of garnet in the eastern, lower parts of the KC also implies a generally lower grade of metamorphism in the Middle Allochthon relative to the higher units of the Upper Allochthon.

DISCUSSION

The major tectonostratigraphic results are summarized in Table I. Owing to lack of biostratigraphic or radiometric information, they had to be based on lithological comparison and correlation, supported by structural and metamorphic criteria. The Middle Allochthon is thus restricted to essentially metasedimentary units. The KSTU is relatively weakly deformed and metamorphosed and therefore obviously better preserved than any other Middle Allochthon unit in the Swedish Caledonides (e.g., Gee and Zachrisson, 1979; Greiling, 1985). The KSTU floor thrust southwest of Mount Njannja (Figs. 1 and 9) merges with the KC floor thrust towards SSW. It occupies the same topographic level as the KC floor thrust (except for the lateral ramp towards the KSTU) and is part of the same contiguous surface at the base of the Middle Allochthon. Therefore, the KSTU is taken here as part of the Middle Allochthon despite its considerably lower metamorphic grade with respect to the KC. Furthermore, the lower greenschist facies of the KSTU is still distinct from the unmetamorphic Lower Allochthon beneath the KSTU.

The well-preserved KSTU allows a reconstruction of the sedimentary environments (see above). So far, no palaeocurrent indicators have been observed in the KSTU, making the location of the source area and its relation to the depositional site uncertain. Abundance of granitoid and rhyolitic pebbles in the Stuortjåkkå Formation implies an acidic crystalline basement as a source area (Kulling, 1982). Such rock types have been mapped in central and western Norrbotten (northern Sweden) and in west-central Sweden (e.g., Zachrisson, 1981) and it is inferred that this belt continues beneath the Caledonian nappe pile from west central Sweden in the south to the Norrbotten area in the north (e.g., Zachrisson, 1981; Gorbatschev, 1980). We, therefore, speculate that this belt was the main source area of the Stuortjåkkå Formation. The depositional area was perhaps located to the west of that belt, towards the axis of a rift, which later developed into the Iapetus Ocean (Fig. 12).

In contrast to the turbidite dominated Stuortjåkkå Formation, the Upper Proterozoic sedimentary rocks southwards along the Caledonides are dominantly of fluvial to shallow marine origin (Kumpulainen and Nystuen, 1985). The nearest known turbidites of approximately the same age are found in Finnmark, northernmost Norway (Føyn, 1985; Siedlecka, 1985), some 600 km northeast of Kvikkjokk. The lithology and sedimentology of Stuortjåkkå Formation equivalents within a distance of 250 km along the Caledonides north and south of Kvikkjokk are poorly studied. The value of turbidites and debrites of the Stuortjåkkå Formation in the regional Late Proterozoic sedimentary and tectonic context is therefore limited. Until data from the adjacent areas become available, the turbidites from the Kvikkjokk area indicating relatively deep water remain a curiosity.

In contrast to the KSTU, relatively strong deformation

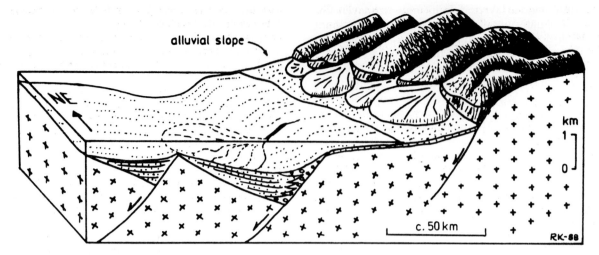

Fig. 12. Interpretation of the geotectonic position of the Stuortjåkkå Formation's depositional site. The probable fan was fed by dominantly sandy debris from a granite–porphyry terrain in the east bypassing a possible narrow shelf. The whole region was subject to rapid subsidence probably due to crustal tension, which later resulted in the birth of the Iapetus Ocean. Scales on the figure are very approximate.

and metamorphism overprinted the sedimentary textures in the overlying KC. Mafic intrusives into Late Proterozoic sediments, sheets or dykes (e.g. Figs. 5 and 7), are elsewhere in the Middle Allochthon restricted to the Särv unit and imply a correlation of the latter with the KC. Unfortunately, deformation and alteration of the metabasites discovered so far in the KC renders a geochemical characterization probably impossible. Intensity of deformation and grade of metamorphism in the KC is comparable with other units of the Middle Allochthon, for example the Stalon Nappe Complex (Kulling, 1942; Greiling, 1985) or the Fjällfjäll unit (Zachrisson, 1964, 1969) farther south. These units and the Särv unit are, similarly to the KC, overlain by Seve units of the Upper Allochthon, which at the eastern Caledonian margin are represented by an "eastern schist and amphibolite belt" (Gee and Zachrisson, 1979) or an "eastern amphibolite area" (Kulling, 1982). Both the mica schists and the metabasite-dominated igneous suite of Ruoutevare overlying the KC relate very well to these eastern Seve units, a view already adopted by Gee *et al.* (1985).

A peculiarity of the KC–Seve boundary is the transitional zone, where garnet prophyroblasts are restricted to the uppermost part of the KC and apparently imply a heating effect of the overriding Seve unit on the underlying KC.

In other areas, the Middle Allochthon frequently contains crystalline basement units, generally imbricated into Late Proterozoic cover sequences, leading to numerous repetitions of basement/cover slices. Examples of these occur in the Akkajaure Complex immediately north of the KSTU (Björklund, 1985; Martna and Hansen, 1986) and to the south (e.g., Kulling, 1942, 1982; Röshoff, 1978, Gee and Zachrisson, 1979; Greiling, 1985). Absence of basement slices in the Middle Allochthon of the Kvikkjokk area points to a relatively high stratigraphic position of the Middle Allochthon floor thrust, i.e. the floor thrust did not cut down into the basement, perhaps owing to the development of a relatively deep sedimentary basin as reflected in the KSTU metasediments. Distinct differences in the structural and metamorphic evolution between KC and KSTU imply that these two units originally occupied slightly different palaeogeographic/tectonic positions prior to and during orogenic shortening and, perhaps, represent a wider area (? a wider continental margin)

than the Middle Allochthon elsewhere in the Swedish Caledonides.

The Middle Allochthon floor thrust is virtually planar in map scale, dipping about 2–5° towards west or northwest. However, imbrication in the underlying Lower Allochthon implies a passive roof deformation of the Middle Allochthon floor. Apparent absence of such a deformation may imply the Middle Allochthon floor thrust to be out-of-sequence. However, such implications will have to be verified by further studies. Recognition of imbrications in the sub-Middle Allochthon sedimentary sequence also requires a revision of the stratigraphy of the Torneträsk Formation as reported by Kulling (1982) from the Kvikkjokk area.

Acknowledgements

Field work was carried out during the summers of 1984 and 1985 in cooperation with the Swedish Geological Survey (S.G.U.) as part of the "Nordkalott" project. We thank S.G.U. for providing field expenses and equipment and sample preparation, in particular E. Zachrisson (Uppsala) for suggesting the Kvikkjokk area for our study and for his support and stimulating discussions. During 1984 E. Ferrow (Uppsala) took part in the field work and his results are included in Fig. 1. R.O.G. prepared his part of the paper when visiting the Department of Geology, University College Cardiff as a Heisenberg research fellow of the German Research Foundation (D.F.G.). He thanks R. A. Gayer (Cardiff) for his patient help and fruitful discussions and, last but not least, for organizing the stimulating Scandinavian Caledonide meeting, at which this paper was presented. R. A. Gayer, J. P. Nystuen (Ås), E. Zachrisson (Uppsala) and an anonymous referee suggested valuable improvements.

REFERENCES

Björklund, L. (1985). The Middle and Lower Allochthons in the Akkajaure–Tysfjord area, northern Scandinavian Caledonides. In: Gee, D. G. and Sturt, B. A. (eds.) *The Caledonide Orogen—Scandinavia and Related Areas*, pp. 515–528. Wiley, Chichester.

Dzulynski, S. and Walton, E. K. (1965). *Sedimentary Features of Flysch and Greywackes.* Elsevier. Amsterdam.

Føyn, S. (1985). The Late Precambrian of northern Scandinavia. In: Gee, D. G. and Sturt, B. A. (eds.) *The Caledonide Orogen—Scandinavia and Related Areas*, pp. 233–245. Wiley, Chichester.

Gavelin, A. (1915). Om den geologiska byggnaden inom Ruoutevare området i Kvikkjokk. *Geol. Fören. Stockh. Förh.* **37** (1), 17–26.

Gee, D. G. and Zachrisson, E. (1979). The Caledonides in Sweden. *Sver. Geol. Unders.* C **769**, 1–48.

Gee, D. G., Kumpulainen, R., Roberts, D., Stephens, M. B., Thon, A. and Zachrisson, E. (1985). Scandinavian Caledonides—Tectonostratigraphic map. *Sver. Geol. Unders.* **Ba35**.

Gorbatschev, R. (1980). The Precambrian development of southern Sweden. *Geol. Fören. Stockh. Förh.* **102**, 129–136.

Greiling, R. (1985). Strukturelle und metamorphe Entwicklung an der Basis grosser, weittransportierter Deckeneinheiten am Beispiel des Mittleren Allochthons in den zentralen Skandinavischen Kaledoniden (Stalon-Deckenkomplex in Västerbotten, Schweden). *Geotekt. Forsch.* **69**, 1–129.

Hamberg, A. (1910). Gesteine und Tektonik des Sarekgebirges nebst einem Überblick der Skandinavischen Gebirgskette. *Geol. Fören. Stockh. Förh.* **32**(4), 681–724.

Hossack, J. R. (1985). The role of thrusting in the Scandinavian Caledonides. In: Gayer, R. A. (ed.) *The Tectonic Evolution of the Caledonide-Appalachian Orogen*, pp. 97–116. Earth Evolution Sciences (Vieweg), Braunschweig.

Kulling, O. (1942). Grunddragen av fjällkedjerandens bergbyggnad inom Västerbottens län. *Sver. Geol. Unders.* C **445**, 1–320.

Kulling, O. (1982). Översikt över södra Norbottensfjällens kaledonberggrund. *Sver. Geol. Unders.* **Ba26**, 1–295.

Kumpulainen, R. and Nystuen, J. P. (1985). Late Proterozoic basin evolution and sedimentation in the westernmost part of Baltoscandia. In: Gee, D. G. and Sturt, B. A. (eds.) *The Caledonide Orogen—Scandinavia and Related Areas*, pp. 213–232. Wiley, Chichester.

Martna, J. and Hansen, L. (1986). Vietas hydroelectric power station headrace tunnel 3, documentation of geology and excavation, final report, July 1986. Vattenfall, Swedish State Power Board, Stockholm, 62 pp.

Pettijohn, F. J., Potter, P. E. and Siever, R. (1972). *Sand and Sandstones*. Springer-Verlag, Heidelberg.

Reading, H. G. (ed.) (1985). *Sedimentary Environment and Facies*. Blackwell, Oxford.

Röshoff, K. (1978). Structures of the Tännäs Augen Gneiss Nappe and its relation to under- and overlying units in the central Swedish Caledonides. *Sver. Geol. Unders.* C **739**, 1–35.

Siedlecka, A. (1985). Development of the Upper Proterozoic sedimentary basins of the Varanger Peninsula, East Finnmark, North Norway. *Geol. Surv. Finland Bull.* **331**, 175–185.

Stow, D. A. V., Howell, D. G. and Nelson, C. H. (1985). Sedimentary, Tectonic and Sea-Level Controls. In: Bouma, A. H., Normark, W. R. and Barnes, N. E. (eds.) *Submarine Fans and Related Turbidite Systems*, pp. 15–22. Springer-Verlag. Heidelberg.

Thelander, T. (1982). The Torneträsk Formation of the Dividal Group, northern Swedish Caledonides. *Sver. Geol. Unders.* C **789**, 1–41.

Trouw, R. A. J. (1973). Structural geology of the Marsfjällen area, Caledonides of Västerbotten, Sweden. *Sver. Geol. Unders.* C **689**, 1–115.

Zachrisson, E. (1964). The Remdalen syncline. *Sver. Geol. Unders.* C **596**, 1–53.

Zachrisson, E. (1969). Caledonian geology of northern Jämtland—southern Västerbotten. *Sver. Geol. Unders.* C **644**, 1–33.

Zachrisson, E. (ed.) (1981). The western Baltic shield within Sweden. In: *Tectonics of Europe and Adjacent Areas, Cratons, Baikalides, Caledonides, Explanatory note to the International Tectonic Map of Europe and Adjacent Areas.* pp. 72–75. Nauka Publishing, Moscow.

Zachrisson, E. and Stephens, M. B. (1984). Mega-structures within the Seve Nappes, southern Norrbotten Caledonides, Sweden. *Meddel. Stockh. Univ. Geol. Inst.* **255**, 241.

8 Age Relationships between Normal and Thrust Faults near the Caledonian Front at the Vietas Hydropower Station, Northern Sweden

Lars Hansen,

Trias Ltd., P.O. Box 16058, S-75016 Uppsala, Sweden.

The Headrace Tunnels 2 and 3 at the Vietas hydropower station, Lapland, Sweden penetrate the middle and lower nappes of the Swedish Caledonides, autochthonous sedimentary Cambrian rocks as well as the Precambrian basement. The latter two rock groups are cut by several faults, which altogether have given the boundary between the Precambrian basement and its cover a vertical displacement of more than 100 m. A detailed geological mapping of the two headrace tunnels and a number of core-loggings have resulted in a longitudinal section, showing that most of the displacement along these fault planes seems to have taken place after the Cambrian, but prior to the Caledonian thrusting.

INTRODUCTION

The Vietas Hydropower Station is situated on the River Lule Älv in the mountains of Swedish Lapland, more than 100 km north of the Arctic Circle. Two of the three headrace tunnels (Tunnels 2 and 3) are roughly parallel, located about 80 m apart and partly at different levels. Tunnel 3 is 16 m high and 15 m wide and Tunnel 2 has approximately the same dimensions. Figure 1 shows the site location and the main geological units. The two tunnels penetrate the Middle and Lower Allochthons (Gee and Zachrisson, 1979), the autochthonous sedimentary rocks as well as the Precambrian basement. Access to the works on Tunnel 3 was obtained through three adits of which the one in the southeast is the most interesting in this context. The Excavation as well as a detailed geological mapping of Headrace Tunnel 3 has been documented in a report published by the Swedish State Power Board (Martna and Hansen, 1986c). In the Swedish version of the report (Martna and Hansen 1986b), a geological map of Headrace Tunnel 2 is also presented. The influence of geology on alignment, excavation and shape of Tunnel 3 has been dealt with in two separate papers (Martna and Hansen 1986a, 1988). The aim of this paper is to present some of the data that led to the geological interpretations presented in the reports.

GENERAL GEOLOGY

The general geology of the area has been treated by Kautsky (1953), Kulling (1964, 1982), Thelander (1982) and Björklund (1985, 1987). Figure 2 shows the mapping of the boreholes and the downstream parts of the tunnels as well as the author's interpretation of the main tectonic and stratigraphic boundaries, based upon mapping of Tunnel 3, Tunnel 2 core logging, and a surface map by Björklund (1987). The rocks encountered in the tunnel can be divided into three main units: (1) Precambrian basement rocks, (2) Palaeozoic sedimentary rocks and (3) Caledonian thrust rocks.

Most of the downstream parts of Tunnels 2 and 3 lie in the Precambrian basement, which consists mainly of red and grey meta-arenites and red metavolcanics belonging to the Snavva–Sjöfall Suite with subordinate sills of greenstone. This rock complex has been subject to folding and subsequent erosion during the Precambrian (Ödman, 1957). In the tunnels, the bedding strikes approximately north to south and dips 55–70° to the west (Fig. 3). Eastwards, at the powerhouse, the bedding dips more gently, 20–30° W and anticlines are found further eastwards.

The Hyolithus Series (or the Torneträsk Formation) was deposited during the early Cambrian upon the

The Caledonide Geology of Scandinavia © R. A. Gayer (Graham & Trotman), 1989, pp. 91–100.

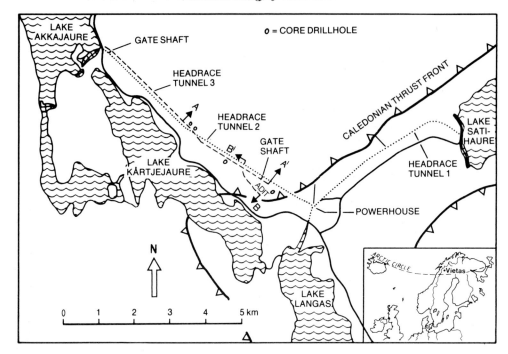

Fig. 1. Map of the Vietas area with main geological units (after Björklund, 1987), and locations of the headrace tunnels and survey boreholes referred to.

Precambrian basement, (Kulling, 1964, 1983; Thelander, 1982). It is represented by a conglomerate with pebbles of Sjöfall rocks, and a sandstone, which are found discordantly bedded upon the steeply westward-dipping Precambrian crystalline Sjöfall Suite (Martna and Hansen 1986c, pp. 26–27). The thickness of the conglomerate and sandstone is overall approximately the same (0.1–0.3 m and 0.5–1 m, respectively). Upon these beds a greenish-grey siltstone is located (Fig. 4). This is probably equivalent to the Upper Siltstone Member (Thelander, 1982). The major part of the beds contains massive siltstone, but upwards it grades into a shale. The thickness of the siltstone/shale member varies in the tunnels and in the survey boreholes between 4 and 40 m. The original horizontal bedding of these rocks is undisturbed except for drag structures close to faults. The upper parts of the shaly feature a diffuse slate cleavage dipping about 30–40° towards the west-northwest.

The Caledonian thrust rocks in the tunnels are represented by the Lower and Middle Allochthons. The Lower Allochthon is separated from the Autochthonous rocks by the Basal Sole Thrust (Björklund, 1985). The thrust is often recognized by imbricate phyllites thrust upon black shale, the thickness of which usually is 1 or 2 m (Fig. 5). On the site, the thickness of the Lower Allochthon varies between 4 and 30 m. It is mainly built up of a grey phyllite and a black, graphite-bearing phyllite. (Phyllite is here used as a field term.) The latter is often crushed and contains loosely embedded layers and lenses of quartzite. These phyllites, together with quartzite layers form duplexes in the tunnels; the roof and floor thrusts are usually almost horizontal, and the imbricate thrusts dip about 30–40° (Fig. 6) northwest.

The rocks of the Middle Allochthon are separated from the Lower Allochthon by the Akkajaure Thrust Fault (Fig. 6). Directly below the Akkajaure Thrust, a thin (0.1 m) graphitic gouge is found. Below the gouge, the imbricate lenses of quartzite and phyllites of the Lower Allochthon are seen. Above the thrust, there is a sheared granitoid.

The Middle Allochthon is divided into six thrust sheets consisting of granitoid gneisses and mylonites with thin sheets of deformed metasediments upon each gneiss unit (Björklund, 1985). In the tunnels, the two lower granitoid sheets are found with metasediments between them. The metasediments are represented by quartz mylonite, quartz phyllite, phyllite and graphitic phyllite. The quartz mylonite is usually horizontally foliated, while the phyllites usually dip about 30° NW. Owing to thrust tectonics, they often thicken into duplexes. The roof and floor thrusts are almost horizontal and the imbricate thrusts dip about 30° NW. There is a repeated geometry of the duplexes. The imbricate lenses within a duplex are built up by smaller duplexes, and so on (Martna and Hansen 1986c, pp. 26–30). At one location, in Tunnel 3, 15–20° forward-dipping thrusts have been found (Martna and Hansen 1986c, App. A).

In the downstream part of the tunnels, several faults are found, which strike north–south and dip 40°–90° westwards and they are usually approximately parallel to the steeply dipping beds of the Sjöfall Suite. The faults displace the boundary between the Precambrian basement and the Cambrian sedimentary rocks altogether for more than 100 m in the vertical direction, the downthrow direction usually being westwards. No horizontal displacement has been observed along the fault planes. Striations have been observed on some of the fault planes within the Sjöfall Suite. They plunge parallel to the dip of the fault planes, i.e. about 60° westwards.

These faults are characteristic for those parts of the tunnels that lie in the Precambrian basement and the Cambrian sediments. No such faults have been recorded in the thrust rocks either during the mapping of the two tunnels or during surface mapping (Björklund, personal communication), although there are a number of horizontal rock boundaries (among them thrusts) that could serve as markers of displacement.

The faults are associated with drag-structures in the siltstone beds and occur as zones of fractured or crushed

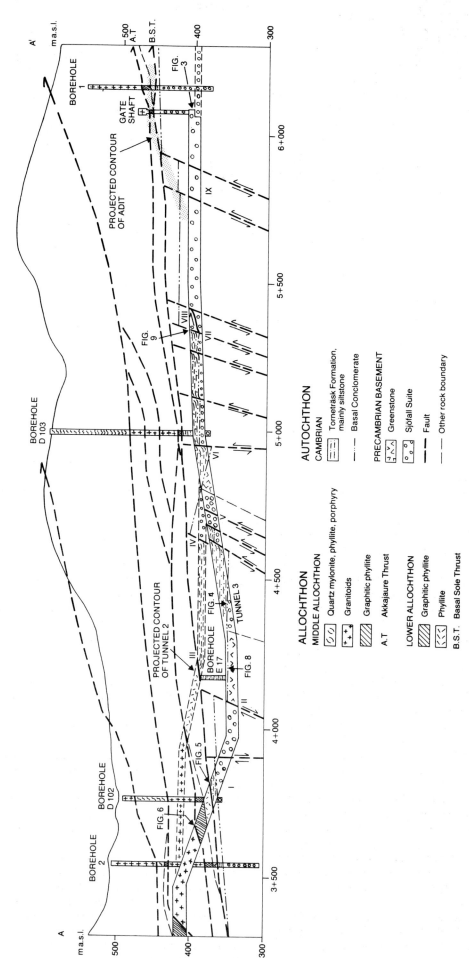

Fig. 2. Geological section of the downstream parts of Headrace Tunnels 2 and 3 between Chainage 3+300 (northwest) and 6+300 (southeast). Vertical exaggeration is 2.5:1. Surface mapping by Björklund (1987). Tunnel 2 mapped by M. Moberg and J. Martna (Martna and Hansen, 1986b) and Tunnel 3 by the author. Boreholes mapped by Moberg (1986) and the author.

Fig. 3. Mud-cracks in beds of the Sjöfall Suite dipping about 70° towards the west (the photograph shows the apparent dip on a tunnel wall striking NE–SW).

Fig. 4. Undisturbed, horizontal beds of Cambrian basal conglomerate, sandstone and siltstone resting upon Precambrian greenstone.

rock within the Precambrian Sjöfall Suite. The fracture zones are sometimes associated with mylonites within the Sjöfall Suite. A number of the fracture zones are associated with breccia. Others feature steeply plunging slickensides or are filled with clay and rock fragments, the latter indicating that some of the faults have subsequently been reactivated.

DESCRIPTION OF LOCATIONS

Figure 7 shows the mapping of the Vietas Adit (mapped by the author) and borehole D 101 (mapped by M. Moberg, Swedish State Power Board). The Akkajaure Thrust is easily recognized near the opening of the adit. The graphitic phyllite below the thrust was unfortunately partly covered by shotcrete before mapping. At the end of shotcrete, another subhorizontal tectonic boundary showing strong deformation was found. Further down in the adit, it passed into the roof of the adit. Below that thrust no more major horizontal tectonic boundaries are found and it has been interpreted as being the Basal Sole Thrust. The diffuse slaty cleavage observed in the shale below the thrust may be due to the thrusting.

The levels of the basal conglomerate correlate well between the borehole and the mapping of the gate shaft as well as with observations along the access road. At the bottom of the adit, however, the conglomerate is shifted about 45 m in the vertical direction in relation to the observation in the borehole. This fact indicates the presence of at least one fault with a downthrow to the west, located between borehole D 101 and Chainage 1200.

At about Chainage 1020 in the adit, drag structures and

Fig. 5. The Basal Sole Thrust viewed towards the southeast in Tunnel 3 at Chainage 3 + 780. Above: grey phyllite with several subhorizontal thrust surfaces causing overbreaks in the tunnel roof. Below: black shale or slate, below which the grey siltstone is located.

Fig. 6. The Akkajaure Thrust as it appears in Tunnel 3 at Chainage 3 + 650. Above: granitic mylonite. Below: imbricated phyllites and quartzite of the Lower Allochthon. The imbricates dip towards the northwest.

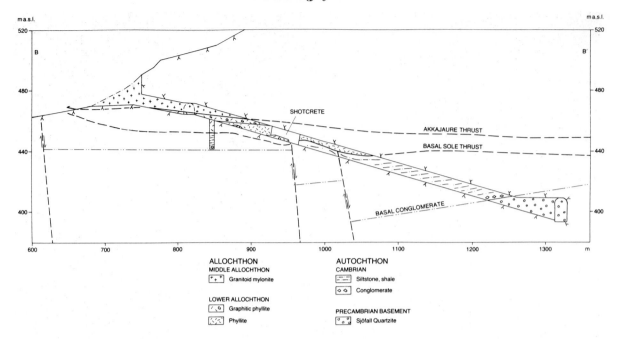

Fig. 7. Author's geological mapping of the Vietas Adit, Chainage 600 (southeast)–1350 (northwest) compiled with borehole D 101 (mapped by M. Moberg). Note the different levels of the Cambrian basal conglomerate in the borehole and at Chainage 1230. Vertical exaggeration is 2:1.

intense fracturing in the shale have been recorded. The structures do not pass into the rock above the flat-lying thrust at this location. They resemble the dragged siltstone beds observed in Tunnel 3 (Martna and Hansen, 1986c, p.29). Some north–south striking fractures at about Chainage 1100 may also represent a fault.

The fault is interpreted to be older than the Akkajaure Thrust because, if younger, the Akkajaure Thrust also would have been displaced about 45 m. As no fault-like structure and no dislocation of the Basal Sole Thrust have been observed in the tunnels, the only possible location for such a fault would be in the unmapped, shotcreted area around Chainage 950. However, a fault there would necessarily have been detected, as it would have placed granitoid mylonite in the adit within Chainage 950–1100. The complete absence of granitoid mylonite within this section indicates that the fault does not cut the Akkajaure Thrust.

Figure 2 shows the levels of the Basal Conglomerate, the Basal Sole Thrust and the Akkajaure Thrust in Tunnels 2 and 3, the adit, the gate shaft and in the drillcores, and also the westward-dipping normal faults. In the following, roman numerals refer to localities on Fig. 2. The Cambrian basal conglomerate is passed by Tunnel 3 at Chainage 3 + 775 − 3 + 865. The contact with the steeply westward-dipping Precambrian Sjöfall Quartzite is weak and some movement may have taken place along the contact. At Chainage 3 + 865 in Tunnel 3 (I), the conglomerate disappears into the tunnel roof, and at Chainage 4 + 140 (II) it reappears at about 35 m below its extrapolated location, which is based upon observations in Tunnel 3 and borehole 2.

The steeply westward-dipping boundary between Sjöfall Quartzite and greenstone at Chainage 4 + 110 contains a chloritic fault gouge, and there is probably a steep reverse fault there. Further evidence of a fault is the difference in thickness of the Cambrian siltstone/shale. In Boreholes 2 and D 102 and in Tunnel 3, Ch 3 + 800, the thickness is about 4 m and in Borehole E 17 drilled from the top heading of tunnel 2, it is 39 m.

The contact between the conglomerate and the underlying Precambrian greenstone is strongly foliated and contains a fault gouge (Fig. 8). This feature has also been observed further north by Kulling (1964, Fig. 26–28). As borehole E 17 almost entirely lies in siltstone, it is most likely that the steep reverse fault is younger than the flat-lying fault in the boundary between the Precambrian and the Cambrian. At least, no major horizontal movement towards the east could have taken place after the formation of the steep fault.

Chainage 4 + 100–4 + 200 of Tunnel 2 lies in phyllites with a foliation dipping about 40° towards the west. Borehole E 17, drilled from the top heading of Tunnel 2, passes through a 1 m thick layer of graphitic phyllite down into horizontally bedded shale and siltstone and ends with a conglomerate containing fragments of Sjöfall rocks. Tunnel 2 cuts the Basal Sole thrust at about Chainage 4 + 250 (III) and downstream of that chainage it is located in horizontally bedded siltstone until the Sjöfall Suite is encountered.

At about Chainage 4 + 600 (IV, still in Tunnel 2), the bedding of the siltstone is bent. This is due to a westward-dipping normal fault that has also been mapped in Tunnel 3, at Chainage 4 + 700 (V). Actually, there are two more faults in Tunnel 3, at Chainages 4 + 600 and 650 which have not been recorded in Tunnel 2. This is possibly due to more time being spent on the mapping of Tunnel 3 than of the corresponding part of Tunnel 2, where only the more obvious faults that form rock boundaries were recorded. It is also possible that the faults anastomose. The vertical displacement of these faults is about 50 m. The downthrow direction is towards the west. The faults are filled with fault gouge. They are interpreted as being older than the Akkajaure Thrust because, if they were younger, they would probably (as would the previously mentioned fault in the Vietas Adit) have displaced this undisturbed thrust about 50 m, resulting in appearance of granitoids in Tunnel 2 around Chainage 4 + 500. No such granitoids have been found there.

In Tunnel 3, Chainage 4 + 950–5 + 350, a number of

Fig. 8.   Cambrian basal conglomerate discordantly bedded upon Precambrian Greenstone. Close to the rock contact, the greenstone is deformed and the rock contact is filled with a 1–2 cm wide chloritic falt gouge.

normal faults are well exposed. Most of them have also been recorded in Tunnel 2 and their orientations correlate well between the tunnels (Martna and Hansen, 1986d, Fig. 6). The basal conglomerate has been observed at several locations and on occasion it has been seen on both sides of a fault, but at different levels. Contrary to the observations at I and II, the boundary between the conglomerate and the Precambrian Sjöfall rocks does not show any sign of having been sheared in these parts of the tunnel.

Particularly important is the fault at Chainage 4 + 950 (VI), which has also been observed in Tunnel 2. It is approximately vertical and the downthrow is about 15 m westwards. Seventeen metres above the tunnel roof, the black phyllite of the Lower Allochthon has been found in Borehole D 103, and the Akkajaure Thrust is located about 20 m above the tunnel. This results in a 90° angle between the thrusts and the steep fault.

Two faults are situated at chainages 5 + 325 and 5 + 400 of which the former is the most spectacular (VII, Fig. 9). Owing to that fault, the tunnel abruptly passes from Cambrian siltstone into metavolcanics belonging to the Precambrian Sjöfall Suite. On the downthrow (west) side of the fault, the basal conglomerate was visible in the lower part of the tunnel. The fault dips 60° westwards and is filled with a 2–5 cm wide fault gouge consisting of silt-, sand- and gravel-sized rock fragments. The rock on both sides of the fault is heavily fractured, and the horizontal bedding of the siltstone is bent upwards towards the fault. This fault also has another interesting character. Another, horizontal fault cuts the steeply dipping beds of the Sjöfall Suite and displaces it about 4 m (VIII). This fault is again bent by the steeply dipping faults (Martna and Hansen, 1988, Fig. 12).

At Chainage 5 + 775, a steeply westward-dipping fault is found (IX). It correlates with the above-mentioned

fractured and bent siltstone in the adit. The fault consists of a half-metre-thick zone of heavily fractured rock with fault gouge and slickensided surfaces. The slickenside striations plunge along the dip of the fault. Adjacent to the fault and parallel to it, there is a 3 m wide breccia of red Sjöfall Quartzite with a matrix of quartz, indicating that the fault has been active at least once before. Also in Tunnel 2, breccias and slickensides plunging steeply westwards have been recorded (Martna, 1970, 1971). Faults with fault gouge are also found at Chainages 5 + 820 and 5 + 850, which may explain the downthrow of the conglomerate observed in the adit and in the gate shaft.

DISCUSSION AND CONCLUSIONS

The faults met with in the downstream parts of Tunnels 2 and 3 have been extrapolated regarding their strikes and dips. This interpretation is shown in Fig. 10. The faults anastomose and the resulting structure resembles an extensional strike-slip duplex (Woodcock and Fischer, 1986).

The westward-dipping, normal faults are often associated with high rock stresses that have caused rock bursting in Tunnel 2 (Martna, 1970, 1971, 1972) as well as in Tunnel 3. Rock stress measurements in the Vietas tunnels have shown magnitudes of up to 55 MPa. Such a high stress cannot have its origin in the overburden alone. The high stresses associated with the faults indicate that there are still active forces in the fault zones. The occurrence of breccias in some of the faults and of fault gouge in others indicates activation of the fault zone at least twice. The stress measurements, their interpretation and their relationships to the faults have been further dealt with in other papers (Martna and Hansen, 1986d, 1988).

Lundqvist and Lagerbäck (1976) have described the

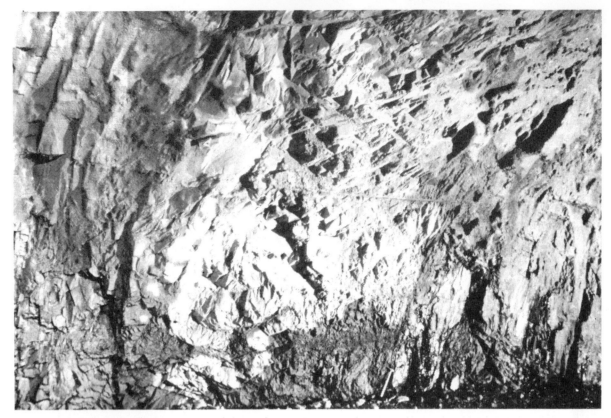

Fig. 9.   Fault (indicated by hammer) dipping about 60° towards the west, containing a 1–5 cm wide fault gouge. To the right (east) of the fault, Precambrian rocks of the Sjöfall Suite are found. Their beds dip 60–70° westwards. To the left (west) horizontal beds of siltstone are located. In the middle of the photograph, the siltstone beds are bent upwards towards the fault.

post-glacial Pärvie Fault system, part of which is located about 20 km east of Vietas. Extrapolation southwards of the main directions of the two fault zones would result in anastomosis somewhere in the area of Lakes Sitojaure and Tjaktjajaure, some tens of kilometres farther south. Talbot (1986) suggests a history of repeated reactivation of an already established fault zone. The recorded signs of reactivation of some of the faults in the Vietas headrace tunnels suggest that the two fault zones may thus be parts of the same, older fault system, but reactivated at different times. The strength of the Caledonian rocks may have prevented the faults beneath the mountain chain from the major post-glacial reactivation, instead favouring the Pärvie fault zone east of the front.

The high stresses in the vicinity of the normal faults and the presence of fault gouge indicate a late, possibly quarternary, movement along the faults. This movement may, however, be very short, which could explain why the allochthonous rocks seem unaffected by the faulting. The graphitic gouges in the Sole and Akkajaure Thrusts may thereby have acted as lubricants. The author's experience from other tunnels is that a movement of only a few metres has been sufficient to produce a fault gouge with a thickness of about 10 cm (unpublished data).

About 160 km farther south-southwest at Lake Laisan, the basement surface is located at different elevations in boreholes (Ljungner, 1950; Kulling, 1982, pp. 32–34). Some of these differences have been interpreted as being caused by steeply westward-dipping thrust faults. With the data from Vietas in mind, the present author would also point out the possibility of some normal faults with a westerly downthrow. It may be of some interest to notice that the locality lies in the extension of the Pärvie fault. Farther south, Kumpulainen and Nystuen (1985) have mentioned block-faulting simultaneously with sediment-ation during Vendian to Cambrian.

The steeply westward-dipping, normal faults observed in the Vietas tunnels cut and displace the Precambrian–Cambrian boundary, which here is identical with the basal conglomerate of the Upper Siltstone member of the Torneträsk Formation (Thelander, 1982). This siltstone has been correlated with the Upper strata of Lower Cambrian (Kulling, 1964, 1982) and consequently, the faults post-date early Cambrian.

The lack of fault observations in the thrust rocks in the area, the westerly downthrow, the absence of Middle Allochthon granitoids in certain parts of the tunnels and the sudden differences in the thickness of the Cambrian siltstone, with a constant thickness of the basal conglo-merate (0.1–0.3 m) and sandstone (0.5–1 m) indicate that the steeply westward-dipping faults encountered in the Vietas tunnels formed before the major Caledonian thrusts.

The tectonic style of the Lower and Middle Allochthons with repetition of duplexes of various scales has not been found in the siltstone and the Sjöfall rocks. The transition occurs abruptly when the tunnel passes the Basal Sole Thrust. The allochthonous rocks are characterized by anastomosing thrusts with low angles between them. The steeply dipping normal faults with their westerly down-throw do not fit into these thrust duplexes. This is particularly clear for the vertical fault at VI, which has a space of only 17 m to turn the 90° which would be necessary to anastomose with the Basal Sole Thrust.

The steeply dipping reverse fault at Chainage 4 + 110, cutting the horizontal tectonic contact at the basal con-glomerate, may or may not be due to Caledonian reactiv-ation of an earlier normal fault. The same uncertainty also applies to the deformation along the basal conglomerate at I and II and the thrust within the Precambrian rocks (IX). The latter shows only a few metres displacement, which could as well be of Precambrian origin. The displacement

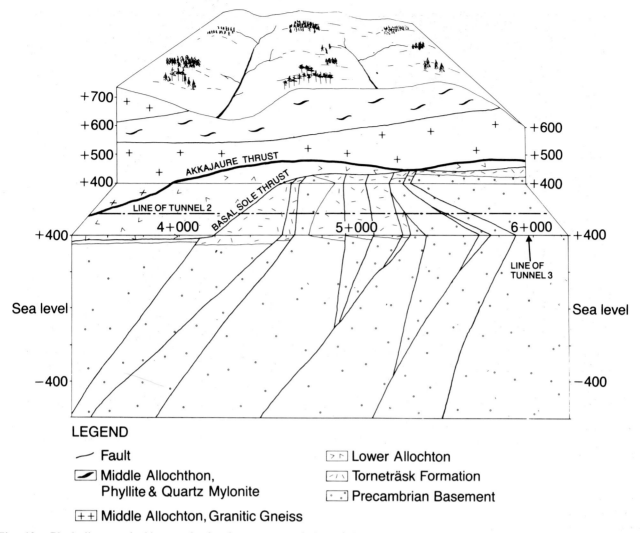

Fig. 10. Block diagram, looking north, showing an extrapolation of the orientations of the faults as observed in Tunnels 2 and 3. Vertical exaggeration is 2:1.

of the fault along the conglomerate is unknown but it does not pass through borehole E17. This movement may have taken place simultaneously with the normal faulting, partly owing to the lesser strength of the Precambrian greenstone compared with that of the Sjöfall rocks.

The faults constitute a graben structure, which may have protected the thicker parts of the siltstone from being eroded by the advancing thrust rocks. The author's conclusion is that there has probably been a shorter or longer period of extension during the Cambrian or Ordovician before the Caledonian thrusting. Whether this was due to simple extension or caused by a strike-slip fault giving rise to an extensional strike-slip duplex at Vietas is uncertain. Further information is needed in the extension of the Vietas fault zone. A Palaeozoic strike-slip fault would possibly also give rise to compressional strike-slip duplexes.

Acknowledgments

The geological mapping of the Vietas tunnels has been financed by the Swedish State Power Board. With Jüri Martna (Swedish State Power Board), David Gee (University of Lund), Ebbe Zachrisson (S.G.U.) and Chris Talbot (Univ. of Uppsala), I have had fruitful discussions about the interpretations of the structures and the rock stresses. Special thanks are due to Reinhard Greiling, Robin Nicholson and Lennart Björklund, who critically read the draft of the manuscript, suggested valuable improvements and corrected my English. The figures have been drawn by Mrs. A. K. Flodén, Mrs. C. Wernström and Mr. J. Carlsson.

REFERENCES

Björklund, L. (1985). The Middle and Lower Allochthons in the Akkajaure-Tysfjord area, Northern Scandinavian Caledonides. In: Gee, D. and Sturt, B. (eds.) *The Caledonide Orogen—Scandinavia and Related Areas.* Wiley, Chichester.

Björklund, L. (1987). Geology of the Akkajaure-Tysfjord area, (maps and sections) Northern Scandinavian Caledonides. From the author, Dept. of Geology, University of Gothenburg, S-412 96 Göteborg, Sweden.

Gee, D. and Zachrisson, E. (1979). The Caledonides in Sweden. *Sver. Geol. Under. C.* **769**, 1–48.

Kautsky, G. (1953). Der geologische Bau des Sulitelma–Salojauregebietes in den Nordskandinavischen Kaledoniden (Eng. summary: Geology of the Sulitelma-Salojaure area, Northern Scandinavian Caledonides) *Sver. Geol. Unders. C.* **528**, 1–228. Stockholm.

Kulling, O. (1964). Översikt över norra Norrbottensfjällens kaledonberggrund (Summary: The geology of Caledonian rocks of the northern Norrbotten Mountains). *Sver. Geol. Unders.* **Ba 19**, 1–166.

Kulling, O. (1982). Översikt över södra Norrbottensfjällens kaledonberggrund (Summary: The geology of Caledonian rocks of the Southern Norrbotten Mountains). *Sver. Geol. Unders.* **Ba 26**, 1–295.

Kumpulainen, R. and Nystuen, J. P. (1985). Late Proterozoic

basin evolution and sedimentation in the westernmost part of Baltoscandia. In: Gee, D. and Sturt, B. (eds.) *The Caledonide Orogen—Scandinavia and Related Areas*. Wiley, Chichester.

Ljungner, E. (1950). Surface form of the crystalline basement at the eastern margin of the Scandes (in Swedish w. English summary). *Q. Geol. Soc. Sweden* **72**, (3), 269–300.

Lundqvist, J. and Lagerbäck, R. (1976). The Pärve Fault: a late glacial fault in the Precambrian of Swedish Lapland. *Q. Geol. Soc. Sweden*. **98**, 45–51.

Martna, J. (1970). Rock bursting in the Suorva-Vietas headrace tunnel. *1st Int. Congr. Int. Ass. Eng. Geol, Paris*, vol. 2, pp. 1134–1139.

Martna, J. (1971). Geologiska synpunkter på smällberget i Suorva–Vietastunneln (Summary: Geological aspects on rock bursting in the Sourva–Vietas tunnel). IVA-rapport no. 38, pp. 141–152.

Martna, J. (1972). Selective overbreak in the Suorva–Vietas tunnel caused by rock pressure. *Int. Symp. Underground Openings, Luzern*. pp. 141–145.

Martna, J. and Hansen, L. (1986a). The influence of rock structure on the shape and the supports of a large headrace tunnel. *Int. ITA Congr. Large Underground Openings, Firenze*, vol. 2, pp. 260–269.

Martna, J. and Hansen, L. (1986b). Vietas Kraftstation. Till-oppstunnel 3 Dokumentation av geologiska förhållanden och tunnelns drivning, pp. 1–138. Swedish State Power Board, Stockholm.

Martna, J. and Hansen, L. (1986c). Vietas Hydro Power Station, Headrace Tunnel 3. Documentation of geology and excavation, pp. 1–62. Swedish State Power Board, Stockholm.

Martna, J. and Hansen, L. (1986d). Initial rock stresses around the Vietas headrace tunnels Nos. 2 and 3, Sweden. *Proc. Int. Symp Rock Stress, Stockholm*, pp. 605–613.

Martna, J. and Hansen, L. (1988). Tunnelling in a complex rock mass. Experience from Vietas, Swedish Lapland. *Tunnelling and Underground Space Technology*. Pergamon Press, in press.

Moberg, M. (1986). Core drilling and rock forecase. In: Martna and Hansen (1986c), Appendix 2.

Ödman, O. H. (1957). Description to map of the Precambrian rocks of Norrbotten County, N Sweden (in Swedish with English summary). *Sver. Geol. Unders.* **Ca 41**, 1–151.

Talbot, C. (1986). A preliminary structural analysis of the pattern of post-glacial faults in northern Sweden. Tech. Rep. no. 86–20, 1–36. Swedish Nuclear Fuel and Waste Management Co. Stockholm.

Thelander, T. (1982). The Törneträsk Formation of the Dividal Group, northern Swedish Caledonides. *Sver. Geol. Unders.* **C789**, 41 pp.

Woodcock, N. H. and Fischer, M. (1986). Strike-slip duplexes. *J. Struct. Geol.* **8**, (7), 725–735.

9 Basement-cover evolution during Caledonian Orogenesis, Troms, N Norway

M. W. Anderson

University of Wales, Cardiff, P.O. Box 914, Cardiff CF1 3YE, UK

Illite crystallinity combined with mineralogical information has been used to detect grade differences between cover sequences around several windows in Troms, N Norway, and the autochthonous sequence of the foreland. Combined with structural observations, this information has been used to constrain basement motions during Caledonian orogenesis. The large Rombak window occupies an anomalous metamorphic position with respect to other windows, and is therefore considered to be allochthonous. On metamorphic grounds, the window has been restored approximately 45 km to the west-northwest of its present position, whilst the Rautas Complex (Lower Allochthon) restores a minimum of 25 km to the west-northwest, beyond the autochthonous Mauken and Straumslia windows but not the restored Rombak window. Doming of the Caledonian nappe pile above the Rombak, Salangsdal and Spansdal windows indicates that basement imbrication was an end-Caledonian event, and may be related to the development of more regional structures such as the Ofoten synform.

INTRODUCTION

Several large areas of Precambrian crystalline basement, often with a thin veneer of cover sediments, are exposed as windows through the nappes of Nordland and Troms (Vogt, 1941; Wilson and Nicholson, 1973) (Fig. 1). They have been interpreted in the past as autochthonous Baltic shield basement, activated in a dominantly vertical sense during Caledonian nappe emplacement (Ramberg, 1980; Cooper and Bradshaw, 1980). Subsequently, the recognition of Caledonian structural and metamorphic features has led to reinterpretation of some basement areas as being allochthonous (e.g., Thelander et al., 1980; Nicholson, 1985; Hossack and Cooper, 1986).

Samples were collected from the autochthonous cover along the Caledonian front, and from correlated cover sequences around the Rombak, Straumslia and Mauken basement windows (Fig. 2). This study aims (i) to establish metamorphic grade of the cover sequences using illite crystallinity, a technique applied successfully to "low-grade" clastic sequences in many other areas (Frey, 1970; Kisch, 1980; Robinson and Bevins, 1986, Bevins et al., 1986), and (ii) to examine the variation in metamorphic grade across the area and the constraints this places on basement evolution during Caledonian orogenesis.

GEOLOGICAL SETTING OF STUDY AREA

The Caledonian Allochthon and Foreland

In the Luopakte area, south of Lake Torneträsk (Fig. 2), the autochthonous crystalline Baltic Shield is overlain by a succession of weakly deformed sedimentary rocks, the Dividal Group (Lindström et al., 1985). This comprises a thin basal conglomerate passing upwards into inter-bedded red and green shales and sandstone horizons, of Vendian to Lower Cambrian age (Kulling, 1964; Ahlberg, 1980), overlain by a penetratively tectonized black shale analogous to the Middle to Late Cambrian "Alum Shale" (Fig. 4) (Lindström et al., 1985; Kulling, 1960, 1964).

Above the autochthonous rocks of the Caledonian front lie a thick pile of nappes, tectonically emplaced during the late Silurian Scandian event of the Caledonian orogeny (Fig. 3) (Barker, this volume; Andresen et al., 1985; Tull et al., 1985; Stephens et al., 1985). In the Torneträsk area, rocks of the Lower Allochthon ("Rautas Complex" of Kulling (1950, 1960)) are overthrust onto the Dividal Group. This comprises of a lower unit of Precambrian crystalline rocks overlain by a sedimentary succession of quartzites, shales and dolomites (Kulling, 1950). A conglomerate is often present at the contact between the two, and the sequence probably represents an imbricated equivalent of the underlying autochthonous succession (Lindström et al., 1985).

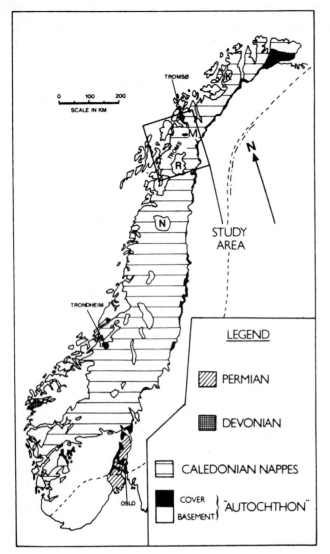

Fig. 1. Simplified geology of Scandinavia, showing location of the study area (Fig. 2) and the central chain of basement windows, in particular the Nasafjell window (N), Rombak window (R), Mauken window (M) and Komagfjord window (K). Also shown are the positions of the pre-erosion thrust front, as estimated by Hossack and Cooper (1986) (dashed) and this study (dashed and dotted).

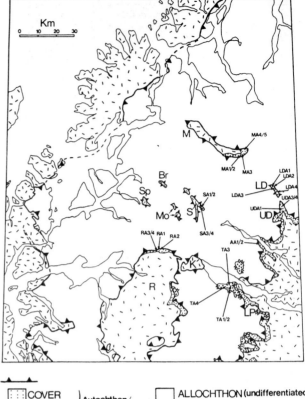

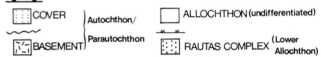

Fig. 2. Location of samples from the various cover sequences exposed in the study area. The various basement windows shown are: Rombak (R), Kuokkel (K), Straumslia (S), Mauken (M), Moholt (Mo), Brandvoll (Br) and Spansdalen (Sp). Other localities shown are: Lower Dividalen (LD), Upper Dividalen (UD) and Luopakte (LP).

Basement "Windows" West of the Caledonian Front

Crystalline basement rocks are exposed in "tectonic windows" through the Caledonian nappes (Figs. 2 and 3). The Rombak window, which is the largest, is dominated by granitic gneisses, granites, gabbroic rocks and supracrustal sequences, exposed over 250 km^2 (Vogt, 1941; Kulling, 1960; Gustavson, 1966; Tull *et al.*, 1985). Around the northern margin of the window (Figs. 2 and 3), a 0–40 m clastic sequence of conglomerates, cross-bedded sandstones and pelites, correlated with the Dividal Group, rests unconformably on Precambrian crystalline basement (Fig. 4) (Vogt, 1941; Tull *et al.*, 1985). The sediments show an intense subhorizontal foliation, parallelling that in the overlying allochthon (dominated by biotite-grade phyllonites) that is domed over the window. This Caledonian fabric also extends into the underlying granites and greenstones of the crystalline basement, contrasting markedly with the steep Precambrian foliation seen in the basement away from the contact.

The crystalline rocks of the Straumslia window (Fig. 2) resemble the granite gneisses and greenstones of the

Rombak Window, and also display a steeply dipping foliation. Above the basement rocks, a 2 m sequence of cleaved, green-black shales with thin siltstone horizons (<1.5 cm thick) is exposed in roadcuts southwest of Straumsmoen. The foliation dips shallowly westwards, subparallel to that in the overlying allochthon. Elsewhere, Gustavson (1966) described a sequence of "white and grey massive sandstones, possibly belonging to the autochthonous Dividal Group sequence", between the basement and overlying allochthon, which was considered to belong to the same cover sequence as the green-black shales (Fig. 4).

The Mauken window, north-northeast of Straumslia (Fig. 2), is dominated by a NW–SE-trending zone of amphibolites, intruded by granodiorite (Landmark, 1967). Along the southeastern margin of the window, an approximately 20 m thick cover sequence of conglomerates, sandstones and green-black shales with thin siltstones, similar to those at Straumslia, overlies the basement (Fig. 4). This has been correlated with the Dividal Group (Gustavson, 1966; Landmark, 1967; Andresen *et al.*, 1985). Northwards, the sequence becomes attenuated beneath, and foliated subparallel to, the subhorizontal mylonites of the overlying allochthon. Unlike the Rombak Window, the allochthon here, and at Straumslia, shows no marked doming over the basement. Indeed, Gustavson (1963, 1972) demonstrated that "updoming of Precambrian rocks and its autochthonous Hyolithus Zone cover predates movement along the basal thrust plane" in the Mauken and Dividalen areas.

Another series of basement windows, including the

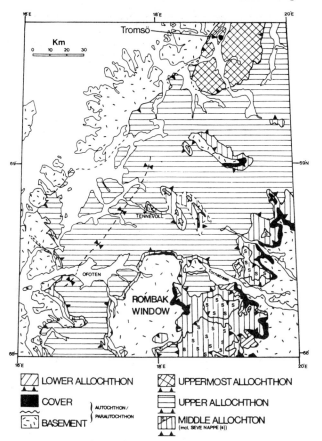

Fig. 3. Generalised Precambrian and Caledonian tectonostratigraphy of the study area, showing numerous basement windows to the north and northeast of the Rombak window, and the trace of the regional Ofoten synform.

Brandvoll and Moholt windows, lies in the north–south trending Salangselv valley (Fig. 2). This is dominated by gabbroic-diorite, augen gneisses, granitic gneisses, amphibolites and various phases of granitic pegmatites, all having a steep, westerly dipping foliation. Towards the contact with the overlying allochthon, the steep pre-Caledonian foliation in the basement has been transposed into a mylonitic foliation subparallel to that in the allochthon, generally dipping gently outwards from the windows (Lindahl *et al.*, 1985; Barker, 1986).

Approximately 10 km west of Moholt, previously unrecorded basement has been found along Spansdalen (Fig. 2). Lithological and structural similarities with other areas indicate that the granite gneisses, augen gneisses, amphibolites and metasediments of the Spansdal window are of pre-Caledonian age. The main structural feature of the basement is a steep, often vertical, foliation that progressively shallows into concordance with the Caledonian foliation in the allochthon above. The overall geometry is one of the basement duplex, with mylonites of the allochthon (in the roof zone) being gently domed above the imbricate stack. Detailed relationships between basement imbrication and evolution of the Caledonian nappe pile in this area will be presented in a future paper.

PEAK METAMORPHISM OF BALTIC COVER SEQUENCES

Little work has been presented on the metamorphic grade of the autochthonous Dividal Group or correlatives around basement windows. Kulling (1964) and Lindström *et al.* (1985) qualitatively describe the essentially unmetamorphosed Dividal Group, and equivalents in the Lower and Middle Allochthons (metamorphosed under chlorite-

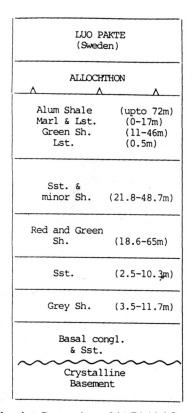

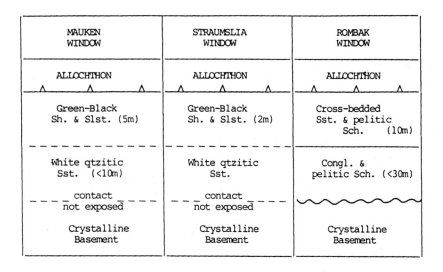

Fig. 4. Comparison of the Dividal Group sequences seen at the Caledonian front (after Petersen, 1878; Gustavson, 1963; Kulling, 1964; Foyn, 1967; Ahlberg, 1980; Lindström *et al.*, 1985) and the generalized sequences around the Mauken, Straumslia and Rombak windows (after Gustavson, 1966; Andresen *et al.*, 1985; Tull *et al.*, 1985).
Abbreviations: Lst. = limestone; Sh. = shale; Sch. = schist; Slst. = siltstone; Sst. = sandstone; Congl. = Conglomerate. --- = uncertain contact; —— = observed contact; ⊿⊿ = tectonic contact; ∿∿ = unconformity.

sericite facies conditions), around Torneträsk. Andresen *et al.* (1985) briefly discuss the biotite-grade cover sequence exposed on the northern edge of the Rombak window. Gustavson (1966) and Landmark (1967) refer to the "low-grade, sercite-chlorite schists" around the Mauken window, considering them to be a "metamorphosed variant" of the Dividal Group shales.

Timing and nature of metamorphism

The sequences studied are believed to range in age from late Precambrian to mid-Cambrian. Therefore, any metamorphism and deformation they record, occurred during Caledonian orogenesis. Dallmeyer and Gee (1986) outline a model for the tectonothermal evolution of the Baltic margin during the Caledonian Orogeny. They suggest that by late Cambrian times, the Dividal Group had developed as a foreland shelf sequence (see also Føyn, 1985), away from synchronous (subduction-related), outboard orogenic activity. Metamorphism recorded in the Dividal Group relates to burial beneath a Scandian nappe pile (containing older Caledonian elements) during accretion onto the Baltic margin, in the mid to end Silurian. Dallmeyer *et al.* (1987) support this model with $^{40}Ar/^{39}Ar$ metamorphic muscovite mineral ages of ca. 420–425 Ma from Dividal Group equivalents in Finnmark. Accepting their model, the possibility of polyorogenic metamorphism in the cover sediments presently being studied seems unlikely, and the results presented below can be interpreted in terms of a late Caledonian (Scandian) metamorphic evolution.

Illite crystallinity

The technique relies on the lattice reorganization that occurs in illite during prograde metamorphism (Kisch, 1983).Quantification of the "sharpness" of the 001 (ca.10 Å) illite peak, an indirect measure of lattice reorganization, is obtained in two ways: (1) Kubler Index— the peak width of the 10 Å peak at half the peak height above background (Kubler, 1967); and (2) Weaver Index—the ratio of the maximum 10 Å peak height minus background against the 10.5 Å peak height minus background (Weaver and Broekstra, 1984). As the 10 Å illite peak progressively narrows with increasing metamorphism, Kubler values decrease and Weaver values increase. Limits of Kubler and Weaver Indices for the very low-grade metamorphic zones have been taken from Robinson and Bevins (1986) (cf. Figs. 5(a) and (b)). Details of sample preparation, analytical techniques used and operating conditions are shown in Appendix A.

RESULTS

Autochthon

Thirteen samples were analysed from the Dividal Group sequences exposed at Dividalen and Altevatn (Fig. 2). Illite crystallinity values indicative of lower to middle anchizone are recorded (Table I and Fig. 5(a)), suggesting metamorphic temperatures of 200–250°C (Frey *et al.*, 1980). A markedly higher crystallinity (metamorphic grade) is apparent for the autochthon of this study area when compared with that of E Finnmark, to the north (data from Bevins *et al.*, 1986).

Basement windows

The presence of chlorite within the shale sequences of the Straumslia and Mauken windows prompted several authors to suggest low greenschist facies (=epizone) conditions of metamorphism (Gustavson, 1966; Andresen *et al.*, 1985). However, recent studies on mineralogical characteristics of very low-grade metamorphism, suggest that chlorite can be present as a significant component throughout the anchizone (Reuter, 1987) and in the upper diagenesis grade (Frey, 1986).

Twelve samples were analysed from the Mauken and Straumslia windows (Fig. 2). The results (Table I(b) and Fig. 5(b) indicate slightly higher illite crystallinity compared with the autochthon, and values generally suggest middle to upper anchizone metamorphism. This suggests a temperature of metamorphism for the windows less than the approximate 300°C value indicated for the anchizone/epizone boundary (Kubler, 1967; Robinson and Bevins, 1986). By comparison with the maximum temperatures suggested for the autochthon, a lower limit of approximately 250°C would seem realistic for the cover sediments of the windows.

The mineralogy of the clastic cover sequence around the Rombak window is strikingly different from the correlated sequences elsewhere. There, quartz + plag-feldspar + K-feldspar + biotite + white mica + chlorite is the dominant assemblage. The appearance of biotite along with phengitic muscovite in the metapelite assemblage (often > 30%), indicates biotite grade metamorphism, suggesting temperatures in excess of approximately 425°C (Nitsch, 1970; Winkler, 1979; Wang *et al.*, 1986).

Lower Allochton (Rautas Complex)

Sediments and Precambrian crystalline rocks constituting the Rautas Complex (Lower Allochthon) around Torneträsk, have tentatively been correlated with the autochthonous foreland succession (Lindström *et al.*, 1985). Four samples were collected from this sequence, yielding a mean crystallinity value and spread of data indicative of true epizone metamorphism (Table I(c) and Fig. 5(b), suggesting temperatures above 300°C (anchizone/epizone boundary of Kubler, 1967).

Discussion of regional metamorphic pattern

The Straumslia and Mauken windows have metamorphic grades similar to that of the foreland sequence, suggesting that they are essentially autochthonous (Fig. 5(a), (b)). Results from the Rombak window and the Rautas Complex indicate significantly higher (anomalous) metamorphic grades when compared with adjacent areas. The Rautas Complex, being part of the Lower Allochthon (Kulling, 1950), has been transported east-southeast from deeper levels within the Caledonian belt and, as expected, displays a higher metamorphic grade than equivalent autochthonous sequences that it now overlies (Fig. 3).

The Rombak window lies a maximum of 15 km across strike from the Straumslia and Mauken windows (Fig. 7), yet the metamorphic grade of the Rombak window cover is much higher than the equivalent sequences to the northeast (Fig. 5b). If the Rombak window were autochthonous, a rapid thickness change in the nappe pile would be required to produce such a variation. Assuming a moderate geothermal gradient (25°C per km) during metamorphism and roughly constant lateral wedge thickness, an across-strike increase in wedge thickness of 5 km is required to produce the observed (minimum) grade difference of 125°C. This corresponds to an increase in wedge taper angle of approximately 19°. Critical wedge taper angles (topographical slope + decollement dip) for presently active wedges (Davis *et al.*, 1983), rarely exceed approximately 13°. It seems unlikely, therefore, that a 19°

Table I. Illite crystallinities for various structural settings examined (see Fig. 2). Grid references in (*a*) and (*b*) are taken from Norwegian 1:50000 topographical maps, series M711 (various sheets). Those in (*c*) are taken from the Swedish 1:100000 topographical map series, second edition (sheet Abisko 301).

(a) Autochthonous Dividal Group, at Caledonian Front

Grid reference	Sample no.	Kubler ($\mathring{A}\ 2\theta$)	Weaver Index
(04487 76269)	UDA1A/86	0.27	7.0
(04487 76269)	UDA1B/86	0.32	5.1
(04486 76274)	UDA3B/86	0.26	6.0
(04486 76272)	UDA4A/86	0.28	7.3
(04439 76366)	LDA1/86	0.30	3.2
(04446 76363)	LDA2A/86	0.23	9.5
(04446 76363)	LDA2B/86	0.26	9.5
(04452 76355)	LDA3A/86	0.24	5.6
(04452 76355)	LDA3B/86	0.24	6.4
(04458 76345)	LDA4A/86	0.27	5.4
(04458 76345)	LDA4B/86	0.28	6.1
(04353 76067)	AA1/86	0.14	10
(04352 76066)	AA2/86	0.24	8.3

(b) Cover sediments to Basement "windows"

Grid reference	Sample no.	Kubler ($\mathring{A}\ 2\theta$)	Weaver Index
(04315 76563)	MA1A/86	0.30	6.4
(04315 76563)	MA1B/86	0.28	6.0
(04315 76562)	MA2/86	0.27	7.1
(04313 76564)	MA3A/86	0.22	8.0
(04313 76564)	MA3B/86	0.19	6.7
(04318 76593)	MA4/86	0.28	7.1
(04318 76593)	MA5/86	0.29	4.1
(04050 76274)	SA1A/86	0.19	6.2
(04050 76274)	SA1B/86	0.22	10.6
(04049 76275)	SA2/86	0.17	15.1
(04049 76274)	SA3/86	0.24	8.6
(04048 76273)	SA4/86	0.26	5.6
(06181 76033)	RA1/86	Biotite-grade	
(06182 76033)	RA2/86	Biotite-grade	
(06179 76032)	RA3/87	Biotite-grade	
(06179 76032)	RA4/87	Biotite-grade	

(c) Dividal Group equivalents within Lower Allochthon (Rautas Complex)

Grid Reference	Sample no.	Kubler ($\mathring{A}\ 2\theta$)	Weaver Index
(16328 75878)	TA1/86	0.14	34.1
(16328 75878)	TA2/86	0.17	12.0
(16318 75878)	TA3/86	0.15	10.3
(16307 75876)	TA4/86	0.17	21.4

increase could be maintained for a sufficiently long period to produce the metamorphic grade variation seen. Similarly, considering the rapidity of the grade change (125°C minimum over a maximum lateral distance of 35 km), a lateral (southwesterly) thickening of the nappe pile from the Straumslia window to the Rombak window does not satisfactorily explain the observed pattern. It is more feasible that the Rombak basement is allochthonous with respect to some adjacent basement windows.

MODELS TO ACCOUNT FOR OBSERVED METAMORPHIC PATTERN

(1) Vertical basement motions

Many of the basement domes in the Scandinavian Caledonides have been interpreted as areas of buoyant continental crust that became active during crustal overloading (by emplacement of the denser allochthon: Ramberg, 1980; Cooper and Bradshaw, 1980). The model suggests development of ductile granite–gneiss domes that arch the overlying allochthon and display steep marginal foliations. However, although, the western margin of the Rombak window shows a steep, westerly dipping (pre-Caledonian) foliation, this becomes progressively more mylonitic and rotated into parallelism with the shallow, westerly dipping foliation in the overlying allochthon (Hodges, 1985). Thus the marginal basement foliation at this contact is apparently associated with emplacement of the allochthon rather than late vertical motions in the basement.

Another model involves reactivation of pre-existing extensional faults and basement lineaments as high-angle reverse faults during compressional orogenesis (Winslow, 1981; Bamford and Higgs, in prep.). However, the model of Winslow (1981) again suggests steeply dipping boundaries, which does not seem applicable to the Rombak Window, particularly the western margin. Furthermore, these models still require a major thickening of the nappe pile to produce the observed grade increase, prior to activation of the basement and its cover.

(2) Horizontally dominated motions

Basement slices are well documented within the Lower and Middle Allochthon, but the large basement areas that are exposed as windows through the allochthon remain enigmatic. Recent work in other regions has shown that many of these windows are major duplex culminations, bounded above and beneath by low-angle Caledonian thrusts (Thelander *et al.*, 1980; Nicholson, 1985; Hossack and Cooper, 1986; Gee, 1986; Gayer *et al.*, 1987). Several lines of evidence indicate that this model is applicable to the Rombak Window. Around the margins of the window, a steep pre-Caledonian fabric becomes rotated parallel with the mylonitic foliation of the overlying allochthon (Hodges, 1985; Tull *et al.*, 1985). Along the northern edge, the thin veneer of clastic sediments between crystalline basement and allochthon also exhibits a well-developed, parallel foliation dipping shallowly to the west. Tight to isoclinal, steeply inclined folds have developed, infolding basal conglomerates and crystalline basement, suggesting the presence of a detachment at a lower structural level. Furthermore, Tull *et al.* (1985) and Bax (1986) describe features on the northeastern and eastern margins of the window, including a series of thrust faults that imbricate Rombak basement and cover. These thrusts strike NE–SW, and contain a lineation approximately parallel to that in the Caledonian allochthon. They dip to the west, and seemingly root beneath the main part of the basement window. High-angle reverse faults within the eastern part of the window are also described by these authors. They show a spatial relationship to linear belts of cover rocks, trending approximatey NEE–SSW, parallel to the strike of the reverse faults, described as large synclines in the basement by Tull *et al.* (1985). They could relate to tip-line folds above blind thrusts in the basement, forming as part of the reverse fault system described in the window elsewhere. These lines of evidence, regardless of precise interpretation, indicate the presence of a detachment

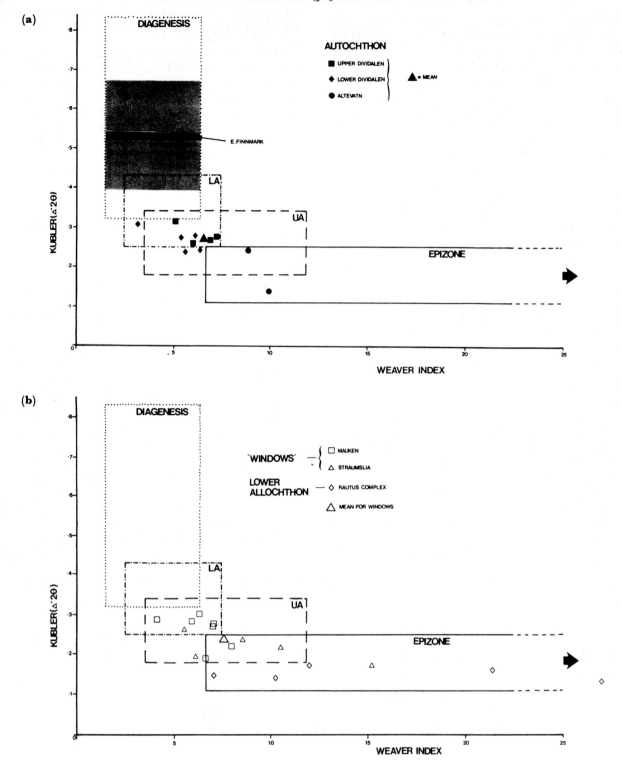

Fig. 5. (*a*) Illite crystallinities as an indication of metamorphic grade for samples from the Caledonian front (Fig. 2). Limits of diagenesis, anchizone and epizone are from Robinson and Bevins (1986). Mean and spread of crystallinity values for the Dividal Group of E Finnmark are also shown (data from Bevins *et al.*, 1986). (*b*) Illite crystallinity plot indicating the metamorphic grade of samples from around tectonic windows in Troms (Fig. 2). Also shown are values obtained from correlated sediments of the Lower Allochthon (Rautas Complex) in the Torneträsk area of Sweden.

beneath the Rombak window. Similarly, the Kuokkel window in Sweden is characterized by imbrication of basement, passing upwards into broad, open folds in the Caledonian allochthon (Bax, 1986).

RESTORATION OF ALLOCHTHONOUS BASEMENT

This study allows a tentative restoration for the Rautas Complex and the Rombak Window. Several assumptions

are made. (1) A smooth and constant wedge taper is maintained throughout metamorphism of the cover sediments (Davis *et al.*, 1983). (2) Estimation of basement shortening requires no absolute values for geothermal gradients during metamorphism, but requires that similar gradients are attained throughout the wedge (Thompson and England, 1984).

Figure 6 shows variation of inferred peak metamorphic temperatures with horizontal distance along a WNW–ESE composite line from the Caledonian front at Dividalen, through the Mauken window to the Rombak window. All distances are projected along strike onto this line and measured along the Caledonian transport direction (ESE). The distance between the Dividalen sequence and the Mauken window is a minimum and the temperature difference between the two is a maximum estimate, hence maximizing the horizontal temperature gradient. These localities are autochthonous, and the gradient should be representative of all other autochthonous basement. Additionally, the distance between the Mauken and Rombak window samples is a maximum whilst the difference between estimated peak metamorphic temperatures is a minimum. Therefore, the horizontal temperature gradient between these points is a minimum. If the Rombak window is autochthonous, then it should lie on a similar temperature–distance line to those of other autochthonous sequences. Figure 6 shows this is not the case. The increase in temperature with distance between the Rombak and Mauken windows is markedly greater than that between the two autochthonous localities. This indicates telescoping of the distance between the Rombak and Mauken windows in the Caldonian transport direction. Extrapolating the temperature–distance line for the autochthon to the 425°C temperature line, and assuming that the increase of temperature with horizontal distance beneath the wedge was linear, gives an approximation of where the Rombak window was situated prior to basement shortening. The present position (R) restores to a position approximately 45 km to the WNW (R') (Figs. 6 and 7). This is a minimum estimate, since the metamorphic grade differences between autochthonous and allochthonous basement cover sequences have been minimized.

The present position of the leading edge of the Rautas Complex (RC) is essentially at the present Caledonian front, where it tectonically overlies the autochthonous Dividal Group sequence (Fig. 2). This restores to a minimum positon (RC') approximately 25 km to the WNW (using the intersection of the peak metamorphic temperature line for the Rautas Complex—minimum 300°C—

and the temperature–distance line constructed for the autochthon) (Fig. 6).

The model also suggests that the Caledonian front must have been approximately 120 km farther east-southeast prior to erosion (see Fig. 6), assuming a uniform taper of the wedge to zero, and that cover sediments at the front were accordingly unmetamorphosed. This estimate compares very favourably with that of approximately 110 km proposed by Hossack and Cooper (1986) on completely independent grounds (Figs. 1 and 6), and suggests that the estimated restorations are also reasonable.

TIMING AND NATURE OF BASEMENT ACTIVATION

The well-exposed basement section along Spansdalen (Fig. 2) allows examination of the nature and timing of basement activation with respect to the emplacement of the Caledonian nappe pile. Detailed mapping has revealed at least ten basement slices within a duplex structure. The geometry of the imbricate stack within the duplex controls the shape of broad, open flexures and monoclines in the allochthon above. Similar features have been described from other areas by many authors (Nicholson, 1985; Hossack and Cooper, 1986; Gee, 1986; Bax, 1986), including the Rombak window and basement areas along Salangsdalen (Gustavson, 1972; Hodges, 1985; Lindahl *et al.*, 1985). It is accepted that some doming is due to late-Caledonian basement motions (Gustavson, 1972), but the relationship to emplacement of the allochthon is poorly documented. However, since the nappe pile generally drapes over, rather than roots under, the large allochthonous basement windows, it seems likely that the basement and the overriding nappe pile were only loosely coupled during basement activation. This possibly caused distinct and contrasting deformation styles to develop under the same bulk stress regime. Shortening was

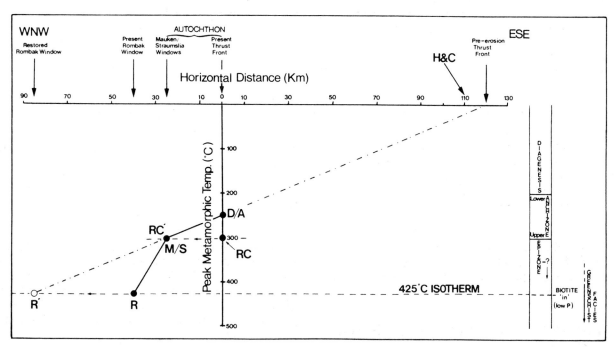

Fig. 6. Estimated peak metamorphic temperature of cover sequences plotted against horizontal distance from the Caledonian front. Distances are shown in Fig. 7. Locality abbreviations as in Fig. 2 except: R and R' = present and restored positions of the Rombak window; RC and RC' = present and restored positions of the Rautas Complex; and H & C = position of the Caledonian front as estimated by Hossack and Cooper, 1986.

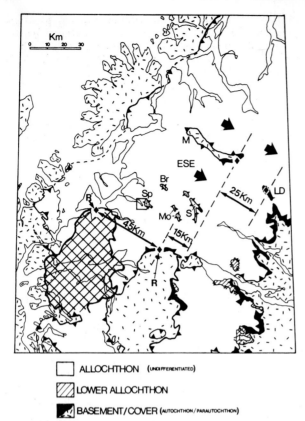

ALLOCHTHON (UNDIFFERENTIATED)

LOWER ALLOCHTHON

BASEMENT/COVER (AUTOCHTHON/PARAUTOCHTHON)

Fig. 7. Distances used in the estimation of basement shortening. All distances are measured along the regional Caledonian
transport direction (ESE), between sample localities projected
along strike. Locality abbreviations as in Fig. 2. Also shown is the
restored position (R′) of the Rombak window sample locality
(R). This is a minimum restoration based upon metamorphism in
the cover sequences (see text).

accommodated in the crystalline basement by imbrication, whilst the nappe pile above deformed internally in
a more homogeneous manner (folding, minor discrete
thrusting and development of extensive S–C mylonites).
Interaction of the two systems is limited to the development of a mutual Caledonian foliation at the contact, and
to the ("passive") doming of the nappe pile as basement
imbrication proceeded.

In the roof zone above the Spansdal duplex, the regional
Caledonian fabric is gently domed and has a variable
low-angle dip. Westwards, at the trailing edge of the
duplex, the dip steepens to the west-northwest, marking
the eastern limb of the regional Ofoten synform (Fig. 3).
Since basement imbrication in the Spansdal, Rombak and
possibly Salangsdal areas shows a spatial relationship to
the eastern limb of this synform, it seems likely that
their structural evolution was mutually dependent. It is
probable that imbrication occurred in response to major
footwall ramps in the basement, inherited features that
possibly control the overall geometry of the Ofoten
synform. Unfortunately, relationships within the Western
Gneiss Terrane, on the western limb of the synform (Fig.
2), are poorly understood. Therefore, it is not yet possible
to determine whether this large basement area has a
similar allochthonous status to the Rombak–
Salangsdal basement on the east side of the synform.

CONCLUSIONS

The Dividal Group and its correlatives in the Straumslia
and Mauken windows show illite crystallinities and

mineral assemblages indicative of sub-greenschist facies
metamorphic conditions. Exceptions to this pattern are the
sequences within the Rautas Complex (lower greenschist
facies metamorphism), and around the Rombak Window
(mid-greenschist facies metamorphism). The Rombak
basement window is at least partly allochthonous on the
basis of anomalous metamorphic grade compared with
adjacent areas. Assuming simple geometric relationships
between depth of burial and peak metamorphic grade,
a restoration of approximately 45 km to the west-northwest
is required to explain the metamorphic and structural
features of the window rocks. Similarly, the Rautas
Complex, which is clearly allochthonous, restores a minimum of 25 km to the west-northwest.

Basement activation was one of the latest Caledonian
events under waning metamorphic conditions, occurring
beneath the accreting nappe pile. The formation of
imbricate systems in the basement controlled the development of broad, open folds in the overlying nappes, and may
relate to much larger structural features such as the Ofoten
synform. However, until relationships within the complex
Western Gneiss Terrane are more fully understood,
the overall structure of the Ofoten region will remain
enigmatic.

Acknowledgements

Special thanks to Hugh Rice, Andy Barker and Rod Gayer
for their advice and encouragement. Illite crystallinity
samples were prepared in the National Museum of Wales,
Cardiff and analysed at the University of Bristol. For their
help on this subject, Richard Bevin, Doug Robinson and
Antje Reuter are particularly thanked, and also Magne
Gustavson for supplying additonal samples. This research
was funded by a NERC studentship (GT4/85/GS/104),
which is gratefully acknowledged. Finally, thanks to Dave
Hunt for help in the field and Nicky Anderson for advice
and support.

APPENDIX A

Analytical procedure

Size fractions <2 m, separated from disaggregated sample
chips by centrifugation and subsequent filtration, were
analysed using the procedure outlined in Bevins *et al.*
(1986). A Phillips PW 1790 series instrument was used with
an automated 0.1 receiving slit, graphite monochromator,
and a tube setting of 40 kV and 30 mA. Crytallinity values
(Weaver and Kubler) were measured manually from
charts and also by computer manipulation of digitally
stored data.

REFERENCES

Ahlberg, P. (1980). Early Cambrian trilobites from northern
Scandinavia. *Norsk Geol. Tidsskr.,* **60**, 153–159.
Andresen, A., Fareth, E., Bergh, S., Kristensen, S.E. and Krogh,
E. (1985). Review of Caledonian lithotectonic units in Troms,
north Norway. In: Gee, D. G. & Sturt, B. A. (eds.) *The
Caledonide Orogen—Scandinavia and Related Areas,* pp. 569–578.
Wiley, Chichester.
Barker, A. J. (1984). The geology between Gratangenfjord and
Salangsdalen, S. Troms, Norway and its regional significance.
Unpublished Ph.D Thesis, University of Wales, 419pp.
Barker, A. J. (1986). The geology between Salangsdalen and
Gratangenfjord, Troms, Norway. *Norges Geol. Unders. Bull.*
405, 41–56.
Barker, A. J. (1988). Metamorphic evolution of the Caledonian

nappes of north-central Scandinavia, *this volume.*

Bax, G. (1986). Basement involved Caledonian nappe tectonics in the Swedish part of the Rombak-Sjangeli Window. *Geol. Fören. Stockh. Förh.* **108**, 3.

Bevins, R. E., Robinson, D., Gayer, R. and Allman, S. (1986). Low-grade metamorphism and its relationship to thrust tectonics in the Caledonides of Finnmark, North Norway. *Norges Geol. Under. Bull.* **404**, 31–42.

Björklund, L. J. O. (1987). Basement-cover relationships and regional correlations of the Caledonian nappes, eastern Hinnoy, N. Norway. *Norsk Geol. Tidsskr.* **67**, 3–14.

Clark, A. H., Dallmeyer, R. D. and Andresen, A. (1985). Basement/Cover relations along the eastern margin of the western gneiss terrane, Senja, Troms, Norway. *Geol. Soc. Amer.,* **17**, 11.

Cooper, M. A. Bradshaw, R. (1980). The significance of basement gneiss domes in the tectonic evolution of the Salta region, Norway. *J. Geol. Soc. Lond.* **137**, 231–240.

Cumbest, R. J. (1986). Tectonothermal Overprinting of the Western Gneiss Terrane, Senja, Troms, Northern Norway. Unpublished M.Sc Thesis, University of Georgia, 179pp.

Dallmeyer, R. D. and Gee, D. G. (1986). $^{40}Ar/^{39}Ar$ mineral dates from retrogressed eclogites within the Baltoscandian miogeocline: Implications for a polyphase Caledonian orogenic evolution. *Geol. Soc. Amer. Bull.,* **97**, 26–34.

Dallmeyer, R. D., Mitchell, J. G., Pharaoh, T. C., Reuter, A. and Andresen, A. (1987). K-Ar and $^{40}Ar/^{39}Ar$ Whole-Rock Ages of Slate/Phyllite from Autochthonous-Parautochthonous sequences in the Komagfjord and Alta-Kvaenangen tectonic windows, northern Scandinavian Caledonides: Evaluating the timing of Caledonian tectonothermal activity. *The Caledonian and Related Geology of Scandinavia.* Abstracts, 13 University College, Cardiff.

Davis, D., Suppe, J. and Dahlen, F. A. (1983). Mechanics of fold-and-thrust belts and accretionary wedges. *J. Geophys. Res.* **88**, 1153–1172.

Føyn, S. (1967). Dividal-gruppen ('Hyolithus-sonen') i Finnmark og dens forhold til de eokamrisk-kambriske formasjoner. *Norges Geol. Unders.,* **249**, 1–84.

Føyn, S. (1985). The late Precambrian in northern Scandinavia. In: Gee, D. G. and Sturt, B. A. (eds.) *The Caledonide Orogen—Scandinavia and Related Areas,* pp. 233–245. Wiley, Chichester.

Frey, M. (1970). The step from diagenesis to metamorphism in pelitic rocks during Alpine orogenesis. *Sedimentology* **15**, 261–279.

Frey, M. (1986). Very low-grade metamorphism of the Alps—an introduction. *Schweiz. Mineral. Petrogr. Mitt.* **66**,13–27.

Frey, M., Teichmuller, M., Teichmuller, R., Mullis, J., Kunzi, B., Breitschmid, A., Gruner, U. and Schwizer, B. (1980). Very low-grade metamorphism in the external parts of the Central Alps: Illite crystallinity, coal rank and fluid inclusion data. *Eclogae Geol. Helv.* **73**, 173–203.

Gayer, R. A., Rice, A. H. N., Townsend, C., Roberts, D. and Welbon, A. (1987). Restoration of the Caledonian Baltoscandian margin from balanced cross-sections: the problem of excess continental crust. *Trans. Roy. Soc. Edinb.* **78**, 197–217.

Gee, D. G. (1986). Middle and upper crustal structure in the central Scandinavian Caledonides. *Geol. Fören. Stockh. Förh.,* **108**, 280–283.

Gunner, J. D. (1981). A reconnaissance Rb-Sr study of Precambrian rocks from the Sjangeli-Rombak window and the pattern of initial $^{87}Sr/^{86}Sr$ ratios from northern Scandinavia. *Norsk Geol. Tidsskr.* **61**, 281–290.

Gustavson, M. (1963). Grunnfjellsvinduer i Dividalen, Troms. *Norges Geol. Unders.* **223**, 92–105.

Gustavson, M. (1966). The Caledonian mountain chain of the Southern Troms and Ofoten areas; Part I: Basement rocks and Caledonian metasediments. *Norges Geol. Unders.* **239**, 162 pp.

Gustavson, M. (1972). The Caledonian mountain chain of the Southern Troms and Ofoten areas; Part III: Structures and structural history. *Norges Geol. Unders.* **283**, 56pp.

Gustavson, M. (1974). Beskrivelse til det berggrunnsgeologiske gradteigskart 1:100,000. *Norges Geol. Unders.* **308**, 34p.

Heier, K. S. and Compston, W. (1969). Interpretation of Rb-Sr age patterns in high grade metamorphic rocks, North Norway. *Norsk Geol. Tidsskr.* **49**, 257–283.

Hodges, K. V. (1985). Tectonic stratigraphy and structural evolution of the Efjord-Sitasjaure area, northern Scandinavian Caledonides. *Norges Geol. Unders. Bull.* **399**, 41–60.

Hossack, J. R. and Cooper, M. A. (1986). Collision tectonics in the Scandinavian Caledonides. In: Coward, M. P. and Ries, A. C. (eds.) *Collision Tectonics. Geological Society Special Publication* **19**, 287–304.

Kisch, H. J. (1980). Incipient metamorphism of Cambro-Silurian clastic rocks from the Jamtland Supergroup, central Scandinavian Caledonides, western Sweden: illite crystallinity and vitrinite reflectance. *J. Geol. Soc. Lond.* **137**, 271–288.

Kisch, H. J. (1983). Mineralogy and petrology of burial diagenesis (burial metamorphism) and incipient metamorphism in clastic rocks. In: Larsen, G. and Chilingar, C. V. (eds.) *Diagenesis in Sediments and Sedimentary Rocks,* (vol. 2), pp. 289–493. Elsevier, Amsterdam.

Kubler, B. (1967). La crystallinite de l'illite et les zones tout a fait superieures du metamorphisme. In: *Etages Tectoniques,* pp. 105–121. A la Baconniere, Neuchatel, Suisse.

Kulling, O. (1950). Berggrunden soder om Torneträsk. *Geol. Fören. Stock. Förh.* **52**, 647–673.

Kulling, O. (1960). On the Caledonides of Swedish Lapland. In: Kulling, O. and Geijer, P. (eds) *Guide to Excursions Nos A25 and C20, International Geological Congress,* 21 Session Norden 1960, pp. 18–39. Sveriges Geologiska Undersökning, Stockholm.

Kulling, O. (1964). Oversikt over norra Norrbottensfjallens kaledonberggrund. *Sver. Geol. Unders.* **Ba. 19**, 1–166.

Landmark, K. (1967). Description of the geological maps "Tromso" and "Malselv", Troms: I. The Precambrian window of Mauken-Andsfjell. *Norges Geol. Unders.* **247**, 172–207.

Lindahl, I., Andresen, A., Rindstad, B. I. and Rundberg, Y. (1985). Age and tectonic setting of the uraniferous Precambrian rocks at Orrefjell, Salangen, Troms. *Norsk Geol. Tiddskr.* **65**, 167–178.

Lindström, M., Bax, G., Dinger, M., Dworatzet, M., Erdtmann, W., Fricke, A., Kathol, B., Klinge, H., Von Pape, P. and Stumpf, U. (1985). Geology of a part of the Torneträsk section of the Caledonian front, northern Sweden. In: Gee, D. G. and Sturt, B. A. (eds.) *The Caledonide Orogen—Scandinavia and Related Areas,* pp. 507–513. Wiley, Chichester.

Nicholson, R. (1985). The structure of a section across the Swedish Caledonides from Nasafjell to the thrust front. In: Gee, D. G. and Sturt, B. A. (eds.) *The Caledonide Orogen—Scandinavia and Related Areas.* Wiley, Chichester.

Nitsch, K. H. (1970). Experimentelle bestimmung der Oberen Stabilitatsgrenze von stilpnomelan (Experimental determination of the upper stability limit of stilpnomelane). *Abstr. Fortsch. Mineral.* **47**, 48–49.

Petersen, K. (1878). Det nordlige Sveriges og Norges Geologi. *Arch. Math. og Naturv.* **3**, 1–38.

Ramberg, H. (1980). Diapirism and gravity collapse in the Scandinavian Caledonides. *J. Geol. Soc. Lond.* **137**, 261–270.

Reuter, A. (1987). Implications of K–Ar ages of whole-rock and grain-size fractions of metapelites and intercalated metatuffs within an anchizonal terrane. *Contrib. Mineral. and Petrol.* **97**, 105–115.

Robinson, D. and Bevins, R. E. (1986). Incipient metamorphism in the Lower Palaeozoic marginal basin of Wales. *J. Metam. Geol.* **4**, 101–113.

Stephens, M. B., Gustavson, M., Ramberg, I. B. and Zachrisson, E. (1985). The Caledonides of central-north Scandinavia—a tectonostratigraphic overview. In: Gee, D. G. and Sturt, B. A. (eds.) *The Caledonide Orogen—Scandinavia and Related Areas,* pp. 135–162. Wiley, Chichester.

Thelander, T., Bakker, E. and Nicholson, R. (1980). Basement-cover relationships in the Nasafjallet Window, central Swedish Caledonides. *Geol. Fören. Stockh. Förh.* **102**, 569–580.

Thompson, A. B. and England, P. C. (1984). Pressure-temperature-time paths of regional metamorphism II. Their inference and interpretation using mineral assemblages in metamorphic rocks. *J. Petrol.* **25**, 929–955.

Tull, J. F., Bartley, J. M., Hodges, K. V., Andresen, A., Steltenpohl, M. G. and White, J. M. (1985). The Caledonides in the Ofoten region (68–69° N), north Norway: key aspects of tectonic evolution. In: Gee, D. G. and Sturt, B. A. (eds.) *The*

Caledonide Orogen—Scandinavia and Related Areas, pp. 553–568. Wiley, Chichester.

Vogt, T. (1941). Trekk av Narvik-Ofoten-Traktens Geologi. *Norsk Geol. Tidsskr*, **21**, 198–213.

Vogt, T. (1967). Fjellkjdeestudier i den oslige del av Troms. *Norges Geol. Unders.*, **248**, 59pp.

Wang, G. F., Banno, S. and Takeuchi, K. (1986). Reactions to define the biotite isograd in the Ryoke metamorphic belt, Kii Peninsula, Japan. *Contrib. Mineral. and Petrol.* **93**, 9–17.

Weaver, C. E. and Broekstra, B. R. (1984). Illite-mica. In: Weaver, C. E. and associates (eds.) *Shale-slate Metamorphism in Southern Appalachians. Developments in Petrology* **10**, 67–97. Elsevier, Amsterdam.

Williams, G. and Chapman, T. (1983). Strain developed in the hangingwalls of thrusts due to their slip/propagation rate: a dislocation model. *J. Struct. Geol.* **5**, 563–571.

Wilson, M. R. and Nicholson, R. (1973). The structural setting and geochronology of basal granitic gneisses in the Caledonides of part of Nordland, Norway. *J. Geol. Soc. Lond.* **129**, 365–387.

Winkler, H. G. F. (1979). *Petrogenesis of metamorphic rocks.* Springer-Verlag, Berlin.

Winslow, M. A. (1981). Mechanisms for basement shortening in the Andean foreland fold belt of southern South America. In: McKlay, K. R. and Price, N. J. (eds.) *Thrust and Nappe Tectonics. Geological Society Spec. Publ.* **9**, 513–528.

10 The structure and stratigraphy of the southwestern portion of the Gaissa Thrust Belt and the adjacent Kalak Nappe Complex, Finnmark, N Norway

C. Townsend[1], A. H. N. Rice[2] and A. Mackay[3]

[1]Cambridge Arctic Shelf Programme, Department of Earth Sciences, West Building, University of Cambridge, Gravel Hill, Huntingdon Road, Cambridge CB3 0DJ, UK

[2]Department of Geology, National Museum of Wales, Cathays Park, Cardiff CF1 3NP, UK

[3]Department of Geology, University of Wales, Cardiff, P.O. Box 914, Cardiff CF1 3YE, UK

In the southwest part of the Gaissa Thrust Belt, a new, arkosic, stratigraphic unit, the Ai'roai'vi Group, has been recognized and informally divided into the Bål'dneraš'ša, Adnevarri, Nav'kaoi'vi and Åbbardasraš'ša Formations. These rocks are structurally overlain by sediments of the Tanafjord Group. Deformation in the Gaissa Thrust Belt was directed towards the east-southeast. The Tanafjord Group has been imbricated into the Munkavarri Imbricate Zone; this has been domed by the Betusordda Antiformal Stack, a duplex developed in the Ai'roai'vi Group. Restoration of the Munkavarri Imbricate Zone suggests that early sedimentation in the Tanafjord Group may have been controlled by extensional faulting.

Two major nappes have been identified in the Kalak Nappe Complex. The Nalganas Nappe, composed of flaggy psammites, and the Gargia Imbricate Stack, composed of four or more minor imbricates repeating a coarse quartzofeldspathic and pelitic sequence. The stratigraphy of the Kalak Nappe Complex has been revised; the sediments within each major thrust sheet have been divided into members that can be correlated with the six formations described in the type stratigraphy, the Sørøy Group.

The Kalak Nappe Complex and Gaissa Thrust Belt overlie the autochthonous Dividal Group, in which minor thrusts and folds have been found in the upper 30 m. The sole thrust cuts down from within Member VI to the lower part of Member II.

INTRODUCTION

The Caledonides of Finnmark were formed by thin-skinned compressional tectonics (Chapman et al., 1985; Townsend et al., 1986; Gayer et al., 1987) during the early Palaeozoic Caledonian Orogeny. Thrust sheets were displaced initially towards the southeast and later towards the east-southeast (Townsend, 1987), with a foreland directed thrust propagation. Five major tectonic units have been identified; the Kalak Nappe Complex, the Laksefjord Nappe Complex, the Barents Sea Caledonides, the Komagfjord Antiformal Stack and the Gaissa Thrust Belt (see Gayer et al., 1987; Rice et al., 1989 for details). The Kalak Nappe Complex is part of the Middle Allochthon of the Scandinavian Caledonides, which were deformed during early Caledonian events between 540 and 490 Ma (Dallmeyer et al., 1988a,b; Pedersen et al., 1988): the other units are all thought to be part of the Lower Allochthon, which was deformed in the Scandian event at 430–410 Ma (Dallmeyer et al., 1988a,b), with the emplacement of the Kalak Nappe Complex.

The Gaissa Thrust Belt, the lowest tectonic unit in the Caledonides of Finnmark, is structurally overlain by the Laksefjord Nappe Complex, which crops out in the

The Caledonide Geology of Scandinavia © R. A. Gayer (Graham & Trotman), 1989, pp. 111–126.

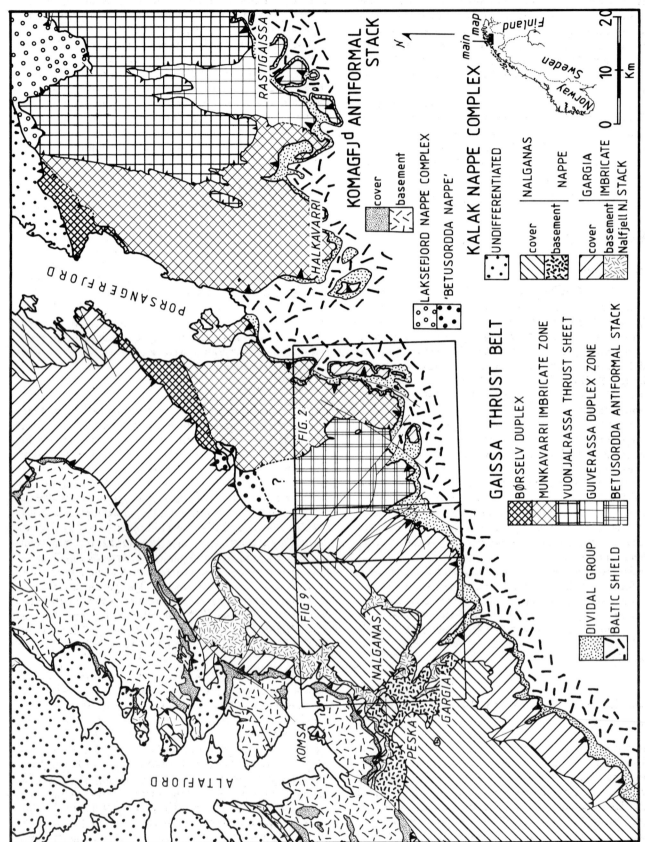

Fig. 1.   Simplified geological map of the Altafjord – Porsangerfjord part of the Caledonides of Finnmark, showing the areas covered by Figs. 2 and 9. (Modified from Sigmond *et al.*, 1984.)

Laksefjord region, the "Betusordda Nappe", which lies to the west of Porsangerfjord and the Kalak Nappe Complex (Fig. 1). The "Betusordda Nappe" was correlated with the Laksefjord Nappe Complex by Williams (1976a) on the basis of their similar tectonostratigraphic position and relatively similar lithologies. Mapping east of Stabbursdalen showed that the "Betusordda Nappe" extended southwards, apparently with a mylonitic and westward-dipping contact over the Gaissa Thrust Belt (Zwaan and Roberts, unpublished mapping, 1983); Sigmond *et al.* (1984) inferred that it extended southwards to the sole thrust.

The Caledonian sole thrust is called the Gaissa Thrust in the Porsangerfjord–Stabbursdal region. To the west of Stabbursdalen, where the Kalak Nappe Complex is the lowest tectonic unit (Figs. 1 and 2), the sole thrust is referred to as the Kalak Thrust. The autochthonous sandstones and shales of the Dividal Group are exposed in the footwall to the sole thrust.

STRATIGRAPHY

Autochthon—Dividal Group

The Dividal Group is a sequence of shales, siltstones, sandstones and conglomerates, up to 260 m thick, resting unconformably on the Baltic Shield (Føyn, 1967). The age of these sediments is constrained by the glacial diamictites, of inferred Varangian (early Vendian) age, underlying the Dividal Group southwest of Lakselv (Figs. 1 and 2; Siedlecka, 1987) and by the early Cambrian fossils found in the middle part of the group (Hamar, 1967; Føyn and Glaessner, 1979).

Føyn (1967) divided the Dividal Group into six members, on the basis of field work in the area southeast of Porsangerfjord; subsequently, Roberts (1974) argued that the two upper members were part of the Gaissa Thrust Belt. However, our mapping, especially in the area to the west of Stabburshaugen (Fig. 2), has clearly shown that there are six members, as initially described by Føyn (1967). Apart from finding glauconite in Members I and VI, no new work on the sedimentology of the autochthon has been undertaken.

Particularly good exposures of the Dividal Group have been found southwest of Jårgastatvarri (Mbr I), at Vuolanjunnasgai'sa (Mbr I–VI, section described by Holtedahl, 1931), on the north side of Vuorji (Mbr II–III) and at Jametgårsa (Mbr III–VI). The unconformity with the Baltic Shield is exposed in the waterfall east of Balgesoai'vi.

Allochthon—Gaissa Thrust Belt

Tanafjord Group

A stratigraphy for the Gaissa Thrust Belt in the Porsangerfjord area was published by Roberts (1974) and considerably amended by Williams (1976b), who suggested a correlation with the Tanafjord Group in East Finnmark (Siedlecka and Siedlecki, 1971). In the present area of the Gaissa Thrust Belt, four out of five of the lower formations of the Tanafjord Group have been recognized (the Stangenes Formation has not been found west of Laksefjord; Føyn *et al.*, 1983), as well as the lowest member described by Williams (1976b; see Fig. 2).

The Grønnes Formation is the most commonly exposed unit of the Tanafjord Group in this region. Two lithologies have been recognized. The upper part of the formation is composed of quartzites that can be distinguished from the massive grey quartzites of the Gamasfjell Formation by their white coloration, often tinted lemon-green, thin and irregular bedding (< 1 m thick) and, in the area studied, by the presence of grits composed of rounded spheroidal quartz grains set in a quartz matrix. Feldspar is generally absent, although small white flecks of ?sericite may represent weathered feldspars.

The lower part of the Grønnes Formation, called the Brennelvfjord Member (Brennelvfjord Interbedded Member of Williams (1976b)), is a variable unit, between 0 and 400 m thick, comprising pale yellow to lemon green sandstones, siltstones and quartzites, with some darker shales and white quartzites. The white quartzites can form units up to 20 m thick, when the rock is essentially indistinguishable from the overlying quartzites in areas of poor exposure. In the lower part of the cliff section east of Divgagai'sa, rust-spotted sandstones, similar to those described by Roberts (1974), were found. Good exposures of the Brennelvfjord Member can be seen in the cliffs east of Divgagai'sa, although more accessible river sections have been found northeast of Stalluskai'di and southwest of Åbbardatčåk'ka (Fig. 2).

The Dakkovarre, Gamasfjell and Vagge Formations crop out in the northern part of the area (Fig. 2). A good section through the Dakkovarre Formation is exposed in the cliffs west of Divgagai'sa. No differences in lithology were observed compared to the descriptions of Roberts (1974) and Williams (1976b) except that the sandstones of the Gamasfjell Formation were not stained red along joint and bedding surfaces.

Åiroai'vi Group

In the extreme western part of the Gaissa Thrust Belt, a sequence of dominantly feldspathic sandstones were mapped. These rocks are equivalent to the sediments of the "Betusordda Nappe" from near the Kalak Thrust west of Porsangerfjord (Fig. 1), which have been correlated with the Laksefjord Group (Føyn *et al.*, 1983), primarily because of their tectonostratigraphic position (Williams, 1976a). However, this correlation is of doubtful validity for three reasons; firstly, the Laksefjord Group is characterized by quartz sandstones (Landersfjord Formation; Føyn *et al.*, 1983), not arkosic rocks; secondly, the Laksefjord Nappe Complex suffered an epizone metamorphism, whilst pelitic rocks within the feldspathic sandstones record an essentially anchizone metamorphism (Rice *et al.*, 1988—this volume); finally, the Laksefjord Nappe Complex suffered significant penetrative ductile strains (Chapman *et al.*, 1979) that have not been found in this area.

These rocks also differ from the Tanafjord Group, in which quartz sandstones are interbedded with pelitic units on a large scale (Williams, 1976b). The only rocks with any similarities at all are the dominantly arkosic to sub-arkosic lithologies (with subordinate conglomerates, quartz-arenites and siltstones) of the Vadsø Group (Banks *et al.*, 1974; Banks and Røe, 1974; Hobday, 1974) that crop out in the autochthon of easternmost Finnmark and with which no detailed correlation can be established. Since no clear correlatives for the feldspathic sandstones in the Stabbursdal area have been found, we have placed these rocks in a new group, the Ai'roai'vi Group, within which four formations have been informally identified. However, the investigations undertaken so far are not sufficient to formalize the stratigraphy with type sections and thicknesses.

The Bål'dneraš'ša Formation is a dark-grey to purple, fine- to medium-grained, massive and often indistinctly

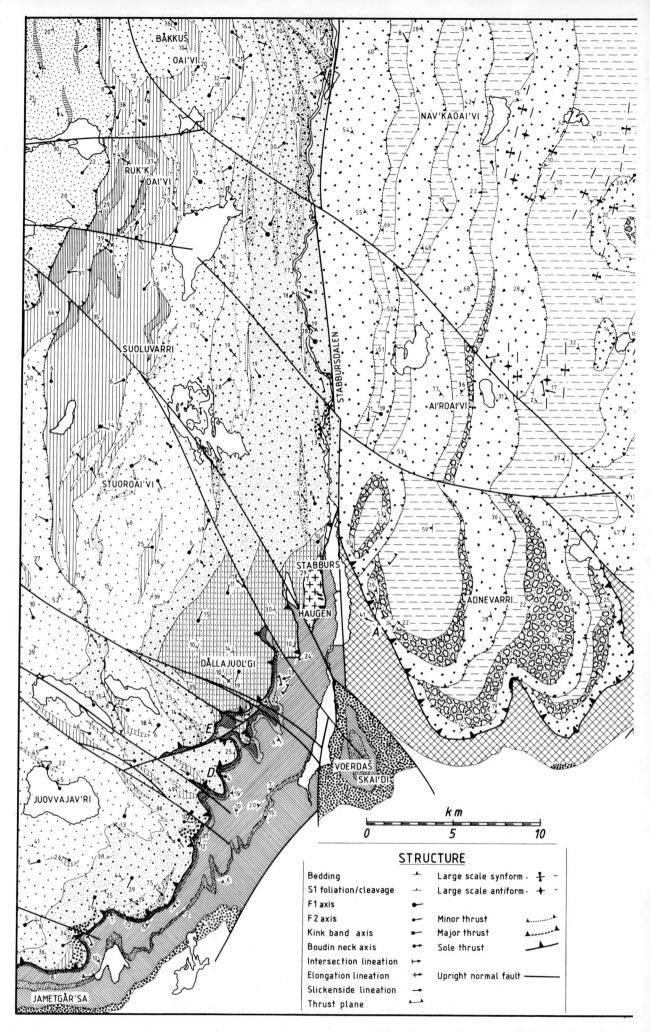

Fig. 2. Geological map of the Caledonides in the western Skoganvarre and Cåkkaraš'sa map sheet areas (1:50,000). Data for the Dividal Group for Voerdašskai'di taken from Zwaan (unpublished data, 1981) and for the area to the south and east of the eastern klippe

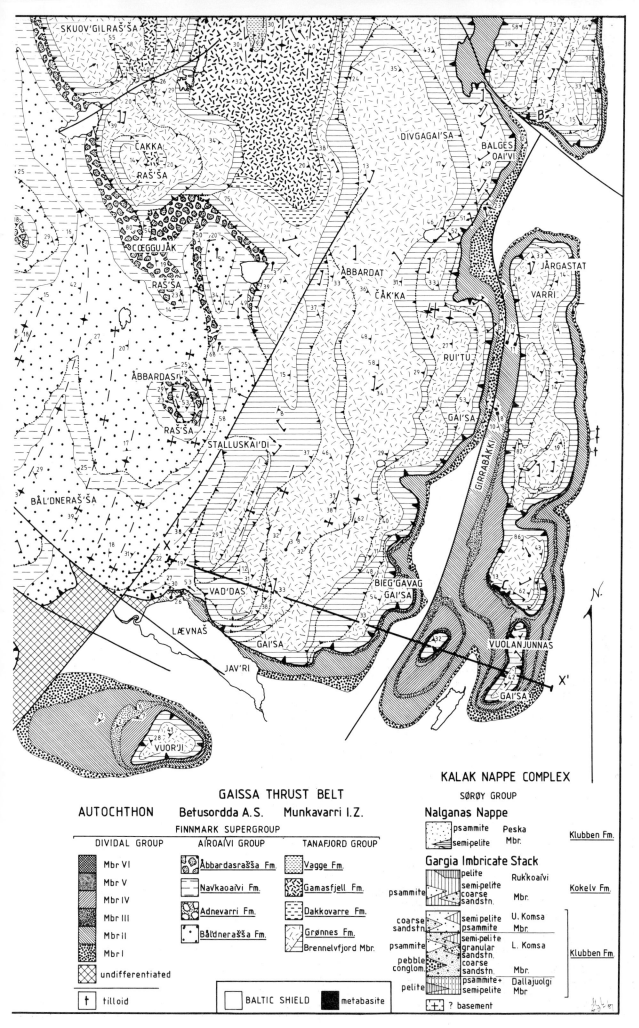

of the Gaissa Thrust Belt from Siedlecka (1987). Other areas modified after Rice and Townsend (1985) and Rice and Mackay (1986). A–E refer to localities where the sole thrust is exposed. X–X' is section line of Fig. 7.

bedded sandstone, with a low quartz:feldspar ratio. Detrital white-mica grains are abundant and biotite and jasper are locally common. Sedimentary structures are absent, except for rare foresets. Bedding surfaces are uneven and spaced at over 1 m. This formation is well exposed in the Bål'dneraš'ša area and to the north of Åbbardasraš'ša. Rarely, well laminated red/green mudstones are present; these are exposed to the east of Coeggujåkraš'ša and northwest of Laevnašjav'ri.

The Adnevarri Formation is a coarse-grained, clast supported, arkose of variable hue (green/pink/brown), with a low quartz: feldspar ratio. The grains, including jasper, are poorly sorted, sub-rounded to sub-angular and occasionally cemented by iron oxides. Bedding planes are fairly smooth, with beds typically less than 0.5 m thick. This formation is well exposed northwest and southeast of Adnevarri.

The Nav'kaoai'vi Formation is very similar in many respects to the Bål'dneraš'ša Formation except that it has abundant thin (<2 mm), discontinuous bands of pelite that often contain detrital mica grains up to 3 mm in diameter. Occasionally, pelite/semi-pelite becomes the dominant lithology, with abundant sedimentary structures such as mud rip-up clasts and ripple laminations. The more pelitic lithologies are well exposed north of Nav'kaoai'vi and east of Åbbardasraš'ša.

The Åbbardasraš'ša Formation is similar to the Adnevarri Formation in that it is a coarse-grained, variably tinted, pale-coloured rock with a low quartz:feldspar ratio. Clasts, which are rounded to sub-spherical and which may be up to 40 cm long, are composed of white quartz and ?jasper. Beds are less than 0.25 m thick and the rock can sometimes be mistaken for parts of the Brennelvfjord Member (see above). However, this formation has the high feldspar content typical of the Ai'roai'vi Group.

Large-scale symmetrical ripples (20–30 cm wavelength) have been seen at the top of Åbbardasraš'ša, where the formation is best exposed.

Petrography of the Ai'roai'vi and Tanafjord Groups

Table I shows the dominant petrographic characteristics of thin-sections of seven samples from the Ai'roai'vi Group and of two samples of grits from the Tanafjord Group. Despite the small data set, two features support field observations. Firstly, although all the sediments are submature, those from the Ai'roai'vi Group have angular to sub-angular clast shapes (except 3324), whereas those from the Tanafjord Group have sub-rounded clast shapes. The coarser grains, usually composed of polycrystalline quartz (Fig. 3A) tend to be more rounded; for example, in sample 3325 the coarse sand/granule components are well rounded whilst the medium/fine sandstone grains tend to be sub-rounded. Secondly, the samples from the Tanafjord Group have much lower feldspar and higher quartz contents than the samples from the Ai'roai'vi Group, although the lithic fragment content is the same in both groups. The samples from the Ai'roai'vi Group plot dominantly in the feldspar-rich fields of the wackes and arkose-arenite diagrams of the classification of Pettijohn *et al.* (1973). The two samples from the Tanafjord Group plot in or close to the sub-lithic arenite field and this may reflect a different provenance for these sediments.

The accessory minerals of these samples are dominated by small (very fine sand size) epidotes and green amphiboles with rare tourmaline and zircon. Many of the ferromagnesian minerals show extensive diagenetic replacement to iron oxide clay minerals, such as goethite. These replacement minerals tend to be found in and around zones of high porosity.

The dominant type of mica found in these samples is

Table I. Petrographic characteristics of thin-sections from the Ai'roai'vi and Tanafjord Groups. QM, monocrystalline quartz; QP, polycrystalline quartz; QT, total quartz; F, feldspar; L, lithic fragments.

Sample number	QM	QP	QT	F	L	Grain shape	Maturity index	Minerals[a]	Cement[a]	Mica[a]
TANAFJORD GROUP										
Grønnes Formation, Brennelvfjord Member										
3325	60	40	85	5	10	Sub-round	Submature	Epidote, Green amph.	Clay Fe ox.	Musc.
3328	70	30	85	10	5	Sub-round	Submature	—	Fe ox.	—
AI'ROAI'VI GROUP										
Åbbardasraš'ša Formation										
3324	60	40	50	30	20	Sub-round	Submature	Epidote	Clay	Musc.
3329	95	5	65	30	5	Sub-ang.	Submature	Epidote Green amph.	Clay	Musc.
Nav'kaoai'vi Formation										
3367	80	20	60	30	10	Sub-ang.	Submature	Zircon, Epidote	Clay	Musc.
3372	80	20	65	30	5	Sub-ang.	Submature	—	Clay	Musc. Sericite
Bål'dneraš'ša Formation										
3154	85	15	60	30	10	Sub-ang.	Submature	Amph., epidote	Mont. Goethite	Musc. Chlorite
3321	70	30	60	30	10	Angular	Submature	Tourmaline Green amph.	Clay Fe ox.	Musc. Chlorite
3333	75	25	60	30	10	Angular	Submature	FeMg minerals Amph.	Clay Fe ox.	Sericite

[a]Musc. = muscovite; Amph. = amphibole; Fe ox. = iron oxides; Mont. = montmorillonite

Fig. 3. (A) Large clast of strained quartz within an essentially unstrained matrix. Bål'dneraš'ša Formation, SE of Bål'dneraš'ša; 3299/85/HR; XPL. (B) Pebble conglomerate from the Lower Komsa Member (Klubben Formation) of the Gargia Imbricate Stack. Taken from within a few metres of the sole thrust near Jametgårsa. (C) Well-foliated (protomylonitic) coarse sandstone from the Lower Komsa Member (Klubben Formation) of the Gargia Imbricate Stack. Southern Stabbursdalen; 4163/86/HR; PPL. (D) Typical minor thrust in the Dividal Group (Mbr IV); from east of Dållajuol'gi.

muscovitic, with a primary diagenetic phase as well as a later metamorphic phase. In rare cases, both types can be seen in the same grain, with the metamorphic type developed around an older, slightly altered, sedimentary core. Associated with muscovite are chlorite clays, probably forming an original grain rimming cement; note that the presence of chlorite does not necessarily indicate greenschist facies metamorphism (Frey, 1986). Where it has been possible to distinguish clay minerals, illite tends to be the dominant variety, as indicated by the extreme wispyness of the crystal terminations in pore spaces.

Clay minerals form the main cement, although syntaxial quartz overgrowths and iron oxides have helped to bind the sediments. As the matrix appears to have been produced by the breakdown of labile grains and minerals, it should be termed a pseudomatrix (Dickinson, 1970), but owing to the subsequent diagenesis and anchizone metamorphism (Rice *et al.*, 1988, this volume) it has not been possible to determine the amount of sedimentary matrix.

Allochthon – Kalak Nappe Complex

Sørøy Group

Work in the Sørøy area established a stratigraphy of five groups (Ramsay, 1971) and this, although modified to six groups by Gayer *et al.* (1985; Klubben, Kokelv, Storelv, Falkenes, Åfjord and Hellefjord Groups), has effectively become a type stratigraphy for the Kalak Nappe Complex. Correlations of lithologies between major imbricates within the complex have been achieved by equating broad lithological units (psammites, muscovite schists, etc). Finer-scale correlations between imbricates has been possible only where minor (late?) thrusting has occurred within a major nappe, forming an imbricate stack (Gayer *et al.*, 1985).

In the following we have used a proposed, revised stratigraphic terminology for the Kalak Nappe Complex. At present all rocks belonging to one group need to be referred to by their tectonic unit, as well as by detailed lithology. Further, the existing system implies that the correlated rocks have a far greater similarity (in terms of lithology and stratigraphic thickness) than is actually the case. We hope that the suggestion outlined will provoke discussion and lead eventually to a formal revision by the Norwegian Stratigraphic Committee.

Since the six previously described stratigraphic groups form the basic mapping units within the Kalak Nappe Complex, we suggest that they be downgraded to formation status. Previously these have been combined into the Kalak Group (Føyn, 1960), but the term Sørøy Group seems more applicable, not least because Kalak lies on the east side of Laksefjord, where the metasediments cannot be directly correlated with the rocks in the Sørøy area (Roberts and Andersen, 1985). To enable correlations to be described more easily, and to facilitate internal division, we suggest that each formation within the Sørøy Group be given one or more distinct member names in each different major tectonic unit (see the key on Fig. 2). Thus, many of the previously used local stratigraphic names for metasediments within the Kalak Nappe Complex, which have previously been discarded, can be reinstated as members (e.g., Hooper and Gronow, 1969; Armitage *et al.*, 1971, Williams *et al.*, 1976).

On the basis largely of work by Zwaan (Zwaan and Roberts, 1978, Zwaan and Gautier, 1980) in the area of the Gargia map sheet immediately to the west of the present region, an outline of the geology to the west of

Stabbursdal has been inferred (Sigmond *et al.*, 1984; Ramsay *et al.*, 1985). Essentially three thrust sheets were recognized, the Nalganas, Nålfjell (Skillefjord) and Gargia Nappes. The present study indicates that this threefold division cannot be substantiated and the Nålfjell and Gargia Nappes have been combined into the Gargia Imbricate Stack, which is formed of at least four significant thrust sheets (and probably seven); the term Nålfjell Nappe has been retained for the thrust sheet composed of basement rocks within the imbricate stack, although this does not crop out in the study area (see below).

Nalganas Nappe

Klubben Formation. Peska Member: This lies in the northwest of the area and is composed almost entirely of white weathering pale grey to buff-coloured flaggy psammites, with rare, thin and impersistent bands of fine-grained biotite–quartz schist. Abundant small white micas (?muscovite), weathered pale brown, often give the rocks a speckled appearance. Bedding was seen only rarely, subparallel to the main tectonic foliation. Previously, these rocks have been called the Peska meta-arkose and correlated with the lower part of the Sørøy Group; we suggest they be renamed the Peska Member of the Klubben Formation.

Gargia Imbricate Stack

Two formations of the Sørøy Group have been recognized in the Gargia Imbricate Stack, the Klubben and Kokelv Formations. The former has been divided into three members, two of which, the Upper and Lower Komsa Members, equate with the larger part of the Komsa Group of Zwaan and Gautier (1980).

Klubben Formation. Lower Komsa Member: This unit, which forms the lower part of the stratigraphy except in the south, where it passes laterally into the Dållajuol'gi Member (see below), is largely composed of pale grey to white coarse-grained sandstones with some granular sandstones and rarely matrix supported pebble conglomerates (Fig. 3B) with rounded white to pink quartzofeldspathic clasts up to 3 cm in diameter. The granular sandstones often have a low matrix content and most grains are composed of translucent quartz, although some ?jasper grains were seen. Bedding in the north and central parts of the area is relatively well developed; further south the member is too massive for bedding to be detected at outcrop. Thin semi-pelitic and psammitic horizons similar to lithologies observed in the higher members were also mapped (Fig. 2).

Dållajuol'gi Member: This is a lateral facies variation of the Lower Komsa Member, cropping out in the Dållajuol'gi area (Fig. 2) with an apparently gradational boundary. The main lithology is a dark blue-grey, coarse- to medium-grained sandstone and psammite, with a variable but often quite high pelitic content. Thin dark "phyllitic" pelites have also been mapped. These rocks have similarities with the Jerta (Tierta) Nappe exposed some 70 km to the southwest and correlated with the Gaissa Thrust Belt (Skjerlie and Tan 1961; Fareth *et al.*, 1977; Binns, 1978; Binns, personal communication 1988). However, although a thrust contact between the Dållajuol'gi and Lower Komsa Members was considered in the field, no evidence to support this idea was found.

Upper Komsa Member: This is dominantly composed of well-bedded (< 1 m thick) psammites in which no relic sedimentary grains were visible in the field and with < 5% pelitic material. The rocks are pale grey-blue to brown-coloured on fresh surfaces, weathering to a paler colour.

Kokelv Formation. Ruk'koai'vi Member: This is mostly composed of well-foliated semi-pelitic and pelitic biotite–quartz schists. Pelitic bands, often lensoid, are more common towards the top of the member. In the lower part of the member finely interbanded (<4 mm) psammites and semi-pelites are common. Quartz veins are locally abundant. No porphyroblastic phases were seen.

Basement lithologies. Underlying the Dållajuol'gi Member at Stabburshaugen are rocks thought to be pre-Caledonian basement (Fig. 2). These are weakly foliated, unbedded and pale-coloured. Coarse feldspar grains without an obvious matrix could be discerned, giving a granitic appearance. These rocks may be similar to some of the basement lithologies exposed in the Alta area (Zwaan and Gautier, 1980). Two weakly foliated metadolerite sheets were found within these "basement" rocks.

STRUCTURE

Structure of the Gaissa Thrust Belt

In the Gaissa Thrust Belt it has been assumed that a layer-cake stratigraphy existed prior to deformation, except where it can be demonstrated otherwise. The Brennelvfjord Member, although considerably thinned and actually missing in some areas, has been assumed to be present at the base of the Tanafjord Group above most of the outcrop of the sole thrust. In the Ai'roai'vi Group, minor imbricate thrusts have been inferred to preserve a consistent stratigraphy wherever possible, although variations in the thickness of the Adnevarri and Åbbardasraš'ša Formations made this more difficult.

The Gaissa Thrust Belt has been divided into several tectonic zones in the area east of Porsangerfjord (Fig. 1; Townsend *et al.*, 1986). The most westerly of these zones, the Munkavarri Imbricate Zone, extends west as far as the outcrop of the Ai'roai'vi Group, which has been deformed into a large-scale duplex structure, here referred to as the Betusordda Antiformal Stack (Fig. 4).

Munkavarri Imbricate Zone

The thrust-related deformation in the region immediately to the east of Porsangerfjord (Townsend *et al.*, 1986) has also been recorded an Oldereidneset (Townsend, 1986) and on the west coast of Porsangerfjord (Zwaan and Roberts, unpublished mapping, 1983). The rocks of the Tanafjord Group in the area under consideration have the same deformation style as, and are thus considered to be a part of, the Munkavarri Imbricate Zone. Major thrusts, which generally dip westwards, detach along the essentially flat-lying Gaissa Thrust and repeat the lower formations of the Tanafjord Group. Excellent exposure of thin-skinned structures along cliff faces, especially south of Vad'dasgai'sa, shows that major thrusts have a spacing of over 1 km and have a smooth trajectory (Cooper and Trayner, 1986). Many minor thrusts, with displacements in the order of a few metres, were observed, particularly within the Brennelvfjord Member (Fig. 5). However, on higher ground, which is extensively covered with glacial boulder fields, only the more obvious stratigraphic repetitions could be identified, except in a few river sections.

North–south trending folds, often with a weak westward-dipping cleavage, and with amplitudes of up to 100 m are often associated with thrust faults. The fold vergence and slickenside striae indicate that the thrusting was directed east to east-southeastwards, as in the Porsangerfjord area (Townsend *et al.*, 1986). Deep erosion of the Gaissa Thrust Belt in the southeast of the area has resulted in the development of several large klippe (Fig. 2).

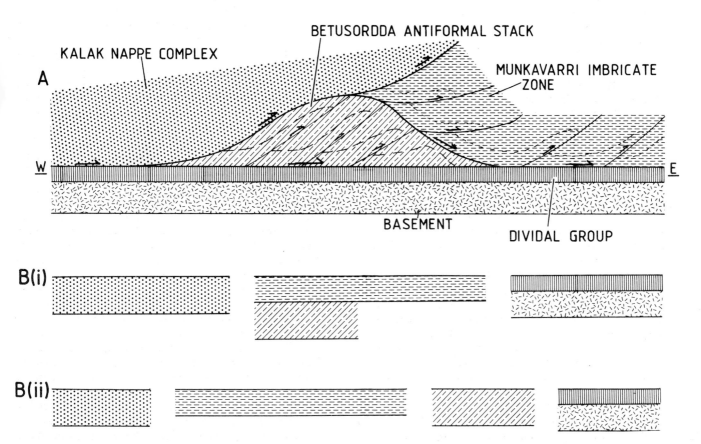

Fig. 4. Schematic cross-section through the area showing the general structure and the relationship of the Betusordda Antiformal Stack to the Munkavarri Imbricate Zone. Two alternative schematic restorations are shown; see text for details.

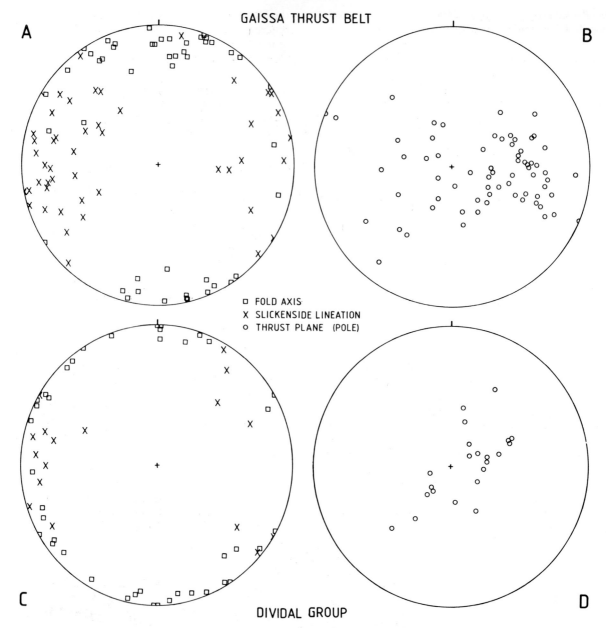

Fig. 5. Equal-area nets showing structural data from the Gaissa Thrust Belt and Dividal Group.

Betusordda Antiformal Stack

Small-scale deformation in the Ai'roai'vi Group was much less commonly observed than in the Tanafjord Group, almost certainly owing to its more competent lithologies. The arkosic rocks do not have a noticeable cleavage (Fig. 3A) but a weak fabric may be present in pelitic rocks. Folding was rarely observed except in the more pelitic parts of the Nav'kaoi'vi Formation, where a crenulation cleavage may also have formed. The most prominent outcrop-scale deformation are quartz veins, with a dominant NE–SW trend. Major thrusts have been inferred from large-scale stratigraphic repetitions and associated folds; very few small-scale thrusts were seen.

Restoration of the Gaissa Thrust Belt

The age of the Ai'roai'vi Group sediments is unknown at present and this has caused problems in constructing restored cross-sections. Two restorations are possible, depending on the assumed relative ages of the Tanafjord and Ai'roai'vi Groups. However, in both restorations the Tanafjord and Ai'roai'vi Groups must both be restored to a position west-northwest of the late Vendian to early

Cambrian-aged rocks of the Dividal Group. This itself causes difficulties because the buried extent of the Dividal Group is unconstrained; Gayer *et al.* (1987) inferred that the buried northern limit was a line running from Andabaktoaivi to the southern tip of Oldereidneset (line A, Fig. 6). In this article we have assumed that the northern limit continues to the southwest on a line between Oldereidneset and the Dividal Group exposed east of Gargia (line B, Fig. 6). This necessitates a minimum restoration of 45 km for the rocks of the Gaissa Thrust Belt in the Halkavarri–Rastigai'sa area (Fig. 6B(i)).

If the Ai'roai'vi Group is older than the Tanafjord Group, no further displacement is required in the restoration, since the Ai'roai'vi Group could underlie any part of the restored Tanafjord Group (Fig. 4B(i)). However, if the Tanafjord Group is inferred to be of an older, or of an equivalent age to the Ai'roai'vi Group, then a further restoration is required to take the Tanafjord Group to the west-northwest of the Ai'roai'vi Group, which would be restored to a position immediately west-northwest of the limit of the Dividal Group (Figs. 4B(ii) and 6B(ii)). In this model, the Ai'roai'vi Group could be a close time correlative of the autochthon and would have been

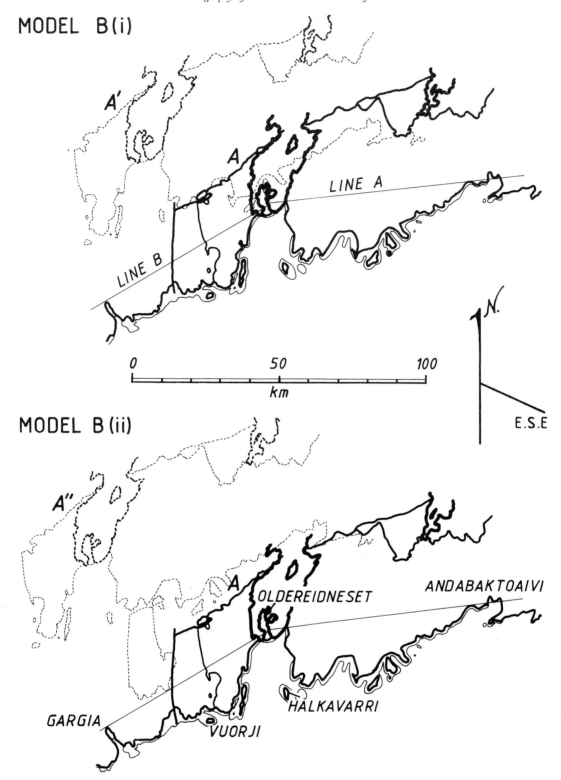

Fig. 6. Plan view of the two possible restorations of the Betusordda Antiformal Stack and Munkavarri Imbricate Zone. See text for details.

incorporated into the Gaissa Thrust Belt by out-of-sequence thrusting. The first model gives a displacement of 45 km and the second a displacement of 65 km. Note that neither model takes account of the *internal* deformation within the Gaissa Thrust Belt.

Two criteria suggest that the former model is more likely to be correct, although definitive evidence is lacking; firstly, since the sole thrust underlies the Ai'roai'vi Group and has a similar appearance there as elsewhere under the Gaissa Thrust Belt, it can be inferred that the Ai'roai'vi Group has been displaced by a comparable distance to that of the Tanafjord Group (see Townsend *et al.*, 1986);

secondly the metamorphism of the Ai'roai'vi Group shows features identical to other parts of the Gaissa Thrust Belt that directly underlie the Kalak Nappe Complex, rather than to the autochthonous Dividal Group (Rice *et al.*, 1988).

Restoration of the Munkavarri Imbricate Zone in this region results in an interesting pre-Caledonian structure. The restored cross-section (Fig. 7) shows that the thickness of the Brennelvfjord Member in the hangingwall increases towards thrust ramps and is considerably thinner in the footwall. For example, to the west of Girrabåkki the Brennelvfjord Member is over 300 m thick in the hanging-

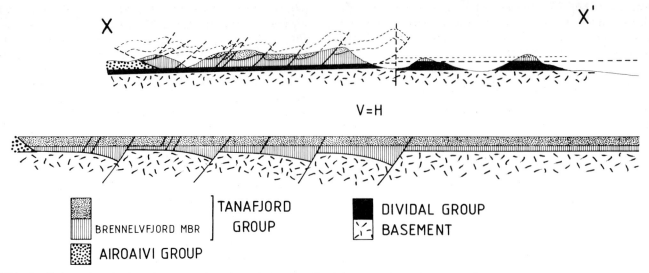

Fig. 7. Balanced and restored cross-sections through the Tanafjord Group rocks in the Munkavarri Imbricate Zone. This shows the variation in thickness of the Brennelvfjord Member across thrusts.

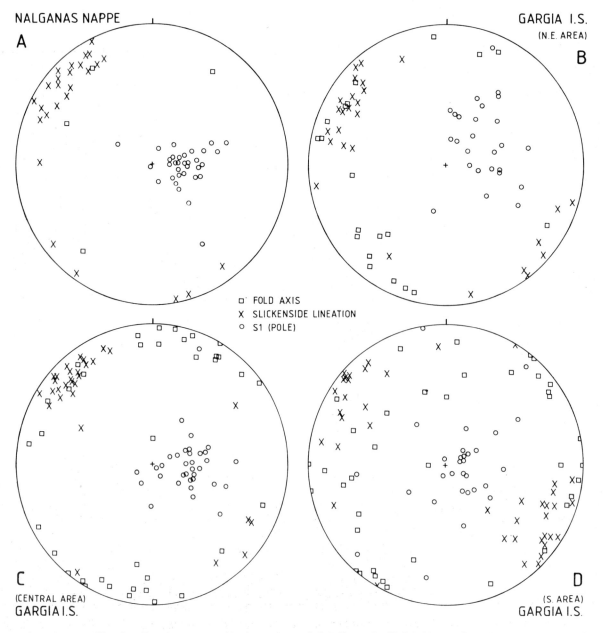

Fig. 8. Equal-area nets showing structural data from the Kalak Nappe Complex.

wall but is only 100 m thick in the footwall to the east. Such thrust faults are interpreted as being inverted extensional faults that were active during the desposition of the lower part of the Tanafjord Group.

Structure of the Kalak Nappe Complex

All the metasediments of the Kalak Nappe Complex have a penetrative foliation (S1) that lies (sub-)parallel to the compositional banding, taken to be the relic bedding (S0) and which is wrapped around relic sedimentary grains (Fig. 3C). In some of the coarser lithologies, this fabric is not immediately obvious, but its development is a critical field criterion for distinguishing the Ai'roai'vi and Sørøy Groups (Figs. 3A and 3C).

In the extreme northeastern part of the Kalak Nappe Complex, S1 strikes NNW–SSE, but further south the regional strike is N–S. South of Dållajuol'gi S0/S1 swings towards the northwest again and appears to intersect the sole thrust at a high angle.

Most of the quartzofeldspathic rocks also contain a mineral stretching lineation of quartz (L1), although porphyroclasts are not noticeably elongate (Fig. 3B). Within the Nalganas Nappe, L1 plunges towards 320° (Fig. 8A), similar to directions in other parts of the Kalak Nappe Complex. In the Gargia Imbricate Stack, L1

plunges towards 290–300° (Fig. 8B–D), again typical of the lower part of the Kalak Nappe Complex (Townsend, 1987). This change in stretching lineation orientation has been used to suggest a change in thrusting direction from southeastwards to east-southeastwards (Townsend, 1987).

Many folds were observed, especially in the Ruk'koai'vi Member. With few exceptions these structures fold the S1 fabric, indicating a D2 fold age. Fold styles vary from upright open to recumbent tight to isoclinal, although in some cases a kink-band geometry was found. In the tighter folds, a weak S2 crenulation cleavage has formed.

Fold-axis orientations show two maxima (Fig. 8); a major one between N–S and NE–SW, typical for most of the Kalak Nappe Complex to the northeast (Gayer *et al.*, 1985) and a second, generally smaller, maxima around NW–SE, subparallel to the L1 stretching lineation direction. This parallelism of L1 and F2 could be taken to infer that the NW–SE-oriented folds developed by rotation of earlier NE–SW-oriented structures, but two interrelated features suggest that the oblique trend is a primary feature; firstly, the rotation of folds implies a high strain, which is not consistent with folds that only have a weakly developed associated crenulation cleavage; secondly, many of the oblique folds are upright open structures, again incompatible with high shear strains.

The Nalganas Thrust is the only major thrust within this

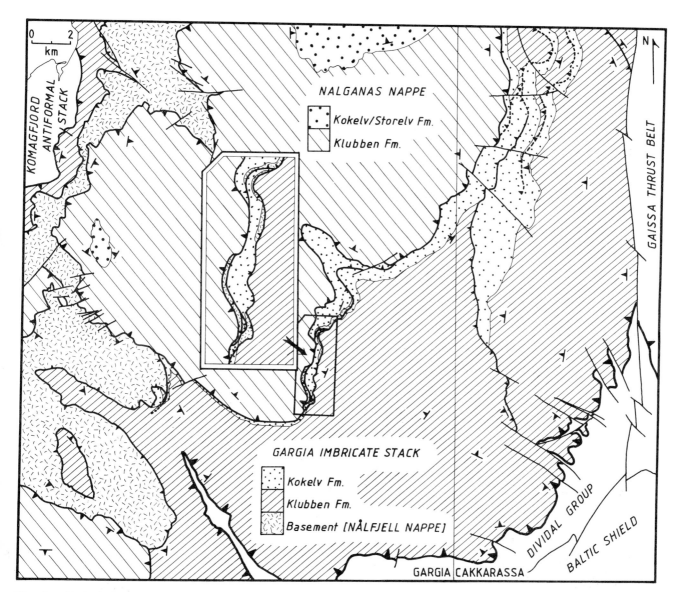

Fig. 9. Geological map of the Kalak Nappe Complex between Altafjord and Stabbursdalen; the area covered is shown in Fig. 1. Simplified from Zwaan and Gautier (1980) and Fig. 2.

part of the Kalak Nappe Complex. Little deformation has been found related to the thrust, although thin mylonitic zones were found at a few localities near the thrust, and most evidence for the thrust lies outside the area (cf. Zwaan and Gautier, 1980).

Within the Gargia Imbricate Stack two minor thrusts have been identified, although it is probable that the threefold repetition of the Ruk'koai'vi and Upper Komsa Members in the Båkkušoai'vi area are of tectonic origin (Fig. 9). The most significant thrust repeats the Ruk'koai'vi and Upper Komsa Members to the west of Suoluvarri (Fig. 2); farther north, the location of the thrust plane becomes uncertain and the displacement may die out. To the south of Stuoroai'vi the thrust swings round to the southeast, parallel to L1 and S0/S1, and probably intersects the sole thrust at a relatively high angle. However, owing to the interference of late upright faulting in the area northeast of Juovvajav'ri (Fig. 2) the location of the thrust is again uncertain. In the overlying imbricate the Lower Komsa Member is thinner than to the northeast, in the Stabbursdal area.

The rocks in the northeast of the Kalak Nappe Complex preserve a higher strain than in other parts of the area. These rocks are probably the upper part of the thick mylonitic sequence mapped in the area between Porsangerfjord and the northern end of Stabbursdal (Townsend, 1987; Zwaan and Roberts, unpublished mapping, 1983). In other parts of this area the basal mylonites have been excised by later, probably out-of-sequence, thrusting and relatively low-strain rocks lie close to the sole thrust (Fig. 3B).

Although previous work had shown that sheared basement rocks produce "pseudo-psammites" (Ramsay and Sturt, 1977), the pelites of the Ruk'koai'vi Member have also been interpreted as sheared basement (Zwaan and Roberts, 1978; Ramsay et al., 1985). This suggestion is not supported by field work, which shows that pelitic rocks entirely similar to those in the Ruk'koai'vi Member have been mapped in the Lower and Upper Komsa Members, with gradational contacts with the quartzofeldspathic sediments. However, it is clear that in the area to the

west there are basement lithologies within the uppermost part of the Gargia Imbricate Stack, interleaved with pelitic schists of the Kokelv Formation. Still further west, near Alta, the pelite thins out entirely, whilst the basement rocks become increasingly thicker (Fig. 9). This relationship has been interpreted as reflecting a dismembered half-graben (Fig. 10); to the west of the extensional fault the Peska Member was deposited on the subsiding hangingwall block, whilst to the east of the uplifted footwall block the Lower Komsa and Dållajuol'gi Members were deposited. With time, subsidence of the footwall allowed westward transgression, with the deposition of the pelitic Ruk'koai'vi Member directly onto the basement.

During the Finnmarkian collision, it is probable that reactivation of the early extensional fault occurred; when this could no longer accommodate the shortening, footwall short-cut thrusting developed across the extensional fault (Fig. 10). Subsequent thrusting imbricated a small basement-cover duplex (inset of Fig. 9) beneath a large basement horse (Nålfjell Nappe) and above thrust sheets composed of the "typical" Sørøy Group stratigraphy. This model implies that the coarser sediments of the Lower and Upper Komsa Members should thin out to the west: whether this is the case cannot be established from published maps (Zwaan and Gautier, 1980). The small sliver of basement rocks beneath the Dållajuol'gi Member has been interpreted as a basement slice above another footwall short-cut fault. The different lithologies in the Dållajuol'gi and Lower Komsa Members may have developed in response to the earlier minor extensional fault.

The sole thrust and its footwall deformation

The sole thrust is a planar horizon with a dip of 1–5° to the north-northwest, across which deformation increases. The Gaissa Thrust has been seen at three localities (A, B and C; Fig. 2). At A and B the thrust is a discrete, narrow, zone containing 1–2 cm of ultracataclasite, whereas at C the fault zone is thicker, with up to 1 m of foliated fault rock

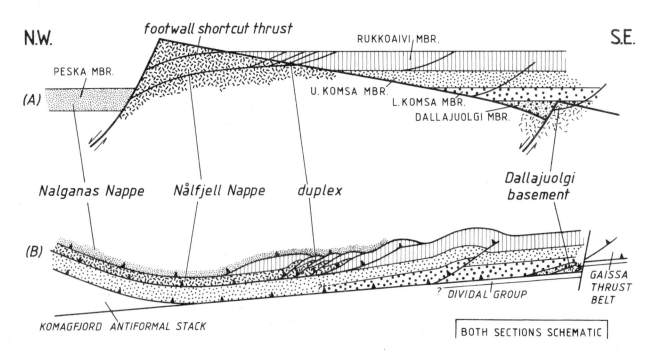

Fig. 10. Schematic sections through the Kalak Nappe Complex in the Alta-Stabbursdal area. (A) illustrates the envisaged restored palaeogeography, with a major half-graben. (B) shows the deformed state, with the development of a basement-cover duplex at the base of the Nålfjell Nappe.

(?ultracataclasite). This is immediately overlain by a metadolerite sheet, part of which has been deformed by the extreme brittle deformation, resulting in a prominent fabric.

The Kalak Thrust is exposed at two localities (D and E; Fig. 2). At D the thrust is overlain by 3 cm to 1 m of foliated fault rock (?ultracataclasite) cut by numerous high-angle brittle fractures. In the underlying Dividal Group the upper 10 cm is an uncleaved quartzofeldspathic cataclasite. This is underlain by 25 cm of foliated green "phyllite" that is folded into close to isoclinal folds with axes parallel to the thrust transport direction and with a weak axial planar cleavage.

The Dividal Group that immediately underlies the Gaissa Thrust in this region is deformed in the top 10–15 m, where folds and slickenside striae are common along bedding surfaces and joints. Under the Kalak Thrust, numerous minor thrusts, associated folds and slickensides were found in the upper 30 m of the autochthon (Fig. 3D). These have a wide range of orientations, reflecting forward, backward and lateral thrusting directions (Fig. 5C & D).

To the west of Stabburshaugen, the sole thrust lies within Member VI of the Dividal Group. Between Stabburshaugen and north of Vuor'ji, the Dividal Group forms a distinct break in topography but is poorly exposed and the position of the sole thrust within the stratigraphy is uncertain. However, the estimated thickness suggests that the thrust lies within Member IV. Between Vuor'ji and Vuolanjunnasgai'sa the autochthon is well exposed and the sole thrust has been mapped lying in higher stratigraphic units further to the southeast. North of Vuolanjunnasgai'sa and Bieg'gai'sa the sole thrust cuts down section rapidly to lie near the base of Member II. Northwest of Balgesoai'vi, the Dividal Group has not been mapped in detail; the outcrop pattern has been inferred from the terrain.

Late vertical faults

Several steeply dipping normal faults have been mapped in the area, most with a NW–SE trend. The displacement vector on these faults has not been directly measured and may contain a component of strike-slip movement.

The N–S trending fault in Stabbursdalen, which has been reported in several publications (Zwaan and Roberts, 1978; Sigmond *et al.*, 1984) is not a major fault, since the sole thrust has been displaced by less than 25 m.

SUMMARY

This work has revealed several significant features of the structure and stratigraphy of the Caledonides of Finnmark. The most important finding is that of the previously unrecorded Ai'roai'vi Group, within which four formations of dominantly feldspathic sediments have been identified. The age of this group is uncertain at present as no fossils have been found.

The Ai'roai'vi Group has been imbricated into a duplex, the Betusordda Antiformal Stack, whilst the Tanafjord Group rocks form the western limit of the Munkavarri Imbricate Zone. The thrust contact between the Betusordda Antiformal Stack and the Munkavarri Imbricate Zone clearly dips towards the foreland; for that reason the Ai'roai'vi Group must either have lain beneath or to the east of the Tanafjord Group prior to deformation. Restoration of the Munkavarri Imbricate Zone suggests that the Tanafjord Group was deposited in a series of extensional basins. With this possibility in mind, previously published balanced and restored sections that have assumed a layer-cake stratigraphy for the Tanafjord Group (Townsend *et al.*, 1986; Gayer *et al.*, 1987) need to be carefully re-examined.

Within the Kalak Nappe Complex, some evidence for synsedimentary extensional faulting has been found, with the basement rocks of the Nålfjell Nappe forming a basement pip developed by footwall short-cut thrusting after inversion of the earlier extensional faults. Beneath this, a minor basement–cover duplex developed. These rocks, all part of the Gargia Imbricate Stack, overlie up to seven imbricates composed of the Klubben and Kokelv Formation of the Sørøy Group. The Klubben Formation in the Gargia Imbricate Stack has been divided into three members, the lower two of which, the Lower Komsa and Dållajuo'gi Members are typically coarse-grained compared to other metasediments in the Kalak Nappe Complex. These coarse sediments have been described in limited regions further to the west.

Acknowledgements

The fieldwork for this article was done under contract to Norges Geologiske Undersøkelse (Rice and Townsend, 1985; Rice and Mackay, 1986) whilst at University College Cardiff (all authors) and University College Galway (AHNR). We thank Anna Siedlecka, N.G.U. staff at Karasjok, Borgne Wood and Une and Per-Øla Gjøvik for their help and hospitality during the fieldwork. Anna Siedlecka, Rod Gayer, David Roberts, Bouke Zwaan and Mike Williams are thanked for discussions on the regional geology.

REFERENCES

Armitage, A. H., Hooper, P. R., Lewis, D. and Pearson, D. E. (1971). Stratigraphic correlations in the Caledonian rocks of S.W. Finnmark and North Troms. *Norges Geol. Unders.* **269**, 318–321.

Banks, N. L. and Røe, S.-L. (1974). Sedimentology of the Late Precambrian Golneselv Formation, Varangerfjorden, Finnmark. *Norges Geol. Unders.* **303**, 17–38.

Banks, N. L., Hobday, D. K., Reading, H. G. and Taylor, P. N. (1974). Stratigraphy of the Late Precambrian 'Older Sandstone Series' of the Varangerfjord area, Finnmark. *Norges Geol. Unders.* **303**, 1–16.

Binns, R. E. (1978). Caledonian nappe correlation and orogenic history in Scandinavia north of lat 67° N. *Geol. Soc.Amer. Bull.* **89**, 1475–1490.

Chapman, T. J., Milton, N. J. and Williams, G. D. (1979). Shape fabric variations in deformed conglomerates at the base of the Laksefjord Nappe, north Norway. *J. Geol. Soc. Lond.* **136**, 683–689.

Chapman, T. J., Gayer, R. A. and Williams, G. D. (1985). Structural cross-sections through the Finnmark Caledonides and timing of the Finnmarkian event. In: Gee, D. G. and Sturt, B. A. (eds.) *The Caledonide Orogen—Scandinavia and Related Areas*, pp. 593–609. Wiley, Chichester.

Cooper, M. A. and Trayner, P. M. (1986). Thrust surface geometry: implications for thrust belt evolution and section balancing techniques. *J. Struct. Geol.* **8**, 305–312.

Dallmeyer, R. D., Mitchell, J. G., Pharaoh, T. C., Reuter, A. and Andresen, A. (1988a). K–Ar and ^{40}Ar/^{39}Ar whole rock ages of slate/phyllite from allochthonous basement and cover in the tectonic windows of Finnmark, Norway: Evaluating the extent and timing of Caledonian tectonothermal activity. *Geol. Soc. Amer. Bull.*, in press.

Dallmeyer, R. D., Reuter, A. and Clauer, N. (1988b). Age of cleavage formation in the lower allochthons of Finnmark, Norway. *Geol. Soc. Amer. Bull.*, in press.

Dickinson, W. R. (1970). Interpreting detrital modes of a grey-wacke and arkose. *J. Sediment. Petrol.* **40**, 695–707.

Fareth, E., Gjelsvik, T. and Lindahl, I. (1977). Tierta. Beskrivelse til det berggrunnsgeologiske kart 1733 II—1:50,000. *Norges Geol. Unders.* **331**, 1–28.

Føyn, S. (1960). Tanafjord to Laksefjord. Guide to Excursion A3. In: J. A. Dons (ed.) *International Geological Congress* **21**, 45–55.

Føyn, S. (1967). Dividal Gruppen ("Hyolithus Zonen") i Finnmark og dens forhold til de eokambriske-kambriske formasjoner. *Norges Geol. Unders.* **249**, 1–84.

Føyn, S. and Glaessner, M. F. (1979). Platysolenites, other animal fossils and the Precambrian–Cambrian transition in Norway. *Norsk Geol. Tidsskr.* **59**, 25–46.

Føyn, S., Chapman, T. J. and Roberts, D. (1983). Adamsfjord og Ul'lugia'sa. Beskrivelse til de berggrunnsgeologiske kart 2135 I og 2135 IV—M 1:50,000. *Norges Geol. Unders.* **381**, 1–78.

Frey, M. (1986). Very low grade metamorphism of the Alps. *Schweitz. Mineral. Petrogr. Mitt.* **66**, 13–27.

Gayer, R. A., Hayes, S. J. and Rice, A. H. N. (1985). The structural development of the Kalak Nappe Complex of Eastern and Central Porsangerhalvøya, Finnmark, Norway. *Norges Geol. Unders. Bull.* **400**, 67–87.

Gayer, R. A., Rice, A. H. N., Roberts, D., Townsend, C. and Welbon, A. (1987). Restoration of the Caledonian Balto-scandian margin from balanced cross-sections: the problem of excess continental crust. *Trans. R. Soc. Edinburgh, Earth Sci.* **78**, 197–217.

Hamar, G. (1967). Platysolenites Antiquissimus Eichw. (Vermes) from the Lower Cambrian of northern Norway. *Norges Geol. Unders.* **249**, 87–95.

Hobday, D. K. (1974). Interaction between fluvial and marine processes in the lower part of the Late Precambrian Vadsø Group, Finnmark. *Norges Geol. Unders.* **303**, 39–56.

Holtedahl, O. (1931). Additional observations on the rock formations of Finnmarken, northern Norway. *Norges Geol. Unders.* **11**, 241–279.

Hooper, P. R. and Gronow, C. W. (1969). The regional significance of the Caledonian structures of the Sandland Peninsula, West Finnmark, northern Norway. *Q. J. Geol. Soc. Lond.* **125**, 193–217.

Pedersen, R. B., Dunning, G. and Robins, B. (1988). The U–Pb (zircon) age of a nepheline syenite pegmatite within the Seiland Magmatic Province, North Norway, *this volume.*

Pettijohn, F. J., Potter, P. E. and Sieverk, K. (1973). *Sand and Sandstones.* Springer-Verlag, Berlin.

Ramsay, D. M. (1971). The stratigraphy of Sørøy. *Norges Geol. Unders.* **269**, 314–322.

Ramsay, D. M. and Sturt, B. A. (1977). A sub-Caledonian unconformity within the Finnmarkian nappe sequence and its regional significance. *Norges Geol. Unders.* **334**, 107–116.

Ramsay, D. M., Sturt, B. A., Zwaan, K. B. and Roberts, D. (1985). Caledonides of northern Norway. In Gee, D. G. and Sturt, B. A. (eds.) *The Caledonide Orogen—Scandinavia and Related Areas*, pp. 163–184. Wiley, Chichester.

Rice, A. H. N. and Mackay, A. (1986). The autochthonous and allochthonous Caledonian rocks west of Stabburselva, Cak-

karassa map sheet 1934 I (1:50,000). Unpublished report to Norges Geologiske Undersøkelse.

Rice, A. H. N. and Townsend, C. (1985). The Gaissa Nappe and Dividal Group of the Skoganvarre and Cakkarassa map sheets, Finnmark. Unpublished report to Norges Geologiske Undersøkelse.

Rice, A. H. N., Bevins, R. E., Robinson, D. and Roberts, D. (1988). Evolution of low grade metamorphic zones in the Caledonides of Finnmark, North Norway, *this volume.*

Rice, A. H. N., Gayer, R. A., Robinson, D. and Bevins, R. E. (1989). Strike-slip restoration of the Barents Sea Caledonides terrane, Finnmark, North Norway. *Tectonics*, in press.

Roberts, D. and Andersen, T. B. (1985). Nordkapp. Beskrivelse til det berggrunnsgeologiske kart, M 1:250,000. *Norges Geol. Unders. Skr.* **61**, 1–49.

Roberts, J. D. (1974). Stratigraphy and correlation of Gaissa Sandstone Formation and Børselv Subgroup (Porsangerfjord Group), South Porsanger, Finnmark. *Norges Geol. Unders.* **303**, 57–118.

Siedlecka, A. (1987). Skoganvarre berggrunnskart 2034 4, 1:50,000. Forelopig utgave. Norges Geologiske Undersøkelse.

Siedlecka, A. and Siedlecki, S. (1971). Late Precambrian sedimentary rocks of the Tanafjord–Varangerfjord region of Varanger Peninsula. *Norges Geol. Unders.* **269**, 246–294.

Sigmond, E. M. O., Gustavson, M. and Roberts, D. (1984). Berggrunnskart over Norge. M 1:1 million. Norges Geologiske Undersøkelse.

Skjerlie, F. J. and Tan, T. H. (1961). The geology of the Caledonides of the Reisa Valley area, Troms–Finnmark, northern Norway. *Norges Geol. Unders.* **213**, 175–196.

Townsend, C. (1986). Thrust tectonics within the Caledonides of northern Norway. Unpublished Ph.D. thesis, University of Wales, Cardiff.

Townsend, C. (1987). Thrust transport directions and thrust sheet restoration in the Caledonides of Finnmark. *J. Struct. Geol.* **9**, 345–352.

Townsend, C., Roberts, D., Rice, A. H. N. and Gayer, R. A. (1986). The Gaissa Nappe, Finnmark, North Norway: an example of a deeply eroded external imbricate zone within the Scandinavian Caledonides. *J. Struct. Geol.* **8**, 431–440.

Williams, D. M. (1976a). A possible equivalent of the Laksefjord Nappe, southwest of Porsangerfjorden, Finnmark. *Norsk Geol. Tidsskr.* **56**, 7–13.

Williams, D. M. (1976b). A revised stratigraphy of the Gaissa Nappe, Finnmark. *Norges Geol. Unders.* **324**, 79–115.

Williams, G. D., Rhodes, S., Powell, D., Passe, C. R., Noake, S. and Gayer, R. A. (1976). A revised tectonostratigraphy for the Kalak Nappe in Central Finnmark. *Norges Geol. Unders.* **324**, 47–61.

Zwaan, K. B. and Gautier, A. M. (1980). Alta og Gargia. Beskrivelse til de berggrunnsgeologiske kart 1834 I og 1934 IV. *Norges Geol. Unders.* **354**, 1–47.

Zwaan, K. B. and Roberts, D. (1978). Tectonostratigraphic successions and development of the Finnmarkian nappe sequence, North Norway. *Norges Geol. Unders.* **343**, 53–71.

11 Palaeogeographic reconstruction of the pre- to syn-Iapetus rifting sediments in the Caledonides of Finnmark, N Norway

R. A. Gayer[1] and A. H. N. Rice[2]

[1]Department of Geology, University of Wales, Cardiff, P.O. Box 914, Cardiff
CF1 3YE, UK
[2]Department of Geology, National Museum of Wales, Cathays Park, Cardiff
CF1 3NP, UK

The Barents Sea Caledonides, previously described as a suspect terrane, have been restored as an integral part of the Finnmark Caledonides to a position along strike from the restored positions of the Komagfjord Antiformal Stack and Laksefjord Nappe Complex. This has allowed modelling of the development of the late Precambrian sedimentary basins in the reconstructed north Norwegian area during pre- to syn-Iapetus rifting times. The simplest model reinforces earlier suggestions that the ENE–WSW trending Gaissa Basin was separated from the NE–SW trending Baltoscandian miogeocline by a ca. 221 km wide topographic high, the Finnmark Ridge. The Gaissa Basin was > 397 km wide and received up to 5 km of mainly fluvial to shallow marine clastic sediments of late Riphean to early Ordovician age. The age of rifting and of the oldest sediments within the Baltoscandian miogeocline is not precisely known but has been inferred to be of pre-mid Vendian age; up to 8 km of fluvial to shallow marine sediments were deposited on its southeast margin. A third basin, the Timanian aulacogen, was oriented WNW–ESE and was juxtaposed against the northwest end of the Baltoscandian miogeocline and Finnmark Ridge, across a major WNW–ESE trending extensional fault. Approximately 15 km of late Riphean to Vendian sediments were deposited in this basin, which may have been connected to the Baltoscandian miogeocline. The southeastern extent of the Timanian aulacogen is not constrained.

Within all three basins an early phase of extension resulted in the deposition of late Riphean to late Sturtian sediments, followed by relative uplift and erosion prior to a period of Vendian to Tremadoc fault-controlled sedimentation. During Caledonian compression, ENE–WSW extensional faults were reactivated as frontal ramps, whilst WNW–ESE trending extensional faults were reactivated as either lateral ramps or major strike-slip faults.

INTRODUCTION

The Caledonides of Finnmark consist of five major tectonic units that thrust southeastwards and east-southeastwards across the Baltic Shield and its unconformable Vendian to lower Cambrian cover (Gayer *et al.*, 1987). The uppermost unit, the Magerøy Nappe (Fig. 1), which lies to the north of the Magerøysundet Fault (Andersen *et al.*, 1982) is composed of Silurian rocks deposited during Iapetus closure and is not considered further in this article. The Kalak and Laksefjord Nappe Complexes have been incorporated into the Middle Allochthon of the Scandinavian Caledonides (Gee *et al.*, 1985; but see below), whilst the Gaissa Thrust Belt (Townsend *et al.*, 1986) has been correlated with the Lower Allochthon. In addition, a line of northeasterly trending large-scale basement horses (Revsbotn and Hatteras Basement Horses and Komagfjord Antiformal Stack) that lie beneath the Kalak Nappe Complex are exposed in at least three tectonic windows (Gayer *et al.*, 1987). The thrust between each unit represents a metamorphic break, the overlying unit showing an upgrade step in peak metamorphic conditions (Rice *et al.*, 1988–this volume).

The Middle Allochthon (Gayer *et al.*, 1985a; Ramsay *et*

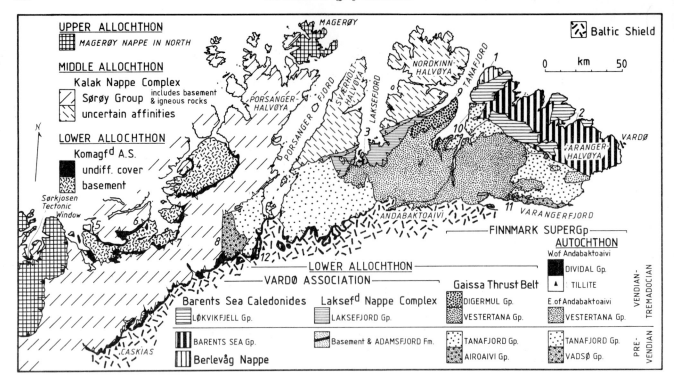

Fig. 1.   Simplified geological map of the Caledonides of Finnmark, showing the distribution of the principal tectonic and sedimentary units.

al., 1985) was metamorphosed and deformed before ca. 531–523 Ma, during the Finnmarkian event (cf. Pedersen *et al.*, 1988–this volume), although hornblende cooling ages of ca. 490 Ma have been recorded (Dallmeyer, 1988). $^{40}Ar/^{39}Ar$ data from muscovites from the Kalak Nappe Complex reflect either slow cooling of, or Scandian reheating of, the Middle Allochthon at 420–430 Ma, whilst similar data from the underlying Lower Allochthon suggest peak metamorphism around 410–420 Ma (Dallmeyer *et al.*, 1988a,b).

The earliest attempt to take account of thrust shortening in this area was made by Gayer and Roberts (1973), who derived a model with two basins separated by an elevated basement area, the Finnmark Ridge. This has been refined by recent structural investigations that have emphasized the importance of thrusting in the deformation (Chapman *et al.*, 1985; Townsend *et al.*, 1986).

The distinctive lithostratigraphy of the Barents Sea Caledonides (Johnson *et al.*, 1978), thought to have been deformed during a pre-Caledonian event (Beckinsale *et al.*, 1976), together with palaeomagnetic data suggesting 500–1000 km displacement dextrally along the Trollfjord–Komagelv (T–K) Fault (Kjøde *et al.*, 1978), led Stephens and Gee (1988) to argue that this unit is a Caledonian strike-slip suspect terrane. In recent palinspastic reconstructions of the Caledonides of the northern Atlantic region, the Barents Sea Caledonides have been placed in a wholly isolated position relative to other units (Winchester, 1988). Reappraisal of the age of deformation and recognition that the Finnmark thrust sheets have been translated by distances comparable to displacements along the T–K Fault have led to a model in which the Barents Sea Caledonides have been restored to a position northeast of the Finnmark Ridge (Rice *et al.*, 1989). The aim of this paper is to outline the palaeogeographic implications of the structural model and to describe possible tectonic controls on the deposition of the late Riphean to Tremadoc sedimentary successions in this region.

REGIONAL GEOLOGY

Autochthon

The autochthon occurs in two zones (Fig. 1). East of Andabaktoaivi (E Finnmark autochthon) it forms a thick succession of fluvial to shallow marine, dominantly siliciclastic, sediments (Fig. 2) that have been divided into three groups (Vadsø, Tanafjord and Vestertana Groups), each separated by a slight unconformity that dips southwards relative to the bedding. These three groups and their lithostratigraphic correlatives in other tectonic units, plus the Ai'roai'vi and Digermul Groups in the Gaissa Thrust Belt (see below) form the Finnmark Supergroup (Gayer and Roberts, 1973).

The sediments of the Vadsø and Tanafjord Groups reflect a series of small scale, and rapid, transgressions and regressions, in which the former group is dominated by deltaic and fluvial deposits and the latter group by shallow marine and intertidal deposits (Banks and Røe, 1974; Banks *et al.*, 1971, 1974; Siedlecka and Siedlecki, 1971; Hobday, 1974; Johnson *et al.*, 1978). The uppermost part of the Tanafjord Group is composed of sabkha facies and stromatolitic dolomites (Grasdal Formation; Siedlecka and Siedlecki, 1971; Siedlecka, 1976). Palaeocurrent directions are variable but are dominantly from the west in the Vadsø Group, changing to southeasterly in the Tanafjord Group (cf. Johnson *et al.*, 1978). Acritarch assemblages indicate a late Riphean to Sturtian age for the Vadsø Group and a mid to late Sturtian age for the Tanafjord Group (Vidal and Siedlecka, 1983; Fig. 3). Dating of shales in the Vadsø Group indicates deposition around 807 Ma (Pringle, 1973).

The Vadsø and Tanafjord Groups are unconformably overlain by the Vestertana Group, the lower part of which is dominated by up to 450 m of glacial sediments of the Varanger Glaciation (Reading and Walker, 1966; Edwards and Føyn, 1981; Figs. 2 and 3). The Smalfjord

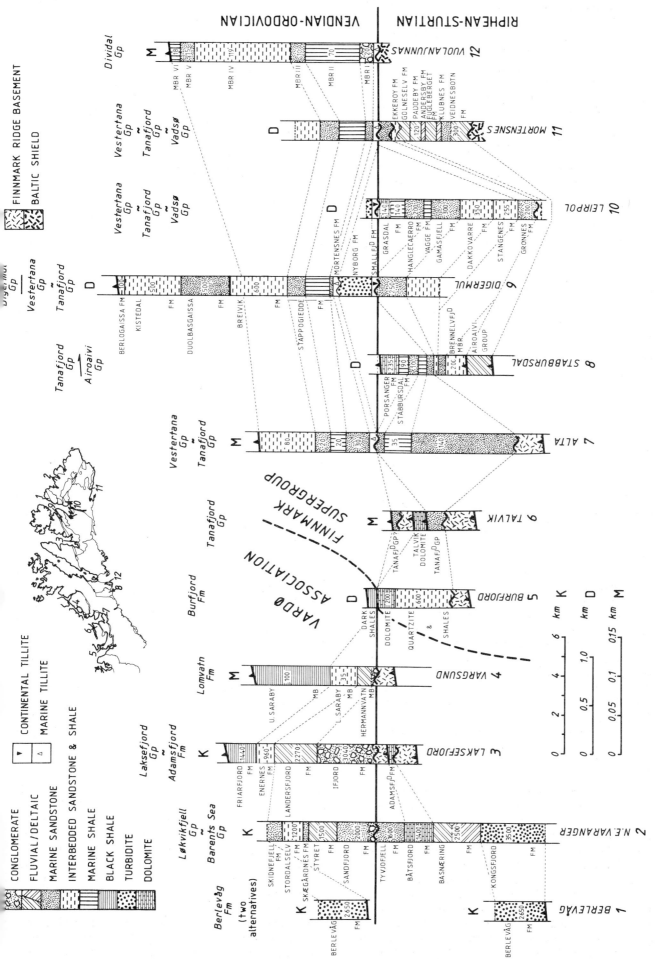

Fig. 2. Summary correlation chart for the major stratigraphic units in the Lower Allochthon and autochthon (and basal Middle Allochthon in column 6). Locations of the columns are shown in Figs. 1, 4 and 5. Note that three different scales have been used (K, D and M). (Data from Milnes and Ritchie, 1962; Williams, 1976; Levell and Roberts, 1977; Johnson *et al.*, 1978; Siedlecki and Levell, 1978; Føyn *et al.*, 1983; Føyn, 1985; Pharaoh, 1985; Siedlecka, 1987.)

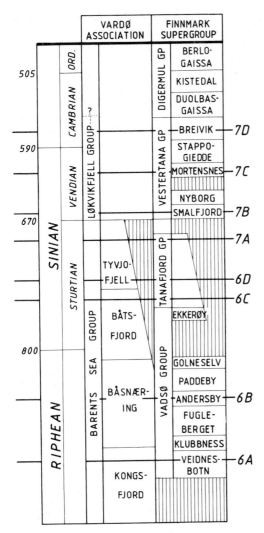

Fig. 3. Simplified chronostratigraphic correlation chart for the Finnmark Supergroup and Vardø Association. 6A–D and 7A–D refer to the time intervals of the palaeogeographic maps of Figs. 6 and 7. (Modified after Vidal and Siedlecka, 1983. Time intervals after Harland *et al.*, 1982.)

Formation is composed of mainly continental glacial deposits and is overlain by the interglacial turbidites of the Nyborg Formation. The Mortensnes Formation unconformably overlies the Nyborg Formation and has a high proportion of marine glacial sediments. These pass upwards conformably into shallow marine/shelf clastic deposits of the Breivik and Stappogiedde Formations (Fig. 2).

East of Andabaktoaivi the autochthon consists of a thin succession (up to 260 m) of siliciclastic sediments (the Dividal Group; Føyn, 1967) resting unconformably on the Baltic Shield. The Dividal Group is correlated with 1100 m of the Breivik and Stappogiedde Formations in the autochthon east of Tanafjord and in the Gaissa Thrust Belt (Fig. 2). Based on limited fossil assemblages, Føyn and Glaessner (1979) have placed the Cambrian–Precambrian boundary in the upper 100 m of the Stappogiedde Formation (Manndreperelv Member).

Southwest of Porsangerfjord, thin lenses of diamictite of presumed glacial origin, correlated with the Mortensnes Formation, have been mapped beneath the Dividal Group (Siedlecka, 1987). Farther to the west, in the Caskias area (Fig. 1) thin autochthonous glacial deposits have also been mapped under the Dividal Group (Skjerlie and Tan, 1960). The condensed nature of the Dividal Group and the absence of pre-Vendian sedimentation has been taken to indicate marine transgression at the basin margin (Føyn, 1985) that may have been fault-controlled (Gayer *et al.*, 1987).

Allochthon

The Gaissa Thrust Belt is the most external tectonic unit and consists of a series of relatively small-scale imbricates, developed within the Finnmark Supergroup, with a cumulative displacement of 55%, formed during east-south-east-directed brittle thrusting (Townsend *et al.*, 1986). Although there are major changes in thickness and detailed lithostratigraphy, the formations comprising the Tanafjord Group have been traced with little difficulty from their type area through to the southeast of Porsangerfjord (cf. Føyn, 1985; Figs. 1 and 2). The glacial and marine sediments of the Vestertana Group have been mapped as far west as to the south of Laksefjord (Fig. 1), where detailed glacial geomorphological features have been described (Føyn and Siedlecki, 1980).

In addition to the stratigraphic groups seen in the autochthon of east Finnmark, the Gaissa Thrust Belt contains the Digermul and Ai'roai'vi Groups (Figs. 1 and 2). The Digermul Group that overlies the Vestertana Group to the east of Tanafjord is dominated by shallow marine clastic sediments reflecting rapid transgressions and regressions (Reading, 1965; Johnson *et al.*, 1978) with an early Cambrian to Tremadocian fauna (Nicolaisen and Henningsmoen, 1985). The Ai'roai'vi Group, which has been mapped to the southwest of Porsangerfjord, is a sequence of immature arkosic sandstones and conglomerates (Townsend *et al.*, 1988—this volume) that are here inferred to be broadly time and facies correlatives of the Vadsø Group, although it is unlikely that a detailed lithostratigraphic correlation can be made.

With the exception of some of the cover successions exposed in the southern part of the Komagfjord Antiformal Stack (see below), the remaining sediments in the Lower Allochthon cannot be correlated with the Finnmark Supergroup and have been combined informally here into the "Vardø Association". These sediments crop out in the Barents Sea Caledonides, the Laksefjord Nappe Complex and in the northern part of the Komagfjord Antiformal Stack; although precise litho- and chronostratigraphic correlations between the sediments in these three units cannot be established, they do have similarities.

Lying immediately above the Gaissa Thrust Belt, the Laksefjord Nappe Complex (Chapman, 1980; Føyn *et al.*, 1983) is composed of three major imbricates. The lowest is composed of basement lithologies (Storfjord Basement Complex) and a thin (< 145 m) succession of dolomites and subordinate siltstones, here informally called the Adamsfjord Formation, that may have an unconformable contact with the basement complex (Føyn *et al.*, 1983). This sedimentary unit is of unknown affinity, but here has been correlated with the dolomites in the Båtsfjord Formation of the Barents Sea Group (Fig. 2).

The Middle and Upper Laksefjord Nappes are composed of the four formations of the Laksefjord Group (Føyn *et al.*, 1983; see also Fig. 2), with small slivers of basement lithologies. The Laksefjord Group represents a transition from alluvial fan and debris flow (Ifjord Formation) through braided stream and shallow marine (Landersfjord and Enernes Formations) to shelf marine deposits (Friarfjord Formation; Føyn *et al.*, 1983). Palaeocurrent directions suggest a proximal source area to the southeast, with sedimentation probably controlled by extensional faulting (Laird, 1972b; Chapman 1980).

Clasts in the Ifjord Formation have been correlated with lithologies in the basement complex (Føyn et al., 1983). Although the composite thickness of the Laksefjord Group is 8000 m, each formation is wedge-shaped and transgressive, so that the total thickness in any area is considerably less. No direct evidence for the age of the Laksefjord Nappe Complex has been found. However, correlation with the Løkvikfjell Group in the Barents Sea Caledonides (Føyn, 1969; Laird 1972a; and see below) suggests a Vendian age (Vidal and Siedlecka, 1983; Fig. 3).

The Komagfjord Antiformal Stack (Figs. 1 and 4), which crops out in the Komagfjord, Altenes and Alta–Kvaenangen (and possibly the Sørkjosen; Fig. 1) Windows, forms one of the most crucial areas for palinspastic reconstruction of the northern part of the Caledonian chain.

Two distinct sedimentary successions have been described from the eastern part of the antiformal stack, both unconformably overlying Raipas Supergroup basement lithologies (Pharaoh, 1985; Pharaoh et al., 1983; Fareth, 1979; Føyn, 1985; Zwaan and Gautier, 1980). South of a line from Korsfjord to Gukkesgurra (Fig. 4), rocks of the Finnmark Supergroup are exposed. In the Alta area the base of the Caledonian succession is a series of quartzites and shales correlated with the middle part of the Tanafjord Group (Føyn, 1985; Fig. 2), but these have not

been found farther north in the Altenes and Komagfjord Windows. The Tanafjord Group correlatives in the Alta–Kvaenangen Window and the basement in the Altenes and southern part of the Komagfjord Tectonic Window are overlain by a thin and locally intermittent glacial diamictite, correlated with the Mortensnes Formation in the autochthon and by a condensed unit correlated with the Stappogiedde and Breivik Formations in the Vestertana Group (Føyn, 1985; Pharaoh, 1985). Metamorphic criteria suggest that a considerable thickness (possibly as much as 500 m) of sediment has been removed by out-of-sequence thrusting (Rice et al., 1988—this volume).

In the Talvik area, quartzites correlated with the Tanafjord Group have been overthrust by dolomite (Talvik Dolomite) that in turn is overlain by a series of minor basement imbricates containing thin slivers of a quartzite, which again has been correlated with the Tanafjord Group (Bowden, 1981). This suggests that the Tanafjord Group quartzites were deposited to the northwest of the restored position of the Talvik Dolomite, in which case it is reasonable to infer that the Talvik Dolomite is a correlative of the Porsanger Dolomite, at the top of the Tanafjord Group (Tucker, 1977).

Farther to the west, the Tanafjord and Vestertana Group equivalents have been mapped along the south and west margins of the window (Fig. 4). However, in the northwest margin of the window a series of interbedded

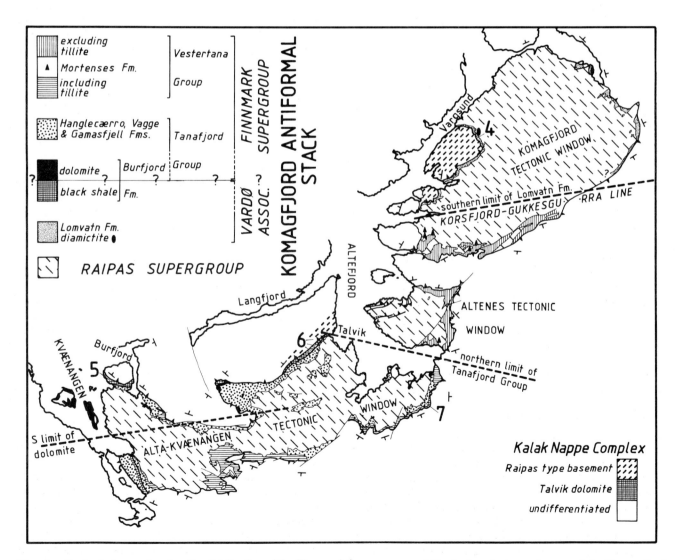

Fig. 4. Geological map showing the distribution of the Finnmark Supergroup and Vardø Association rocks around the Komagfjord Antiformal Stack.

quartzites and shales, overlain by stromatolitic dolomites and then dark shales, here informally called the Burfjord Formation, has been described (Milnes and Ritchie, 1962). These dolomites have been correlated with the Talvik Dolomite (Bowden, 1981) and the whole succession has been correlated with the Tanafjord Group (Zwaan, 1977). However, the status of the dark upper shales is in doubt, with no obvious correlative in either the Tanafjord or Vestertana Groups. The nearest dark shales form the uppermost part of the Lomvatn Formation (see below). Note that the figures quoted for the thickness of the Burford Formation (Fig. 2) are probably overestimates; Zwaan (1977) has marked imbricate thrusts within the Burford Formation.

North of the Korsfjord–Gukkesgurra line, rocks of the Lomvatn Formation have been mapped at the margin of the Komagfjord Antiformal Stack (Fig. 4). These rocks have been correlated with the Laksefjord Group and, although the succession is considerably thinner, the four major lithologies within the Laksefjord Group have been found in the Lomvatn Formation (Rhodes, 1976; Pharaoh, 1985). The thin diamictite found at the base of the formation in the Hermannvatn area, for which Pharaoh (1985) suggested a glacial origin, has been reinterpreted as a fluvial debris flow deposit, based on a similar reinterpretation of the Ifjord Formation at the base of the Laksefjord Group (Laird, 1972b; Chapman, 1980). Similar sedimentary rocks have been mistakenly interpreted as tilloids in other parts of the Vardø Association (Siedlecka and Roberts, 1972). No overlap has been found between the sediments to the north and south of the Korsfjord–Gukkesgurra line (Fig. 4), except possibly in the Burfjord area, where black shales overlying dolomites have been very tentatively correlated with the upper part of the Lomvatn Formation. Reitan (1963) postulated a gradual facies change between the two areas, but Pharaoh (1985) rejected this notion in favour of a model with deposition in two separate and unrelated basins.

The age of the sediments within the Kalak Nappe Complex is poorly constrained. The nappes to the west of Porsangerfjord (Gayer *et al.*, 1985a), Ramsay *et al.*, 1985) imbricate a dominantly siliciclastic succession (Sørøy Group), passing from fluviatile and shallow marine sandstones through mudstones to deeper water turbidites (Ramsay, 1971; Roberts 1968). The lower part has been cut by dolerite dykes (Gayer *et al.*, 1985b; Rice, 1987) that have similar compositions to those described from farther south in Scandinavia (Solyom *et al.*, 1979), which have been dated at ca. 640 Ma, implying a pre-mid Vendian age of deposition. Similarly, the youngest rocks were deformed in the Finnmarkian event at 540–530 Ma (Pedersen *et al.*, 1988—this volume) indicating a pre-mid Cambrian deposition age.

The rocks of the Kalak Nappe Complex in the Svaerholt and Nordkinn Peninsulas (Fig. 1) are similar to those in the Berlevåg Formation (Roberts and Andersen 1985; Levell and Roberts, 1977), which has recently been placed in the Lower Allochthon (Rice *et al.*, this volume 1988). However, the possible correlation of the rocks of the Kalak Nappe Complex on the Svaerholt and Nordkinn peninsulas with the Berlevåg Formation, and thus with the sediments of the Lower Allochthon, may be of considerable importance in that it may provide a sedimentological link between the Middle and Lower Allochthons. The age relationship of the sediments on the two peninsulas with those in the Kalak Nappe Complex west of Porsangerfjord is uncertain but since the rocks on the peninsulas have been intruded by dolerite dykes of similar composition to those

described above (Townsend, 1986), a pre-mid Vendian age, equivalent to the lower part of the Sørøy Group, is probable.

BARENTS SEA CALEDONIDES

In this article the term Barents Sea Caledonides includes all the rocks lying to the north of the T–K fault. Two main groups have been recognized within this region, the Barents Sea Group and the unconformably overlying Løkvikfjell Group (cf. Johnson *et al.*, 1978).

Dolerite dykes in the Barents Sea Group and lowest part of the Løkvikfjell Group have compositions similar to those in the Kalak Nappe Complex and to those from central Scandinavia (Roberts, 1975; Gayer *et al.*, 1985b; Rice, 1987; Solyom *et al.*, 1979), which have been interpreted as being syn-Iapetus rifting intrusions. However, since the dykes in the Barents Sea Caledonides lie parallel to the axial surface of folds but contain a penetrative axial planar fabric, their intrusion has been interpreted previously as post-dating the initial development of the folds (Roberts, 1972). We prefer to interpret the dykes as pre-dating the folds, which subsequently developed at the dyke margin anisotropy (cf. Gayer *et al.*, 1978; Krill, 1986). Thus, the ca. 640 Ma age of the dykes (Beckinsale *et al.*, 1976) indicates a major extensional event in the Barents Sea Caledonides, contemporaneous with extension farther south in the Caledonides (Claesson and Roddick, 1983), and which post-dated the onset of sedimentation and pre-dated Caledonian deformation.

The Barents Sea Group is a 9 km thick succession that has been divided into four formations (Fig. 2). The group represents a regressive sequence from deep marine turbidites (Kongsfjord Formation) through deltaic facies (Båsnaering Formation) to shallow marine and intertidal facies (Båtsfjord and Tyvjofjell Formations), including the stromatolitic dolomites in the Båtsfjord Formation (Siedlecka, 1978). Palaeocurrent directions suggest a southwesterly source, with the shoreline prograding to the north and northeast (Pickering, 1984); the great thickness of the Barents Sea Group suggests tectonic control with major subsidence influencing sedimentation (Siedlecka and Edwards, 1980). Radiometric dating of the lower part of the group indicates deposition around 807 Ma (Råheim, personal communication cited in Siedlecka and Edwards, 1980), confirmed by acritarch assemblages (Vidal and Siedlecka, 1983).

The Løkvikfjell Group unconformably overlies the Barents Sea Group, lying on stratigraphically older sediments farther to the west. The 5.6 km thick group has been divided into five formations (Siedlecki and Levell, 1978) that reflect a transgressive–regressive–transgressive sequence of clastic sediments (cf. Johnson *et al.*, 1978). Although the Løkvikfjell Group has been correlated with the lower part of the Laksefjord Group, it seem unlikely that a detailed lithostratigraphic correlation is possible (Føyn, 1969; Laird, 1972a,b). Acritarch assemblages cannot give a precise age for the sediments in the Løkvikfjell Group, but do indicate that deposition was contemporary with the Varanger glaciation or younger. Only the Sandfjord Formation (Fig. 2) has been cut by the ca. 640 Ma dolerite dykes (Beckinsale *et al.*, 1976).

The Berlevåg Formation, which crops out in the northwest corner of Varangerhalvøya (Fig. 1), comprises ca. 2.65 km of dominantly distal turbidites with subordinate sandstones and conglomerates (Levell and Roberts, 1977). One dolerite dyke cutting these sediments gave a K–Ar age of 542 Ma. This age is difficult to interpret if the Berlevåg Formation is part of the Lower Allochthon (Scandian deformation; cf. Rice *et al.* 1988–this volume), but could represent either a Finnmarkian metamorphic age, suggesting that the northwest margin of the Lower Allochthon was deformed in the Finnmarkian event, or a partial Scandian resetting of a ca. 640 Ma age of emplacement, which has resulted in a purely fortuitous Finnmarkian age. However, although the age of the metamorphism cannot be constrained from the available radiometric data, the presence of the dyke, thought to be of the same origin as those in the Barents Sea Caledonides, suggests a pre-mid Vendian age of deposition, making correlation with the upper part of the Laksefjord Group (Laird, 1972a) untenable. The alternative suggestion, that the Berlevåg Formation is a correlative of the Kongsfjord Formation (cf. Siedlecka and Siedlecki, 1972; Pickering, 1979) is structurally and sedimentologically possible, although there are significant differences in lithologies.

RESTORATION OF FINNMARK AND BARENTS SEA CALEDONIDES

The restoration of the Caledonian deformation in Finnmark has been documented by Gayer *et al.* (1987) and Rice *et al.* (1989). Although the former account assumed an early Caledonian age for the main thrusting event, only minor modifications to the restoration are required in view of the late Caledonian age of deformation demonstrated in the external part of the orogen by Dallmeyer *et al.* (1988b). Gayer et al. (1987) constructed a 169 km long balanced cross-section from Sørøy to 46 km east of Porsangerfjord, which restores to ca. 800 km. Within this restoration, the trailing branch line of the Gaissa Thrust Belt restores to ca. 240 km, which, including the area of the Finnmark Supergroup to the east of the section line, gives a ca. 400 km wide Gaissa Basin. The trailing branch lines of the Laksefjord Nappe Complex and the Revsbotn Basement Horse restore to 460 km, indicating that the positive basement area, the Finnmark Ridge, that separated the Gaissa Basin from the Baltoscandian miogeocline was up to ca. 220 km wide.

The restoration of the Barents Sea Caledonides along the T–K Fault involves a model that requires the same displacement rates for strike-slip faulting and for thrusting, such that rocks which overthrust the line of the T–K Fault Zone prior to the onset of strike-slip movement were not displaced differentially. This has been achieved by arguing that both rates were controlled by regional plate motions. Strike-slip movement along the T–K fault was initiated after the emplacement of the Berlevåg Formation and 18 km of shortening in the Barents Sea Caledonides. During the imbrication of the Laksefjord Nappe Complex and the basement horses to the south of the fault, ca. 220 km of southeast directed dextral strike-slip occurred followed by further ca. 170 km of east-southeast directed faulting at the same time that imbrication of the Gaissa Thrust Belt occurred (Fig. 5).

PALAEOGEOGRAPHIC MODEL

The restoration proposed by Rice *et al.* (1989; Fig. 5) allows a more detailed palaeogeography for the late Riphean to early Cambrian to be inferred. However, the basic model, in which a topographic high (Finnmark Ridge) separated areas of different sedimentation is similar to that proposed by Gayer and Roberts (1973), whilst the development of a thick sedimentary sequence in a basin to

the north (Timanian aulacogen) follows the model of Siedlecka (1975).

Three "basins" have been recognized. To the south of the T–K Fault, rocks of the Finnmark Supergroup were deposited in the Gaissa Basin. These are thickest in the east and thin both to the west, onto the Finnmark Ridge, and to the south, onto the Baltic Shield. To the north of the T–K Fault, rocks belonging to the Vardø Association were deposited in the Timanian aulacogen; these are exposed in the Barents Sea Caledonides as well as in the northern part of the Komagfjord Antiformal Stack and in the Laksefjord Nappe Complex; farther east from the restored position of the Barents Sea Caledonides, the precise location of the boundary between the Gaissa Basin and Timanian aulacogen is unknown; almost certainly the T–K Fault does not represent the boundary, but probably was the most southerly of a series of major faults dividing the two basins (cf. Lipard and Roberts, 1987; Townsend, 1988). Finally, to the northwest of the Finnmark Ridge, sediments were deposited in the Baltoscandian miogeocline. The proposed palaeogeography clearly indicates that the sediments of the Baltoscandian miogeocline and Timanian aulacogen overlapped and merged (Fig. 5); this is supported by the preliminary geochemical work of Ottesen *et al.* (1985). In this context it is possible that the rocks of the Berlevåg Formation form an important sedimentary link between the Barents Sea Caledonides (Lower Allochthon) and the Kalak Nappe Complex (Middle Allochthon).

Sedimentation in the Timanian aulacogen and Gaissa Basin occurred in two distinct episodes. A late Riphean to late Sturtian period of sedimentation was followed by uplift and erosion. This was followed by an extensive period of deposition in the Gaissa Basin, from early Vendian to early Tremadoc times, although in the Timanian aulacogen no rocks of demonstrably Cambrian or younger age have been found (Fig. 3). Owing to major problems of dating the sediments of the Kalak Nappe Complex, this article deals only with the most south-easterly part of the Baltoscandian miogeocline (Laksefjord Group and correlatives). Thus, a somewhat unbalanced view of the development of the main Caledonian ocean has resulted. However, since the sediments at the base of the Kalak Nappe Complex have been cut by metadolerite dykes that have similar geochemical composition to those found in several other parts of the Caledonides (Roberts, 1975; Solyom *et al.*, 1979; Gayer *et al.*, 1985b; Rice, 1987) for some of which an emplacement date of ca. 640 Ma has been determined (Beckinsale *et al.*, 1976; Claesson and Roddick, 1983; Sundvoll, 1987), it has been inferred that at least the oldest sediments in the Kalak Nappe Complex are of pre-mid Vendian age.

The considerable thickness of the Barents Sea Group suggests that deposition occurred in a fault-controlled basin; Pickering (1984) has demonstrated that the "axis" of sedimentation during the deposition of the lower part of the Barents Sea Group had an orientation essentially parallel to the T–K Fault. Consequently, the early history of the T–K Fault line has been interpreted as a major extensional fault that controlled deposition in this region during the later part of the Riphean and early Sturtian. However, since sedimentation also occurred immediately to the south of the T–K Fault (cf. Pickering, 1984) during deposition of the Kongsfjord Formation, little or no footwall uplift was taking place at this time; this may imply that extension occurred considerably earlier and that thickening of the isotherms under the footwall had occurred, leading to subsidence south of the T–K Fault.

During deposition of the turbitites of the Kongsfjord

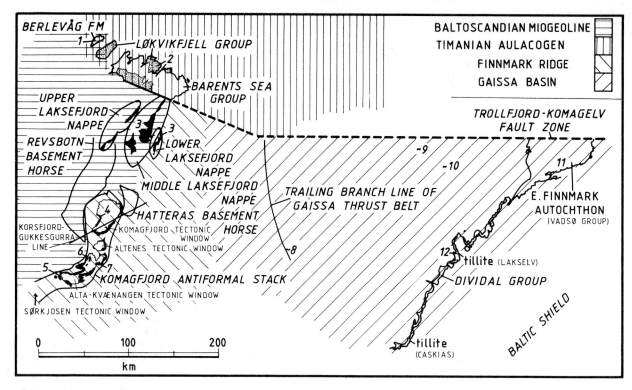

Fig. 5. Palinspastic reconstruction of the rocks now comprising the Lower Allochthon and autochthon in the Caledonides of Finnmark. The distribution of the three major sedimentary basins and of the Finnmark Ridge is also shown. For details of this restoration see Gayer *et al.* (1987) and Rice *et al.* (1989).

Formation, Pickering (1984) has inferred that a major deltaic system developed to the south, in the area that later formed the Finnmark Ridge (Fig. 6A). About 400 km to the southeast, the deltaic sediments of the Vadsø Group (Banks *et al.*, 1974) were deposited in a tectonic position (relative to the T–K Fault) compatible with the inferred deltaic sediments deposited to the south of the Kongsfjord Formation (Fig. 6B). Subsequently, north-wards progradation of the deltaic system occurred, with the deposition of the Båsnaering Formation. During this period, the ?fluviatile sediments of the Airoaivi Group were deposited in the western side of the Gaissa Basin.

This suggests that, during the deposition of the lower part of the Barents Sea Group, deltaic systems were fed by rivers running off the Baltic Shield to the south, whilst to the north a deep water basin developed. With time, infilling of the basin margin occurred, leading the north-wards progradation and fluviatile sedimentation (upper Båsnaering Formation).

Following the deposition of the Båsnaering Formation, subsidence and southwards transgression occurred in the Timanian aulacogen, with the development of a variety of tidal environments (Båtsfjord Formation), including dolomitic stromatolites (Siedlecka, 1978; Fig. 6C). By correlating the dolomites in the Adamsfjord Formation with those in the Båtsfjord Formation, it is implicit that marine sedimentation transgressed across the T–K Fault line. At the same time a major interval in sedimentation occurred in the Gaissa Basin, reflected in the eastern areas by the unconformity between the Golnaselv and Ekkerøy Formations (Fig. 3).

Sedimentation in the Tanafjord Group reflects a grad-ual westwards transgression. Not only are the formations thickest in the east, thinning considerably between Tana-fjord and Porsangerfjord (Fig. 2, columns 8 and 10) and still further onto the basement rocks incorporated in the Komagfjord Antiformal Stack, but younger formations

overlie the basement in the west (Gamasfjell Formation rather than Grønnes Formation). These sedimentation patterns are interpreted to reflect positive areas both east and southeast of the area of the Adamsfjord Formation, possibly controlled by extensional faulting (Fig. 6C).

The deposition of siliciclastic sediments in the Tyvjofjell Formation, concurrent with the sediments of the middle part of the Tanafjord Group, may reflect a regional subsidence, with an influx of coarser detritus derived from the Baltic Shield (Fig. 6D). Up to this period, prior to any development of an Iapetus rift zone, there is no compelling evidence for the development of a Finnmark Ridge.

Evidence for subsequent eastward regression in the Gaissa Basin can be found in the latter part of the deposition of the Tanafjord Group. Sabkha facies and stromatolitic dolomites have been described from both the eastern and central parts of the Gaissa Basin (columns 8 and 10, Fig. 5; Tucker, 1976, 1977; Siedlecka, 1976) and possibly from the westernmost part of the Komag-fjord Antiformal Stack, if the correlation of the Burfjord Dolomites with the Porsanger and Grasdal Formations is accepted (Fig. 7A). These sediments reflect a shallow marine environment with a minimal clastic sediment input. Comparison of the stromatolite faunas between the Porsangerfjord and Tanafjord areas indicate that those in the east are somewhat younger (Bertrand-Sarfarti and Siedlecka, 1980), indicating that uplift af-fected western areas earlier. Contemporaneous with this uplift was the deposition of the shallow marine siliciclastic sedimentation (upper part of the Tyvjofjell Formation) in the Timanian aulacogen. In Fig. 7A it has been inferred that these deposits were derived from a positive basement source to the south developed during the earliest Iapetus rifting extension. Southward directed palaeocurrents have been recorded in the upper part of the Tanafjord Group in the Porsangerfjord area (Tucker, 1976, 1977).

The boundary between the Sturtian and the Vendian

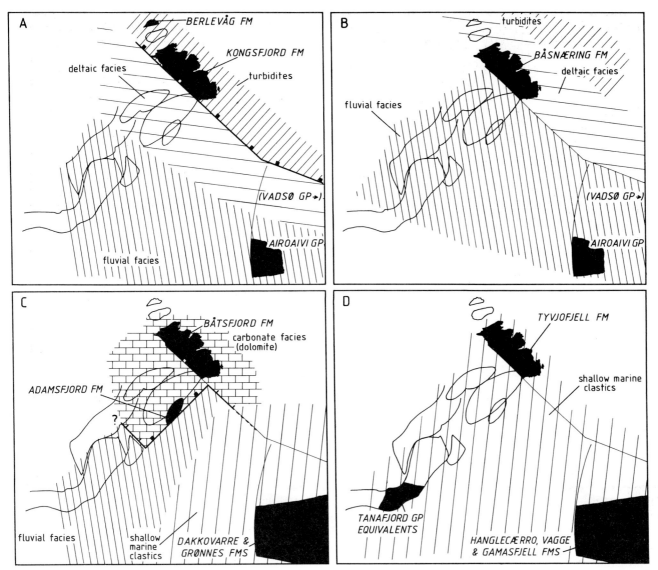

Fig. 6. Restored palaeogeography for the pre-Vendian sediments in the western part of the Lower Allochthon in Finnmark. Time intervals of the diagrams are shown in Fig. 3. See text for details. Square blocks on faults indicate extensional faults; small ticks indicate transfer faults.

phases of deposition is marked by a significant, although gentle, unconformity that has been described from both the Timanian aulacogen and the Gaissa Basin (Johnson *et al.*, 1978; Siedlecki and Levell, 1978). In the eastern part of the Gaissa Basin > 1500 m of sediment were eroded; in the west the amount cannot be estimated. In the Timanian aulacogen the lowest beds in the Løkvikfjell Group rest on older beds in the Barents Sea Group further to the west, with in the order of 8 km eroded in some areas.

Our understanding of the pattern of sedimentation during the Vendian is less clear than for the Riphean and Sturtian periods. This is because in addition to the tectonic effects caused by the rifting of Iapetus, major eustatic changes in sea level resulted from the Varanger glaciation during the early part of the Vendian. Further problems are also caused by the lack of precise age data from the sediments of the Løkvikfjell and Laksefjord Groups. The coarse alluvial deposits at the base of the Laksefjord Group were deposited in a rapidly subsiding, fault-controlled basin; the recognition of clast lithologies in the Ifjord Formation similar to those in the Storfjord Basement Complex (Føyn *et al.*, 1983) suggests a proximal source. Restoration of the Caledonides of Finnmark (Gayer *et al.*, 1987) also revealed that major NE–SW oriented extensional faulting in four zones (Fig. 7B) was

responsible for the early development of the three basement horses, reflecting the initial extensional phase in the rifting of Iapetus that may correlate with the ca. 640 Ma extensional dyke emplacement event in the Barents Sea Caledonides (Beckinsale *et al.*, 1976). However, the differences in thickness between the Laksefjord Group and Lomvatn Formation (Fig. 2, columns 3 and 4) demonstrates that significant faulting also occurred on a NW–SE line, downthrowing to the northeast, and this may reflect the development of transfer faults associated with northwesterly extension.

The amount of erosion that occurred prior to the deposition of the Smalfjord Formation tillite was extremely variable, being greatest in the east (cf. Fig. 2, columns 9–11) and apparently less over the area of the basement horses, since much of the thinned Tanafjord Group is preserved (Fig. 2, columns 5–7). This ammount of erosion would appear to be glacially, rather than tectonically controlled. Deposits resulting from the early glacial advance (Smalfjord Formation) and subsequent interglacial marine sediments are absent from the Finnmark Ridge and western part of the Gaissa Basin (Fig. 7B), whilst deposits from the dominantly marine later glacial advance (Mortensnes Formation) have been found over the Finnmark Ridge (Fig. 7C). The lack of glacial deposits, including the

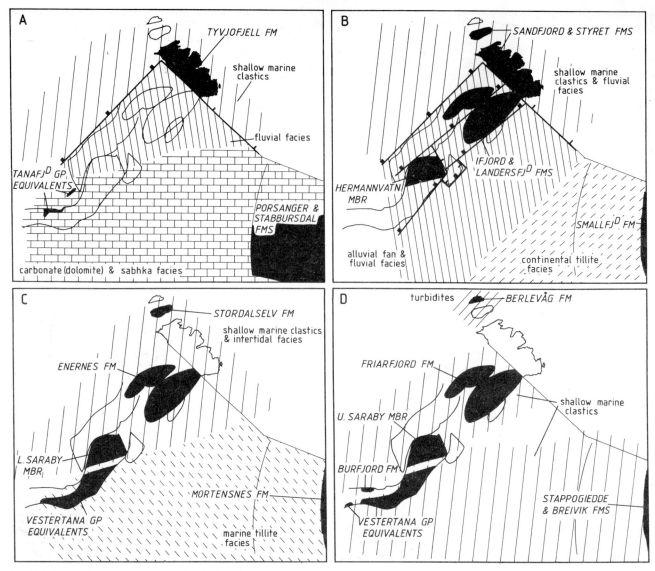

Fig. 7. Restored palaeogeography for the late Sturtian to lower Cambrian sediments in the western part of the Lower Allochthon of Finnmark. Time interval of the diagrams is shown in Fig. 3. See text for details.

younger, dominantly marine unit, in the Baltoscandian miogeocline and Timanian aulacogen may reflect a wide variety of processes, including contemporaneous oceanographic and geomorphological influences.

However, it is clear that the marine deposits of the Mortensnes Formation were followed by those of the Stappogiedde and Breivik Formations (Vestertana Group) and these are preserved as a thinned sequence in the Komagfjord Antiformal Stack (Fig. 2, columns 7 and 9). Thus even ca. 50 Ma after Iapetus rifting (taken at ca. 640 Ma) the Finnmark Ridge was still a significant positive area, controlling sedimentation. Direct correlation of the sediments on the northwest flank of the ridge with those in the Gaissa Basin is not possible, but the deposition of marine sediments in the Gaissa Basin implies that deeper water marine sedimentation was probably occurring at the same time in the Baltoscandian miogeocline and Timanian aulacogen, seen as the black shales of the upper part of the Laksefjord Group and in the Lomvatn and Burfjord Formations (Fig. 7D).

CONCLUSIONS

Three sedimentary basins developed in the region now preserved as the Lower Allochthon and autochthon of the

Caledonides of Finnmark, prior to Caledonian thrusting and strike-slip faulting. These are: (1) the ENE–WNW-oriented Gaissa Basin in which up to 5 km of late Riphean to early Ordovician mainly fluvial or shallow marine clastic sediments accumulated; (2) the WNW–ESE-trending Timanian aulacogen containing up to 14 km of late Riphean to Sturtian dominantly clastic marine sediments derived in part from the Finnmark Ridge area to the southwest; and (3) the NE–SW-oriented Baltoscandian miogeocline with up to 8 km of pre-mid Vendian to early Cambrian fluvial to shallow marine clastic sediments.

The status of the metasediments of the Sørøy Group (Ramsay, 1971; Roberts, 1968), interpreted generally as the more distal facies of the Baltoscandian miogeocline is uncertain. However, preliminary geochemical work by Ottesen *et al.* (1985) indicates similarities between the Barents Sea Caledonides and the rocks of the Kalak Nappe Complex, suggesting that the Kalak Nappe Complex can be "tied" to the autochthon.

Sedimentation in the first two basins outlined occurred in two episodes: late Riphean to middle/late Sturtian sedimentation ended in uplift; regression and erosion before major transgression occurred during the early Vendian. Sedimentation then continued to the Tremadoc.

Deposition was controlled by two sets of major extensional faults; one set oriented NE–SW and downthrowing to the northwest, and the other set oriented NW–SE and downthrowing to the northeast. Reactivation of these syndepositional faults may have controlled the location of major Caledonian thrust faults. The NE–SW striking set were inverted to form frontal thrusts, with associated footwall shortcut faulting, whilst the other set were reactivated as either lateral thrust ramps or, in the case of the T–K Fault, as a major dextral strike-slip fault.

Acknowledgements

We thank the Royal Society for a research travel grant and our colleagues for discussions. We are grateful to Kevin Pickering for suggesting improvements to an earlier version of this article.

REFERENCES

Andersen, T. B., Austrheim, H. B., Sturt, B. A., Pedersen, S. and Kjaesrud, K. (1982). Rb–Sr whole rock ages from Magerøy, North Norwegian Caledonides. *Norsk Geol. Tidsskr.* **62**, 79–85.

Banks, N. L. and Røe, S.-L. (1974). Sedimentology of the late Precambrian Golneselv Formation, Varangerfjorden, Finnmark. *Norges Geol. Unders.* **303**, 17–38.

Banks, N. L., Edwards, M. B., Geddes, W. P., Hobday, D. K. and Reading, H. G. (1971). Late Precambrian and Cambro-Ordovician sedimentation in East Finnmark. *Norges Geol. Unders.* **269**, 197–236.

Banks, N. L., Hobday, D. K., Reading, H. G., and Taylor, P. N. (1974). Stratigraphy of the late Precambrian "Older Sandstone" Series of the Varangerfjord area, Finnmark. *Norges Geol. Unders.* **303**, 1–16.

Beckinsale, R. D., Reading, H. G. and Rex, D. C. (1976). Potassium-argon ages for basic dykes from East Finnmark: stratigraphical and structural implications. *Scott. J. Geol.* **12**, 51–65.

Bertrand-Sarfarti, J. and Siedlecka, A. (1980). Columnar stromatolites of the terminal Precambrian Porsanger Dolomite and Grasdal Formations of Finnmark, north Norway. *Norsk Geol. Tidsskr.* **60**, 1–27.

Bowden, P. L. (1981). Basement-cover distribution within the Caledonian nappe sequence in the northwestern envelope of the Alta-Kvaenangen Window. *Terra Cognita* **1**, 35–36.

Chapman, T. J. (1980). The geological evolution of the Laksefjord Nappe Complex, Finnmark, North Norway. Unpublished Ph.D. thesis, University of Wales, Cardiff.

Chapman, T. J., Gayer, R. A. and Williams, G. D. (1985). Structural cross-sections through the Finnmark Caledonides and timing of the Finnmarkian event. In: Gee, D. G. and Sturt, B. A. (eds.) *The Caledonide Orogen—Scandinavia and Related Areas*, pp. 593–609. Wiley, Chichester.

Claesson, S. and Roddick, J. C. (1983). ^{40}Ar/^{39}Ar data on the age and metamorphism of the Ottfjället dolerites, Sarv Nappe, Swedish Caledonides. *Lithos* **16**, 61–73.

Dallmeyer, R. D. (1988). Polyorogenic ^{40}Ar/^{39}Ar mineral age record within the Kalak Nappe Complex, northern Scandinavian Caledonides. *J. Geol. Soc. Lond.*, in press.

Dallmeyer, R. D., Reuter, A. and Clauer, N. (1988a). Age of cleavage formation in the lower allochthons of Finnmark, Norway. *Geol. Soc. Amer. Bull.* **145**, 705–716.

Dallmeyer, R. D., Mitchell, J. G., Pharaoh, T. C., Reuter, A. and Andresen, A. (1988b). K–Ar and ^{40}Ar/^{39}Ar whole rock ages of slate/phyllite from allochthonous basement and cover in the tectonic windows of Finnmark, Norway: Evaluating the extent and timing of Caledonian tectonothermal activity. *Geol. Soc. Amer. Bull.*, in press.

Edwards, M. B. and Føyn, S. (1981). Late Precambrian tillites in Finnmark, North Norway. In: Hambrey, M. J. and Harland, W. B. (eds.), *Earth's Pre-Pliestocene Glacial Record*, pp. 605–610. Cambridge University Press, Cambridge.

Fareth, E. (1979). Geology of the Altenes area, Alta-Kvaenangen Window, North Norway. *Norges Geol. Unders.* **351**, 13–30.

Føyn, S. (1967). Dividal-gruppen ("Hyolithus-sonen") i Finn-mark og dens forhold til de eokambriske-kambriske formasjoner. *Norges Geol. Unders.* **249**, 1–84.

Føyn, S. (1969). Laksefjord-gruppen ved Tanafjord. *Norges Geol. Unders.* **258**, 5–16.

Føyn, S. (1985). The late Precambrian in northern Scandinavia. In Gee, D. G. and Sturt, B. A. (eds.) *The Caledonide Orogen—Scandinavia and Related Areas*, pp. 233–245. Wiley, Chichester.

Føyn, S. and Glaessner, M. F. (1979). Platysolenites, other animal fossils and the late Precambrian-Cambrian transition in Norway. *Norsk Geol. Tidsskr.* **59**, 25–46.

Føyn, S. and Siedlecki, S. (1980). Glacial stadials and interstadials of the late Precambrian Smalfjord tillite on Laksefjordvidda, Finnmark, North Norway. *Norges Geol. Unders.* **358**, 31–45.

Føyn, S., Chapman, T. J. and Roberts, D. (1983). Adamsfjord og Ul'lugaisa. Beskrivelse til de berggrunnsgeologiske kart 2135 I og 2135 II – M 1:50.000. *Norges Geol. Unders.* **381**, 1–78.

Gayer, R. A. and Roberts, J. D. (1973). Stratigraphic review of the Finnmark Caledonides with possible tectonic implications. *Proc. Geol. Assoc.* **84**, 405–428.

Gayer, R. A., Powell, D. B. and Rhodes, S. (1978). Deformation against metadolerite dykes in the Caledonides of Finnmark, Norway. *Tectonophysics* **46**, 99–115.

Gayer, R. A., Hayes, S. J. and Rice, A. H. N. (1985a). The structural development of the Kalak Nappe Complex of Eastern and Central Porsangerhalvøya, Finnmark, Norway. *Norges Geol. Unders. Bull.* **400**, 67–87.

Gayer, R. A., Humphreys, R. J., Binns, R. E. and Chapman, T. J. (1985b). Tectonic modelling of the Finnmark and Troms Caledonides based on high level igneous rock geochemistry. In: Gee, D. G., and Sturt, B. A. (eds.) *The Caledonide Orogen—Scandinavia and Related Areas*, pp. 931–951. Wiley, Chichester.

Gayer, R. A., Rice, A. H. N., Roberts, D., Townsend, C. and Welbon, A. (1987). Restoration of the Caledonian Baltoscandian margin from balanced cross-sections: the problem of excess continental crust. *Trans. R. Soc. Edinburgh* **78**, 197–217.

Gee, D. G., Kumpulainen, R., Roberts, D., Stephens, M. B., Thon, A. and Zachrisson, E. (1985). Scandinavian Caledonides: Tectonostratigraphic map. *Sver. Geol. Unders. Ser.* **BA**, No 35.

Harland, W. B., Cox, A. V., Llewellyn, P. G., Pickton, C. A. G., Smith, A. G. and Walters, R. (1982). *A Geologic Time-scale.* Cambridge University Press.

Hobday, D. K. (1974). Interaction between fluvial and marine processes in the lower part of the late Precambrian Vadsø Group, Finnmark. *Norges Geol. Unders.* **303**, 39–56.

Johnson, H. D., Levell, B. K. and Siedlecka, A. (1978). Late Precambrian sedimentary rocks in East Finnmark, North Norway and their relationship to the Trollfjord-Komagelv Fault. *J. Geol. Soc. Lond.* **135**, 517–533.

Kjøde, J. K., Storvedt, M., Roberts, D. and Gidskehaug, A. (1978). Palaeomagnetic evidence for large-scale dextral movement along the Trollfjord–Komagelv Fault, Finnmark, North Norway. *Phys. Earth Planet. Int.* **161**, 132–144.

Krill, A. G. (1986). Eidsvoll Quarry, Oppdal, South Norway: a one-outcrop model for some aspects of Trollheimen-Dovre-fjell tectonics. *Norges Geol. Unders. Bull.* **404**, 21–30.

Laird, M. G. (1972a). Sedimentation in the ?late Precambrian Raggo Group, Varanger Peninsula. *Norges Geol. Unders.* **278**, 1–12.

Laird, M. G. (1972b). Stratigraphy and sedimentology of the Laksefjord Group, Finnmark. *Norges Geol. Unders.* **278**, 13–40.

Levell, B. K. and Roberts, D. (1977). A re-investigation of the geology of northwest Varanger Peninsula, East Finnmark, North Norway. *Norges Geol. Unders.* **334**, 83–90.

Lipard, S. J. and Roberts, D. (1987). Fault systems in Caledonian Finnmark and the southern Barents Sea. *Norges Geol. Unders. Bull.* **410**, 55–64.

Milnes, A. G. and Ritchie, A. (1962). Contribution to the geology of the Kvaenangen Window, Burfjord, Troms, Norway. *Norsk Geol. Tidsskr.* **42**, 77–102.

Nicolaisen, F. and Henningsmoen, G. (1985). Upper Cambrian and lower Tremadoc Olenelid trilobites from the Digermul Peninsula, Finnmark, northern Norway. *Norges Geol. Unders. Bull.* **400**, 1–49.

Ottesen, R. T., Bolviken, B. and Volden, T. (1985). Geochemical provinces in the northern parts of the Baltic Shield and Caledonides: preliminary results. *Norges Geol. Unders. Bull.* **403**, 197–208.

Pedersen, R. B., Dunning, G. and Robins, B. (1988). The U–Pb (zircon) age of a nepheline syenite pegmatite within the Seiland Magmatic Province, North Norway, *this volume*.

Pharaoh, T. (1985). The stratigraphy and sedimentology of autochthonous metasediments in the Repparfjord-Komagelv Tectonic Window, west Finnmark. In: Gee, D. G. and Sturt, B. A. (eds.) *The Caledonide Orogen—Scandinavia and Related Areas*, pp. 347–357. Wiley, Chichester.

Pharaoh, T. C., Ramsay, D. M. and Jansen, Ø. (1983). Stratigraphy and structure of the northern part of the Repparfjord–Komagfjord Window, Finnmark, northern Norway. *Norges Geol. Unders.* **377**, 1–45.

Pickering, K. T. (1979). A Precambrian submarine fan and upper basin-slope succession in the Barents Sea Group, Finnmark, North Norway. Unpublished D.Phil. thesis, University of Oxford, England.

Pickering, K. T. (1984). Kongsfjord turbidite system, Norway. In: Bouma, A. H., Normark, W. R. and Barnes, N. E. (eds.) *Submarine Fans and Related Turbidite Systems*, pp. 237–244. Springer-Verlag, New York.

Pringle, I. R. (1973). Rb–Sr age determinations on shales associated with the Varanger Ice Age. *Geol. Mag.* **109**, 465–472.

Ramsay, D. M. (1971). The stratigraphy of Sørøy. *Norges Geol. Unders.* **269**, 314–322.

Ramsay, D. M., Sturt, B. A., Zwaan, K. B. and Roberts, D. (1985). Caledonides of northern Norway. In: Gee, D. G. and Sturt, B. A. (eds.). *The Caledonide Orogen—Scandinavia and Related Areas*, pp. 163–184. Wiley, Chichester.

Reading, H. G. (1965). Eocambrian and Lower Palaeozoic geology of the Digermul Peninsula, Tanafjord, Finnmark. *Norges Geol. Unders.* **234**, 167–191.

Reading, H. G. and Walker, R. G. (1966). Sedimentation of Eocambrian tillites and associated sediments in Finnmark, Northern Norway. *Palaeogeog. Palaeoclimatol. Palaeoecol.* **2**, 177–212.

Reitan, P. H. (1963). The geology of the Komagfjord Tectonic Window of the Raipas Suite, Finnmark, Norway. *Norges Geol. Unders.* **221**, 1–71.

Rhodes, S. (1976). The geology of the Kalak Nappe and its relation to the northeast margin of the Komagfjord Tectonic Window. Unpublished Ph.D. thesis, University of Wales, Cardiff.

Rice, A. H. N. (1987). A tectonic model for the evolution of the Finnmarkian Caledonides of North Norway. *Can. J. Earth Sci.* **24**, 602–616.

Rice, A. H. N., Bevins, R. E., Robinson, D. and Roberts, D. (1988). Evolution of low-grade metamorphic zones in the Caledonides of Finnmark, North Norway, *this volume*.

Rice, A. H. N., Gayer, R. A., Robinson, D. and Bevins, R. E. (1989). Strike-slip restoration of the Barents Sea Caledonides terrane, Finnmark, North Norway. *Tectonics*, in press.

Roberts, D. (1968). Hellefjord Schist Group—a probable turbidite formation from the Cambrian of Sørøy, West Finnmark. *Norsk Geol. Tidsskr.* **48**, 231–244.

Roberts, D. (1972). Tectonic deformation in the Barents Sea region of Varanger Peninsula, Finnmark. *Norges Geol. Unders.* **282**, 1–39.

Roberts, D. (1975). Geochemistry of dolerite and metadolerite dykes from Varanger Peninsula, Finnmark, North Norway. *Norges Geol. Unders.* **322**, 55–72.

Roberts, D. and Andersen, T. B. (1985). Nordkapp, beskrivelse til det berggrunnsgeologiske kartbladet, M. 1:250,000. *Norges Geol. Unders. Skr.* **61**, 1–49.

Siedlecka, A. (1975). Late Precambrian stratigraphy and structure of the northeastern margin of the Fennoscandian Shield (East Finnmark–Timan Region). *Norges Geol. Unders.* **316**, 313–348

Siedlecka, A. (1976). Silicified Precambrian evaporite nodules from northern Norway: a preliminary report. *Sedimentary Geol.* **16**, 193–204.

Siedlecka, A. (1978). Late Precambrian tidal flat deposits and algal stromatolites in the Båtsfjord Formation, East Finnmark, North Norway. *Sedimentary Geol.* **21**, 277–310.

Siedlecka, A. 1987. Skoganvarre berggrunnskart 2034 4, 1:50,000. Forelopig utgave. *Norges Geol. Unders.*

Siedlecka, A. and Edwards, M. B. (1980). Lithostratigraphy and sedimentology of the Riphean Båsnaering Formation, Varanger Peninsula, North Norway. *Norges Geol. Unders.* **355**, 27–48.

Siedlecka, A. and Roberts, D. (1972). A late Precambrian tilloid from Varangerhalvøya—evidence of both glaciation and subaqueous mass movement. *Norsk Geol. Tidsskr.* **52**, 135–141.

Siedlecka, A. and Siedlecki, S. (1967). Some new aspects of the geology of Varanger Peninsula (northern Norway). *Norges Geol. Unders.* **247**, 288–306.

Siedlecka, A. and Siedlecki, S. (1971). Late Precambrian sedimentary rocks of the Tanafjord-Varangerfjord area of Varanger Peninsula, Northern Norway. *Norges Geol. Unders.* **269**, 246–294.

Siedlecka, A. and Siedlecki, S. (1972). Lithostratigraphical correlation and sedimentology of the late Precambrian of Varanger Peninsula and neighbouring areas of East Finnmark, Northern Norway. *XXIV International Geological Congress*, Montreal, Vol. 6, pp. 349–358.

Siedlecki, S. and Levell, B. K. (1978). Lithostratigraphy of the Late Precambrian Løkvikfjell Group on Varanger Peninsula, East Finnmark, North Norway, *Norges Geol. Unders.* **343**, 73–85.

Skjerlie, F. and Tan, T. H. (1960). The geology of the Caledonides of the Reisa Valley area, Troms-Finnmark, North Norway. *Norges Geol. Unders.* **213**, 175–197.

Solyom, Z., Gorbatschev, R. and Johansson, I. (1979). The Ottfjäll dolerites: geochemistry of the dyke swarm in relation to the geodynamics of the Caledonide orogen of Central Scandinavia. *Sver. Geol. Unders.* **C786**, 1–38.

Stephens, M. B. and Gee, D. G. (1988). Terranes and polyphase accretionary history in the Scandinavian Caledonides. In: Dallmeyer, R. D. (ed.) *Terranes in the Circum-Atlantic Palaeozoic Orogens*. Geological Society of America Special Publication, in press.

Sundvoll, B. (1987). The age of the Egersund dyke-swarm, SW Norway: some tectonic implications. *Terra Cognita* **7**, 180.

Townsend, C. (1986). Thrust tectonics within the Caledonides of northern Norway. Unpublished Ph.D. thesis, University of Wales, Cardiff.

Townsend, C. (1988). The inner shelf of North Cape, North Norway and its implications for the Barents Shelf–Finnmark Caledonian boundary. *Norsk Geol. Tidsskr.* **67**, 151–153.

Townsend, C., Roberts, D., Rice, A. H. N. and Gayer, R. A. (1986). The Gaissa Nappe, Finnmark, north Norway: an example of a deeply eroded external imbricate zone. *J. Struct. Geol.* **8**, 431–440.

Townsend, C., Rice, A. H. N. and Mackay, A. (1988). The structure and stratigraphy of the southwestern portion of the Gaissa Thrust Belt and adjacent Kalak Nappe Complex, Finnmark, North Norway, *this volume*.

Tucker, M. (1976). Replaced evaporites from the late Precambrian of Finnmark, Arctic Norway. *Sedimentary Geol.* **16**, 193–204.

Tucker, M. (1977). Stromatolite biostromes and associated facies in the late Precambrian Porsanger Dolomite Formation of Finnmark, Arctic Norway. *Palaeogeog. Palaeoclimatol. Palaeoecol.* **21**, 55–83.

Vidal, G. and Siedlecka, A. (1983). Planktonic, acid-resistant microfossils from the upper Proterozoic strata of the Barents Sea region of Varanger Peninsula, East Finnmark, North Norway. *Norges Geol. Unders.* **382**, 45–79.

Williams, D. M. (1976). A revised stratigraphy of the Gaissa Nappe, Finnmark, *Norges Geol. Unders.* **324**, 63–78.

Winchester, J. A. (1988). Late Proterozoic environments and tectonic evolution in the northern Atlantic lands. In: Winchester, J. A. (ed.) *Later Proterozoic Stratigraphy of the Northern Atlantic Lands*, pp. 253–270. Blackie, London.

Zwann, K. B. (1977). Berggrunnsgeologiske kart Nordreisa NR33 34-9-M. 1:250,000. Preliminaer utgave. Norges Geologiske Undersøkelse.

Zwaan, K. B. and Gautier, A. M. (1980). Alta og Gargia. Beskrivelse til de berggrunnskart 1834 I og 1934 IV, M. 1:50,000. *Norges Geol. Unders.* **357**, 1–47.

Part III

Igneous Geology

12 Basic igneous rocks from a portion of the Jotun Nappe: evidence for Late Precambrian ensialic extension of Baltoscandia?

Trevor F. Emmett

Geology Section, Cambridgeshire College of Arts and Technology,
East Road, Cambridge, CB1 1PT, UK

A small gabbroic body and associated dolerite dykes, the Leirungsmyran Gabbroic Complex, were intruded into the Jotun Nappe of central S Norway, apparently some time in the period 700–900 Ma ago. The presence of garnet coronas in some of the dykes suggests that they were emplaced into relatively hot crust. Twenty-one basic igneous rocks have been analysed and they appear to be within-plate alkaline rocks with tholeiitic tendencies. They are more alkaline in character than the Ottfjället Dolerites, but closely resemble the alkaline metadolerites from the Leksdal Nappe of the Trondheim region. Petrogenetically, the rocks represent liquids evolved by the extraction of plagioclase and clinopyroxene; the absence of coexisting plagioclase and olivine on the liquidus and their mildly alkaline character suggests depths of magma genesis in excess of 30 km. From this and other contextural evidence, it is concluded that these igneous rocks represent the initial deep-seated stages of rift magmatism and that they were emplaced soon after the beginning of extension of the Laurentian–Baltoscandian craton, but before the onset of sea-floor spreading that eventually generated the Iapetus Ocean.

INTRODUCTION AND GEOLOGICAL SETTING

The earliest recognizable event in the evolution of the Scandinavian Caledonides is the extension and eventual division of the Laurentian–Baltoscandian craton in late Riphean to early Vendian times (Kumpulainen and Nystuen, 1985). Igneous activity associated with this extension is now widely recognized in the Lower and Middle Allochthons of the orogen (Andréasson *et al.*, 1979; Furnes *et al.*, 1983; Stephens *et al.*, 1985; Andréasson and Gee, 1987), and some dyke swarms in the autochthon of Sweden (e.g., the Blekinge–Dalarna Dolerites) may be of similar provenance (Solyom *et al.*, 1984). In the southern portion of the orogenic belt, the most extensive suite of such igneous rocks is the Ottfjället Dolerites of the Särv Nappe (Solyom *et al.*, 1979). These have a broadly MORB geochemistry, and are interpreted as portions of the Iapetus oceanic magmas intruded into the sedimentary rocks lying on the trailing edge of the Baltoscandian craton.

During this initial Late Proterozoic extension, the deep continental crustal rocks that now make up most of the Jotun Nappe lay ca. 300 km northwest of their present position (Hossack, 1978, and Fig. 1), in which position

they may have been quite close to the axis of extension. This nappe, which shows only minor effects of Caledonian deformation and metamorphism (Battey and McRitchie, 1973; Emmett, 1982b; Milnes and Koestler, 1985), may well, therefore, preserve some evidence of the igneous activity associated with the initiation of the Iapetus Ocean.

The Jotun Nappe (JN) was emplaced as a southeastward-translated fold nappe during the Scandian Orogeny, at ca. 395 Ma (Hossack and Cooper, 1986). The high-grade meta-igneous (Jotun kindred) rocks of the nappe had formed the basement to a series of generally arkosic sediments and these, the Valdres Group (VG) were transported with it. The basement and cover together form the regionally inverted Jotun–Valdres Complex (JVC) of Milnes and Koestler (1985). This paper notes the occurrence within the JN of a suite of undeformed and apparently rift-related igneous rocks, the Leirungsmyran Gabbroic Complex (LGC) and the external dyke set (EDS). These occur in the vicinity of Leirungsmyran, which is located at the southern end of Sjodalen (Fig. 1). The country rock is Svartdalen Gneiss, an S-tectonite pyriclasite Jotun kindred orthogneiss (Emmett, 1989b). Though direct radiometric evidence is lacking, Emmett (1987) argues that this unit has a magmatic age of ca. 1300 Ma, and that it was metamorphosed to low-

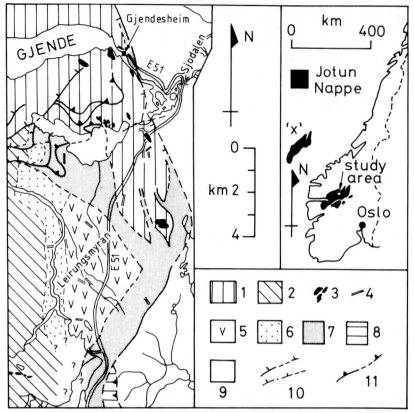

Fig. 1. Location and geological map of the study area. The inset shows the approximate pre-thrusting position 'x' of the Jotun Nappe (after Hossack, 1978). Key: 1, Storådalen Complex; 2, Svartdalen Gneiss; 3, ultrabasic bodies; 4, basic dykes; 5, Leirungsmyran Gabbroic Complex; 6, Knutshø Amphibole-gneiss; 7, retrograde rocks of major fault and shear zones; 8, mylonitic syenite-gneiss; 9, (unshaded) metasedimentary rocks of the Valdres Nappe; 10, steep faults, ticks on down-thrown side; 11, thrusts, barbs on superimposed side.

pressure granulite facies grade some 50–100 Ma after consolidation.

The LGC consists of a steep-sided stock of medium- to coarse-grained gabbro, the Main Gabbro, cut by a number of steeply dipping basic dykes, the internal dyke set (IDS).

Note that the terms "internal" and "external" refer to the relationship of the dykes to the margins of the Main Gabbro, which outcrops as a number of isolated knolls as shown in Fig. 2. The gabbro grades into patches of very coarse gabbro-pegmatite, which itself is occasionally seen

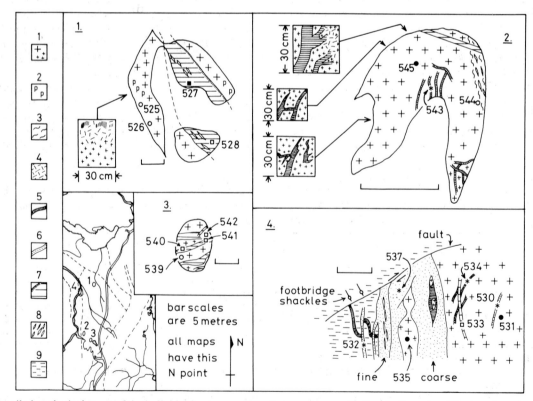

Fig. 2. Detailed geological maps of the individual outcrops of the Leirungsmyran Gabbroic Complex. The inset location map is from Fig. 1. Key: 1, Main Gabbros types I and II; 2, patches and veinlets of gabbro-pegmatite; 3, shear zones; 4, coarse gabbro-pegmatite; 5, acidic leucocratic vein; 6, intermediate leucocratic vein, often with cataclastic texture; 7, dolerite with chilled margin; 8, feldspathic streaks; 9, Knutshø Amphibole-gneiss. Sample point symbols as Fig. 5.

to grade into fine-grained leucocratic material (Fig. 2, exposure 1). The latter may also be found as patches within the Main Gabbro, but it is usually systematized into planar veins (exposures 2 and 3 in Fig. 2) or into localized zones of net-veining (exposure 2 in Fig. 2). The country rock is not hornfelsed, but it does show a degree of amphibolitization that diminishes with distance from the contact with the gabbro. The IDS comprises a number (the seven mapped are shown in Fig. 2) of steeply inclined dykes of medium- to fine-grained dolerite that vary in thickness from a few centimetres to several metres, and all except the thinnest have well-developed chilled margins (e.g., exposure 1 in Fig. 2).

The EDS consists of subvertical basic dykes (seven have been mapped, see Fig. 1) that are rather variable in trend and are usually between 1 and 2 m in thickness. Unsheared examples have no apparent chilled margins, and clearly cross-cut the foliation in the host gneiss, which shows no apparent thermal alteration effects. These dykes may have some affinities with those described from elsewhere in the JN (e.g., Battey and McRitchie, 1973; Battey *et al.*, 1979; Emmett, 1982a), a point that will be discussed later.

Most of the dykes shown in Figs. 1 and 2 have exposed lengths of <10 m. They occupy an exceedingly small fraction of the total volume of rock exposed, and they show rather inconsistent trends. In only one instance (Fig. 2 exposure 3) is a cross-cutting relationship between dykes observed, and, with the exception of some members of the EDS, none of the dykes is sheared or foliated. The Main Gabbro is cut by a number of narrow shear zones, but the effects of these are minor and restricted to the immediate vicinity of the shear.

AGE OF THE LEIRUNGSMYRAN GABBROIC COMPLEX

There is no unambiguous evidence regarding the age of these intrusive rocks. Sparagmitic metasediments of the VN can lie no more than a few hundred metres below the currently exposed level of the Main Gabbro, and these contain no igneous rocks comparable to those so far described. Synsedimentary igneous rocks that are reported from the VN near Espedalen may be comparable, but petrographic and geochemical data are lacking (Kumpulainen and Nystuen, 1985). Accordingly, the LGC can be no younger than the depositional age of the lower part of the Valdres Group sediments (Riphean–Vendian,

ca. 700 Ma; see Kumpulainen and Nystuen, 1985). Since the LGC is undeformed and unmetamorphosed, it must, therefore, have been intruded after the youngest regional tectonothermal event recorded in the Svartdalen Gneiss. This would appear to be the ca. 900 Ma retrograde event recorded by Schärer (1980), and referred to as the "Main Orogeny" by Milnes and Koestler (1985). An age for the LGC and associated external dyke set in the region of ca. 700–900 Ma is therefore indicated.

PETROGRAPHY

Representative samples of the LGC and the EDS were examined in thin section, but the following descriptions concern only those rocks of basic composition. Quantitatively determined modes are listed in Table I.

(1) Main Gabbro

The Main Gabbro consists of two petrographic types. Type I is a biotite gabbro, and type II a biotite–amphibole gabbro. The two types show no obvious systematic distribution in the field, except that possibly type II has a tendency to occur marginally to type I (see Fig. 2). The general characters shared by the two types are coarse, undeformed ophitic to subophitic textures, sparsely developed phenocrysts of complexly intergrown clinopyoxene and orthopyroxene, a predominance of clinopyroxene over orthopyroxene, no modal olivine, and the presence of late interstitial quartz. The plagioclase is free from antiperthite, and the opaques may be interstitial or skeletal. Grain size varies from 500 μm to 5 mm, with a median ca. 3 mm. In type I, the amphibole is always subordinate to interstitial, apparently primary igneous biotite, and does not exceed ca. 5 vol.%. Texturally, this amphibole appears to be mainly secondary in origin. In type II, euhedral primary amphibole is dominant over biotite. The secondary amphibole in both types is sporadically developed, and it grows at the expense of biotite. The development of the secondary amphibole is thought to be due to hydrous fluids circulating through the gabbro soon after consolidation.

As will be demonstrated later, these gabbros have a somewhat alkalic geochemical character, and it is apparent from the modes that the main repository of sodium would be plagioclase and of potassium would be biotite. The primary and secondary amphiboles in both

Table I. Representative modal analyses, in vol.%, of the rocks of this study. Most sample points are shown in Fig. 2, with details of the others being available on request from the author. 1, external dyke set; 2, internal dyke set; 3, Main Gabbro, type I; 4, Main Gabbro, type II.

	1	*2*	*2*	*2*	*2*	*3*	*3*	*3*	*3*	*3*	*3*	*3*	*3*	*4*	*4*	*4*	*4*
Sample	510	527	528	540	550	525	526	538	539	544	546	548	531	535	545	549	
Quartz	–	–	–	–	–	0.8	0.9	3.3	0.8	0.7	0.4	1.1	1.0	0.8	1.3	0.3	
Plag.	44.3	29.5	32.0	35.0	42.9	56.6	29.7	58.2	42.5	34.0	53.4	53.0	46.1	33.8	23.0	37.9	
Clinopyrx.	33.9	33.2	18.7	16.3	15.4	24.1	32.9	17.8	32.4	27.8	33.5	10.1	20.4	18.0	46.4	18.9	
Orthopyrx.	–	0.5	1.6	2.9	3.7	1.0	2.3	1.6	3.2	1.2	1.8	0.6	2.1	1.7	3.0	2.1	
Prim. amph.	–	16.1	18.0[a]	17.9[a]	15.3[a]	–	–	–	2.5[a]	5.2[a]	tr.[a]	4.0[a]	15.2[a]	29.5[a]	19.0[a]	13.2[a]	
Secd. amph.	1.1	–	–	–	–	1.3	1.9	3.6	–	–	–	–	–	–	–	–	
Prim. bio.	–	16.1	20.8[a]	20.3[a]	17.6[a]	12.8	28.7	10.6	10.3	25.5	6.6	20.1[a]	9.7[a]	12.1[a]	4.3[a]	23.0[a]	
Secd. bio.	2.3	–	–	–	–	–	–	–	–	–	–	–	–	–	–	–	
Opaques	6.6	4.6	8.9	7.6	5.1	2.6	3.3	2.3	7.1	5.1	3.4	4.3	4.3	4.0	2.7	3.4	
Others[c]	11.7[b]	–	–	–	–	0.8	0.3	2.6[c]	1.2[c]	0.5	0.8[c]	6.8[d]	0.8[c]	0.1	0.3	1.2[c]	

Key: tr. = trace amounts; prim. = primary; secd. = secondary; amph. = amphibole; bio. = biotite; plag. = plagioclase.
[a] Total amount, [b] garnet 11.4, [c] K-feldspar, [d] apatite. [e] "Others" are mostly sphene, apatite, and, in 528, allanite.

types of gabbro are pleochroic in shades of green. A small number of electron-microprobe analyses have been completed and, whilst hardly exhaustive, they show that the hornblende in type I gabbros is generally less magnesian and more calcic than those in type II. These figures seem to apply to coarser varieties of secondary amphibole in both types, and to the primary variety in type II. The primary biotite in both types is pleochroic in shades of brown. That in type I has a Fe^{2+}/Mg^{2+} ratio of ca. 1 and contains 4–5 wt.% TiO_2. The figures for primary biotite in type II are ca. 0.7 and 5–5.5 respectively. These preliminary data on the amphiboles and biotites suggest that type II gabbros are less magmatically evolved than those of type I.

(2) Internal dyke set

The samples examined are all broadly doleritic in composition, but they are rather variable in their petrography. Most carry corroded phenocrysts of clear plagioclase, and have matrices of subophitically intergrown plagioclase and clinopyroxene. Green amphibole and brown biotite occur within the matrix of all samples, whilst samples 533, 541, 542, and 543 all have phenocrysts of complexly intergrown clinopyroxene and orthopyroxene similar to those seen in the Main Gabbro. No coronas are seen and no garnet was noted.

(3) External dyke set

With only one exception, these basic dykes all show some evidence of shearing and recrystallization. With grain sizes between 50 and 100 μm, granular (blasto-mylonitic) textures predominate. The sheared rocks are usually 60–80 vol.% matrix, which is composed mainly of plagioclase, epidote *s.l.*, green amphibole, and brown biotite. Some samples contain patches of scapolite, whilst calcite and a white mica are abundant in sample 519. The common porphyroclast phase is plagioclase, of which crystals may be up to 750 μm across and which usually show a dense pinkish clouding. Clinopyroxene and, rarely, orthopyroxene, may also occur in porphyroclasts up to 500 μm across. Garnet occurs as either granular trains within the matrix, or as euhedral porphyroblasts up to 500 μm across. The one unsheared example is 510 (Table I), and it consists of an intimate interlocking framework of clear plagioclase laths, the interstices so formed being filled with clinopyroxene, opaques, and a little green amphibole. The clinopyroxene crystals are marginally occluded with a fine opaque dust, and occasionally have a thin rind of pale-green uralite. Coronas of garnet are developed between opaque grains, mainly ilmenite and plagioclase. Dykes petrographically similar to 510 have been described by Emmett (1982a), and there may be some affinities to the example described by Battey *et al.* (1979). Both of these papers suggest that the formation of garnet is promoted by abnormally slow cooling of the dykes after their intrusion into relatively hot country rock. In the context of this work, it is tempting to speculate that these elevated crustal temperatures may be due to the crust undergoing extension and thinning contemporaneously with dyke emplacement.

GEOCHEMISTRY

Twenty-seven samples were analysed, of which six were of material from leucocratic veins. FeO was determined colorimetrically by the method of Wilson (1960), and loss on ignition (L.O.I.) gravimetrically. All other elements were determined by XRF methods using international standard rock samples for calibration. A representative number of analyses, the corresponding C.I.P.W. norms, and selected element ratios, are given in Table II. Fifteen of the twenty-one basic rocks show olivine in the norm, whilst the remainder show small amounts of quartz. Normative nepheline is absent, and so these rocks have the character of quartz- or olivine-tholeiites. There is a distinct Daly gap apparent between the basic rocks and the leucocratic veins (Fig. 3); this may be a real feature, but field evidence strongly suggests that the leucocratic veins are late differentiates from the gabbros. It is thought that the apparent gap reflects the absence of analyses from the material intermediate, both spatially and geochemically, between the Main Gabbro and the leucocratic veins. Type II gabbros are slightly impoverished in incompatible elements compared to type I, and they tend to lie on the low D.I. (=differentiation index, after Thornton & Tuttle, 1960) part of the differentiation trends; hence, they may represent the least evolved rocks in the sequence. Overall, however, the four groups of basic rocks have a remarkable uniformity in both composition and differentiation character, and it is concluded that they represent slightly different expressions of the same magmagenetic process.

Most of the major element oxides show well-defined linear trends against D.I., as do some trace elements, notably Rb, Y and Zr. Ni and Cr show much scatter (see Fig. 3). These features suggest that the rocks represent the trend of liquid composition evolving by the extraction of mafic mineral phases. However, it must emphasized that the basic rocks analysed were not entirely aphyric. Clinopyroxene phenocrysts are present in the Main Gabbro, and samples from both dyke sets may have phenocrysts of plagioclase present. The analyses cannot therefore truly represent liquids, a conclusion that may account for the poorly defined trends for FeO, TiO_2, and Cr and Ni, elements which tend to be compatible with mafic mineral phases.

The erratic behaviour of Ni and the diminishment of Cr/Ni ratio with D.I. suggest that olivine has not played a major role in the evolution of the magmas, a conclusion supported by the Ni–Ti diagram, constructed after the method of Pearce and Flower (1977), shown in Fig. 4. However, the predominance of olivine-bearing normative compositions (Table II) may suggest that the parental magma was in equilibrium with an olivine-bearing residue (Yoder, 1976). These arguments, together with the presence of plagioclase phenocrysts and the absence of modal olivine, could indicate final fractionation at pressures >7 kb (i.e., depths >25 km; see Pressnall *et al.*, 1978).

Figure 5 shows the rocks of this study plotted on several widely used major-element discriminant diagrams (see figure caption for details). They show mixed characteristics, these being transitional between alkalic and tholeiitic bulk composition and ambiguous oceanic and/or continental provenance. The immobile trace elements present a less-ambiguous set of relations. Figure 6 shows the discriminant diagrams of Floyd and Winchester (1975) and Winchester and Floyd (1976). The diagrams with Zr/P_2O_5 as abcissae again illustrate the transitional nature of the rocks, perhaps indicating a stronger alkalic character than the major element discriminants, a feature also apparent from the TiO_2–Y/Nb diagram. The popular Pearce and Cann diagram (Fig. 7) classifies the samples as within-plate basalts, and overall they appear to represent magmas of a transitional alkalic–tholeiitic character, generated in an intraplate environment. Figures 5 and 7 compare their composition in terms of immobile minor

Table II. Geochemical analyses of C.I.P.W. norms of selected rocks from this study. Oxides in wt.%, trace elements in p.p.m. Column headings as Table I. The differentiation index, D.I., is that of Thornton and Tuttle (1960), and An% = 100 An/(Ab + An).

	1	2	2	2	2	3	3	3	3	3	3	3	4	4	4	4
Sample	510	527	528	540	550	525	526	538	539	544	546	548	531	535	545	549
SiO_2	48.72	48.04	47.87	48.87	48.15	51.03	48.88	50.80	48.36	48.68	52.03	51.05	50.78	48.82	50.71	48.76
Al_2O_3	14.39	13.35	13.72	14.47	14.65	15.51	12.46	13.92	10.46	12.64	15.44	15.33	15.23	15.29	8.82	14.05
Fe_2O_3	2.41	3.32	4.26	0.47	2.41	1.13	4.63	3.10	1.22	3.48	1.89	1.23	3.80	4.73	1.63	1.21
FeO	12.10	9.39	9.14	12.20	10.64	9.05	8.61	9.28	14.84	10.53	7.55	9.85	8.48	7.66	10.01	11.45
MgO	5.92	7.28	6.48	5.76	5.48	6.23	7.43	6.27	7.67	6.81	6.68	4.39	6.11	6.77	11.34	5.93
CaO	9.92	8.87	8.21	8.11	8.17	8.26	7.90	8.39	8.71	7.95	10.44	6.87	8.27	10.27	12.56	8.03
Na_2O	2.66	2.86	3.05	3.11	3.04	3.66	2.35	3.23	2.16	2.39	2.95	3.65	3.12	2.70	1.53	2.93
K_2O	0.79	1.96	1.91	2.03	1.83	1.38	3.05	1.22	1.03	2.74	0.66	1.85	1.78	1.65	0.44	2.18
MnO	0.23	0.16	0.16	0.15	0.15	0.13	0.15	0.16	0.20	0.16	0.13	0.13	0.26	0.22	0.18	0.15
TiO_2	2.25	2.70	3.95	3.15	3.38	1.94	2.53	2.55	4.30	2.62	1.38	4.33	1.01	0.87	1.46	3.08
P_2O_5	0.22	0.46	0.44	0.53	0.53	0.29	0.45	0.36	0.22	0.52	0.15	0.27	0.55	0.38	0.10	0.46
LOI[a]	1.11	1.57	1.20	1.81	1.96	1.75	1.48	2.30	2.39	1.60	1.43	2.05	1.30	1.22	1.80	2.63
Total	100.72	99.96	100.39	100.66	100.39	100.36	99.92	101.58	101.56	100.12	100.73	101.00	100.69	100.58	100.58	100.86
Cr	62	211	77	71	57	129	286	94	116	203	66	c.10	90	88	350	92
Ni	78	198	157	135	120	144	208	128	143	178	113	74	55	54	371	142
Zn	120	134	140	147	144	112	158	132	184	182	89	120	136	130	107	142
Cu	169	67	65	56	72	64	67	60	85	67	35	47	130	118	89	60
Pb	7	9	7	8	7	8	9	6	6	7	<5	8	11	13	6	10
Sr	266	691	854	772	780	699	612	923	684	634	662	1091	872	939	329	716
Rb	18	39	37	37	32	22	72	14	16	78	<5	32	22	45	9	46
Ba	325	650	793	696	668	538	1167	643	571	957	295	502	1518	589	221	776
Y	33	20	21	21	21	16	21	17	19	24	14	14	26	23	16	21
Zr	134	210	196	201	195	122	116	108	146	215	79	143	60	68	70	180
Nb	11	37	39	38	43	21	28	31	46	35	12	34	<5	<5	9	37
V	311	204	241	175	190	157	184	143	185	181	146	136	304	308	268	192
q	–	–	–	–	–	–	–	0.41	–	–	0.92	–	–	–	–	–
c	–	–	–	–	–	–	–	–	–	–	–	0.27	–	–	–	–
Or	4.67	11.58	11.29	12.00	10.81	8.16	18.02	7.21	6.09	16.19	3.90	10.93	10.52	9.75	2.60	12.88
Ab	22.51	24.20	25.81	26.32	25.72	30.97	19.89	27.33	18.28	20.22	24.96	30.89	26.40	22.85	12.95	24.79
An	25.00	17.81	18.11	19.53	20.93	21.82	14.45	19.89	15.81	15.68	26.95	19.99	22.30	24.74	15.90	18.75
Di	18.93	18.85	15.93	14.35	13.30	14.19	17.59	15.84	21.50	16.68	19.44	10.18	12.40	19.26	37.11	14.97
Hy	14.47	3.35	7.80	3.08	8.74	7.11	10.28	18.44	24.53	11.21	17.42	16.07	15.50	4.75	23.96	5.82
Ol	5.76	11.59	5.56	15.68	7.78	10.37	5.65	–	2.53	7.32	–	–	3.57	8.63	0.90	12.34
Mt	3.49	4.81	6.18	0.68	3.49	1.64	6.71	4.49	1.77	5.05	2.74	1.78	5.51	6.86	2.36	1.75
Il	4.27	5.13	7.50	5.98	6.42	3.68	4.81	4.84	8.17	4.98	2.62	8.22	1.92	1.65	2.77	5.85
Ap	0.52	1.09	1.04	1.26	1.26	0.69	1.07	0.85	0.52	1.23	0.36	0.64	1.30	0.90	0.24	1.09
An%	52.6	42.4	41.2	42.6	44.9	41.3	42.1	42.1	46.4	43.7	51.9	39.3	45.8	52.0	55.1	43.1
D.I.	27.2	35.8	37.1	38.3	36.5	39.1	37.9	35.0	24.4	36.4	29.8	42.1	36.9	32.6	15.6	37.7
Cr/Ni	0.79	1.07	0.49	0.53	0.48	0.90	1.38	0.73	0.81	1.14	0.58	0.14	1.64	1.63	0.94	0.65
Rb/Sr	0.068	0.056	0.043	0.048	0.041	0.031	0.118	0.015	0.023	0.123	*[b]	0.029	0.025	0.048	0.027	0.064
K/Rb	364	417	429	455	475	521	352	723	534	292	*[b]	480	672	304	406	393
Ba/Sr	1.22	0.94	0.93	0.90	0.86	0.77	1.91	0.70	0.83	1.51	0.45	0.46	1.74	0.63	0.67	1.08
Zr/Y	4.06	10.5	9.33	9.57	9.29	7.63	5.52	6.35	7.68	8.96	5.64	10.2	2.31	2.96	4.38	8.57
Ti/Zr	100.7	77.1	120.8	94.0	103.9	95.3	130.8	141.5	176.6	73.1	104.7	181.5	100.9	76.7	125.0	102.6
Y/Nb	3.00	0.54	0.54	0.55	0.49	0.76	0.75	0.55	0.41	0.69	1.17	0.41	6.50	5.75	1.78	0.57
Ti/V	43.4	79.3	98.3	107.9	106.6	74.1	82.4	106.9	139.3	86.8	56.7	190.9	19.9	16.9	32.7	96.2

[a] Loss on ignition (L.O.I.) figure corrected for oxidation of $Fe^{2+} \rightarrow Fe^{3+}$.

*[b] Indicates one or both parameters in a ratio below detection limit.

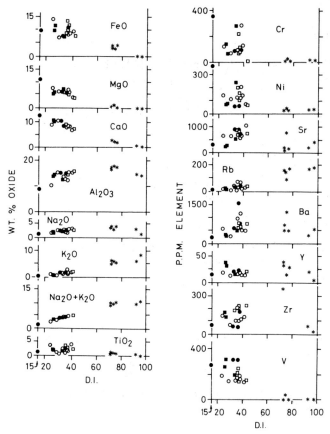

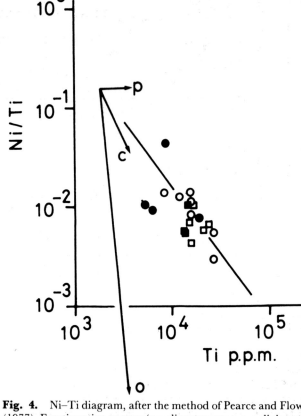

Fig. 3. Geochemical differentiation trends in the rocks of this study. The differentiation index is from Thornton and Tuttle (1960). Symbols as Fig. 5.

Fig. 4. Ni–Ti diagram, after the method of Pearce and Flower (1977). Fractionation vectors (c = clinopyroxene, o = olivine, p = plagioclase) are calculated for an arbitrary 50% fractionation. The distribution coefficients used were taken from Pearce & Norry (1979) and Cox *et al.* (1979). Symbols as Fig. 5.

and trace elements with Phanerozoic igneous rocks of known tectonic provenance, i.e. ensialic alkali basalts from Mull (Beckinsale *et al.*, 1978), and Gondwanan flood basalts from the Serra Geral subprovince of the Paraná suite (Fodor *et al.*, 1985), and the similarities are obvious. There are some differences, e.g. the Paraná basalts have a much stronger "oceanic" character in the Pearce

and Cann diagram, but these serve to emphasize the transitional nature of the Jotunheimen rocks.

Within the Norwegian Caledonides, the rocks of the LGC and EDS have possible correlatives in the alkaline metadolerites of the Leksdal Nappe (Andréasson *et al.*, 1979), the Ottfjället Dolerites of th Särv Nappe (Solyom *et al.*, 1979), and lavas low in the Hedmark Group

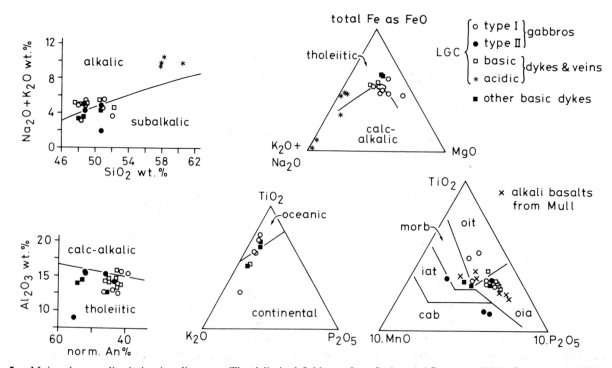

Fig. 5. Major-element discrimination diagrams. The delimited fields are from Irvine and Baragar (1971), Pearce *et al.* (1975), and Mullen (1983).

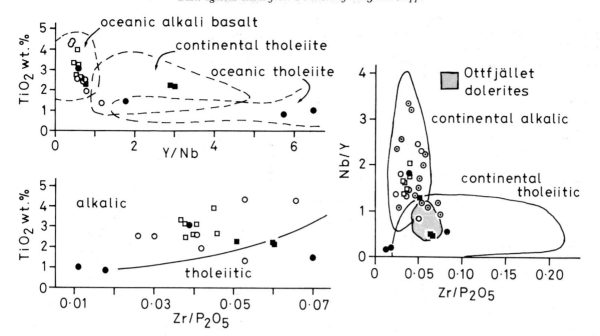

Fig. 6. Discrimination diagrams, after Floyd and Winchester (1975) and Winchester and Floyd (1976). Alkaline metadolerites from the Leksdal Nappe are shown by bullseyes, and the other symbols are as Fig. 5.

succession (Furnes *et al.*, 1983); at least, all these appear to be in the same Late Proterozoic age bracket as the rocks of this study. Important compositional parameters of the four units are listed in Table III, wherein strong similarities between the rocks of this study and the metadolerites from the Leksdal Nappe are apparent. Similar trace element geochemistries are also indicated on the Zr/P_2O_5 vs. Nb/Y diagram (Fig. 6). Andréasson *et al.* (1979) regard these alkaline rocks as being emplaced during "the earliest stages of continental distension . . . leading to the development of the Caledonian Iapetus Ocean" (their abstract). The Ottfjället rocks are generally poorer in Fe and Sr (Table III, column 2), and show unambiguous tholeiitic tendences on Floyd and Winchester's Zr/P_2O_5 vs. Nb/Y diagram (Fig. 6), and a clear ocean floor affinity on the Pearce and Cann diagram (Fig. 7). The Hedmark Group

lavas also show very little comparability to the rocks of this study (Table III, column 4).

DISCUSSION AND CONCLUSIONS

The igneous rocks described in this paper are small in their extent and volume, but their interest lies in their significance in understanding the late Precambrian evolution of the Baltoscandian margin. They were intruded into a portion of the continental crust that was probably at elevated temperature during their emplacement. Evidence for this lies in the presence of coronas in the external dyke set and other dykes (Emmett, 1982a, and cf. Battey *et al.*, 1979), and in the lack of hornfelsing of the country rock. The author prefers to explain these elevated crustal

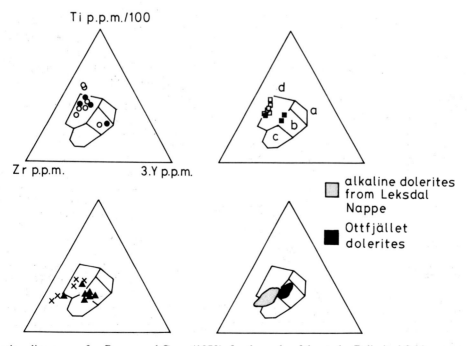

Fig. 7. Discrimination diagrams, after Pearce and Cann (1973), for the rocks of the study. Delimited fields represent low-K tholeiites (a + b), calc-alkaline basalts (b + c), ocean floor basalts (b), and within-plate basalts (d). Symbols as Fig. 5, but including triangles for Paraná continental flood basalts.

Table III. Selected comparative analyses: 1, mean of all *basic* rocks of this study; 2, mean of all Ottfjället Dolerites (Solyom *et al.*, 1979); 3, mean of alkaline metadolerites from the Leksdal Nappe Andréasson *et al.*, 1979); 4, mean of basaltic lavas from the Hedmark Group (Furnes *et al.*, 1983). (*n* = number of analyses averaged.)

	1	2	3	4
SiO_2	49.55 (1.46)	50.46 (0.86)	47.31 (1.84)	47.04 (0.58)
FeO	9.96 (1.84)	7.72 (1.10)	8.66 (1.01)	14.05 (1.15)
Na_2O	2.83 (0.50)	2.43 (0.30)	3.08 (0.84)	2.79 (0.63)
K_2O	1.68 (0.66)	0.58 (0.33)	1.37 (0.35)	3.82 (0.68)
TiO_2	2.59 (0.95)	1.28 (0.25)	2.15 (0.65)	3.18 (0.12)
Cr	125 (91)	–	158 (193)	57 (13)
Ni	140 (74)	–	124 (88)	–
Rb	33 (18)	16 (16)	38 (15)	148 (63)
Sr	692 (212)	255 (46)	612 (146)	118 (24)
Y	22 (6)	26 (4)	32 (8)	24 (2)
Zr	154 (54)	96 (34)	232 (78)	106 (7)
Nb	27 (13)	–	53 (17)	12 (1)
n	21	123	19	12

temperatures by invoking crustal stretching (MacKenzie, 1978), but it is possible that the crust was still cooling down after regional metamorphism (Dallmeyer, personal communication). This matter can only be resolved by obtaining reliable radiometric dates on the Leirungsmyran igneous rocks. The rocks also have an obvious alkaline imprint, and also appear to lack olivine on the liquidus; these observations lead to the conclusion that the magmas originated at some depth. Yoder (1976, chap. 8) and Pressnall *et al.* (1978) both present arguments that this depth must, at least, be the equivalent of some 10–15 kb pressure (i.e., depths of the order of 30–45 km), figures comparable with those suggested by Harris (1969). If continental lithosphere is distended, it undergoes a general rise in temperature and the upwelling of the upper mantle/asthenosphere generally accompanies basin formation at the surface (MacKenzie, 1978). The deep-seated nature of the LGC–EDS magmatism, and its failure to penetrate into the basal sediments of the VG, indicates that the magmas described here are representatives of the earliest phase of rift-related igneous activity (cf. Andréasson *et al.*, 1979).

If the arguments presented above are true, one would expect igneous intrusions related to the LGC and EDS to be widespread in the JN. Emmett (1982a) has summarized the geographical occurrences of undeformed basic dykes in the JN, and there are doubtless a great many more to be discovered. However, there is a general absence of detailed geochemical and isotopic studies, and so it is not possible to integrate these occurrences with those described in the current work. Certainly the small number of incomplete analyses published by Battey *et al.* (1979) and Emmett (1982a) differ in many significant respects from those presented here. A detailed appraisal of this matter is not possible here, but it is doubtless a study that could produce many interesting and important data.

Acknowledgements

Field work in Norway was financed by grants from the Royal Society of London and C.C.A.T. The analytical work was completed whilst the author was on a Cambridgeshire County Council Industrial Secondment to Aston University. Professor Hawkes is thanked for making this secondment possible, and technical assistance from R. Wightman and M. Cowley is gratefully acknowledged. P. Banham, I. Bryhni, D. C. Green, and A. J. R. Rickard are thanked for their helpful and much-appreciated comments on earlier versions of this paper.

REFERENCES

Andréasson, P.-G. and Gee, D. G. (1987). Baltoscandian rift magmatism. In: *The Caledonian and Related Geology of Scandinavia*, Gayer, R. and Townsend (eds.) C. Abstract. University College, Cardiff.

Andréasson, P.-G., Solyom, Z. and Roberts, D. (1979). Petrochemistry and tectonic significance of basic and alkaline-ultrabasic dykes in the Leksdal Nappe, northern Trondheim region, Norway. *Norges Geol. Unders.* **348,** 47–71.

Battey, M. H. and McRitchie, W. D. (1973). A geological traverse across the pyroxene-granulites of Jotunheimen in the Norwegian Caledonides. *Norsk Geol. Tidsskr.* **53,** 237–265.

Battey, M. H., Davison, W. and Oakley, P. J. (1979). Almandine pseudomorphous after plagioclase in a metadolerite dyke from the Jotunheim, Norway. *Mineral. Mag.* **43,** 127–130.

Beckinsale, R. D., Pankhurst, R. J., Skelhorn, R. R. and Walsh, J. N. (1978). Geochemistry and petrogenesis of the early Tertiary lava pile of the Isle of Mull, Scotland. *Contrib. Mineral. Petrol.* **66,** 415–427.

Cox, K. G., Bell, J. D. and Pankhurst, R. J. (1979). *The Interpretation of Igneous Rocks.* George Allen & Unwin, London.

Emmett, T. F. (1982a). The petrography and geochemistry of coronite dolerites from the Jotun Nappe, central southern Norway. *Mineral. Mag.* **46,** 43–48.

Emmett, T. F. (1982b). Structure and petrology of the Bergen-Jotun kindred rocks from the Gjendebu region, Jotunheimen, central southern Norway. *Norges Geol. Unders.* **373,** 1–32.

Emmett, T. F. (1987). A reconnaissance study of the distribution of Ba, Nb, Y, and Zr in some Jotun kindred gneisses from central Jotunheimen, southern Norway. *J. Metam. Geol.* **5,** 41–50.

Floyd, P. A. and Winchester, J. A. (1975). Magma type and tectonic setting discrimination using immobile elements. *Earth Planet. Sci. Lett.* **27,** 211–218.

Fodor, R. V., Corwin, C. and Roisenberg, A. (1985). Petrology of Serra Geral (Paraná) continental flood basalts, southern Brazil: crustal contamination, source material, and South Atlantic magmatism. *Contrib. Mineral. Petrol.* **91,** 54–65.

Furnes, H., Nystuen, J. P., Brunfelt, A. O. and Solheim, S. (1983). Geochemistry of Upper Riphean-Vendian basalts associated with the 'sparagmites' of southern Norway. *Geol. Mag.* **120,** 349–361.

Harris, P. G. (1969). Basalt type and African Rift Valley tectonism. *Tectonophysics* **8,** 427–436.

Hossack, J. R. (1978). The correction of stratigraphic sections for tectonic finite strain in the Bygdin area, Norway. *J. Geol. Soc. Lond.* **135,** 229–241.

Hossack, J. R. and Cooper, M. A. (1986). Collision tectonics in the Scandinavian caledonides. In: Coward, M. P. and Ries, A. C. (eds.) *Collision Tectonics. Geol. Soc. London. Spec. Publ.* **19,** 287–304.

Irvine, T. N. and Baragar, W. R. A. (1971). A guide to the chemical classification of the common volcanic rocks. *Can. J. Earth Sci.* **8,** 523–548.

Kumpulainen, R. and Nystuen, J. P. (1985). Late Proterozoic basin evolution and sedimentation in the westernmost part of Baltoscandia. In: Gee, D. G. and Sturt, B. A. (eds.) *The Caledonide Orogen: Scandinavia and Related Areas*, pp. 213–232. Wiley, Chichester.

MacKenzie, D. (1978). Some remarks on the development of sedimentary basins. *Earth Planet. Sci. Lett.* **40,** 25–32.

Milnes, A. G. and Koestler, A. G. (1985). Geological structure of Jotunheimen, southern Norway (Sognefjell-Valdres cross-section). In: Gee, D. G. and Sturt, B. A. (eds.) *The Caledonide Orogen: Scandinavia and Related Areas*, pp. 457–474. Wiley, Chichester.

Mullen, E. D. (1983). $MnO/TiO_2/P_2O_5$: a minor element discriminant for basaltic rocks of oceanic environments and its implications for petrogenesis. *Earth Planet. Sci. Lett.* **62**, 53–62.

Pearce, J. A. and Cann, J. R. (1973). Tectonic setting of basic volcanic rocks determined using trace element analyses. *Earth Planet. Sci. Lett.* **19**, 290–300.

Pearce, J. A. and Flower, M. F. J. (1977). The relative importance of petrogenetic variables in magma genesis at accreting plate margins: a preliminary investigation. *J. Geol. Soc. Lond.* **134**, 103–127.

Pearce, J. A. and Norry, M. J. (1979). Petrogenetic implications of Ti, Zr, Y, and Nb variation in volcanic rocks. *Contrib. Mineral. Petrol.* **69**, 33–47.

Pearce, T. H., Gorman, B. E. and Birkett, T. C. (1975). The TiO_2–K_2O–P_2O_5 diagram: a method of discriminating between oceanic and non-oceanic basalts. *Earth Planet. Sci. Lett.* **24**, 419–426.

Pressnall, D. C., Dixon, S. A., Dixon, J. R., O'Donnell, T. H., Brenner, R. L., Schrock, R. L. and Dycus, D. W. (1978). Liquidus phase relationships on the join Di–Fo–An from 1 atm–20 kb and their bearing on the generation and crystallisation of basaltic magma. *Contrib. Mineral. Petrol.* **66**, 203–220.

Schärer, U. (1980). U–Pb and Rb–Sr dating of a polymetamorphic nappe terrain: the Caledonian Jotun Nappe, Southern Norway. *Earth Planet. Sci. Lett.* **49**, 205–218.

Stephens, M. B., Furnes, H., Robins, B. and Sturt, B. A. (1985). Igneous activity within the Scandinavian Caledonides. In: Gee, D. G. and Sturt, B. A. (eds.) *The Caledonide Orogen: Scandinavia and Related Areas*, pp. 623–656. Wiley, Chichester.

Solyom, Z., Andréasson, P.–G., Johansson, I. and Hedvall, R. (1984). Petrochemistry of Late Proterozoic rift volcanism in Scandinavia I: The Blekinge–Dalarna Dolerites (BDD)—volcanism in a failed arm of Iapetus? *Lund Publ. Geol.* **23**, 56 pp.

Solyom, Z., Gorbatschev, R. and Johannson, I. (1979). The Ottfjället Dolerites. *Sver. Geol. Unders.* C**756**, 1–38.

Thornton, C. P. and Tuttle, O. F. (1960). Chemistry of igneous rocks: part I. Differentiation index. *Amer. J. Sci.* **258**, 664–684.

Wilson, A. D. (1960). The micro-determination of ferrous iron in silicate minerals by a volumetric and a colorimetric method. *Analyst* **85**, 823–827.

Winchester, J. A. and Floyd, P. A. (1976). Geochemical magma type discrimination: application to altered and metamorphosed basic igneous intrusions. *Earth Planet. Sci. Lett.* **28**, 459–469.

Yoder, H. S. Jnr. (1976). *Generation of Basaltic Magma*. National Academy of Sciences, Washington, D.C., 265 pp.

13 The geochemistry of the Sulitjelma ophiolite and associated basic volcanics: tectonic implications

A. P. Boyle

Department of Earth Sciences, University of Liverpool, P.O. Box 147, Liverpool L69 3BX, UK

The Sulitjelma ophiolite is part of a volcanosedimentary succession in the Köli Nappes of north Central Scandinavia. Field relations suggest formation in an ensialic marginal-basin formed by rifting of the Skaiti Supergroup. The basin was infilled by the Furulund and Sjønstå Groups. The geochemistry of Sulitjelma ophiolite metabasalts is consistent with their origin as MORB at a spreading centre in a rifted continental margin. A suite of intrusions in the Furulund Group, the Kjeldvann metadolerites, also have MORB affinities, whereas the basic members of the younger basalt to rhyolite Lomivann volcanics towards the top of the Furulund Group have CAB affinities. These relations, together with the sedimentology of the Furulund Group, are interpreted as representing formation of the ophiolite in an ensialic back-arc marginal-basin. The inferred polarity of the back-arc basin suggests it formed outboard of Baltica during westward-directed subduction, possibly on the Laurentian side of Iapetus.

INTRODUCTION

The Sulitjelma region of arctic Scandinavia lies within the Köli Nappes of the Upper Allochthon (Stephens *et al.*, 1985) and is famous for its stratabound copper ore-bodies mined since 1887. These have been variously interpreted as syngenetic sediments (Krause, 1956), magmatic segregations (Vogt, 1927), epigenetic thrust-related deposits (Kautsky, 1953), and syngenetic volcanic-exhalites (Oftedahl, 1958; Hansen, 1980). The last of these models is consistent with the interpretation that associated basic rocks are ophiolitic in character (Boyle, 1980; Boyle *et al.*, 1985), forming in a rifted continental margin environment (Stephens, 1986). A general outline of the geology of the Sulitjelma region is given in Figs. 1 and 2. The latter illustrates primary igneous and stratigraphic relations between the component parts of the Sulitjelma ophiolite and associated rock units. Note that although Stephens (1986) envisages formation of the basic igneous rocks at Sulitjelma in a rifted continental margin setting during a "major extensional event", he does not think the term "ophiolite" appropriate. I and other workers think the term "ophiolite" appropriate and regard the intrusive relations between the gabbros and dykes of the Sulitjelma

ophiolite and the metasediments of the Skaiti Supergroup (Mason, 1971; Boyle, 1980, 1987; Brekke *et al*, 1984; Boyle *et al.*, 1985; Furnes *et al.*, 1985; Billett, 1987) as indicating formation of the ophiolite in an ensialic marginal-basin analogous to the Mesozoic "Rocas Verdes" ensialic marginal-basins of Chile (e.g., Saunders *et al.*, 1979). The massive copper ore-bodies formed on the marginal-basin floor by exhalation of metal-rich brines derived by hydrothermal leaching of the ophiolite (Hansen, 1980). The basin was infilled by graphitic shales, greywackes, volcanics and fossiliferous limestones of the Furulund Group and by the Sjønstå Group.

The stratigraphically lower part of the Furulund Group is intruded by a suite of basic igneous sheets, the Kjeldvann metadolerites. The upper part of the Furulund Group includes a sequence of volcanics, here termed the Lomivann volcanics, and fossiliferous marbles (Nicholson, 1966; Kekwick, 1988). The Lomivann volcanics and associated marbles pass stratigraphically up into the younger Sjønstå Group. Much of the sequence in Fig. 2 is now found on the inverted lower limb of a megascopic recumbent fold structure, the Sulitjelma Fold-Nappe (Boyle, 1987), produced by closure and obduction of the basin during the Scandian orogenic phase

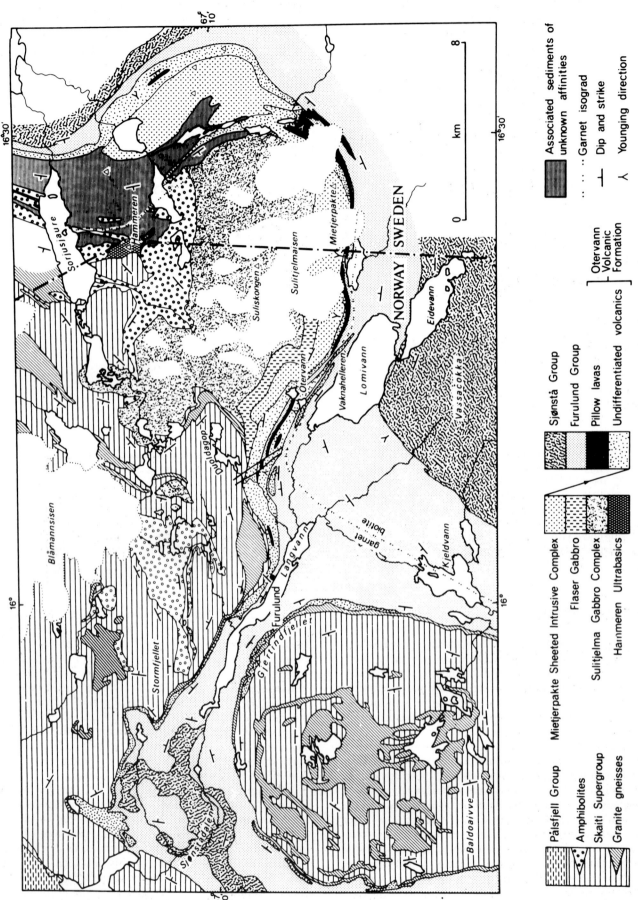

Fig. 1. Summary geological map of the Sulitjelma region (after Boyle 1987).

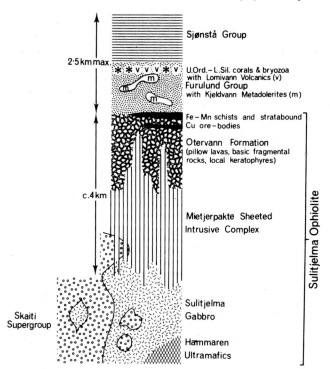

Fig. 2. Palinspastic column section summarizing inferred primary relations between the lithologies of Fig. 2 (after Boyle *et al.*, 1985).

(Roberts and Sturt, 1980). The rocks comprising the Sulitjelma ophiolite and the Furulund Group underwent variable degrees of deformation together with greenschist to amphibolite facies metamorphism. Thus, although primary igneous features such as agglomerates, pillow lavas, dykes, porphyritic and ophitic textures, and phase layering are widely recognizable, preservation of primary igneous minerals is effectively limited to the gabbroic part of the ophiolite (Mason, 1971). Therefore, any treatment of the geochemistry of the Sulitjelma ophiolite, the Kjeldvann metadolerites and the Lomivann volcanics is dependent on the investigation of whole-rock compositions rather than primary igneous phases, and the degree to which the rocks have retained their primary geochemical imprints during a variety of processes that could lead to alteration.

SAMPLES AND METHODS

All samples discussed in this paper preserve enough of their original igneous character to be recognized as gabbros (13 gabbro and 8 flaser gabbro samples), dykes (11 samples) or pillows (17 samples) in the ophiolite, or pillows (Kekwick, 1988, 6 samples) in the Lomivann volcanics. Kjeldvann metadolerites (6 samples) are easily recognized as such. Ophiolite samples whose igneous origin is unclear (essentially schistose amphibolites) have also been analysed (Boyle, 1982), but the results are not presented here because they generally appear to have been more affected by the alteration processes discussed below, and also because it is unclear to what extent they represent igneous liquids (as opposed to tuffs or tuffites, for example). Samples have been analysed by a combination of: (1) x-ray fluorescence (XRF) methods at Bedford College London (now part of Egham) and the University of Liverpool on duplicate fusion discs for the major elements Si, Ti, Al, Fe, Mn, Mg, Ca, Na, K, and P, and powder pellets for the trace elements Rb, Sr, Y, Zr, Nb, Cr, and Ni; (2) atomic

absorption spectrophotometry (AAS) methods at University College London for Cu and Zn; (3) instrumental neutron activation analysis (INAA) methods at the University of London Reactor Centre and the Risley Reactor Centre for rare-earth elements (REE). H_2O^+ and CO_2 were analysed simultaneously by gravimetric methods at University College London and the University of Liverpool. Representative analyses are given in Tables I and II (the full data set is available on request).

GEOCHEMISTRY

Sources of alteration

There are two main sources of potential whole-rock geochemical alteration for the rocks studied: (1) sea water–basalt interaction; (2) Scandian metamorphism and deformation. Sea water interaction can lead to effects such as increased K, Na, large ion lithophile (LIL), and light

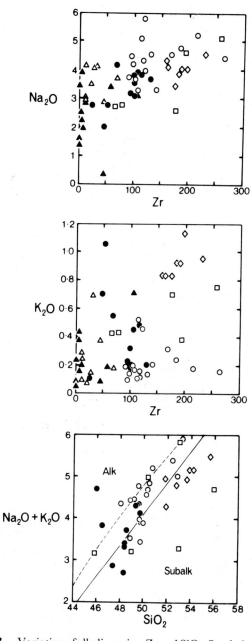

Fig. 3. Variation of alkalis against Zr and SiO_2. Symbols: filled triangles = gabbros; open triangles = flaser gabbros; filled-circles = dykes; open circles = pillows; open squares = Kjeldvann metadolerites; open diamonds = Lomivann volcanics pillows.

rare-earth elements (LREE) contents of basalts (Ludden and Thompson, 1979; Saunders and Tarney, 1984). Figure 3 illustrates the generally low K_2O contents but rather high Na_2O contents of Sulitjelma ophiolite pillow lavas, suggesting that sea water interaction may have increased Na_2O but decreased or not changed K_2O content. Many samples are Ne-normative and appear alkaline on the SiO_2–$(Na_2O + K_2O)$ plot, whereas "immobile" trace element ratios (such as $Y/Nb = 4$–8) indicate a tholeiitic character. It is suggested that the apparent alkaline character is a function of major element alteration, especially Na_2O increase. Note the higher K_2O contents, but lower Na_2O contents of Lomivann

volcanic pillows compared to ophiolite pillows. Is this variation a primary feature, or a function of sea water interaction? Further evidence for sea water interaction is provided by the development of massive, volcanic-exhalative, Fe–Cu–Zn–S ore-bodies and the wide scatter in Cu and Zn contents (Fig. 4). This scatter is more pronounced in rocks close to ore zones (data not presented here, but available on request).

Studies of alteration related to regional metamorphism are less abundant, but Hellman *et al.*, (1979), for example, have demonstrated various types of REE enrichment and depletion during burial and orogenic metamorphism. It is clear therefore that, in studying rocks that have potentially

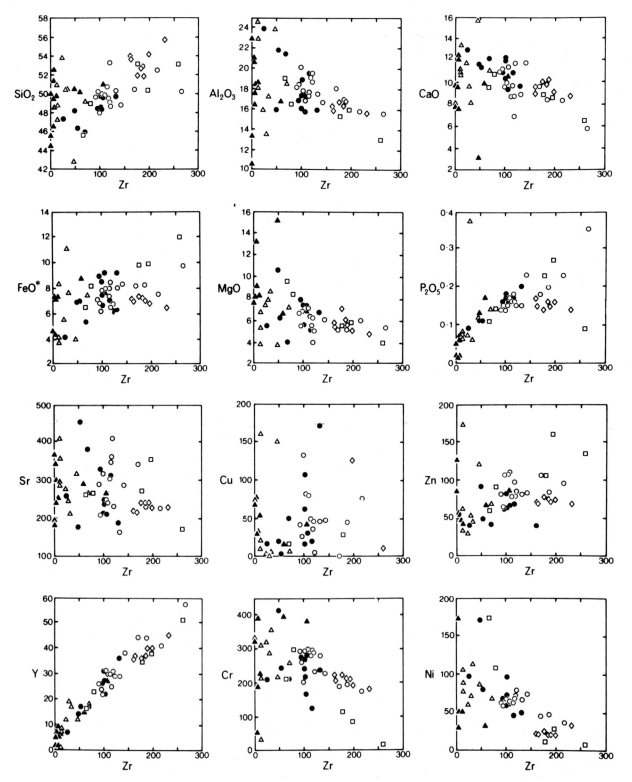

Fig. 4. Variation of selected major and trace elements. Symbols as in Fig. 3.

undergone sea water alteration and metamorphic alteration, care must be taken when evaluating the significance of whole-rock chemical variations. In view of this, the element Zr is used as a fractionation index, rather than for example SiO_2, MgO, FeO*/MgO or differentiation index (D.I.), because Zr is relatively abundant, precisely determined by XRF techniques, has a low bulk distribution coefficient in basaltic systems, and, perhaps most importantly, is believed to be immobile during weathering and metamorphism (Pearce and Cann, 1973; Weaver *et al.*, 1979).

Major- and trace-element variation

Representative analyses are given in Tables I and II. The three igneous suites exhibit different trends on an AFM plot: dykes and pillows of the Sulitjelma ophiolite have a poor iron-enrichment trend (probably owing to Na_2O enrichment during sea water alteration and FeO* mobility during ore formation); the Kjeldvann metadolerites have a pronounced iron-enrichment trend; and the Lomivann volcanics have no iron-enrichment

trend (Fig. 5). This difference between the three suites is also borne out by Fig. 4: the Lomivann volcanics show an iron-depletion trend when Zr is used as a fractionation index. In general, the gabbros show a wide scatter of concentrations (Fig. 4, excepting highly incompatible elements such as yttrium and phosphorous), reflecting the cumulate processes active during their formation.

Ophiolite dykes and pillows show trends of increasing SiO_2, FeO*, Na_2O, P_2O_5, Y and Ti, and decreasing Al_2O_3, CaO, MgO, Cr and Ni with increasing Zr (Figs. 3, 4 and 8), suggesting fractionation of olivine, plagioclase and clinopyroxene. The scatter present in these data is partly attributed to alteration processes discussed earlier, but also to variable amounts of phyric plagioclase now pseudomorphed by metamorphic plagioclase, quartz, epidote and mica assemblages. The Kjeldvann metadolerites differ mainly in having higher Ti, FeO* and K_2O contents, but lower Cr and Ni contents in more fractionated rocks (Figs. 3, 4 and 8). The Lomivann volcanics have higher SiO_2 and K_2O contents and distinctive FeO*, P_2O_5 and Zn trends, together with slightly higher Cr and lower Ni in the range 150–250 ppm

Table I. Representative analyses of Sulitjelma ophiolite rocks

Sample	T158	B112	B84	B137	A14	B128	B129	A24	A62	D1	A78	D2
Type	G[a]	G[a]	G[a]	FG[a]	D[a]	D[a]	D[a]	D[a]	P[a]	P[a]	P[a]	P[a]
SiO_2	44.53	48.58	50.15	50.50	48.29	45.98	48.38	49.71	49.66	49.52	48.84	50.24
TiO_2	0.15	0.31	3.16	0.71	0.80	0.87	1.27	1.54	1.18	1.38	1.41	2.22
Al_2O_3	13.41	18.42	16.76	13.59	15.94	21.37	17.36	16.03	18.17	17.61	16.63	15.57
FeO*	9.89	6.97	8.72	11.08	6.88	5.38	7.46	8.52	7.09	8.02	8.34	9.73
MnO	0.16	0.12	0.16	0.13	0.14	0.09	0.13	0.15	0.13	0.16	0.14	0.15
MgO	21.92	9.07	6.68	7.96	10.64	4.11	7.44	7.34	6.70	5.59	6.14	5.41
CaO	8.00	12.06	9.90	9.66	11.63	12.17	12.03	10.35	10.99	9.75	11.83	5.81
Na_2O	1.36	1.98	3.38	4.02	2.00	4.14	3.09	3.81	3.76	4.25	3.32	4.41
K_2O	0.88	0.21	0.19	0.15	0.70	0.55	0.21	0.32	0.19	0.52	0.21	0.16
P_2O_5	0.05	0.07	0.17	0.37	0.11	0.14	0.17	0.17	0.14	0.18	0.23	0.35
H_2O^+	0.35	1.52	1.20	1.36	2.25	3.01	1.87	1.35	1.01	1.01	1.49	3.63
CO_2	0.20	0.38	0.15	0.25	0.52	2.68	0.66	0.42	0.70	1.75	1.29	2.72
Total	100.10	99.69	100.62	99.78	99.90	100.49	100.07	99.71	99.7	99.4	99.87	100.40
Ba	nd[b]	nd	nd	nd	nd	nd	nd	nd	nd	nd	nd	nd
Ce	2.3	3.0	16.3	24.6	7.7	12.0	16.7	16.3	14.1	15.0	21.9	40.6
Cr	333	388	393	287	409	211	242	267	292	nd	229	nd
Cu	74	34	17	1	20	50	107	62	42	nd	48	nd
Dy	nd	nd	nd	nd	nd	nd	nd	nd	nd	5.2	nd	nd
Er	nd	nd	nd	nd	nd	nd	nd	nd	nd	3.8	nd	nd
Eu	0.4	0.9	1.4	1.4	0.9	1.1	1.3	1.4	1.2	1.3	1.6	2.4
Gd	0.8	0.9	2.6	3.2	2.4	3.1	3.4	4.5	3.8	4.4	5.2	8.3
Ho	nd	nd	1.1	n.d	n.d	n.d	n.d	n.d	1.1	1.1	nd	nd
La	1.0	1.9	7.4	10.0	2.9	3.8	4.6	5.8	4.2	4.4	8.1	13.7
Lu	0.1	0.1	0.2	0.3	0.3	0.3	0.4	0.6	0.5	0.4	0.7	0.9
Nb	2	2	7	4	3	5	5	2	1	7	4	11
Nd	1.3	3.2	9.0	12.6	5.2	6.8	10.3	10.2	9.6	11.0	14.2	23.1
Ni	428	51	32	71	170	69	72	59	62	nd	74	nd
Pr	nd	nd	2.1	nd	nd	nd	nd	nd	nd	2.1	nd	nd
Rb	1	2	1	1	7	10	1	2	2	9	1	2
Sc	nd	nd	nd	nd	nd	nd	nd	nd	nd	nd	nd	nd
Sm	0.3	0.8	2.2	2.9	1.8	2.0	3.0	3.0	2.9	3.4	3.9	6.3
Sr	196	253	289	245	178	382	236	213	289	348	287	105
Tb	0.1	0.2	0.5	0.7	0.4	0.6	0.7	0.8	0.7	0.7	1.0	1.5
Th	nd	nd	nd	nd	nd	nd	nd	nd	nd	nd	nd	nd
Tm	0.0	0.1	0.2	0.2	0.3	0.2	0.3	0.3	0.3	0.3	0.5	0.7
V	nd	nd	nd	nd	nd	nd	nd	nd	nd	nd	nd	nd
Y	2	5	15	19	14	17	22	31	26	25	38	57
Yb	0.3	0.7	1.6	1.8	1.8	2.1	2.5	3.1	3.1	3.0	4.2	6.7
Zn	84	48	67	53	91	42	63	81	81	nd	83	nd
Zr	1	8	59	28	49	69	102	101	92	116	142	264

[a] G = gabbro, FG = flaser gabbro, D = dyke, P = pillow.

[b] nd = not determined.

* total Fe as FeO.

Table II. Representative analyses of Kjeldvann metadolerites (KM) and Lomivann volcanics (LV)

Sample	B118	X6	Giles1	Giles5	Giles7
Type	KM	KM	LV	LV	LV
SiO_2	48.99	53.08	51.90	51.84	55.62
TiO_2	1.23	3.55	1.10	1.11	1.01
Al_2O_3	16.52	13.04	16.67	16.75	15.64
FeO*	8.18	12.00	7.35	7.06	6.46
MnO	0.15	0.19	0.13	0.13	0.12
MgO	8.42	4.03	7.12	5.58	4.90
CaO	10.77	6.54	10.11	10.26	8.81
Na_2O	2.77	5.06	3.46	4.03	4.54
K_2O	0.43	0.75	0.83	0.91	0.95
P_2O_5	0.14	0.09	0.14	0.15	0.14
H_2O^+	2.37	1.01	nd	nd	nd
CO_2	0.17	0.88	nd	nd	nd
Total	100.14	100.22	98.81	97.82	98.19
Ba	nd	nd	168	211	144
Ce	10.0	33.3	44.0	37.0	44.0
Cr	297	19	223	193	182
Cu	222	12	nd	nd	nd
Dy	nd	nd	6.5	6.3	5.1
Er	nd	nd	4.6	4.7	4.5
Eu	1.2	1.9	1.7	1.4	1.5
Gd	4.0	6.8	6.5	5.2	6.8
Ho	nd	nd	1.3	1.3	1.5
La	3.4	11.3	18.0	14.0	18.0
Lu	0.4	0.8	0.71	0.49	0.57
Nb	5	6	nd	nd	nd
Nd	8.8	20.8	20.0	19.0	22.0
Ni	108	9	26	21	34
Pr	nd	nd	5.1	4.2	4.9
Rb	10	25	21	34	31
Sc	nd	nd	37	33	30
Sm	2.8	5.4	5.5	4.6	5.4
Sr	267	174	242	229	230
Tb	0.7	1.4	0.86	0.84	0.97
Th	nd	nd	7	7	9
Tm	0.2	0.7	nd	nd	nd
V	nd	nd	179	167	157
Y	23	51	36	40	45
Yb	2.6	5.3	4.5	3.4	4.1
Zn	91	135	77	74	69
Zr	80	258	175	197	230

Zr. Subjectively, these differences suggest the Kjeldvann metadolerites and the Lomivann volcanics are not co-magmatic with the Sulitjelma ophiolite.

REE patterns

The most notable features of the REE patterns of the Sulitjelma ophiolite dykes and pillows are the flatness of the patterns, and the progressively less-positive Eu anomaly with increasing total REE (Fig. 6(a)). There is also a tendency towards some LREE enrichment with increasing total REE in both dykes and pillows. La/Yb ratios vary from 1.4 to 2.0. Gabbros have similar patterns, but a generally lower total abundance of REE and a larger positive Eu anomaly. Kjeldvann metadolerites also exhibit flat REE patterns, total REE abundances and La/Yb ratios (1.3 to 2.1) similar to ophiolite pillows and dykes, but less pronounced Eu anomalies (Fig. 6(b)). Lomivann volcanics are characterized by LREE enrichment, flat HREE, larger La/Yb ratios (4.0 to 4.4) and small negative Eu anomalies (Fig. 6(b)).

Eruptive setting

Pillows and dykes of the Sulitjelma ophiolite show flat to nearly flat patterns on MORB-normalized spidergrams (Fig. 7). There is slight LIL element enrichment, but

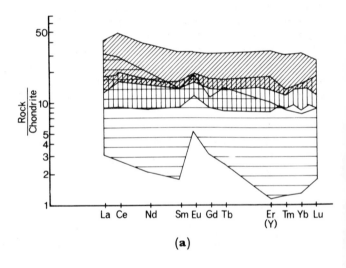

(a)

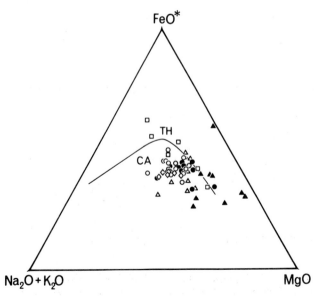

Fig. 5. AFM plot for ophiolite pillows and dykes, Kjeldvann metadolerites and Lomivann volcanics. Symbols as in Fig. 3.

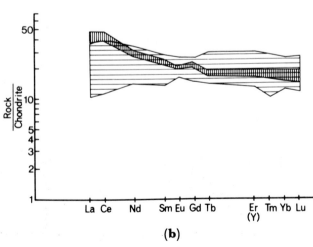

(b)

Fig. 6. Chondrite-normalized REE plots for (a) 4 gabbros (horizontal lines), 4 dykes (vertical lines) and 4 pillows (diagonal lines) of the ophiolite, and (b) 2 Kjeldvann metadolerites (horizontal lines) and 3 Lomivann volcanics (vertical lines). Actual values are given in Tables I and II.

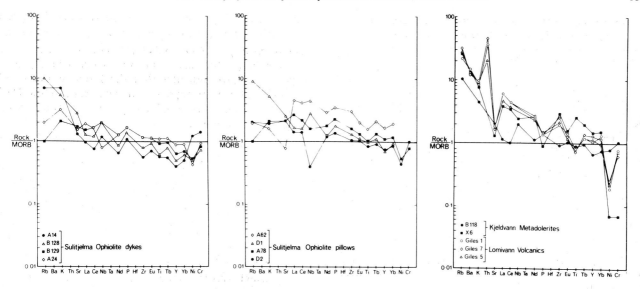

Fig. 7. Spidergrams of whole-rock chemistry normalized to MORB (values of Saunders and Tarney, 1984).

close correspondence of HFS element concentrations to typical MORB (Saunders and Tarney, 1984). In contrast, the Kjeldvann metadolerites and Lomivann volcanics show more pronounced LIL element enrichment and a tendency for relative depletion in Ni and/or Cr. The variation in LIL element enrichment may not be significant in view of the potential for sea water interaction.

Eruptive-setting discriminant diagrams (Figs. 8 and 9) clearly show the MORB-like character of the Sulitjelma ophiolite dykes and pillows, consistent with the spidergram plots and REE patterns. Conversely, the Lomivann volcanics have CAB-affinities, suggesting the LIL element enrichment in the spidergrams and the LREE enrichment

is primary rather than a function of alteration (see Brekke *et al.*, 1984). The Kjeldvann metadolerites show predominantly MORB-like affinities.

PETROGENESIS

The Sulitjelma ophiolite

The absence of primary igneous phases in ophiolite dykes and pillows precludes modelling of major-element variations using actual mineral compositions. However, the coherent chemical variation of the ophiolite dykes and pillows is here regarded as approximating to a liquid line of

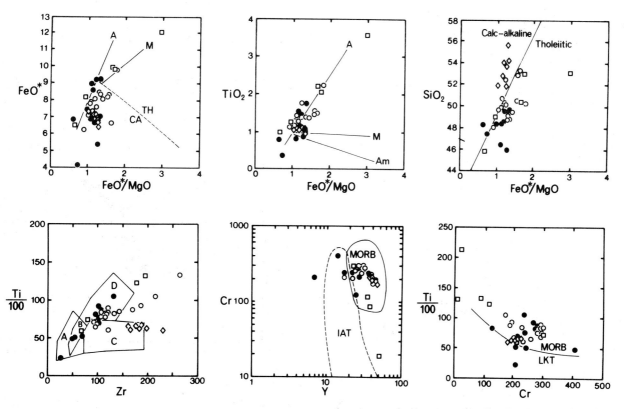

Fig. 8. Eruptive setting discriminant diagrams: FeO*–FeO*/MgO, TiO_2–FeO*/MgO, and SiO_2–FeO*/MgO after Miyashiro (1975) (A, M, Am refer to trend lines for abyssal tholeiite, Macauley Island tholeiitic island-arc series and Amagi calc-alkaline series respectively); Ti/100–Zr after Pearce and Cann (1973) (A+B=LKT, B+D=OFB, B+C=CAB); Cr–Y after Pearce *et al.*, (1984); Ti/100–Cr after Pearce (1975). Symbols as in Fig. 3.

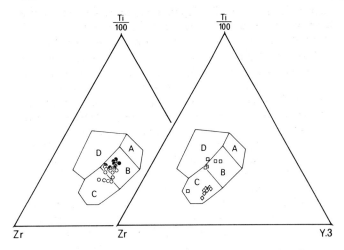

Fig. 9. Eruptive setting discriminant diagrams after Pearce and Cann (1973). Symbols as in Fig. 3. Fields: A + B = LKT, B = OFB, B + C = CAB, D = WPB.

descent, involving increasing SiO_2, FeO*, MnO, Na_2O, P_2O_5 and Ti, and decreasing Al_2O_3, MgO, CaO, Cr and Ni.

Qualitatively, these variations can be attributed to low-pressure fractional crystallization of Mg-rich olivine, calcic plagioclase and clinopyroxene in the source magma. These are the principal cumulate phases in the Sulitjelma gabbro (Mason, 1971). The flat pattern of Al4, the most primitive dyke (presumed to be closest to the parental liquid composition, base line of dyke field in Fig. 6(*a*)), suggests a garnet-free lherzolitic mantle source because HREE partition coefficients in garnet/basaltic-liquid systems are much higher than unity (Nicholls and Harris, 1980) and this would result in a LREE-enriched partial melt. The tendency for LREE enrichment with increasing total REE can be explained by low-pressure fractionation of clinopyroxene or hornblende since both these phases have higher partition coefficients for the HREE than the LREE in basaltic systems (Arth, 1976; Nicholls and Harris, 1980). Hornblende occurs as a primary crystallization product only in localized late-stage gabbro pegmatites, and as a deuteric alteration phase in much of the gabbro (Mason, 1971). It seems more probable, therefore, that the observed LREE enrichment, given that it is not an alteration effect (cf. Ludden and Thompson, 1979; Hellman *et al.*, 1979), is related to clinopyroxene fractionation. This effect will be partly offset by plagioclase fractionation, since plagioclase preferentially incorporates the LREEs. Evidence for plagioclase fractionation is provided by the progressively less positive Eu anomalies. The larger positive Eu anomalies in the gabbros relate to plagioclase accumulation during crystal fractionation. The persistence of the positive Eu anomaly, even in the most fractionated pillow lavas, may indicate the retention of Eu by residual clinopyroxene in the source region (Leeman, 1976). Alternatively, it may reflect the presence of xenocrystic plagioclase in dykes and pillows, though the aphyric nature of some of the samples argues against this explanation.

Quantitative modelling of the REE patterns in Fig. 6(a) (data in Table I) indicates that the range of pillow lava compositions (A62, D1, A78, D2) can be obtained from the dyke compositions (A14, B128, B129, A24) by 60–70% Rayleigh fractional crystallization of 50% plagioclase, 25% clinopyroxene, 15% olivine and 10% orthopyroxene using the partition coefficients of Arth (1976). This is the average mode of the Sulitjelma gabbro (Mason, 1971) and

confirms the field evidence for the co-magmatic nature of the gabbros, dykes and pillow lavas of the Sulitjelma ophiolite (Boyle, 1980). However, this fractionation series is not a simple time series. Field evidence for the relative ages of samples indicates that the fractionation trends discussed above are repeated in time (e.g., pillow D2 is older but more fractionated than pillow Dl). This could be explained in a number of ways:

(1) an open-system fractionation in which a single magma chamber is periodically replenished causing fractionation trends to be repeated;

(2) a closed-system fractionation in which a zoned magma chamber is periodically tapped at different levels enabling, under certain circumstances, more fractionated higher levels to be tapped before less fractionated lower levels,

(3) a series of essentially closed-system fractionations in which discrete magna chambers are supplied by batches of magma from the same source, such that each chamber undergoes a similar fractional crystallization resulting in fractionation trends in the dyke and pillow lava pile being periodically reproduced.

Unpublished mapping (Craig Lindsay and Ruth Manson, unpublished B.Sc. mapping theses, University of Liverpool, 1980, and Roger Mason personal communication) indicates that the Sulitjelma gabbro is a multiple intrusion, comprising an as-yet unknown number of discrete magma pulses. These observations favour model 3 above.

To sum up, the geochemical evidence favours the production of batches of parental magma of tholeiitic composition by high-pressure partial melting of a garnet-free lherzolitic mantle source, followed by a low-pressure fractional crystallization processes in discrete high-level magma chambers that repeatedly produced characteristic differentiation trends through a range of basaltic compositions owing to crystallization of olivine, plagioclase, clinopyroxene and orthopyroxene.

Kjeldvann metadolerites and Lomivann volcanics

Data on these two suites are less extensive and hence more difficult to interpret. The Kjeldvann metadolerites have sufficiently different geochemistry, particularly in terms of their Fe-enrichment, Cr and Ni trends, and LIL enrichment (Figs. 4, 5, 7 and 8), to indicate that they are not co-magmatic with the ophiolite. However, on REE abundance and eruptive setting plots (Figs. 6, 8 and 9) they are virtually indistinguishable from the ophiolite dykes and pillows, with the possible exception of the Ti–Zr–Y plot (Fig. 9), where the generally higher Ti contents of the Kjeldvann metadolerites push the analyses towards the within-plate field. Thus the metadolerites have MORB affinities, suggesting a similar source to the ophiolite, but a different geochemical evolution in detail.

The Lomivann volcanics show calc-alkaline affinities (Figs. 5, 6, 7, 8 and 9), apart from their MORB-like Y and Cr contents. Trends for FeO*, P_2O_5 and Ti are suggestive of hornblende fractionation. In the field the Lomivann volcanic pillow lavas are associated with andesites, dacites and rhyolites (Nicholson, 1966; Findlay, 1980). The association and its geochemistry are consistent with derivation from a volcanic arc.

DISCUSSION

The three igneous suites indicate a progression with time from MORB geochemistry to calc-alkaline chemistry. How can this be related to the field relationships of the three suites? There can now be little doubt that the Sulitjelma ophiolite formed in some sort of ensialic marginal-basin related to Iapetus. All of the ophiolite pseudostratigraphy from ultramafites to cover sediments is present, and the gabbro and dyke parts intrude continental crust in the form of Skaiti Supergroup (Fig. 2; Boyle, 1980; Billett, 1987). The Kjeldvann metadolerites intrude the sedimentary cover of the ophiolite, suggesting that they are younger than the ophiolite pillows and dykes. The Lomivann volcanics lie in levels of the Furulund Group stratigraphically above those containing the metadolerites, and are not cut by the metadolerites, suggesting they were erupted after emplacement of the metadolerites. The recorded arc-activity would therefore appear to be younger than construction of the ophiolite. However, if the Sulitjelma ophiolite does represent marginal-basin crust, then the boundary between it and the Furulund Group must be diachronous, not a time plane. As the basin extended and grew with time, the Furulund Group would have

prograded across it. The Kjeldvann metadolerites and the Lomivann volcanics need not therefore be younger than *all* of the ophiolite, but must be younger than *some* of it.

Can the Lomivann volcanics be precluded on age grounds from representing the products of an arc active during the formation of the Sulitjelma ophiolite in a marginal-basin? There are no relevant radiometric age constraints, but some constraints can be made based on the type of infill and rates of extension and sedimentation in present-day marginal-basins (Carey and Sigurdsson, 1984; Leitch, 1984). The thickness of the Furulund Group is not precisely known because it has undergone thinning due to flattening strains, but also thickening due to repetition by folding. Maximum structural thickness on Vogt's (1927) structural cross-sections are 1300 m. Hydrothermal, pelagic and hemipelagic deposits are common at the stratigraphic base of the Furulund Group (Boyle, 1980). These pass up into calcareous greywackes showing abundant evidence of grading in packets of several centimetre thick beds to metre-plus-thick beds (dominated by Bouma intervals A and E with sporadic development of

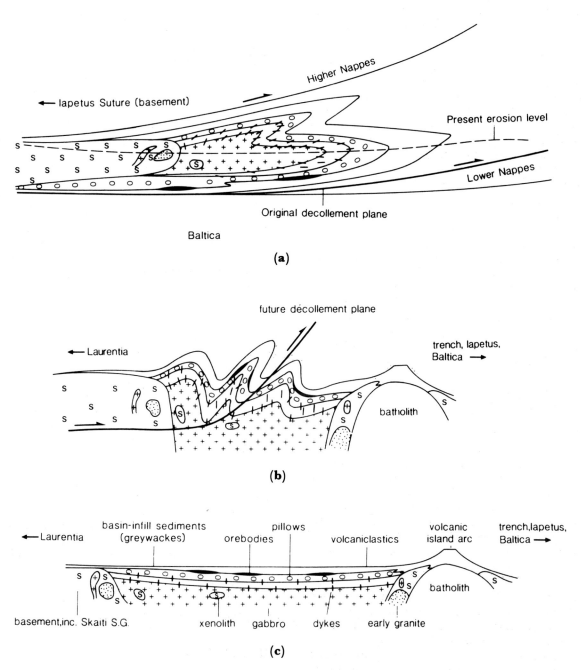

Fig. 10. Model showing the sequential restoration of the Sulitjelma ophiolite ensialic, back-arc, marginal basin (partly after Boyle, 1987). See text for discussion.

interval B) with local graphitic pelites, tuffites, meta-andesites, and conglomerates containing basic and acid fragments (Nicholson, 1966; Findlay, 1980; Kirk and Mason, 1984). The greywackes in turn pass up into the Lomivann volcanics, which show local evidence of slumping on volcanic slopes (Nicholson, 1966) and pillow lava eruption (Kekwick, 1988), and associated fossiliferous marbles, with minor calcareous greywackes. This sequence is analogous to that proposed by Carey and Sigurdsson (1984) in a model for marginal-basin sedimentation. In this context the greywackes represent the dominant clastic/epiclastic influx formed by erosion of the adjacent arc and continental crust; the Lomivann volcanics represent the later progradation of a pyroclastic apron (with intermittent clastic/epiclastic deposits and lava flows) formed when the arc and basin had matured (stages 2 to 3 of Carey and Sigurdsson (1984)); and the pelagic facies, in the form of graphitic pelites, represent the background sedimentation interrupted by the clastic/epiclastic and pyroclastic events. The sedimentology and volcanology of the Furulund Group are consistent with the formation of the Sulitjelma ophiolite in a back-arc marginal-basin. Carey and Sigurdsson report sedimentation rates of $0.1–0.3 \, \text{km} \, \text{Ma}^{-1}$ in back-arc basins. The thickness of the Furulund Group from the ophiolite to the fossil occurrences (Upper Ordovician–Lower Silurian) need not therefore represent more than about 3–10 Ma, constraining the age of the ophiolite to be Ordovician or younger. Average half spreading rates in SW Pacific marginal-basins are about $3 \, \text{cm} \, \text{a}^{-1}$ (Leitch, 1984), but generally slower in ensialic marginal-basins (M. Atherton, personal communication). Using this value, together with the sedimentation rate data, the Sulitjelma ophiolite marginal-basin would have attained a width of 180–600 km, intermediate in size between the Lau and South Fiji basins (Leitch, 1984), at the faster spreading rate, but perhaps 60–200 km at a slower half-spreading rate of $1 \, \text{cm} \, \text{a}^{-1}$ in the time it took the Furulund Group to be deposited. In this context, the Lomivann volcanics could be contemporaneous with the later-formed parts of the Sulitjelma ophiolite (i.e., the central part of the mature basin), but younger than the earlier formed parts (i.e., the present margin of the basin). It is therefore suggested that the Lomivann volcanics represent the products of an arc active during the construction of the Sulitjelma ophiolite marginal-basin, which is thus interpreted as forming in a back-arc environment. An alternative explanation is that the Lomivann volcanics simply represent the apron from a younger, unrelated arc.

The tectonic setting of the Sulitjelma ophiolite is restored in Fig. 10. Figure 10a shows the geometry of the Sulitjelma Fold-Nappe prior to late stage D3 upright folding (Boyle *et al.*, 1985). Unfolding and restoration of the fold-nappe structure suggests that the arc producing the Lomivann volcanics lay to the present day east of the ophiolite (Fig. 10(c)). This polarity of the back-arc basin suggests westward-directed subduction and favours formation of the basin outboard of Baltica and possibly on the Laurentian side of Iapetus. Stephens and Gee (1985) presented a model for the evolution of Central Scandinavian eugeoclinal terranes in which westward-directed subduction beneath Laurentia began in the early Silurian, but raised the possibility that it may have started earlier in the Ordovician. Conversely, Brekke *et al.* (1984) favour westward-directed subduction during the Cambrian and Lower Ordovician followed by eastward-directed subduction beneath Baltica during the Middle/Upper Ordovician and Silurian, the eastward-directed subduction being responsible for the Sulitjelma ophiolite

ensialic marginal-basin. The present work favours westward-directed subduction, contrary to Brekke *et al.*'s eastward-directed subduction model. However, it also suggests westward-directed subduction began earlier than the Silurian age, as put forward by Stephens and Gee (1985), though this may be a function of along-strike variations in the timing of subduction initiation; westward-directed subduction was perhaps earlier in north-Central Scandinavia than in Central Scandinavia.

Acknowledgements

This work has benefited from discussions on marginal-basins with Mike Atherton and Pat Brenchley, and discussions on Sulitjelma geology with Roger Mason, Wendy Kirk, Mike Billett and Tor Søyland Hansen. Mike Stephens and an anonymous reviewer are thanked for their constructive comments. Giz Marriner, Dee Stewart and Julie Sharman are thanked for assistance with XRF analyses, and Sue Parry and Mike Brotherton for assistance with INAA work. Financial assistance for field and laboratory work came from a NERC studentship, Sulitjelma Bergverk A. S. and Liverpool University.

REFERENCES

Arth, J. G. (1976). Behaviour of trace elements during magmatic processes–a summary of theoretical models and their applications. *J. Res. U.S. Geol. Surv.*, **4**, 41–47.

Billett, M. F. (1987). The geology of the northern Sulitjelma area and its relationship to the Sulitjelma Ophiolite. *Norsk Geol. Tidsskr.* **67**, 71–83.

Boyle, A. P. (1980). The Sulitjelma amphibolites, Norway: part of a Lower Palaeozoic ophiolite complex? In: Panayiotou, A. (ed.) *Ophiolites: Proceedings International Ophiolite Symposium, Cyprus, 1979*, pp. 567–575. Cyprus Geological Survey Department.

Boyle, A. P. (1982). Structure, petrology and ophiolitic affinities of the Sulitjelma amphibolities, Central Scandinavian Caledonides, Unpublished PhD. thesis, University of London.

Boyle, A. P. (1987). A model for stratigraphic and metamorphic inversions at Sulitjelma, Central Scandes. *Geol. Mag.* **124**, 451–466.

Boyle, A. P., Mason, R. and Hansen, T. S. (1985). A new tectonic perspective of the Sulitjelma Region. In: Gee, D. G. and Sturt, B. A. (eds.) *The Caledonide Orogen—Scandinavia and Related Areas*, pp. 529–541. Wiley, Chichester.

Brekke, H., Furnes, H., Nordås, J. and Hertogen, J. (1984). Lower Palaeozoic convergent plate margin volcanism in Bømlo, SW Norway, and its bearing on the tectonic environments of the Norwegian Caledonides. *J. Geol. Soc. Lond.* **141**, 1015–1032.

Carey, S. and Sigurdsson, H. (1984). A model of volcanogenic sedimentation in marginal basins. In: Kokelaar, B. P. and Howells, M. F. (eds.) *Marginal Basin Geology: Volcanic and Associated Sedimentary and Tectonic Processes in Modern and Ancient Marginal Basins*, pp. 37–58. Geol. Soc. Lond. Spec. Publ.

Findlay, R. H. (1980). A regional lithostratigraphy for southern and eastern Sulitjelma, north Norway. *Norsk Geol. Tidsskr.* **60**, 233–234.

Furnes, H., Ryan, P. D., Grenne, T., Roberts, D., Sturt, B. A. and Prestvik, T. (1985). Geological and geochemical classification of the ophiolite fragments in the Scandinavian Caledonides. In: Gee, D. G. and Sturt, B. A. (eds.), *The Caledonide Orogen—Scandinavia and Related Areas*, pp. 657–670, Wiley, Chichester.

Hansen, T. S. (1980). Some guide-lines to ore in the Sulitjelma ore field. *Norges Geol. Unders.* **360**, 237–238.

Hellman, P. L., Smith, R. E. and Henderson, P. (1979). The mobility of rare earth elements: evidence and implications from selected terrains affected by burial metamorphism. *Contrib. Mineral. Petrol.* **71**, 23–44.

Kautsky, G. (1953). Der geologische Bau des Sulitjelma–Salojaure – gebietes in den nordskandinavischen Kaledoniden. *Sver. Geol. Under. C* 528, 1–233.

Kekwick, G. R. P. (1988). Inverted pillow lava at the base of the Furulund Group, Sulitjelma. *Norsk Geol. Tidsskr.*, in press.

Kirk, W. and Mason, R. (1984). Facing structures in the Furulund Group, Sulitjelma, Norway. *Proc. Geol. Assoc.* **95**, 43–50.

Krause, H. (1956). Zur Kenntnis der metamorphen Kieslagerstätte von Sulitjelma (Norwegen). *Neues Jahrbuch für Mineralogie Abhandlungen* **24**, 151–155.

Leitch, E. C. (1984). Marginal basins of the SW Pacific and the preservation and recognition of their ancient analogues: a review. In: Kokelaar, B. P. and Howells, M. F. (eds.), *Marginal Basin Geology: Volcanic and Associated Sedimentary and Tectonic Processes in Modern and Ancient Marginal Basins*, pp. 97–108. *Geol. Soc. Lond. Spec. Publ.*

Leeman, W. P. (1976). Petrogenesis of McKinney (Snake River) olivine tholeiite in the light of rare earth element and Cr/Ni distributions. *Geol. Soc. Amer. Bull.* **87**, 1582–1586.

Ludden, J. N. and Thompson, G. (1979). An evaluation of the behaviour of the rare earth elements during the weathering of sea-floor basalt. *Earth Planet. Sci. Lett.* **43**, 85–92.

Mason, R. (1971). The chemistry and structure of the Sulitjelma gabbro. *Norges Geol. Unders.* **269**, 108–141.

Miyashiro, A. (1975). Classification, characteristics and origin of ophiolites. *J. Geol.* **83**, 249–281.

Nicholls, I. A. and Harris, K. L. (1980). Experimental rare earth element partition coefficients for garnet, clinopyroxene and amphibole coexisting with andesitic and basaltic liquids. *Geochim. Cosmochim. Acta* **44**, 287–308.

Nicholson, R. (1966). On the relations between volcanic and other rocks in the fossiliferous east Lomivann area of Norwegian Sulitjelma. *Norges Geol. Unders.* **260**, 143–156.

Oftedahl, C. (1958). A theory of exhalative-sedimentary ores. *Geol. Fören. Stockh. Förh.* **80**, 1–19.

Pearce, J. A. (1975). Basalt geochemistry used to investigate past tectonic environments on Cyprus. *Tectonophysics* **25**, 41–67.

Pearce, J. A. and Cann, J. R. (1973). Tectonic setting of basic volcanic rocks determined using trace element analyses. *Earth Planet. Sci. Lett.* **19**, 290–300.

Pearce, J. A., Lippard, S. J. and Roberts, S. (1984). Characteristics and tectonic significance of supra-subduction zone ophiolithes. In: Kokelaar, B. P. and Howells, M. F. (eds.), *Marginal Basin Geology: Volcanic and Associated Sedimentary and Tectonic Processes in Modern and Ancient Marginal Basins*, pp. 77–94. *Geol. Soc. Lond. Spec. Publ.*

Roberts, D. and Sturt, B. A. (1980). Caledonian deformation in Norway. *J. Geol. Soc. Lond.* **137**, 241–250.

Saunders, A. D. and Tarney, J. (1984). Geochemical characteristics of basaltic volcanism within back-arc basins. In: Kokelaar, B. P. and Howells, M. F. (eds.), *Marginal Basin Geology: Volcanic and Associated Sedimentary and Tectonic Processes in Modern and Ancient Marginal Basins*, pp. 59–76. *Geol. Soc. Lond. Spec. Publ.*

Saunders, A. D., Tarney, J., Stern, C. R. and Dalziel, I. W. C. (1979). Geochemistry of Mesozoic marginal basin floor igneous rocks from southern Chile. *Geol. Soc. Amer. Bull.* **90**, 237–258.

Stephens, M. B. (1986). Terrane analysis of Sulitjelma, Upper Allochthon, Scandinavian Caledonides. *Geol. Fören. Stockh. Förh.* **108**, 303–304.

Stephens, M. B. and Gee, D. G. (1985). A tectonic model for the evolution of the eugeoclinal terranes in the central Scandinavian Caledonides. In: Gee, D. G. and Sturt, B. A. (eds.) *The Caledonide Orogen—Scandinavia and Related Areas*, pp. 953–77. Wiley, Chichester.

Stephens, M. B., Gustavson, M., Ramberg, I. B. and Zachrisson, E. (1985). The Caledonides of Central-North Scandinavia – a tectonostratigraphic overview. In: Gee, D. G. and Sturt, B. A. (eds.) *The Caledonide Orogen—Scandinavia and Related Areas*, pp. 135–162. Wiley, Chichester.

Vogt, T. (1927). Sulitelmafeltets geologi og petrografi. *Norges Geol. Unders.* **121**, 1–560.

Weaver, S. D., Saunders, A. D., Pankhurst, R. J. and Tarney, J. (1979). A geochemical study of magmatism associated with the initial stages of back-arc spreading. *Contrib. Mineral. Petrol.* **68**, 151–169.

Note: The Norwegian spelling of the place names Lomivann and Kjeldvann has recently been changed to Låmivatn and Kjeldvatn.

14 Xenolithic dykes on Seiland and preliminary observations on the lithospheric mantle beneath the Seiland Province, W Finnmark, Norway

J. M. Leaver,[1] *M. C. Bennett*[1] *and B. Robins*[2]

[1]Geology Department, University of Bristol, Wills Memorial Building,
Queen's Road, Bristol BS8 1RJ, UK
[2]Geologisk Institutt, Avd. A, Universitetet i Bergen, Allégt 41, 5007 Bergen, Norway

Swarms of ENE-trending dykes associated with a late phase in the evolution of the Melkvann ultramafic complex on the island of Seiland contain ultramafic xenoliths. These represent a sample of the lithospheric mantle beneath the Seiland Igneous Province at the time of dyke emplacement, between 531 Ma and (?)600 Ma. The dykes document a period of local crustal extension accompanied by intrusion of hydrous alkali olivine basalt or basanitic magmas. Distribution of xenoliths within the dykes requires xenolith accumulation via gravitational settling from magmas with low effective viscosity and yield strength. Calculations using reasonable rheological assumptions and maximum xenolith diameter suggest minimum magma ascent velocities of 2–3 km h^{-1}.

The xenolith suite contains Type I, Type II and composite xenoliths. Type I xenoliths are mainly laminated, disrupted, porphyroclastic or granuloblastic Al-spinel and Cr-spinel lherzolites and harzburgites showing variable degrees of depletion and modal metasomatism. The least-depleted xenoliths have mg numbers of ca. 86. Equilibrium temperatures based on several geothermometers are in the range 870–1070° C. The Type II xenoliths are variably recrystallized mantle cumulates, and are common at only one locality, where they occur with composite xenoliths that provide evidence for magma conduit flow and associated modal metasomatism in at least part of the sub-Seiland mantle prior to entrainment.

INTRODUCTION

Swarms of xenolithic ultramafic dykes associated with the later stages in the evolution of the Melkvann ultramafic complex (Bennett *et al.*, 1986) have been discovered recently at several places on the island of Seiland, W Finnmark. The dykes are unusual because they are choked with ultramafic xenoliths. They offer a rare opportunity to examine the physical aspects of magma transport at intermediate–lower crustal levels, as well as providing a sample of the sub-Seiland Province lithospheric mantle. The Seiland Igneous Province of W Finnmark and N Troms (Roberts, 1974; Robins and Gardner, 1975) is confined to the Sørøy Nappe. This is the uppermost tectonic unit within the Kalak Nappe Complex that constitutes the Middle Allochthon in the northern Scan-

dinavian Caledonides (Ramsay *et al.*, 1985; Roberts and Gee, 1985). The tectonic setting for the voluminous basic and ultrabasic magmatism of the province remains the subject of current debate (Stephens *et al.*, 1985; Bennett *et al.*, 1986; Gayer *et al.*, 1987; Krill and Zwaan, 1987; Pedersen *et al.*, 1988), and the timing of the magmatic episode is as yet poorly constrained. It has been shown, however, that in the northern and eastern parts of the province, the magmatism progressed from synkinematic subalkaline activity to the emplacement of alkaline olivine basalt and picritic magmas that gave rise to olivine gabbro–ultramafic complexes through *in situ* fractional crystallization at 6–10 kbar, corresponding to crustal depths of ca. 20–35 km (Robins and Gardner, 1975; Bennett *et al.*, 1986). Further phases of activity are represented by a regional picrite–ankaramite dyke swarm (Robins and Takla, 1979) and the carbonatite-bearing alkaline complexes of Lillebukt (Stjernøy) and Breivikbotn (Sørøy) together with their associated syenite and nepheline syenite pegmatite dykes, recently dated at 531 ± 2 Ma and 523 ± 2 Ma (U–Pb, zircon) (Pedersen *et al.*, 1988, this volume).

The dykes described in this account were emplaced at a late stage in the evolution of the Melkvann ultramafic complex (Fig. 1) and pre-date both the nepheline syenite pegmatites and the regional picrite–ankaramite dyke swarm. They appear to represent a short-lived period of alkali olivine basalt (or basanite) magmatism associated with local crustal extension following the emplacement of the ultramafic complexes, the timing of which is rather poorly constrained by Sm/Nd internal isochron ages for Melkvann olivine gabbro of 559 ± 36 Ma and 513 ± 24 Ma (Snow *et al.*, 1986).

THE XENOLITHIC DYKES

Swarms of dykes containing ultramafic xenoliths occur in two areas (Holmevann and East Melkvann) within the outcrop of the Melkvann ultramafic complex, and in a further area close to its sheeted contact with foliated gabbronorite at Nik'kavarri (Fig. 1). Single dykes are exposed at Store Bekkarfjord (Robins, 1975) and Storfjell, outside the ultramafic complex. Some members of the picrite–ankaramite dyke swarm are also xenolithic (Robins, 1975) but these are excluded from the present account.

All the dykes have an ENE–WSW trend and are either vertical or steeply inclined to the north-northwest. Their widths vary usually between 5 and 20 m, but one multiple dyke at Nik'kavarri (Fig. 2) is ca. 30 m wide. Although exposure is imperfect, some dykes may be traced along strike for 1 km. The most dense swarm is found at Nik'kavarri (Fig. 2), where reasonable exposure allows the following observations.

(1) The dykes occur in a subparallel ENE–WSW swarm and are cut only by rare amphibolitized picritic and ankaramitic dykes and in places by syenite pegmatites.

(2) Dyke widths vary from ca. 5 m to ca. 30 m, but are not usually greater than 20 m. The dykes dip very steeply to the north-northwest.

(3) The host rock is either poikilitic olivine–hornblende clinopyroxenite (varying to cortlandite) or foliated gabbronorite.

(4) The marginal facies of dykes contain dispersed xenoliths in a variably amphibolitized olivine clino-pyroxenite that grades locally into amphibole-bearing wehrlite. In these marginal facies, xenoliths are frequently smaller and their margins more diffuse than in the central facies of dykes (Fig. 3).

(5) The central facies of dykes contain high concentra-tions of xenoliths, and in many instances these facies resemble framework-supported conglomerates, with the "clasts" separated by thin septa of "matrix" (Fig. 4) varying in composition from olivine–horn-blende clinopyroxenite to poikilitic wehrlite.

(6) Dyke margins and, more commonly, dyke cores contain narrow dykes, veins and segregrations of hornblende gabbro pegmatite that intrude both the interstitial matrix *and* xenoliths.

(7) The dykes are all dilational and indicate local SSE–NNW crustal extension of 25–30%.

In dykes at all five localities (Fig. 1) the matrix between xenoliths consists of olivine (ca. Fo_{75-76}) + Al-rich salite $\pm$ pargasitic amphibole $\pm$ green hercynitic spinel $\pm$ Fe–Ti oxide. In most cases, the euhedral or subhedral salite and olivine represent the liquidus phases precipitated from the entraining magma. In a few dykes, the clinopyroxene is interstitial or poikilitic, which suggests that in these cases olivine was the only silicate liquidus phase at the time of xenolith entrapment. The mineralogy of the dyke matrices is consistent with that expected from the frac-tional crystallization of hydrous alkali olivine basalt magma, or basanite (Bennett *et al.*, 1986). It may be concluded that the xenoliths were entrained in a magma of this compositional range, but because some olivine in the matrix is xenocrystic (derived by chemical and mechanical erosion of xenoliths), it is difficult to constrain the entraining magma composition more precisely. The injection of hornblende gabbro pegmatite is attributed to a later magmatic phase involving petrogenetically-related but more evolved magmas during the waning stages of activity.

The xenolith suite consists mainly of spinel lherzolite, harzburgite and minor dunite derived from the litho-spheric mantle above the zone of magma generation. The majority of these xenoliths are characterized by strongly anisotropic fabrics that record deformation and recrystallization in a tectonically active lithospheric mantle. In addition, the suite contains a small proportion (maximum 20% at Holmevann, see below) of meta-cumulate xenoliths with less strongly developed aniso-tropic fabrics. The strong textural anisotropy in the majority of xenoliths has to a significant extent influenced xenolith dimensions, so that in general they have a low degree of sphericity and many approximate to spindle or plate shapes. The xenoliths also possess a low degree of roundness, and the majority of small fragments (<4 cm in diameter) are angular and appear to have resulted from brecciation of large xenoliths during entrainment and/or transport. In the central parts of dykes, xenoliths have sharp margins, and show little evidence of reaction with the entraining magma. In marginal facies however, dispersed xenoliths tend to be more rounded, have more diffuse margins and often show signs of recrystallization, all of which suggest a greater degree of re-equilibration with the entraining magma, and its crystallization products, than is shown by xenoliths in central facies. The size variation of xenoliths is extreme. In cases where brecciation has occurred, angular xenoliths vary in size from ca. 4 cm to <1 cm and occur in a matrix rich in xenocrystic olivine. More usually, however, xenoliths do not show evidence of brecciation and maximum dimensions of individual xenoliths vary between 4 and 20 cm, but with occasional larger blocks up to 63 cm across.

Sorting according to xenolith diameter is poor, and if the extremely high concentration of xenoliths (Fig. 4) is considered, it is difficult to escape the conclusion that the central facies of dykes represent rapid gravita-tionally-controlled accumulation of xenoliths within magma conduits. Several authors, notably Carmichael *et al.* (1977), Sparks *et al.* (1977) and Spera (1980, 1984), have used xenolith dimensions to calculate xenolith-settling velocities (and hence minimum magma-ascent velocities) assuming reasonable rheological properties for the entraining magma.

One very important parameter in such calculations is the crystallinity (and hence yield strength and effective viscosity) of the entraining magma (Spera, 1980, 1984; Marsh, 1981). In the Seiland xenolithic dykes, it may be inferred that the difference between dyke marginal

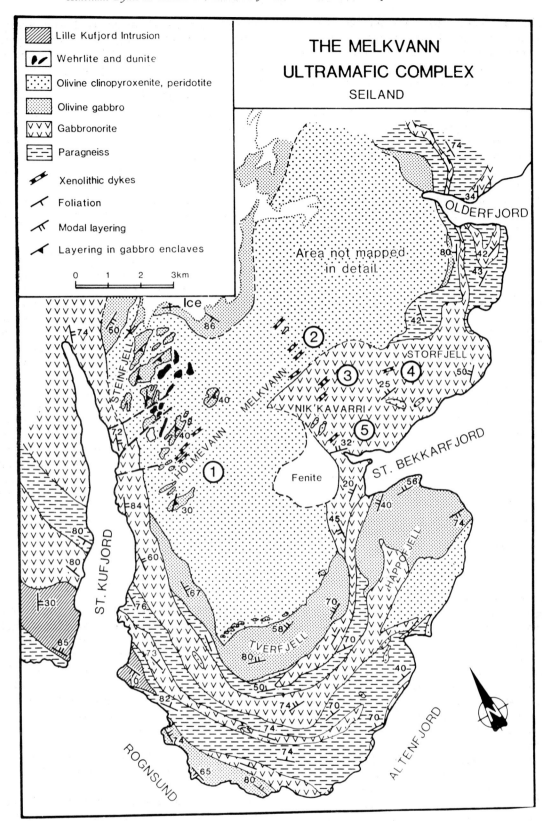

Fig. 1. Geological map of the Melkvann ultramafic complex, Seiland, W Finnmark, Norway showing the main ultramafic xenolithic dyke localities. 1, Holmevann; 2, East Melkvann; 3, Nik'kavarri; 4, Storfjell; 5, Store Bekkarfjord.

facies and central facies (Figs. 3 and 4) derive essentially from differing degrees of crystallinity. Close to the dyke margins, dissipation of heat to wall rocks coupled with increased frictional drag (and hence decrease in velocity) would have been major factors responsible for an increase in magma crystallinity at dyke margins. This is reflected in the higher proportion of matrix—i.e., the crystallization products from the entraining magma, and the relatively low concentration of xenoliths. It is likely that during

magma flow, xenolith concentrations in the central regions of the dykes would have exceeded that near dyke margins owing to flowage segregation (Komar, 1976). It is unlikely, however, that this mechanism alone was responsible for the very high xenolith/matrix ratios observed in central facies, because the percentage of solids (mainly xenoliths, but also xenocrysts and phenocrysts) would have been too high, and the porosity too low, to allow magma flow (Roscoe, 1953; Marsh, 1981, p. 94).

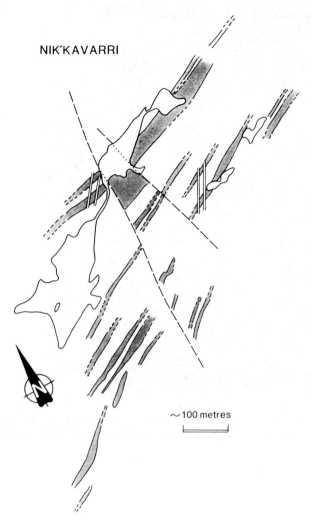

NIK'KAVARRI

~100 metres

Fig. 2. Sketch map showing the outcrop of the xenolithic ultramafic dyke swarm at Nik'kavarri (loc. 3 in Fig. 1). The unshaded dykes with a more northerly strike are later syenite pegmatites.

It may be inferred therefore that gravitational accumulation of xenoliths occurred, and it follows that to allow accumulation the entraining magmas in the central volumes of the conduits must have had relatively low effective viscosities and yield strengths.

Although the magma composition cannot be precisely constrained, the modal and mineralogical compositions described above indicate that the entraining magmas were of compositions close to hydrous alkali olivine basalt or basanite, and the field evidence demonstrates that these magmas had a low effective viscosities and yield strengths.

In the following calculations, several assumptions on magma rheology have been made to derive *first approximations* of minimum magma-ascent velocities using xenolith diameters of 20 cm and 60 cm. The assumptions rely in the first instance on the field evidence for low crystallinity and secondly on published estimates of density and rheology of alkali olivine basalt magmas (Shaw, 1969, 1972; Shaw *et al.*, 1968; Murase and McBirney, 1973; Carmichael *et al.*, 1977; Kushiro, 1980; Marsh, 1981), assuming temperatures of 1300–1400° C at ca. 7 kbar (Yeo, 1984; Bennett *et al.*, 1986).

Xenolith settling velocity (X_n) may be found by balancing inertial, buoyancy and viscous forces, assuming Bingham rheology for sub-liquidus magma using the following expression (Spera, 1984, equation (1)) for

xenoliths with a particle Reynolds number > 2.0,

$$X_n = 0.344 \left(\frac{\Delta \rho g}{\rho_l}\right)^{5/7} \left(\frac{\rho_l}{\eta_l}\right)^{3/7} \left(R_x - \frac{15\delta_0}{4\Delta \rho g}\right)^{8/7}$$

where $\Delta \rho$ = difference in density between xenolith and melt;

g = acceleration due to gravity;
ρ_l = density of melt;
η_l = plastic viscosity for Bingham magma;
δ_0 = magma yield strength;
R_x = xenolith radius.

It is assumed that melt density was $28000 \, kg \, m^3$, the density difference between xenolith and melt was $500 \, kg \, m^{-3}$ and that the crystallinity of the melt was between 15 and 25% by volume, giving approximate values of $\eta_l = 20$–$35 \, kg \, m^{-1} \, s^{-1}$ and $\delta_0 = 50$–$100 \, N \, m^{-2}$. For these ranges of effective viscosity and yield strength, minimum magma-ascent velocities are 5–$18 \, cm \, s^{-1}$ for a $20 \, cm$ diameter xenolith and 60–$92 \, cm \, s^{-1}$ for a $60 \, cm$ diameter xenolith. These calculations imply ascent velocities in the range 2–$3 \, km \, h^{-1}$ for $60 \, cm$ diameter xenoliths. If the depth of entrainment is taken as 60–$65 \, km$ (see below) and the depth of accumulation was 20–$25 \, km$, then residence times were in the order of 12–$18 \, h$.

It must be emphasised that these calculated velocities are *first approximations*. Refinements depend chiefly on the critical factor of magma composition, especially the concentrations of H_2O and CO_2 (Spera, 1987), which exert a strong influence on effective viscosity. The field evidence does, however, provide *prima facie* qualitative evidence for low yield strength and low viscosity and for this reason the calculated velocities are believed to be of the correct magnitude.

It is inferred that because of the high ascent velocities, the intrusion mechanism was one of magma fracture conduit flow (Spera, 1987). It appears likely that the dykes extended to the Earth's surface, producing a pressure gradient concomitant with high magma velocities. If the conduits were blind, then ascent velocities would have been constrained by the rate of crack propagation, which is likely to have been somewhat lower (Spera, 1987, p. 8). It is envisaged that the peculiarly high concentration of xenoliths within the dykes was produced partly by rapid accumulation, probably triggered by a reduction in magma flow below the critical value necessary to sustain buoyancy. The reduction in velocity could be explained by a decrease in magma pressure at the source following eruption at the surface.

THE XENOLITH SUITE

The ultramafic xenolith suite includes a variety of rock types (Fig. 5) but their proportions vary between localities (Fig. 6). Using petrographic and mineralogical observations the xenoliths can be classified as follows.

(1) Type I (Frey and Prinz, 1978). This group includes spinel lherzolites harzburgites, dunites and wehrlites.
(2) Type II (Frey and Prinz, 1978). This group includes olivine websterites, olivine clinopyroxenites, wehrlites and dunites.
(3) Composite xenoliths.

Type I xenoliths

Type I xenoliths are by far the most abundant and widespread on Seiland. They are olivine-rich rocks charac-

Fig. 3. The marginal facies of a 5 m wide xenolithic ultramafic dyke at Nik'kavarri. Somewhat rounded fine-grained Type I harzburgite and lherzolite xenoliths, frequently with diffuse margins, are dispersed in coarsely crystalline (ca. 5 mm) matrix of olivine, sub-poikilitic clinopyroxene and a little amphibole. The cross-cutting dyke at the top of the figure is composed of coarse-grained hornblende melanogabbro.

Fig. 4. The central facies of a 10 m wide xenolithic ultramafic dyke at Nik'kavarri. The xenoliths are Type I porphyroclastic and granuloblastic Cr-spinel lherzolites and harzburgites with low degrees of roundness and sphericity. Maximum xenolith dimensions range from 4 to 20 cm, with a mode of ca. 8 cm. The matrix is olivine clinopyroxenite with a low, but variable amphibole-content and with a crystal size of ca. 5 mm. In places, the matrix is injected by coarse-grained hornblende gabbro that also occurs in narrow dykes throughout the outcrop.

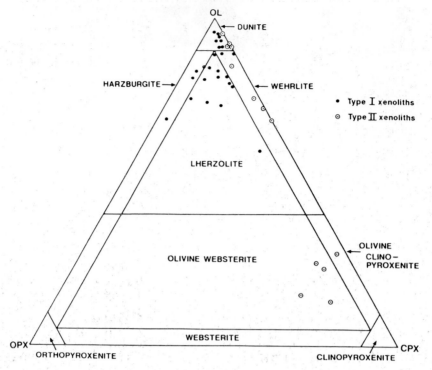

Fig. 5. Modal compositions of Type I and Type II ultramafic xenoliths from Seiland xenolithic dykes projected into the olivine-orthopyroxene-clinopyroxene triangle.

terized by Mg-rich and Cr-rich compositions with mg ($= 100$ Mg/Mg + Fe) ranging from 85 to 90 (Table I). Olivine, orthopyroxene and clinopyroxene together constitute more than 90% of the mode of each xenolith. Spinel is present in most xenoliths, while amphibole, phlogopite, plagioclase and carbonate are also present in some. The microstructures of these xenoliths indicate a wide variation of fabric types from laminated, disrupted, porphyroclastic to granuloblastic (nomenclature of Harte, 1977). Fabrics also vary from locality to locality. For example, mosaic porphyroclastic xenoliths constitute over 50% of all xenoliths in the Nik'kavarri, East Melkvann and Storfjell areas but are absent from Holmevann, where porphyroclastic xenoliths are dominant. The porphyroclasts at Holmevann are of olivine with kink bands set in a matrix of unstrained olivine, orthopyroxene, and clinopyroxene neoblasts. In the other areas, the olivine is almost entirely recrystallized and the porphyroclasts are dominantly of orthopyroxene. The latter display severely-deformed exsolution lamellae of spinel and/or diopside and show recrystallization in the form of small neoblasts <0.1 mm along grain boundaries and fractures. The striking contrast between the amount of strain in the porphyroclasts and the total absence of any optically visible strain in the neoblasts would suggest that they were close to areas of melting and consequently subject to plastic flow at the time of extraction (Mercier and Nicholas, 1975). Following deformation, recovery and recrystallization must have occurred with variable amounts of grain growth. Where a fabric is seen in individual xenoliths, it is caused by a roughly parallel alignment of large ortho-pyroxenes up to 10×2 mm in size and by "stretched" spinels. In some cases the recrystallized olivines have assumed a tabular form, accentuating the alignment of the other minerals.

In the Type I xenoliths, olivine has a compositional range from Fo_{87} to Fo_{91}, with NiO between 0.3 and 0.5 wt.%, there being a positive correlation between NiO and the forsterite content. There appears to be no significant

difference in chemistry between olivine porphyroclasts and neoblasts within individual xenoliths.

Orthopyroxene compositions vary from $Wo_{0.2}En_{84.0}Fs_{15.8}$ to $Wo_{0.7}En_{90.0}Fs_{9.3}$, the average enstatite content being En_{88}. Al_2O_3-contents vary from 0.44 to 3.9 wt.% and Cr_2O_3 is in the range of 0.19–0.55 wt.%. In general, as the En content increases, Cr_2O_3 increases and Al_2O_3 decreases and there is no significant difference in mineral chemistry across the porphyroclasts. The clinopyroxenes are chrome diopsides, characteristic of the Type I series. They vary in composition from $Wo_{44}En_{50}Fs_6$ to $Wo_{49}En_{47}Fs_4$. Al_2O_3 in these diopsides ranges from 2.0 to 4.6 wt.% and Cr_2O_3 from 0.77 to 2.5 wt.%, depending on the composition of the co-existing spinel. As a consequence, and not unexpectedly, there is a negative correlation between Al_2O_3 and Cr_2O_3 in the diopsides but a good positive correlation between Al_2O_3 in the diopside and $Cr/(Cr + Al + Fe^{3+})$ in the spinel. Na_2O-contents are normally between 1 and 2 wt.%.

The spinels in the Type I xenoliths vary considerably in their Al_2O_3- and Cr_2O_3-contents. Al_2O_3 ranges from 12 to 64 wt.% and Cr_2O_3 from 12 to 57 wt.%. A correlation is observed between increasing $Cr/(Cr + Al + Fe^{3+})$ and the development of haloes around the spinel porphyroclasts. The haloes consist mainly of granular plagioclase that is Na-rich, but in many samples pale-green amphibole is also developed. In some, the spinel porphyroclasts are partially replaced by phlogopite, and the residual spinel is granular chromite. The petrogenetic history of the haloes requires detailed analysis, but at this stage it seems likely that the production of plagioclase resulted mainly from decompression through reactions (or partial melting) involving orthopyroxene, clinopyroxene and Al-rich spinel, and that the development of amphibole and phlogopite are probably of metasomatic origin (cf. Robins, 1975).

The amphibole in spinel haloes is pargasitic and similar in composition to the amphibole that replaces clinopyroxene in some xenoliths, particularly those from Holmevann.

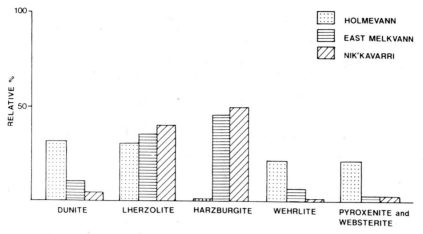

Fig. 6. The relative abundances of the five main modal varieties that comprise the xenolith suite at Holmevann, East Melkvann and Nik'kavarri. Note that at Holmevann, harzburgite xenoliths are scarce, but dunite, wehrlite, pyroxenite and websterite (Type II) xenoliths are relatively abundant compared with the suites at E Melkvann and Nik'kavarri, which are dominated by (Type I) lherzolites and harzburgites.

Table I. Bulk rock analyses of selected xenoliths, Seiland, N Norway

Sample	P55	P104	P125	P87	P90
Type	I	I	I	II	II
SiO$_2$	42.02	38.89	41.35	47.55	45.31
Al$_2$O$_3$	2.98	0.79	1.69	6.83	6.51
TiO$_2$	0.12	0.08	0.01	1.34	1.45
Fe$_2$O$_3$	1.62	3.10	2.52	1.40	1.63
FeO	8.60	8.62	5.76	6.54	7.58
MgO	37.84	40.94	40.24	16.54	19.59
CaO	2.80	2.30	1.85	18.65	15.59
Na$_2$O	0.67	0.01	0.13	0.50	0.47
K$_2$O	0.04	0.02	0.03	0.01	0.06
MnO	0.16	0.17	0.12	0.14	0.16
P$_2$O$_5$	2.20	3.87	3.25	0.63	0.89
CO$_2$	0.35	2.10	2.13	0.39	0.54
Cr$_2$O$_3$	0.40	0.34	0.76	0.28	0.29
NiO	0.26	0.30	0.10	0.08	0.09
H$_2$O	2.20	3.87	3.25	0.63	0.89
Total	100.09	99.48	99.45	100.42	100.19
100 Mg/(Mg+Fe)	86.8	86.3	89.9	78.8	79.1

The phlogopites associated with residual chromite have Cr$_2$O$_3$-contents of 0.82–1.60 wt.%, TiO$_2$-contents of 0.12–0.90 wt.% and relatively high Na$_2$O-contents of 1.60–2.20 wt.%. The carbonate that occurs in some Type I xenoliths is frequently found in association with pargasitic amphibole and is dolomitic in composition.

Type II xenoliths

The xenoliths belonging to this group are much less common than Type I xenoliths and their distribution is uneven. At Holmevann they comprise approximately 20% of the nodule suite, while at the other sampling localities they make up no more than 2% of the entire suite. They typically have higher concentrations of Al and Ti and higher ratios of Fe to Mg than Type I xenoliths (Table I). Most of the textures observed in hand specimens and thin sections can be interpreted as igneous in origin but many show evidence of partial recrystallization.

Although Al-augite is considered to be the characteristic mineral phase (together with Al-spinel) of most Type

II xenoliths world-wide (Kempton, 1987), the clino-pyroxene observed in the Seiland xenoliths is essentially diopsidic and varies in composition from Wo$_{47}$En$_{44}$Fs$_9$ to Wo$_{51}$En$_{40}$Fs$_9$. Compared to the diopsides from the Type I suite, however, the Al$_2$O$_3$-contents are higher (5–8 wt.%) and Cr$_2$O$_3$-contents lower (0.1–0.8 wt.%). The spinels are Al-rich, with Al$_2$O$_3$ ranging from 55 to 62 wt.% with Cr$_2$O$_3$-contents of 2–5 wt.%.

Other mineral phases observed in these xenoliths include olivine (Fo$_{74-81}$), kaersutite $\pm$ phlogopite, ortho-pyroxene, magnetite, ilmenite and iron sulphides.

Composite xenoliths

At Holmevann, xenoliths showing sharp contacts between two or more rock types were collected. These composite xenoliths are not readily apparent at any of the other sampling localities. All the composite xenoliths observed consist of a hydrous lithology crosscutting a Type I lherzolite host. The contacts are sharp, and parallel-sided, the hydrous lithologies occurring as dykes (1–4 cm wide) or veinlets (<1 cm). These have mineral assemblages dominated by clinopyroxene, amphibole with minor amounts of spinel, olivine and plagioclase.

The vein minerals are almost always undeformed and the dykes and veinlets cross-cut mineral layering in the Type I host, implying that they post-date the plastic deformation of the mantle rocks.

Equilibration temperatures

Estimates of equilibration temperatures for a selected sample of spinel lherzolite xenoliths are given in Table II. Average temperatures obtained by the four methods lie between 875 and 1070° C. Such values fall largely within the lower part of the temperature range (900–1150° C) determined from spinel lherzolite xenoliths world-wide (Carswell, 1980) and reflect subsolidus recrystallization. The limitations of geothermometry have been the subject of extensive debate in the literature (e.g., Carswell, 1980; Fuji and Scarfe, 1982), but further discussion of the data in Table II is beyond the scope of this contribution.

Accurate estimates of the depth of origin of the Type I xenoliths are limited by the uncertainties in the slopes of Al$_2$O$_3$ isopleths for orthopyroxene in the spinel lherzolite stability field (Carswell, 1980). The pyroxene thermobarometric method of Mercier (1980) indicates a depth range of 25–55 km (8–18 kbar), which is con-

Table II. Equilibrium temperatures for eight Seiland spinel lherzolite xenoliths from pyroxene geothermometry

Sample[a]	Wood and Banno (1973), two-pyroxene	Wells (1977), solvus	Mercier (1980), single pyroxene	Mysen (1976), Al^{VI}–Cr partition	Average[b]
	Temperature (°C)				
50	775	655	947	1129	876
738	887	757	930	964	885
P99	937	886	950	901	919
P106	905	796	945	1061	927
P55	789	677	924	1438	957
P20	934	825	949	1196	976
P37	1018	932	929	1082	990
P110	1119	1044	1113	977	1063

[a]Samples:

    738: Store Bekkarfjord (5 in Fig. 1)

    P106, P110: E. Melkvann (2 in Fig. 1)

    P20, P37, P50, P55, P99: Holmevann (1 in Fig. 1)

[b]Analyses of pyroxene pairs used in the calculations are available from the authors.

sidered reasonable because these values lie within the spinel lherzolite stability field determined by Green and Ringwood (1967).

It is widely believed that Type I xenoliths represent mantle wall rock, whereas those belonging to Type II originate as high-pressure cumulates (Irving, 1980; Kempton, 1987). Using the Ti/Cr content of the diopside in both suites, a plot similar to that used by Ross (1983) emphasizes the fact that at Holmevann there is a much higher proportion of Type II xenoliths. Furthermore, Ross considered a low Ti-content (<0.008 at./f.u. (atoms per formula unit)) and variable Cr-content (0.020–0.050 at./f.u.) to be representative of undepleted mantle (Fig. 7).

Using similar constraints all the Seiland lherzolite xenoliths that plot in the "undepleted" field have bulk rock values of mg = 86. Those considered as depleted have mg values of 87–90, the value increasing with increasing depletion. Consequently it appears that, compared with pyrolite models for the mantle (Ringwood, 1975), the lithospheric mantle below the Seiland Province was relatively iron enriched. Interestingly, the value of mg = 86 for the least-depleted Seiland lithospheric mantle is the same as that proposed by Wilkinson (1976) for "undepleted" mantle from experimental anhydrous partial melting of spinel lherzolite at elevated pressures.

The presence of amphibole within the Type I xenoliths from Holmevann suggests that some regions of the lithospheric mantle had been modally metasomatized. The composite xenoliths provide further evidence for this. Irving (1980) has interpreted similar dykes and veinlets as precipitates from flowing magma within narrow mantle conduits. Mineralogical evidence from traverses across the veins into the host rock show that there is an approximately 3.5 cm wide reaction zone defined by large variations in spinel composition. The presence of carbonate and mica in some xenoliths would appear to suggest that volatile-rich fluids may also have been metasomatizing agents. However, further constraints on the nature, timing and extent of metasomatism require trace-element and isotopic data, at present not yet available.

CONCLUSIONS

The xenolith-bearing dykes on Seiland are of great interest in that they represent a sample of the lithospheric mantle below the Seiland Province that is at least 531 Ma old. Because the xenoliths were transported in basaltic or basanitic (rather than kimberlitic) hosts, we have also

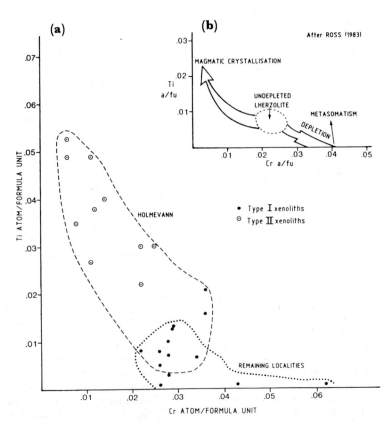

Fig. 7. Ti- and Cr-contents of diopsides in Type I and Type II xenoliths (*a*), compared with (*b*) the fields for undepleted and depleted lithospheric mantle compositions proposed by Ross (1983).

been offered some insight into the physical processes of magma transport. Although interpretation is at present constrained by a limited amount of geochemical data, the following can be inferred with a reasonable degree of certainty.

(1) The entraining magma was hydrous alkali olivine basalt or basanite with low effective viscosity and low yield strength.
(2) Magma ascent velocities were as high as $2-3 \text{ km h}^{-1}$.
(3) The intrusion mechanism was one of magma fracture conduit flow, with the dykes probably connected to the surface.
(4) The least-depleted lithospheric mantle below the Seiland Province was relatively iron rich with an mg value of 86.
(5) Equilibration temperatures estimated by a variety of thermometric methods are in the range from 875 to 1070°C.
(6) The mantle had undergone both depletion and enrichment events and was heterogeneous on a relatively small scale, before entrainment of the mantle sample at some time earlier than 531 Ma but probably later than 600 Ma.

From mineral and petrographic evidence it has become clear that the xenoliths from Holmevann are in many respects different from those from the other sampled areas. This is not only interesting but puzzling, considering that Holmevann lies only 5 km along-strike from E Melkvann, and that field evidence requires that all the xenolithic dykes are essentially the same age. It is possible that the Holmevann sample originated from a higher level in the mantle. The answer to this problem, and that of the geochemical, petrological and dynamic history of the sub-Seiland Province lithospheric mantle will emerge as research continues.

Acknowledgements

The authors gratefully acknowledge the financial support of the following bodies during the course of the research reported in this contribution: The Norwegian Research Council for Science and Humanities (B.R.), The Natural Environment Research Council (Grants GR3/3817 and GR3/4517), The Royal Society, and the University of Bristol (M.C.B.). The line drawings were prepared in the cartographic office of the Geological Institute of the University of Bergen, under the direction of E. Irgens, and Jane Hawker of the University of Bristol typed numerous versions of the manuscript. We also thank Peter Nixon and Eric Condliffe of the University of Leeds for microprobe facilities and much useful advice. This is Norwegian contribution No. 39 to the International Lithosphere Programme.

REFERENCES

Bennett, M. C., Emblin, S. R., Robins, B. and Yeo, W. J. A. (1986). High Temperature ultramafic complexes in the North Norwegian Caledonides: I—Regional setting and field relationships. *Norges Geol. Unders. Bull.* **405**, 1–40.

Carmichael, I. S. E., Nicholls, J., Spera, F. J., Wood, B. J. and Nelson, S. A. (1977). High temperature properties of silicate liquids: applications to the equilibration and ascent of basic magma. *Phil. Trans. R. Soc. Lond.*, A **286**, 373–431.

Carswell, D. A. (1980). Mantle derived lherzolite nodules associated with kimberlite, carbonatite and basalt magmatism: A review. *Lithos* **1** 121–138.

Frey, F. A. and Prinz, M. (1978). Ultramafic inclusions from San Carlos, Arizona. Petrologic and geochemical data bearing on their petrogenesis. *Earth Planet. Sci. Lett.* **38**, 129–176.

Fuji, T. and Scarfe, C. M. (1982). Petrology of ultramafic nodules from West Kettle River near Kelawna, southern British Columbia. *Contrib. Mineral. Petrol.* **80**, 297–306.

Gayer, R. A., Rice, A. H. N., Roberts, D., Townsend, C. and Welbon, A. (1987). Restoration of the Caledonian Baltoscandian margin from balanced cross-sections: the problem of excess continental crust. *Trans. R. Soc. Edinburgh*, Earth Sci. **78**, 197–217.

Green, D. H. and Ringwood, A. E. (1967). The stability fields of aluminous pyroxene peridotite and garnet peridotite and their relevance in upper mantle structure. *Earth Planet. Sci. Lett.* **3**, 151–160.

Harte, B. (1977). Rock nomenclature with particular relation to deformation and recrystallisation textures in olivine-bearing xenoliths. *J. Geol.* **85**, 279–288.

Irving, A. J. (1980). Petrology and geochemistry of composite ultramafic xenoliths in alkalic basalts and implications for magmatic processes within the mantle. *Amer. J. Sci.* **280-A**, 389–426.

Kempton, P. D. (1987). Mineralogic and geochemical evidence for differing styles of metasomatism in spinel lherzolite xenoliths: enriched mantle source regions of basalts. In: Menzies, M. A. and Hawkesworth, C. J. (eds.) *Mantle Metasomatism*, pp. 45–89. Academic Press, London.

Komar, P. D. (1976). Phenocryst interactions and the velocity profile of magma flowing through dykes or sills. *Geol. Soc. Amer. Bull.* **87**, 1336–1342.

Krill, A. G. and Zwaan, B. (1987). Reinterpretation of Finnmarkian deformation on Western Sørøy, northern Norway. *Norsk Geol. Tidsskr.* **67**, 15–24.

Kushiro, A. (1980). Viscosity, density and structure of silicate melts at high pressures, and their petrological applications. In: Hargraves, R. B. (ed.) *Physics of Magmatic Processes*, pp. 93–120. Princeton University Press, Princeton, N.J.

Marsh, B. D. (1981). On the crystallinity, probability of occurrence and rheology of lava and magma. *Contrib. Mineral Petrol.* **78**, 85–98.

Mercier, J. C. C. (1980). Single pyroxene thermobarometry. *Tectonophysics* **70**, 1–39.

Mercier, J. C. C. and Nicolas, A. (1975). Textures and fabrics of upper mantle peridotites as illustrated by basalt xenoliths. *J. Petrol.* **16**, 454–87.

Murase, T. and McBirney, A. R. (1973). Properties of some common igneous rocks and their melts at high temperatures *Geol. Soc. Amer. Bull.* **84**, 3563–3592.

Mysen, B. O. (1976). Experimental determination of some geochemical parameters relating to conditions of equilibration of peridotite in the upper mantle. *Amer. Mineral.* **61**, 677–683.

Pedersen, R. B., Dunning, G. R. and Robins, B. (1988). UPb (zircon) ages of nepheline syenite pegmatites from the Seiland Magmatic Province, North Norway, *this volume*.

Ramsay, D. M., Sturt, B. A., Zwaan, K. B. and Roberts, D. (1985). Caledonides of Northern Norway. In: Gee, D. G. and Sturt, B. A. (eds.) *The Caledonide Orogen—Scandinavia and Related Areas*, pp. 611–622. Wiley, Chichester.

Ringwood, A. E. (1975). *Composition and Petrology of the Earth's Mantle.* McGraw-Hill, New York.

Roberts, D. (1974). Hammerfest. Description of the 1:250000 geological map. *Norges Geol. Unders.* **301**, 1–66.

Roberts, D. and Gee, D. G. (1985). An introduction to the structure of the Scandinavian Caledonides. In: Gee, D. G. and Sturt, B. A. (eds.) *The Caledonide Orogen—Scandinavia and Related Areas*, pp. 55–68. Wiley, Chichester.

Robins, B. (1975). Ultramafic nodules from Seiland, Northern Norway. *Lithos* **8**, 15–27.

Robins, B. and Gardner, P. M. (1975). The magmatic evolution of the Seiland Petrographic Province, Finnmark. *Norges Geol. Unders.* **312**, 91–130.

Robins, B. and Takla, M. A. (1979). Geology and geochemistry of a metamorphosed picrite-ankaramite dyke suite from the Seiland Province, Northern Norway. *Norsk Geol. Tidsskr.* **59**, 67–95.

Roscoe, R. (1953). Suspensions. In: Hermans, J. J. (ed.) *Flow properties of disperse systems*, pp. 1–38. North Holland, Amsterdam.

Ross, J. V. (1983). The nature and rheology of the Cordilleran upper mantle of British Columbia: Inferences from peridotite xenoliths. *Tectonophysics* **100**, 321–357.

Shaw, H. R. (1969). Rheology of basalt in the melting range. *J. Petrol.* **10**, 510–535.

Shaw, H. R. (1972). Viscosities of magmatic silicate liquids: an empirical method of prediction. *Amer. J. Sci.* **272**, 870–893.

Shaw, H. R., Peck, D. L., Wright, T. L. and Okamura, R. (1968). The viscosity of basaltic magma: An analysis of field measurements in Makaopulii lave lake, Hawaii. *Amer. J. Sci.* **266**, 255–264.

Snow, J., Bennett, M. C., Yeo, W. J. A., Basu, A. R. and Tatsumoto, M. (1986). The Melkvann complex, N. Norway: Nd and Sr isotopic evidence for the origin of Alaskan type ultramafic complexes (abstr.) *ICOG IV Symposium* abstracts, *Terra Cognita*.

Sparks, R. S. Pinkerton, H. and MacDonald, R. (1977). The transport of xenoliths in magmas. *Earth Planet. Sci. Lett.* **35**, 234–238.

Spera, F. J. (1980). Aspects of Magma Transport. In: Hargraves, R. (ed.) *Physics of Magmatic Processes*, pp. 265–323. Princeton University Press, Princeton, N.J.

Spera, F. J. (1984). Carbon dioxide in petrogenesis III: role of volatiles in the ascent of alkaline magma with special reference to xenolith-bearing mafic lavas. *Contrib. Mineral. Petrol.* **88**, 217–232.

Spera, F. J. (1987). Dynamics of Translithospheric Migration of Metasomatic Fluid and Alkaline Magma. In: Menzies, M. A. and Hawkesworth, C. J. (eds.) *Mantle Metasomatism*, pp. 1–20. Academic Press, London.

Stephens, M. B., Furnes, H., Robins, B. and Sturt, B. A. (1985). Igneous activity within the Scandinavian caledonides. In: Gee, D. G. and Sturt, B. A. (eds.) *The Caledonide Orogen— Scandinavia and Related Areas*, pp. 623–656. Wiley, Chichester.

Wells, P. R. A. (1977). Pyroxene thermometry in simple and complex systems. *Contrib. Mineral. Petrol.* **62**, 129–139.

Wilkinson, J. F. G. (1976). Some Subcalcic Clinopyroxenites from Salt Lake Crater, Oahu and Their Petrogenetic Significance. *Contrib. Mineral. Petrol.* **58**, 181–201.

Wood, B. J. and Banno, S. (1973). Garnet-orthopyroxene and orthopyroxene-clinopyroxene relationships in simple and complex systems. *Contrib. Mineral. Petrol.* **42**, 109–42.

Yeo, W. J. A. (1984). The Melkvann Ultramafic Complex, Seiland igneous province, North Norway: intrusive mechaniss and petrological evolution. Unpublished Ph.D. thesis, University of Bristol, U.K.

Part IV

Metamorphic Geology

15 Evolution of low-grade metamorphic zones in the Caledonides of Finnmark, N Norway

A. H. N. Rice[1], *R. E. Bevins*[1], *D. Robinson*[2] *and D. Roberts*[3]

[1]Department of Geology, National Museum of Wales, Cathays Park, Cardiff CF1 3NP, U.K.
[2]Department of Geology, University of Bristol, Wills Memorial Building, Queens Road, Bristol BS 1RJ, U.K.
[3]Norges Geologiske Undersøkelse, Postboks 3006 Lade, N-7002, Trondheim, Norway.

Two hundred and three pelite samples from the Lower Allochthon and Autochthon of the Caledonides of Finnmark, North Norway have been analysed for illite crystallinity and 75 for the b_0 lattice parameter. Most samples were collected from the Gaissa Thrust Belt; other samples were obtained from the Berlevåg Formation, Barents Sea Caledonides, Komagfjord Antiformal Stack, Laksefjord Nappe Complex, E Finnmark Autochthon and Dividal Group.

The results indicate the following. (1) The base of the Middle Allochthon was at least at middle greenschist facies when emplaced over the Lower Allochthon during the Scandian event, and caused an inverted thermal gradient in the underlying rocks. (2) In the Gaissa Thrust Belt, high anchizone–low epizone metamorphism occurred near the roof and high diagenetic–low anchizone metamorphism occurred near the floor. Roof metamorphism occurred during a high-intermediate baric type and floor metamorphism during an intermediate baric type. (3) The Laksefjord Nappe Complex, Komagfjord Antiformal Stack and Barents Sea Caledonides are components of the internal part of the Lower Allochthon and suffered epizone/high–intermediate baric type metamorphism. (4) The Berlevåg Formation suffered the same metamorphic history as the Barents Sea Caledonides and is probably part of the Lower Allochthon. (5) In areas of high-intermediate baric type metamorphism, P_{max} and T_{max} were contemporaneous. In areas of intermediate baric type metamorphism, P_{max} and T_{max} were separated by a period of erosional decompression and heating due to thermal relaxation. (6) Imbrication in the Lower Allochthon was syn- to post-peak metamorphism in the roof areas and pre- to syn-peak emplacement of the floor areas. (7) Brittle deformation in the Gaissa Thrust Belt was contemporaneous with ductile deformation in the Middle Allochthon. (8) No major sole thrust has developed between the Lower Allochthon and autochthon in E Finnmark. (9) Metamorphism in the Dividal Group also reflects an inverted thermal gradient.

INTRODUCTION

This paper evaluates the results of an investigation of low-grade metamorphism in the lowest part of the Middle Allochthon, the Lower Allochthon and in the Parautochthon and Autochthon of the Caledonides of Finnmark, North Norway. Both illite crystallinity and the b_0 lattice parameter have been determined and the data used to elucidate relationships between the emplacement of the Middle Allochthon and deformation/metamorphism in the Lower Allochthon and Parautochthon–Autochthon during the Scandian phase of the Caledonian orogeny.

REGIONAL CONTEXT

Of the four major allochthons in the Scandinavian Caledonides (Roberts and Gee, 1985), only the lower two are of any great extent in Finnmark. The Upper Allochthon is present only on Magerøy (Andersen, 1984; Fig. 1) and the Uppermost Allochthon is completely absent.

The Middle Allochthon is composed wholly of rocks of the Kalak Nappe Complex. This is an assemblage of several nappes (cf. Gayer *et al.*, 1987) imbricating a late Proterozoic to earliest Palaeozoic sequence that passes up from fluviatile/shallow marine sandstones through mudstones to turbiditic sediments (Roberts, 1968; Ramsay, 1971) overlying basement rocks of Karelian and pre-

Karelian age (Zwaan and Roberts, 1978; Ramsay *et al.*, 1979). The observed peak metamorphic grade in these rocks decreases from upper amphibolite facies in the highest imbricate on Sørøy, through middle to lower amphibolite facies between the peninsulas of Porsanger and Nordkinn, to epizone (lower greenschist facies) on the northwest part of Varangerhalvøya (Roberts, 1985a; Gayer *et al.*, 1985; Bevins *et al.*, 1986; Rice and Roberts, 1988).

The Lower Allochthon is represented by the Gaissa Thrust Belt, Laksefjord Nappe Complex, Komagfjord Antiformal Stack and Barents Sea Caledonides (Fig. 1). Details of the lithostratigraphies are contained in Føyn (1985), Johnson *et al.* (1978), Pharaoh (1985) and Gayer *et al.* (1987). The principal, and most extensive, units are the Tanafjord, Vestertana and Digermul Groups of late Riphean to Tremadoc age. These dominantly shallow marine sediments and their close correlatives, locally containing the glacigene sediments of the Vestertana Group, crop out mainly in the Gaissa Thrust Belt and Autochthon, but have also been recorded in the southern part of the Komagfjord Antiformal Stack (Føyn, 1985) and in the Kalak Nappe Complex (Bowden, 1981). The lithologies in the Barents Sea Caledonides, Laksefjord Nappe Complex and the northern part of the Komagfjord Antiformal Stack, which together form the most north western parts of the restored palaeogeography of the Lower Allochthon (Rice *et al.*, 1989b) and which have been loosely correlated (Siedlecki and Levell, 1978; Føyn, 1985; Pharaoh, 1985; Gayer and Rice, 1988–this volume), are quite distinct from those found farther southeast in the Lower Allochthon, although of broadly comparable age. The rocks of the Berlevåg Formation, which crop out in the northwest corner of Varangerhalvøya have previously been correlated with the rocks of the Kalak Nappe Complex on the Nordkinn peninsula (Levell and Roberts, 1977). However, for reasons outlined below, the Berlevåg Formation has been included in the Lower Allochthon in this paper.

In the west, the Gaissa Thrust Belt comprises a series of relatively small-scale imbricates from which a cumulative shortening of 55% has been estimated (Townsend *et al.*, 1986). Further east, however, the belt is composed of a single large thrust sheet, extensively folded but with only rare thrusts. Although not interpreted as such previously, it seems possible that this eastern part is a passive roof duplex (Banks and Warburton, 1986). Relatively little deformation has occurred within the other major units in the Lower Allochthon, although restored sections indicate that the westernmost part has been transported at least 300 km (Gayer *et al.*, 1987; Rice *et al.*, 1989b).

Two distinct zones have been identified within the Autochthon (Fig. 1). East of Andabaktoaivi the precise location of the Sole thrust (if emergent) is uncertain but sedimentary rocks entirely equivalent to those found in the Gaissa Thrust Belt are thought to form a broad auto-chthonous to parautochthonous zone on southern Var-angerhalvøya (East Finnmark Autochthon). Figure 1 shows the position of the sole thrust to the northeast of Andabaktoaivi postulated by Chapman *et al.* (1985). West of Andabaktoaivi the Autochthon is composed of rocks belonging to the Dividal Group, a < 260 m thick sequence of shallow-marine sandstones and mudstones (divided into six members: Føyn, 1967) equivalent in age to the upper part of the Vestertana Group. West of Stabbursdal, where the Middle Allochthon overlies the Dividal Group (Figs. 1 and 6), the Kalak Thrust overlies Member VI of the group but east of Stabbursdal the contact between the Lower Allochthon and the Autochthon has cut down section so that in some places the Dividal Group has been

completely excised (Roberts, 1974). Although not investigated in detail, the uppermost 30 m of the Dividal Group is considerably more deformed than the lower parts, with the development of folds, some associated with minor, probably blind, thrusts and the development of slickenside lineations and a weak bedding-subparallel planar fabric (Roberts, 1974; Rice and Mackay, 1986).

The age of deformation within these rocks is currently the subject of debate. Until recently, all the deformation was thought to be of Finnmarkian age—540 ± 17 Ma on Sørøy to 504 ± 7 Ma on Varangerhalvøya (cf. Sturt *et al.*, 1978). However, Pedersen *et al.* (1988—this volume) have shown that the deformation associated with the Seiland Igneous Province, which was considered to be a crucial component of the Finnmarkian event, had finished by 531 Ma. The subsequent ca. 490 Ma event recorded in hornblende cooling ages in some parts of the Kalak Nappe Complex (Dallmeyer, 1988) is, therefore, recording a younger, regionally significant, high-grade metamorphic, event. For this reason, the age of peak metamorphism and main deformation in other parts of the Kalak Nappe Complex is uncertain. Although $^{40}Ar/^{39}Ar$ data from amphiboles in the Kalak Nappe Complex yield an age of ca. 490 Ma, muscovites have given early Scandian ages of ca. 420–430 Ma, reflecting either a slow cooling, or a Scandian reheating, of the Finnmarkian nappe pile. Within the Lower Allochthon, only Scandian ages (ca. 410–420 Ma) have been recorded by $^{40}Ar/^{39}Ar$ techniques, using the finest mica grain fractions (Dallmeyer *et al.*, 1988a,b).

LOW-GRADE METAMORPHISM

Analytical techniques

A total of 203 samples of pelite, which include 60 utilized by Bevins *et al.* (1986), form the data set in this paper.

Small rock chips were disaggregated in an ultrasonic tank and a $< 2\,\mu m$ fraction was separated by centrifugation and subsequent filtration of the supernatant liquid. The separated clay minerals were then mounted as smears on glass slides. A Phillips PW1790 series X-ray diffractometer (with an automated divergence slit, 0.1 μm receiving slit and graphite monochromator) at the University of Bristol was used for clay mineral identification and determination of illite crystallinity. Illite crystallinity values, performed in duplicate, were obtained by scanning the 7–10° 2θ range at 0.5° 2θ/minute using Cu K$_\alpha$ radiation, at a tube setting of 40 kV and 30 mA. Kubler crystallinity values (see Kubler, 1967) were determined from diffraction data collected digitally and stored using on-line computer control of the diffractometer, as well as from data recorded using a conventional chart recorder. The Kubler ($\Delta°2\theta$) index is the parameter used in this study and the ranges utilized are similar to those adopted by Kisch (1980) and Bevins *et al.* (1986), namely; diagenesis grade $\Delta°2\theta > 0.38$; anchizone grade $\Delta°2\theta = 0.38$–0.20; epizone grade $\Delta°2\theta \leqslant 0.21$. Essentially, as the metamorphic grade increases $\Delta°2\theta$ values decrease. Multiple repeat precision measurements on in-house standards gave results in the range of ± 0.02–$0.03°2\theta$.

The b_0 lattice parameter, an indirect measurement of the Fe/Mg substitution in the octahedral site (Sassi and Scolari, 1974; Padan *et al.*, 1982), was determined on powdered whole-rock samples using a special mount to provide oriented samples (see Robinson, 1981). Mineral identification scans from 2 to 35° 2θ were made on all samples to establish that the mineralogical constraints on

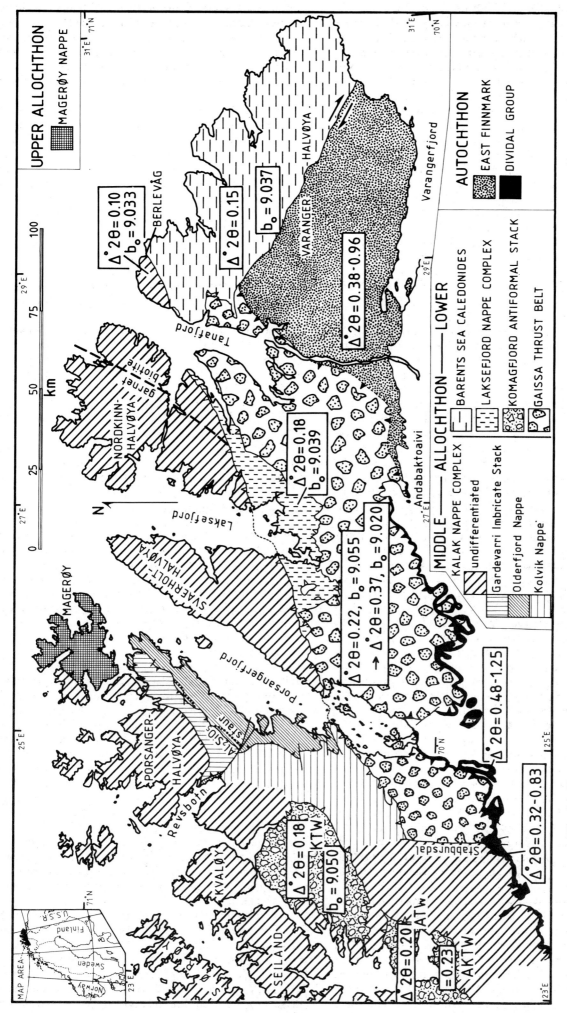

Fig. 1. Simplified geological map of the Caledonides of northern Norway, showing the major tectonostratigraphic units, the average $\Delta°2\theta$ and b_0 values in areas of low metamorphic grades and some of the localities referred to in the text; other localities are shown on Figs. 2, 3 and 6. KTW, ATW and AKTW are the Komagfjord, Altenes and Alta–Kvaenangen Tectonic Windows, the exposed parts of the Komagfjord Antiformal Stack.

the use of the b_0 parameter were met (Sassi and Scolari, 1974). The special sample holder utilized in this study enhanced the white mica (060) reflections (Robinson, 1981) and reduced the overlapping effect of the (331) peak (Guidotti, 1984), thus allowing a truer measurement of b_0. The position of the (060) peak was calculated using a peak-search routine with the quartz (211) peak (at 1.541 Å) as an internal standard. The range 59–63° was scanned at 0.25° 2θ/minute, at a tube setting of 40 kV and 40 mA, with digitized data storage. The b_0 parameter describes the baric type (and thus the geothermal gradient) of the peak metamorphic event. Although precise b_0 values cannot be ascribed to particular baric types, values over 9.035 have been taken as indicating a high–intermediate pressure baric type (Otago type metamorphism) and values of ca. 9.002 have been taken as indicating an intermediate pressure baric type (Barrovian type metamorphism).

RESULTS

The mean values of the analytical results, with standard deviations, are presented in Table I. Detailed analyses and locality maps for the samples can be obtained on request.

Lower Allochthon

(1) *The Berlevåg Formation* (Levell and Roberts, 1977) crops out in the extreme northwest of the Varanger Peninsula (Fig. 1). These rocks have been correlated with the Kalak Nappe Complex (Middle Allochthon) to the west, on Nordkinnhalvøya, where garnet and biotite porphyroblasts are abundant and indicate an upper greenschist facies metamorphism (Rice and Roberts, 1988). However, the results described here indicate that the Berlevåg Formation is probably a part of the Lower Allochthon (see below). Opaque (probably ilmenite) and chlorite porphyroblasts have been identified in our thin sections of phyllites, although biotite has not been found as a *porphyroblastic* phase, as previously described by Teisseyre (1972).

Eight samples of pelitic rocks were analysed for illite crystallinity. These gave a mean value of $\Delta°2\theta = 0.10$ ($\sigma_n = 0.03$, $N = 8$), indicative of an epizone grade of metamorphism. Analysis of the b_0 lattice parameter indicates an intermediate to high–intermediate baric type ($b_0 = 9.033$ Å, $\sigma_n = 0.010$, $N = 6$). The standard deviation is higher and the mean is lower than reported by Bevins *et al.* (1986), but this is a consequence of one sample (63/83/HR) for which $b_0 = 9.018$ Å. If this sample is omitted the mean $b_0 = 9.036$ Å ($\sigma_n = 0.008$, $N = 5$), comparable to the values quoted by Bevins *et al.* (1986).

(2) *Komagfjord Antiformal Stack.* Twenty-two samples have been analysed for illite crystallinity and 11 for the b_0 lattice parameter. Samples from the northern part of the Komagfjord Tectonic Window (Fig. 1) have epizone crystallinity values ($\Delta°2\theta = 0.18$, $\sigma_n = 0.04$, $N = 9$) but farther south samples have higher values, indicative of high anchizone–low epizone metamorphic grades ($\Delta°2\theta = 0.20$, $\sigma_n = 0.04$, $N = 5$ for the Altenes Tectonic Window, and $\Delta°2\theta = 0.23$, $\sigma_n = 0.05$, $N = 8$ for the Alta–Kvaenangen Tectonic Window; Fig. 1).

Associated with this slight (and possibly not significant) increase in $\Delta°2\theta$ towards the south is a decrease in b_0 spacing, from $b_0 = 9.050$ Å ($\sigma_n = 0.006$, $N = 7$) in the north of the Komagfjord Tectonic Window to $b_0 = 9.033$ Å ($\sigma_n = 0.003$, $N = 4$) in the Alta–Kvaenngen Tectonic Window. This represents a change from a high-intermediate towards an intermediate baric type.

(3) *Laksefjord Nappe Complex* (Fig. 1). Seventeen samples from the Laksefjord Group have been analysed for illite crystallinity, all but one of which have also been analysed for the b_0 lattice parameter. No difference between data from the Upper and Middle Laksefjord Nappes has been found (Table I). Combined, the data indicate an epizone grade of metamorphism ($\Delta°2\theta = 0.18$, $\sigma_n = 0.04$) equivalent to that determined in the northern part of the Komagfjord Antiformal Stack. The b_0 lattice parameter value, however, is lower, although still indicative of a high–intermediate baric type ($b_0 = 9.039$ Å, $\sigma_n = 0.007$).

(4) *Barents Sea Caledonides* (Fig. 1). Fifteen samples

Table I. Summary of illite crystallinity and b_0 data used in the estimation of low metamorphic grades in the present article ($\bar{x}$ = mean; σ_n = standard deviation; N = number of samples).

Tectonostratigraphic unit	Kubler value ($\Delta°2\theta$)			Lattice parameter b_0		
	$\bar{x}$	σ_n	N	$\bar{x}$	σ_n	N
Lower Allochthon						
Berlevåg Formation (Kalak Nappe Complex)	0.10	0.03	8	9.033	0.010	6
Laksefjord Nappe Complex (all samples)	0.18	0.04	17	9.039	0.007	16
Upper Laksefjord Nappe	0.18	0.03	4	9.037	0.004	4
Middle Laksefjord Nappe	0.18	0.04	13	9.040	0.007	12
Barents Sea Caledonides (all samples)	0.15	0.03	15	9.037	0.013	11
NW Barents Sea Caledonides	0.13	0.03	6	9.043	–	2
SE Barents Sea Caledonides	0.16	0.03	9	9.036	0.014	9
Komagfjord Antiformal Stack (all samples)	0.20	0.05	22	9.044	0.100	11
Komagfjord Tectonic Window	0.18	0.04	9	9.050	0.006	7
Altenes Tectonic Window	0.20	0.04	5	–	–	–
Alta–Kvaenangen Tectonic Window	0.23	0.05	8	9.033	0.003	4
Gaissa Thrust Belt (all samples)	0.29	0.07	78	9.036	0.023	33
Autochthon/Parautochthon						
E Finnmark	0.48	0.20	28	9.037	–	2
Dividal Group	0.60	0.24	35	–	–	–
Under Kalak Nappe Complex	0.43	0.13	16	–	–	–
Under Gaissa Thrust Belt	0.74	0.22	19	–	–	–

have been analysed for illite crystallinity and 11 for the b_0 lattice parameter. The data indicate an epizone grade of metamorphism ($\Delta°2\theta = 0.15$, $\sigma_n = 0.03$) with a baric type similar to that determined for the Laksefjord Nappe Complex ($b_0 = 9.037$Å, $\sigma_n = 0.013$). However, division of the data into two subsets indicates a slight variation from northwest to southeast, although this may not be significant statistically, particularly in view of the small sample population. Samples from the Løkvikfjell Group (NW Barents Sea Caledonides of Tables I and II) immediately underlying the Berlevåg Formation have an illite crystallinity value of $\Delta°2\theta = 0.13$ ($\sigma_n = 0.03$, $N = 6$) and $b_0 = 9.043$Å ($N = 2$), whilst samples from the underlying Barents Sea Group (SE Barents Sea Caledonides) have an illite crystallinity of $\Delta°2\theta = 0.16$ ($\sigma_n = 0.03$, $N = 9$) and $b_0 = 9.036$ A ($\sigma_n = 0.014$, $N = 9$). These values might suggest a general decrease in peak metamorphic grade towards the southeast, associated with a lower baric type of metamorphism, but more data are required to confirm this.

(5) *Gaissa Thrust Belt.* By far the greatest number of samples analysed have been collected from the Gaissa Thrust Belt. In all, 78 samples have been analysed for illite crystallinity and 33 for the b_0 lattice parameter, although only 21 have been analysed for both. Mean values are $\Delta°2\theta = 0.29$ ($\sigma_n = 0.07$), typical of middle anchizone meta-

morphism, and $b_0 = 9.036$ Å ($\sigma_n = 0.023$), between a high-intermediate and intermediate pressure baric type. However, these averaged results give a misleading view of the metamorphic evolution of the Gaissa Thrust Belt and the data are best considered in terms of three subareas.

(i) *SUBAREA A* utilizes 29 samples taken from the road section between the Laksefjord Thrust exposed near Ifjord and the Tana River (Fig. 2). Essentially, this area contains a threefold stratigraphy (Fig. 2(a)), with the Tanafjord Group (Dakkovarre Formation) at the base, overlain unconformably by the lower Vestertana Group (Mortensnes, Nyborg and Smalfjord Formations), which is in turn overlain by the upper Vestertana Group (Stappogiedde and Breivik Formations). This sequence has been folded into a series of NNE–SSW-trending upright to moderately inclined folds (Fig. 2(a)), probably related to blind thrusting; few thrusts have been mapped in this region (Siedlecki, 1980; Roberts, 1985b).

Figure 2(b) illustrates the distribution of the illite crystallinity results; the highest $\Delta°2\theta$ values (lowest metamorphic grade) coincide with the lowest stratigraphic unit, whilst the lower $\Delta°2\theta$ values tend to be associated with the highest stratigraphic unit. In the east, the metamorphic zonation reflects the antiformal and synformal structural grain of the region.

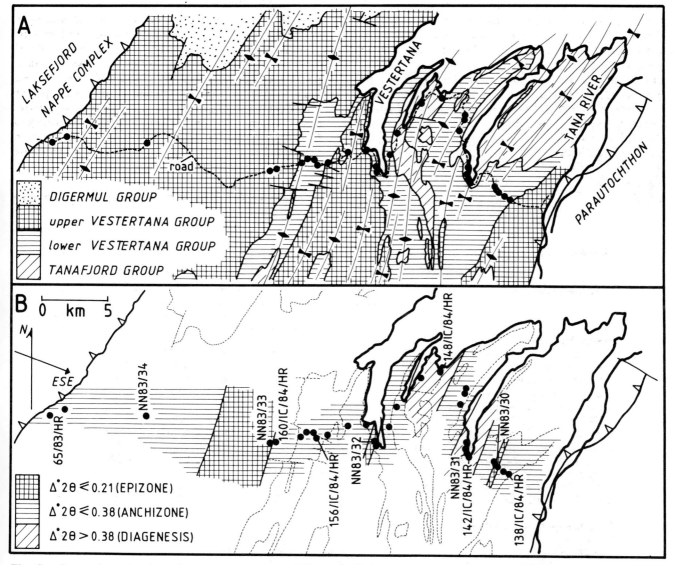

Fig. 2. Comparison of geological outcrop pattern (*a*) and illite crystallinity values (*b*) in the Gaissa Thrust Belt along the road section from the Tana River to the Lakesfjord Nappe Complex.

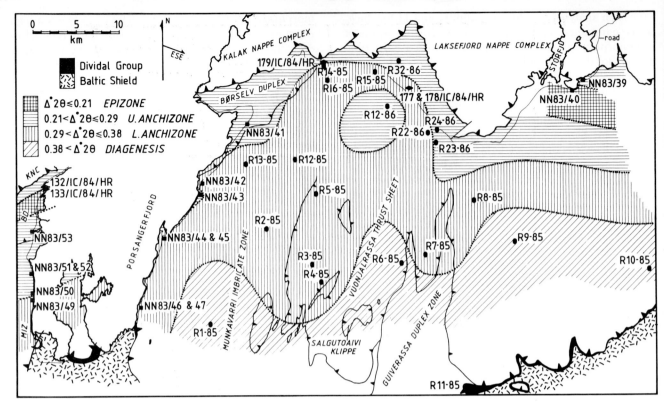

Fig. 3. Map of the Porsangerfjord–Storfjord (southwest Laksefjord) area, showing the main tectonic units within the Gaissa Thrust Belt (modified after Townsend *et al.*, 1986) and the variation in $\Delta°2\theta$ values.

(ii) *SUBAREA B* extends from the west coast of Porsangerfjord to Storfjord, southwest of Laksefjord (Fig. 3). In this large region, 39 samples have been analysed. Figure 3 shows the distribution of the sample localities in relation to the major structural units within the Gaissa Thrust Belt (modified after Townsend *et al.*, 1986). The data have been divided into four ranges of $\Delta°2\theta$ values; the lines delimiting these ranges are not to be considered as indicating the position of metamorphic isograds in view of the small data set. However, they do show an increase in metamorphic grade towards the north, closer both geographically and also structurally to the roof thrust, a trend that is oblique to the dominant NNE–SSW fold and thrust strike trend (Townsend *et al.*, 1986). Note that the upper part of the Børselv Duplex is composed of dolomitic lithologies from which suitable samples for this study cannot be obtained.

Generally, the topography has had little effect on the distribution of metamorphic zones. The only exceptions are an upper anchizone sample (R12-86) that is separated from equivalent-grade rocks by a major valley to the north and northeast that exposes low anchizone rocks, and the close proximity of low anchizone grade rocks to the Kalak Nappe Complex (samples R14 to R16-85), which is a result of the steep topography in that area.

(iii) *SUBAREA C* lies immediately to the east of Stabbursdal (Fig. 1), close to the trailing branch line of the Gaissa Thrust Belt, where the boundary between the Middle and Lower Allochthons trends north–south. Eight samples were collected from shale bands within the Ai'roai'vi Group (Townsend *et al.*, 1988) and again these show a decrease in $\Delta°2\theta$ towards the Kalak Nappe Complex (i.e., closer to the Kalak Thrust). Samples close to the thrust have high anchizone–low epizone crystallinities, whilst samples from at least 300 m below the Kalak Thrust have high diagenesis–low anchizone crystallinities.

In summary, illite crystallinity data from these three subareas within the Gaissa Thrust Belt show similar characteristics; a marked decrease in $\Delta°2\theta$ in passing upwards from the sole thrust, where values typical of low anchizone–high diagenesis metamorphism were recorded to values typical of low epizone–high anchizone metamorphism close to the roof thrust.

Figure 4 shows a cumulative frequency plot for the b_0 data from all parts of Finnmark under discussion, with the range of data from each tectonic unit marked by a bar. From this it may be seen that all units from the upper part of the Lower Allochthon have b_0 spacings indicative of metamorphism at the high-pressure end of intermediate-pressure metamorphism. The data from the Gaissa Thrust Belt, however, show a very wide range, from $b_0 = 8.988$ Å to 9.066 Å, although the lower values are associated with lower anchizone grade samples, for which the technique may not be suitable (see Padan *et al.*, 1982), although several authors have in fact obtained consistent results at such low grades (Kemp *et al.*, 1985; Robinson and Bevins, 1986). The remaining data from middle to upper anchizone grade rocks from the Gaissa Thrust Belt still show a wide range, from $b_0 = 9.015$ Å, to 9.066 Å, suggesting that the baric type of metamorphism varied significantly within the Gaissa Thrust Belt, from an intermediate to at least a high-intermediate baric type. Figure 5 shows the change in $\Delta°2\theta$ with b_0 from other orogenic belts; this illustrates the variation in the b_0 values that can be attributed to changes in peak metamorphic temperature. The data from the Gaissa Thrust Belt, also shown, clearly displays an oblique trend, with low b_0 values associated with high Kubler ($\Delta°2\theta$) values and vice versa. Thus the change in b_0 spacing in the Gaissa Thrust Belt reflects a real change in the baric type of metamorphism during T_{max}, from intermediate pressure in the lower-temperature areas to high-intermediate pressure (at least) in the higher-temperature areas.

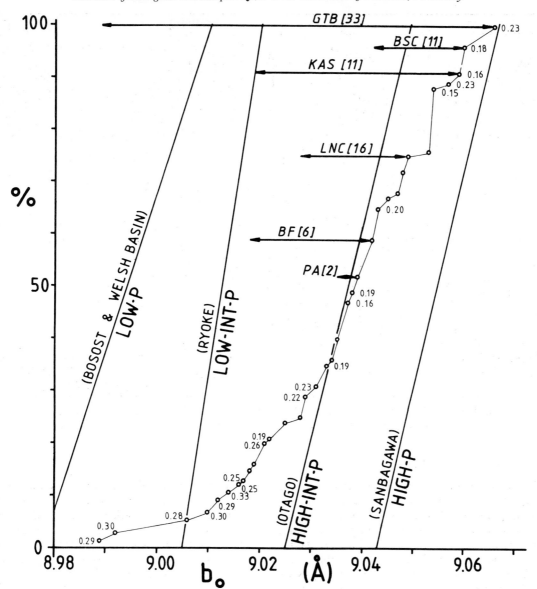

Fig. 4. Cumulative b_0 plot for all samples analysed for this study. The range of b_0 values for each tectonostratigraphic unit is shown by a bar, with the number of samples. Individual $\Delta°2\theta$ values for samples collected from the Gaissa Thrust Belt are also shown. The cumulative frequency curves for other areas have been simplified from Sassi and Scolari (1974) and Bevins *et al.* (1986). GTB = Gaissa Thrust Belt; BSC = Barents Sea Caledonides; KAS = Komagfjord Antiformal Stack; LNC = Laksefjord Nappe Complex; BF = Berlevåg Formation; PA = Parautochthon.

Autochthon

(1) *East Finnmark.* To the east of the postulated sole thrust (cf. Chapman *et al.*, 1985), 28 samples have been analysed for illite crystallinity (Fig. 6). These indicate that the metamorphic grade varies between uppermost anchizone immediately below the sole thrust (as defined by Chapman *et al.*, 1985) to low diagenesis further east (Fig. 6). However, in the extreme east, metamorphic grade increases slightly to upper diagenesis. Two samples of mid to upper anchizone grade have been analysed for b_0 and have an average value of 9.037 Å.

(2) *Dividal Group.* Thirty-five samples have been analysed from the Dividal Group, although only one was collected from east of Porsangerfjord. Sixteen samples were obtained from the area where the Kalak Thrust forms the sole thrust, the remainder were collected from beneath the Gaissa Thrust Belt. Figure 7 shows the variation in $\Delta°2\theta$ with depth below the sole thrust in the two regions. In both areas $\Delta°2\theta$ increases with depth, although the actual values are different, being higher beneath the Gaissa

Thrust Belt. At 0–10 m below the Kalak Thrust, $\Delta°2\theta$ varies between 0.32 and 0.45 (lowest anchizone–high diagenesis), whilst at the same depth below the Gaissa Thrust, $\Delta°2\theta$ varies between 0.48 and 0.73.

Best-fit linear regression lines through the data also show an increase in $\Delta°2\theta$ with depth, both with smaller ranges. In both cases the fit is poor and may be strongly influenced by the relatively few samples collected at depth. The line through the data from samples collected below the Kalak Nappe Complex indicates an increase in $\Delta°2\theta$ from 0.39 to 0.69 over 0–250 m depth, whilst samples from under the Gaissa Thrust Belt have a range from 0.43 to 1.25 over the same depth. However, the data from samples collected under the Gaissa Thrust Belt may show a form of exponential increase, here divided into two segments, with $\Delta°2\theta$ increasing from 0.52 immediately below the Gaissa Thrust to 0.78 at 10 m depth and then to 1.22 at 250 m depth. This variation bears no relation to the increase in deformation observed in the upper 30 m of the Dividal Group.

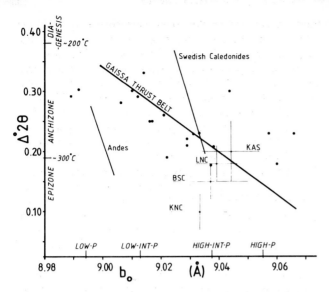

Fig. 5. Variation in b_0 with $\Delta°2\theta$. A linear regression line has been drawn through the data from the Gaissa Thrust Belt. Other data have been shown only by mean values and standard deviations.

DISCUSSION

In both the Gaissa Thrust Belt and the Dividal Group Kubler values are markedly lower in close proximity to the overlying major thrust, indicating that metamorphism occurred at a higher temperature in higher parts of these units. Inverted metamorphic sequences have been described from many parts of the world, commonly, but not always associated with ophiolite obduction (Nicholson and Rutland 1969; Spray and Williams, 1980; St-Onge, 1981; Arita, 1983; Watkins, 1985; Worthing, 1986). The cause of inversion is usually ascribed to the thrust emplacement of a hot slab onto a cold basement (cf. Oxburgh and Turcotte, 1974) although in some instances

shear heating along the thrust may be a significant source of heat (Graham and England, 1975). Inverted metamorphic sequences have been described from low-grade regions by Aprahamian and Pairis (1981), who invoked a shear-heating mechanism in the Alps, and by Kisch (1980), who ascribed thermal inversion in the Lower Allochthon of the southern Scandinavian Caledonides to the emplacement of higher-grade nappes of the Middle Allochthon. The preservation of inverted metamorphic sequences indicates that uplift and cooling occurred before thermal reequilibration; this implies rapid uplift and erosion (Pavlis, 1986).

It is not the purpose of this article to develop a precise thermal model by which the inverted metamorphic sequences described developed; quite probably shear heating may have aided conductive heat transfer (thrusting). However, some general inferences can be made from a structural viewpoint in relation to the development of inverted metamorphic zones. Firstly, the effects of conductive heat transfer would be most significant immediately upon underthrusting, whereas shear heating could continue as long as thrusting. Since the former process makes the greater contribution to the heating process, T_{max} would occur at essentially the same time (and pressure) as the emplacement of the hot slab. Secondly, since orogenic shortening results in the lateral displacement of a wedge of material over the autochthon, which is progressively detached into the allochthon, the maximum pressure achieved by a rock in a simple foreland-propagating thrust system would occur during emplacement of the immediately overlying thrust-sheet. Once detached, the rocks are uplifted with consequent decompression through erosion.

From these two inferences, it is apparent that at the top of a thrust-sheet preserving an inverted metamorphic sequence T_{max} and P_{max} occurred contemporaneously with the emplacement of the immediately overlying thrust-sheet, assuming that no significant out-of-sequence thrusting occurred. Further, it can be deduced that P_{max} at the base of a thrust-sheet would be greater than at the top of

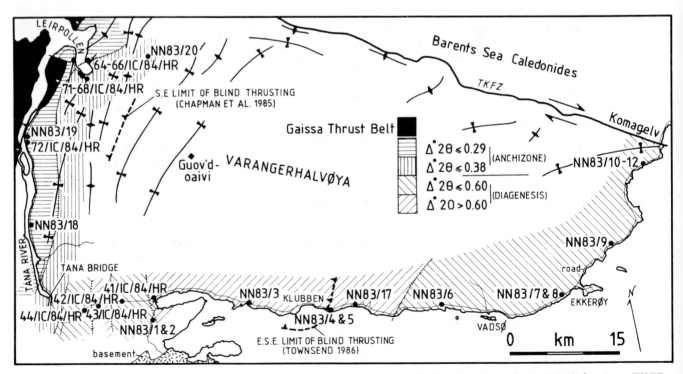

Fig. 6. Simplified map of the Parautochthon and Autochthon of E Finnmark, showing the variation in $\Delta°2\theta$ values. TKFZ = Trollfjord—Komagelv Fault Zone.

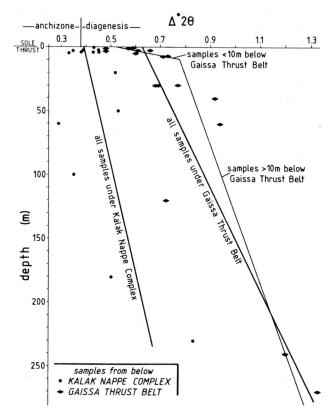

Fig. 7. Variation in $\Delta^0 2\theta$ values with depth below the Caledonian sole thrust for samples from the autochthonous Dividal Group.

the thrust-sheet by the thickness of the thrust-sheet, here termed D_p. P_{max} would develop at the same time throughout the thrust-sheet but this would not be the case for T_{max}. At the base of the thrust-sheet, T_{max} would develop some time after P_{max}, as a consequence of thermal relaxation; this heating would continue until either an equilibrium geo-

therm developed or the rocks were thrust up and over colder rocks. In the latter case, the metamorphic peak (T_{max}) at the base of the thrust-sheet would thus effectively define the temperature at which such thrusting occurred. The reasoning behind these interpretations of the tectono-metamorphic relationships are summarized in Fig. 8.

The b_0 lattice parameter provides an indication of the geothermal gradient of the peak metamorphic conditions, assumed here to be T_{max}. This is the same assumption as made by other workers although no data on the relationship of b_0 to T_{max} have been published. However, cooling must have occurred after the rocks passed through the geotherm recorded by the b_0 data, otherwise the lattice spacing would have re-equilibrated to the new geotherm. For intermediate and high-intermediate baric types geothermal gradients of 100 and 82°C kbar^{-1} respectively have been adopted.

Since the Kubler value provides an estimate of the temperature at the peak of metamorphism, a combination of the b_0 and $\Delta^0 2\theta$ parameters gives a general idea of the peak metamorphic P–T conditions. Note, however, that since both techniques are essentially statistical, relying on large data sets, precise P–T data cannot be determined. Despite this, any attempt to draw P–T–t paths for low-grade rocks will appear to be affixing precise values to parts of the curve. We emphasize, therefore, that although precise P–T values have been quoted, these carry an inherent and unknown uncertainty and should at best be regarded as being relative estimates compared to the other values quoted.

Utilizing the limited metamorphic data available in the low-grade metamorphic region of the Caledonides of Finnmark and using the structural constraints outlined above, a complex P–T–t evolution for the lower part of the orogen has been developed. The crucial unit in this model is the Gaissa Thrust Belt, for which both b_0 and $\Delta^0 2\theta$ values have been obtained from the roof and floor areas. This effectively offers constraints on the emplacement of the Middle and Lower Allochthons.

Table II. Estimated peak metamorphic conditions developed within the Lower Allochthon and Parautochthon/Autochthon in the Caledonides of Finnmark.

Tectonostratigraphic unit	Metamorphic grade	Baric type
Middle Allochthon	Mid-greenschist	?Intermediate
Lower Allochthon		
Berlevåg Formation	Upper epizone	High-intermediate
Laksefjord Nappe Complex	Epizone	High-intermediate
Barents Sea Caledonides	Epizone	High-intermediate
Komagfjord Antiformal Stack		
Komagfjord Tectonic Window	Epizone	High-intermediate
Altenes Tectonic Window	Upper anchizone	?
Alta–Kvaenangen Tectonic Window	Upper anchizone	Intermediate
Gaissa Thrust Belt		
Upper part	Upper anchizone	High-intermediate
Lower part	Lower anchizone	?Intermediate
Autochthon/Parautochthon		
E Finnmark	Lower anchizone–diagenesis	?Intermediate
Dividal Group		
Under Kalak Nappe Complex		
Upper part	Lowest anchizone	?Intermediate
Lower part	Diagenesis	?Intermediate
Under Gaissa Thrust Belt		
Upper part	Diagenesis	?High-intermediate
Lower part	Diagenesis	?Intermediate

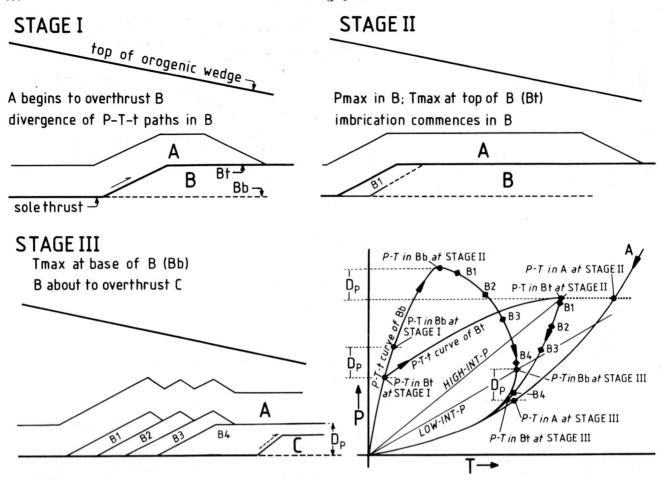

Fig. 8. Schematic diagram to show the relationship between metamorphism (T_{max}, P_{max}) at the top (Bt) and bottom (Bb) of thrust sheet B (composed finally of imbricates B1–B4) and the initiation and development of thrusting. No out-of-sequence thrusting occurs in the model. *Stage I*: Hot nappe A overthrusts B; pressure in A is decreasing owing to erosion and in B increasing owing to emplacement of A. Transfer of thermal energy from A to B causes rapid heating at the roof of B (Bt) at relatively high pressures (and thus relatively a low geothermal gradient), whilst the base heats more slowly as thermal relaxation occurs. Thus the *P–T–t* paths of Bt and Bb diverge when A is overthrust. *Stage II*: Imbrication commences in B. Since overthrusting of A ceases, P_{max} occurs in B, but is higher in Bb by the thickness of B, defined as D_p. Since fresh hot nappe A is no longer being emplaced over B, heating of Bt ceases at the same time. Thus, in Bt, P_{max} and T_{max} are contemporaneous and pre-imbrication. Since overthrusting of A has stopped, decompression through erosion occurs; in Bb thermal relaxation (heating) continues during erosion. *Stage III*: Imbricated B about to be thrust over C, causing an inverted thermal regime in C. Since the imbricates B1–B4 are emplaced over colder rocks, no further heating occurs in Bb. Note the changing relationships between imbrication of B1–B4 and the peak of metamorphism, marked by squares on the *P–T–t* loops.

    In reality, A equates to the Middle Allochthon, B to the Lower Allochthon (with B1 approximating to the Komagfjord Antiformal Stack, Laksefjord Nappe Complex, Barents Sea Caledonides and Berlevåg Formation, and B2–B4 approximating to the Gaissa Thrust Belt) and C to the Autochthon (Dividal Group).

Evolution of Gaissa Thrust Belt

Peak metamorphic conditions at the base and top of the Gaissa Thrust Belt are estimated at low anchizone–intermediate baric type (ca. 210°C/2.1 kbar) and highly-anchizone–high-intermediate baric type (ca. 290°C/3.6 kbar) respectively. For the top part of the Gaissa Thrust Belt this represents T_{max} and P_{max}, whilst towards the base this represents $P_{T max}$. P_{max} for the base is $3.6 + D_p$ for the Gaissa Thrust Belt; utilizing aeromagnetic studies (Åm, 1975), it has been shown that D_p is of the order of 0.3 kbar in the Porsanger area, so that P_{max} was ca. 3.9 kbar.

    The *P–T–t* curves shown for the Gaissa Thrust Belt (Fig. 9) conform to the values derived above. Within these constraints, however, there is a wide range of possible *P–T–t* loops (Thompson and England, 1984); those shown illustrate reasonable possibilities. Note firstly that the temperature at P_{max} for the base of the Gaissa Thrust Belt is not 0°C; in other words, a perfect sawtooth geotherm is not envisaged as the first stage after thrusting. This is because thrusting is not an instantaneous process and consequently

the underthrust sediments would have been subject to tectonic burial (and thus heating) for some time prior to the sole thrust cutting down section to directly overlie them.

    The pressure at which the *P–T–t* curves for the top and base of the Gaissa Thrust Belt begin to diverge represents the time at which the thrust at the base of the Middle Allochthon cut down to lie at the top of the then undeformed rocks now comprising the Gaissa Thrust Belt. This pressure is unconstrained, but would have been only slightly less than the final P_{max} value.

    Between the times of P_{max} and T_{max} at the base of the Gaissa Thrust Belt, ca. 1.8 kbar of erosion occurred (Fig. 9) caused either by tectonic process (e.g., Platt, 1987) or by simple mechanical erosion. The amount of decompression is surprisingly large and remains a problem within the model. During this period of erosion, cooling of the upper part of the Gaissa Thrust Belt occurred, whilst heating occurred at the base; theoretically, some central part could have remained at a constant temperature during part of this decompression.

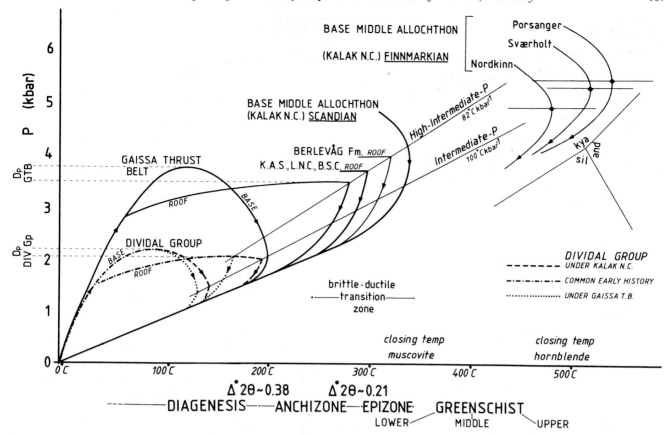

Fig. 9. Inferred *P–T–t* model for the main tectonic units within the Caledonides of Finnmark. Definitions of metamorphic zones are as used throughout the text; Al₂SiO₅ phase diagram after Salje (1986). In all units of the Lower Allochthon peak metamorphic conditions have been estimated by combining the b_0 data (which gives the geothermal gradient) with the illite crystallinity data (which gives the peak temperature). P_{max} for the base of the Gaissa Thrust Belt has been estimated by adding D_p to the peak pressure at the roof of the belt. Inferred *P–T–t* loops for the Kalak Nappe Complex have been shown assuming a Finnmarkian age of deformation for the Gaissa Thrust Belt (old model) and a Scandian deformation age (new model). Note that ductile deformation could have continued in the base of the Middle Allochthon after brittle imbrication commenced in the Lower Allochthon. In the Laksefjord Nappe early deformation could have been ductile, followed by brittle deformation: this agrees with the model of Williams *et al.* (1984).

Since emplacement of the Gaissa Thrust Belt onto the Dividal Group caused the observed metamorphic inversion in the autochthon (Fig. 9), T_{max} at the base of the Gaissa Thrust Belt must have occurred prior to emplacement; indeed, the emplacement itself may have caused the cessation in heating. Thus, the pressure at T_{max} at the base of the Gaissa Thrust Belt a approximately defines the pressure at which this emplacement occurred. However, this does not imply that imbrication and deformation *within* the Gaissa Thrust Belt occurred at this pressure; quite probably, imbrication started soon after the emplacement of the Middle Allochthon, contemporaneous with heating in the lower part and cooling in the upper part of the thrust belt. Thus, in the upper part of the Gaissa Thrust Belt peak metamorphism was pre- to syn-imbrication of the belt, the two events having been more nearly contemporaneous in the earlier-formed imbricates. In the lower parts of the Gaissa Thrust Belt, the relationship depended on the relative velocities of imbrication and the development of the peak of metamorphism. Thus, the peak of metamorphism near the base of the Gaissa Thrust Belt was syn- to post-imbrication, the two events being more nearly contemporaneous in the late-formed imbricates (cf. B1–B4 in Fig. 8). Note, further, that although all the deformation in the Gaissa Thrust Belt was brittle, the simultaneous deformation in the Middle Allochthon and more internal parts of the Lower Allochthon could have been ductile. This has considerable significance in the restoration of the orogen (cf. Williams *et al.*, 1984; Townsend, 1987; Gayer *et al.*, 1987).

Evolution of the Komagfjord Antiformal Stack and Laksefjord Nappe Complex

From structural considerations, Gayer *et al.* (1987) restored the Komagfjord Antiformal Stack to a position 166 km west-northwest of the trailing branch line of the Gaissa Thrust Belt. Further, because of the similar metamorphic conditions in the northern part of the Komagfjord Antiformal Stack and in the Laksefjord Nappe Complex, the latter was restored to a position along tectonic strike from the former. These conditions, estimated at epizone–high-intermediate baric type (ca. 310°C/ 3.8 kbar), are similar if somewhat higher-grade, to those found at the top of the Gaissa Thrust Belt. This similarity, especially in the high b_0 values, has been interpreted as indicating a similar origin—by "contact" metamorphism from the hot Middle Allochthon. The higher metamorphic grade of the Laksefjord Nappe Complex and Komagfjord Antiformal Stack reflects the higher temperatures at the base of the Kalak Nappe Complex when first thrust over the more internal parts of the Lower Allochthon (see Fig. 9).

In the more southerly parts of the Komagfjord Antiformal Stack, b_0 values are lower, whilst illite crystallinity values are higher (lower grade); specifically, in the Altenes Window $\Delta°2\theta = 0.20$ ($\sigma_n = 0.04$, $N = 5$) and in the Alta–Kvaenangen Window $\Delta°2\theta = 0.23$ ($\sigma_n = 0.05$, $N = 8$) with $b_0 = 9.033$ Å ($\sigma_n = 0.03$, $N = 4$). These values indicate a lower baric type than in the northern part of the Komagfjord Antiformal Stack and a high anchizone–low epizone grade of metamorphism. Accepting the argument that metamorphism at the roof of the Komagfjord Antiformal Stack was due to "contact" metamorphism that resulted in an epizone–high-intermediate baric type metamorphism, then the lower-grade metamorphic conditions in the south suggest that these Caledonian cover rocks cannot have been adjacent to the base of the Middle Allochthon during emplacement of the latter. This has two implications. Firstly, it indicates that the sedimentary sequence in the south was originally thicker, a point of interest in palinspastic reconstructions, although how much thicker is uncertain (< 500 m?). Secondly, if the cover sequence was thicker it has subsequently been removed by thrusting after the main emplacement of the Middle Allochthon, with the corollary that little such thrusting occurred towards the north of the antiformal stack. This is in agreement with the observed structural pattern in which the deformation along the northwest margin of the Komagfjord Antiformal Stack increases towards the southwest (Fareth, 1979; Bowden, 1981; Pharaoh *et al.*, 1983).

The timing of the emplacement of the Laksefjord Nappe Complex cannot be precisely constrained from metamorphic criteria. However, since the metamorphic zonal pattern in the Gaissa Thrust Belt is not disturbed by the presence of the Laksefjord Nappe Complex (Fig. 3), it may be inferred that the part of the Gaissa Thrust Belt that underlies the Laksefjord Nappe Complex suffered the same thermal history as other parts. Thus, the hot Middle Allochthon was emplaced over the Gaissa Thrust Belt, developing the inverted metamorphic zones, prior to the imbrication and emplacement of the Laksefjord Nappe Complex.

Scandian metamorphism in the Middle Allochthon

Conditions in the Middle Allochthon during the Scandian event can only be loosely constrained. Peak metamorphic conditions in the Laksefjord Nappe Complex have been estimated at epizone–high-intermediate baric type (ca. 310°C/3.8 kbar; see above); this, therefore, was also the pressure at the base of the Middle Allochthon during its emplacement. $^{40}Ar/^{39}Ar$ hornblende data from some parts of the Kalak Nappe Complex yield ages of ca. 490 Ma (Dallmeyer, 1988), whilst muscovites yield early Scandian ages (420–430 Ma ; Dallmeyer *et al.*, 1988a,b), indicating that peak Scandian temperatures in the Middle Allochthon were > 350°C and < 500°C. Since the muscovite closing temperature is so close to the peak temperature in the Laksefjord Nappe Complex, it seems reasonable to infer that this was associated with the lowest possible pressure. Thus, conditions at the base of the Middle Allochthon were at least middle greenschist/high-intermediate baric type (ca. 350°C/3.8 kbar).

Between Stabbursdalen and the west side of Tanafjord the lowest parts of the Kalak Nappe Complex lie within the upper greenschist facies. Garnet–biotite geothermometry (Ferry and Spear, 1978; Ganguly, 1979) yields peak temperatures of 491°C ($\sigma_n = 43$, $N = 6$) on Nordkinnhalvøya, 530°C ($\sigma_n = 58$, $N = 6$) on southern Svaerholthalvøya and 551°C ($\sigma_n = 78$, $N = 14$) to the south of Porsangerhalvøya (Rice, 1987), indicating little decrease in metamorphic grade along tectonic strike (Fig. 9). In the Berlevåg area, however, $\Delta°2\theta$ and b_0 data are not dissimilar to those from the internal parts of the Lower Allochthon (Komagfjord Antiformal Stack and Laksefjord Nappe Complex), although the grain size is coarser and locally chlorite schists have developed, suggesting a marginally higher peak temperature. Peak metamorphic conditions

have been estimated at high-epizone/high-intermediate baric type (ca. 330°C/3.8 kbar). Although considerable uncertainties are associated with all these temperature estimates, they are consistent with the observed mineral assemblages. Thus, across Tanafjord, from Nordkinnhalvøya to Berlevåg, a distance of < 50 km, the determined peak metamorphic temperature decreases by ca. 140°C. By comparison, between Sørøy, where upper amphibolite facies metamorphism has been reported (Roberts, 1985a) with inferred peak metamorphic temperatures of ca. 690°C (Rice *et al.*, 1989a), and Nordkinnhalvøya, a distance of over 175 km, the determined peak metamorphic temperature falls by only ca. 200°C.

Following early mapping, the Berlevåg Formation was correlated with the Kongsfjord Formation at the base of the Barents Sea Group (Siedlecka and Siedlecki, 1972). This correlation fell into disfavour, partly as a result of the work of Levell and Roberts (1977), who demonstrated a thrust at the base of the Berlevåg Formation and correlated these rocks with those of the Kalak Nappe Complex. This was reinforced when the amount of strike-slip displacement along the Trollfjord–Komagelv Fault was estimated at 500–1000 km (Kjøde *et al.*, 1978). However, recent work has shown that the Barents Sea Caledonides can be restored to a position along strike from the Laksefjord Nappe Complex and Komagfjord Antiformal Stack and are part of the Lower Allochthon (Rice *et al.*, 1989b). Thus, in view of the extreme metamorphic telescoping required by the correlation between the Kalak Nappe Complex and Berlevåg Formation and the similarity of metamorphic conditions in the Berlevåg Formation and other parts of the Lower Allochthon (Table I), it seems more reasonable to infer that the thrust under the Berlevåg Formation is not equivalent to the Kalak Thrust, which thus must lie beneath Tanafjord, and that the Berlevåg Formation is part of the Lower Allochthon.

Metamorphism in the Barents Sea Caledonides

By consideration of structural and metamorphic criteria it has been argued (Rice *et al.*, 1989b) that the Barents Sea Caledonides are an integral part of the Caledonides of Finnmark and are *not* a suspect terrane as previously inferred (Roberts and Gee, 1985). The similarity of metamorphic conditions, both in terms of $\Delta°2\theta$ and b_0 values, suggests that the Barents Sea Caledonides may be restored to a position along tectonic strike from the Komagfjord Antiformal Stack, to the northeast of the Laksefjord Nappe Complex, itself restored partly on metamorphic grounds (Rice *et al.*, 1989b). The Berlevåg Formation/Barents Sea Caledonides package, therefore, is here interpreted as being the upper part of an inverted metamorphic sequence, slightly dismembered by thrusting. Closest to the base of the Middle Allochthon, metamorphic conditions were highest ($\Delta°2\theta = 0.10$, b_0 between 9.036 Å and 9.033 Å; Table II) whilst to the southeast, and structurally lower, metamorphic conditions were somewhat lower (Table I; Fig. 9). However, the very broad area of epizone grade rocks with high b_0 values, both indicative of proximity to the roof thrust of the Lower Allochthon, suggests that the base of the Middle Allochthon was flat-lying and was positioned not far above the present erosion surface.

Metamorphism in the East Finnmark Autochthon

To the south of the Trollfjord–Komagelv Fault Zone the Autochthon and basal part of the Lower Allochthon are exposed. In the Leirpollen area and to the west of the river Tana, large-scale moderately inclined folds have been mapped (Siedlecki, 1980; Roberts, 1985b; Townsend, 1986), occasionally with associated thrusts, although all are presumed to have associated blind thrusts at depth (Chapman *et al.*, 1985). However, no major decollement has been reported. To the east, the limit of blind thrusting has been variously estimated; Chapman *et al.* (1985) suggested that it lies to the northeast of Guovdoaivi, where large-scale gentle folding dies out, whilst Townsend (1986) suggested that it occurs in the Klubben area, where cleavage and intermediate-scale folds have developed (Fig. 6). This indicates that both the position of the sole thrust (*if* emergent) and the limit of blind thrusting are uncertain.

Metamorphic data, unfortunately, do little to resolve these problems. Samples collected from the Vestertana Group near the inferred sole thrust in the Leirpollen–Tana Bridge area have suffered middle to upper anchizone grades of metamorphism (which the very limited b_0 data suggest was of an intermediate to high-intermediate baric type ($b_0 = 9.037$ Å, $N = 2$)). This is entirely comparable to the metamorphism on the west side of the Tana river (Fig. 2), although in the Autochthon comparison of illite crystallinity with stratigraphic position does not reveal any systematic variation (almost certainly owing to the major unconformable relationships between the Tana, Vestertana and Vadsø Groups).

Samples from farther to the east and southeast have values indicative of a diagenesis grade of metamorphism, which may vary with the structural grain immediately east of Tana Bridge (Fig. 6). The uncertainty in the position of the sole thrust and in the limit of blind thrusting under the sole, combined with the lack of a distinct metamorphic break (compared with the Autochthon further west), suggests that deformation died out gradually and that the concept of a major emergent sole thrust is not applicable to this area. However, the essentially homogeneous metamorphic grade across the East Finnmark autochthon does suggest that structural conditions were fairly uniform, implying that the base of the Middle Allochthon, which controlled burial metamorphism, was flat-lying across the region. The slight increase in metamorphic grade in the Vadsø–Ekkerøy region may reflect an increase in temperature from burial metamorphism rather than a greater proximity to the sole of the Lower Allochthon.

Metamorphism in the Dividal Group

Illite crystallinity values in the Dividal Group indicate sub-middle anchizone metamorphic grades. At these metamorphic grades the b_0 technique may not be applicable (Padan *et al.*, 1982), although it was used by Kemp *et al.* (1985) and Robinson and Bevins (1986). Consequently, metamorphic conditions in the footwall to the regional sole thrust cannot be estimated in the same manner as in other units. The illite crystallinity data have, therefore, been interpreted as reflecting the model developed for the Gaissa Thrust Belt, rather than as supporting it.

On the basis of the Gaissa Thrust Belt model, metamorphism in the upper part of the Dividal Group that underlies the Lower Allochthon (Fig. 9) has been inferred to have been of a high-intermediate baric type. Since P_{max} must have been essentially similar to the pressure in the Gaissa Thrust Belt during its emplacement (ca. 2 kbar), conditions have been inferred at upper diagenesis/high-intermediate baric type (ca. 175°C/2 kbar). However, for samples collected close to the Kalak Thrust, lowest anchizone–highest diagenesis illite crystal-

linity values have been recorded, indicating peak metamorphic temperatures of ca. 200°C. With a high-intermediate baric type, this would indicate *P–T* conditions of ca. 200°C/2.3 kbar, suggesting a rapid increase in pressure across the trailing branch line of the Gaissa Thrust Belt (essentially what is now the Stabbursdal Fault; Fig. 1). This seems an unlikely occurrence, since the metamorphic history of all the Dividal Group rocks was essentially the same until the moment the Kalak Nappe Complex was overthrust, i.e. *P–T–t* paths could only diverge after this overthrusting, synchronous with peak conditions in the Autochthon beneath the Gaissa Thrust Belt. A more reasonable solution would have been that there was no significant change in pressure, but that the late emplacement of the hotter Kalak Nappe Complex (as shown by Fig. 9) elevated temperatures at the top of the Dividal Group. This led to a high diagenesis–low anchizone/intermediate pressure metamorphism (ca. 200°C/2 kbar).

P–T–t curves for the lower parts of the Dividal Group have been drawn with a similar shape to that illustrated for the base of the Gaissa Thrust Belt. Since $\Delta°2\theta$ values are higher to the east of Stabbursdal, these rocks have been given a slightly lower temperature.

CONCLUSIONS

(1) Structural, metamorphic and radiometric age constraints indicate that metamorphic conditions within the Kalak Nappe Complex (Middle Allochthon) during Scandian reactivation were at least as high as middle greenschist facies when emplaced over the rocks which now constitute the Lower Allochthon.

(2) Peak metamorphic conditions determined within the Gaissa Thrust Belt record a thermal inversion. High anchizone–low epizone conditions developed near the roof thrust and high diagenesis–low epizone conditions developed near the floor thrust. Values of b_0 indicate a concurrent change from high-intermediate to an intermediate baric type. This inversion resulted from the Scandian emplacement of the hot Middle Allochthon.

(3) The similarity between metamorphic conditions near the roof of the Gaissa Thrust Belt and in the Laksefjord Nappe Complex, Komagfjord Antiformal Stack and Barents Sea Caledonides suggests that the latter three units are part of the Lower Allochthon and that the metamorphism in all these areas resulted from "contact" metamorphism arising from the emplacement of the hot Middle Allochthon.

(4) Metamorphic criteria suggest that the Berlevåg Formation may be a part of the Lower Allochthon, rather than a correlative of the Kalak Nappe Complex (Middle Allochthon).

(5) In the upper part of the Gaissa Thrust Belt and in the Laksefjord Nappe Complex, Komagfjord Antiformal Stack and Barents Sea Caledonides, P_{max} and T_{max} occurred at essentially the same time. Nearer the base of the Gaissa Thrust Belt, P_{max} and T_{max} were separated by a considerable period of decompression.

(6) Deformation *within* the sediments of the Gaissa Thrust Belt probably started soon after the emplacement of the Middle Allochthon. Peak metamorphism in the roof area was pre- to syn-thrusting, whilst in the floor area it was syn- to post-thrusting, depending on the relative velocities of imbrication and thermal relaxation.

(7) Although deformation in the Gaissa Thrust Belt was brittle, it was contemporaneous with ductile deformation in the base of the Middle Allochthon.

(8) No significant break in metamorphic grade has

been found between the Autochthon in the Varanger area (East Finnmark Autochthon) and the Gaissa Thrust Belt. This suggests that deformation died away gradually and that there may not be a *Major* sole thrust (emergent or not) in the eastern part of the Lower Allochthon/Parautochthon.

(9) Metamorphism in the Dividal Group shows a thermal inversion, with higher grades developed closer to the regional sole thrust. Higher conditions also developed, at any given depth below the sole thrust, in areas to the west of the Stabbursdal Fault, under the Kalak Nappe Complex.

Acknowldgements

Funds were provided by the Royal Society, Norges Geologiske Undersøkelse and University College Galway to A.H.N.R. and by the National Museum of Wales to R.E.B. Most of the fieldwork was undertaken while A.H.N.R. was at University College, Galway. We thank Sheila Bevins, Chris Townsend and Andy Mackay for help with sample collection and Mike Lambert and Chris Jarman for assistance with sample preparation and analysis. Discussions with Rod Gayer over a number of years have greatly benefited our understanding of Finnmarkian geology. University College Cardiff is thanked for loan of equipment. We thank Hanan Kisch and Grahame Oliver for thoroughly reviewing the typescript.

REFERENCES

Åm, K. (1975). Aeromagnetic basement complex mapping north of latitude 62° N, Norway. *Norges Geol. Unders.* **316**, 351–374.

Andersen, T. B. (1984). The stratigraphy of the Magerøy Supergroup, Finnmark, North Norway. *Norges Geol. Unders. Bull.* **395**, 25–39.

Aprahamian, J. and Pairis J.-L. (1981). Very low-grade metamorphism with a reverse gradient induced by an overthrust in Haute-Savoie (France). In: McClay, K. R. and Price, N. J. (eds.) *Thrust and Nappe Tectonics.* Geol. Soc. Spec. Publ. Lond. **9**, 159–165.

Arita, K. (1983). Origin of the inverted metamorphism of the lower Himalayas, Central Nepal. *Tectonophysics* **95**, 43–60.

Banks, C. J. and Warburton J. (1986). Passive-roof duplex geometry in the frontal structures of the Kirthar and Sulaimen mountain belts, Pakistan. *J. Struct. Geol.* **8**, 229–238.

Bevins, R. E., Robinson, D., Gayer, R. A. and Allman, S. (1986). Low-grade metamorphism and its relationship to thrust tectonics in the Caledonides of Finnmark, North Norway. *Norges Geol. Unders. Bull.* **404**, 33–44.

Bowden, P. L. (1981). Basement-cover distribution within the Caledonian nappe sequence in the northwest envelope of the Alta-Kvaenangen Window. *Terra Cognita* **1**, 35–36.

Chapman, T. J., Gayer, R. A. and Williams, G. D. (1985). Structural cross-sections through the Finnmark Caledonides and timing of the Finnmarkian event. In: Gee, D. G. and Sturt, B. A. (eds.) *The Caledonide Orogen—Scandinavia and Related Areas,* pp. 593–669. Wiley, Chichester.

Dallmeyer R. D. (1988). Polyorogenic $^{40}Ar/^{39}Ar$ mineral age record within the Kalak Nappe Complex, northern Norwegian Caledonides. *J. Geol. Soc. Lond.* **145**, 705–716.

Dallmeyer, R. D., Mitchell, J. G., Pharaoh, T. C., Reuter, A. and Andresen, A. (1988a). K–Ar and $^{40}Ar/^{39}Ar$ whole rock ages of slate/phyllite from allochthonous basement and cover in the tectonic windows of Finnmark, Norway: Evaluating the extent and the timing of Caledonian tectonothermal activity. *Geol. Soc. Amer. Bull.,* in press.

Dallmeyer, R. D., Reuter, A. & Clauer, N. (1988b). Age of cleavage formation in the lower allochthons of Finnmark, Norway. *Geol. Soc. Amer. Bull.,* in press.

Fareth, E. (1979). Geology of the Altenes area, Alta-Kvaenangen Window, North Norway. *Norges Geol. Unders.* **351**, 13–29.

Ferry, J. M. and Spear F. S. (1978). Experimental calibration of Fe and Mg between garnet and biotite. *Contrib. Mineral. Petrol.* **66**, 113–117.

Føyn, S. (1967). Dividal-gruppen ("Hyolithus-sonen") i Finnmark og dens forhold til de eokambriske-kambriske formasjoner. *Norges Geol. Unders.* **249**, 1–84.

Føyn, S. (1985). The Late Precambrian in northern Scandinavia. In: Gee, D. G. and Sturt, B. A. (eds.) *The Caledonide Orogen—Scandinavia and Related Areas,* pp. 233–245. Wiley, Chichester.

Ganguly, J. (1979). Garnet and clinopyroxene solid solutions and geothermometry based on Fe–Mg distribution coefficients. *Geochim. Cosmochim. Acta* **43**, 1021–1029.

Gayer, R. A. and Rice A. H. N. (1988). Palaeogeographic reconstruction of the pre- to syn-Iapetus rifting sediments in the Caledonides of Finnmark, N Norway, *this volume.*

Gayer, R. A., Hayes, S. J. and Rice, A. H. N. (1985). The structural development of the Kalak Nappe Complex of eastern and central Porsangerhalvøya, Finnmark, Norway. *Norges Geol. Unders. Bull.* **400**, 67–87.

Gayer, R. A., Rice, A. H. N., Roberts, D., Townsend, C. and Welbon, A. I. F. (1987). Restoration of the Caledonian Baltoscandian margin from balanced cross-sections: the problem of excess continental crust. *Trans. R. Soc. Edinburgh* **78**, 197–217.

Graham, C. M. and England, P. C. (1975). Thermal regimes and regional metamorphism in the vicinity of overthrust faults: an example of shear heating and inverted metamorphic zonation from southern California. *Earth Planet. Sci. Lett.* **31**, 142–152.

Guidotti, C. V. (1984). Micas in metamorphic rocks. In: Bailey, S. W. (ed.) *Micas. Reviews in Mineralogy* **13**, pp. 357–467. Mineralogical Society of America.

Johnson, H. D., Levell, B. K. and Siedlecki S. (1978). Late Precambrian sedimentary rocks in East Finnmark, North Norway and their relationship to the Trollfjord-Komagelv fault. *J. Geol. Soc. Lond.* **135**, 517–534.

Kemp, A. E. S., Oliver, G. J. H. and Baldwin, J. R. (1985). Low-grade metamorphism and accretionary tectonics: Southern Uplands terrain, Scotland. *Mineral. Mag.* **49**, 335–344.

Kisch, H. J. (1980). Incipient metamorphism of Cambro-Silurian clastic rocks from the Jamtland Supergroup, Central Scandinavian Caledonides, Western Sweden: illite crystallinity and vitrinite reflectance. *J. Geol. Soc. Lond.* **137**, 217–288.

Kjøde, J., Storetvedt, K. M., Roberts, D. and Gidskehaug, A. (1978). Palaeomagnetic evidence for large-scale dextral movement along the Trollfjord-Komagelv Fault, Finnmark north Norway. *Phys. Earth Planet. Interiors* **16**, 132–144.

Kubler, B. (1967). La cristallinité de l'illite et les zones a fait superieures du metamorphisme. In: *Étages Tectonique,* pp. 105–121. A la Baconnière Neuchatel, Suisse.

Levell, B. K. and Roberts, D. (1977). A re-investigation of the geology of northwest Varanger Peninsula, East Finnmark, North Norway. *Norges Geol. Unders.* **334**, 83–90.

Nicholson, R. and Rutland, R. W. R. (1969). A section through the Norwegian Caledonides; Bodø to Sulitjelma. *Norges Geol. Unders.* **260**, 1–86.

Oxburgh, E. R. and Turcotte, D. L. (1974). Thermal gradients and regional metamorphism in overthrust terranes with special reference to the Eastern Alps. *Schweit. Mineral. Petrogr. Mitt.* **54**, 641–662.

Padan, A., Kisch, H. J. and Shagam, R. (1982). Use of the lattice parameter b_0 of dioctahedral illite/muscovite for the characterization of P/T gradients in incipient metamorphism. *Contrib. Mineral. Petrol.* **79**, 85–95.

Pavlis, T. L. (1986). The role of strain heating in the evolution of megathrusts. *J. Geophys. Res.* **91-B12**, 12407–12422.

Pedersen, R. B., Dunning, G. and Robins, B. (1988). The U–Pb (zircon) age of a nepheline syenite pegmatite within the Seiland Magmatic Province, North Norway, this volume.

Pharaoh T. C. (1985). The stratigraphy and sedimentology of autochthonous metasediments in the Repparfjord–Komagfjord Tectonic Window, west Finnmark. In: Gee, D. G. and Sturt, B.

A. (eds.) *The Caledonide Orogen—Scandinavia and Related Areas*, pp. 347–357. Wiley, Chichester.

Pharaoh, T. C., Ramsay, D. M. and Jansen, O. (1983. Stratigraphy and structure of the northern part of the Repparfjord–Komagfjord Window, Finnmark, northern Norway. *Norges Geol. Unders.* **377**, 1–45.

Platt, J. P. (1987). The uplift of high-pressure low-temperature metamorphic rocks. In: Oxburgh, E. R., Yardley, D. B. W. and England, P. C., (eds.) *Tectonic Setting of Regional Metamorphism. Phil. Trans. R. Soc. Lond. A*321, 87–103.

Ramsay, D. M. (1971). The stratigraphy of Sørøy. *Norges Geol. Unders.* **269**, 314–317.

Ramsay, D. M., Sturt, B. A. and Andersen, T. B. (1979). The sub-Caledonian unconformity on Hjelmsøy—new evidence of primary basement-cover relationships in the Finnmarkian nappe sequence. *Norges Geol. Unders.* **351**, 1–12.

Rice, A. H. N. (1987). Continuous out-of-sequence ductile thrusting in the Norwegian Caledonides. *Geol. Mag.* **124**, 249–260.

Rice, A. H. N. and Mackay, A. (1986). The autochthonous and allochthonous Caledonian rocks west of Stabburselva, Cakkarassa map sheet 134 I (1:50,000). Unpublished report to Norges Geologiske Undersøkelse.

Rice, A. H. N. and Roberts, D. (1988). Multi-textured garnets from a single growth event: an example from northern Norway. *J. Metam. Geol.* **6**, 159–172.

Rice, A. H. N., Bevins, R. E., Robinson, D. and Roberts, D. (1989a). Thrust related metamorphic inversion in the Caledonides of Finnmark, North Norway. In: Daly, S., Yardley, B. W. D. and Cliff, B. (eds.) *Evolution of Metamorphic Belts. Geol. Soc. Lond. Spec. Publ.*, in press.

Rice, A. H. N., Gayer, R. A., Robinson, D. and Bevins, R. E. (1989b). Strike-slip restoration of the Barents Sea Caledonides terrane, Finnmark, North Norway. *Tectonics*, in press.

Roberts, D. (1968). Hellefjord Schist Group—a probable turbidite formation from the Cambrian of Sørøy, West Finnmark. *Norsk Geol. Tidsskr.* **48**, 231–244.

Roberts, D. (1985a). The Caledonian fold belt in Finnmark: a synopsis. *Norges Geol. Unders. Bull.* **403**, 161–167.

Roberts, D. (1985b). Geologiske kart over Norge, berggrunnsgeologiske kart Honningsvåg—M 1:250,000, forelopig utgave. Norges Geologiske Undersøkelse.

Roberts, D. and Gee, D. G. (1985). An introduction to the structure of the Scandinavian Caledonides. In: Gee, D. G. and Sturt B. A. (eds.) *The Caledonide Orogen—Scandinavia and Related Areas*, pp. 55–68. Wiley, Chichester.

Roberts, J. D. (1974). Stratigraphy and correlation of Gaissa Sandstone Formation and Børselv Subgroup (Porsangerfjord Group), South Porsanger, Finnmark. *Norges Geol. Unders.* **303**, 57–118.

Robinson, D. (1981). Metamorphic rocks of an intermediate facies series juxtaposed at the Start boundary, south-west England. *Geol. Mag.* **118**, 297–301.

Robinson, D. and Bevins, R. E. (1986). Incipient metamorphism in the Lower Palaeozoic marginal basin of Wales. *J. Metam. Geol.* **4**, 101–113.

Salje, E. (1986). Heat capacities and entropies of andalusite and sillimanite: the influence of fibrolitization on the phase diagram of the Al_2SiO_5 polymorphs. *Amer. Mineral.* **71**, 1366–1371.

Sassi, F. P. and Scolari, A. (1974). The b_0 value of the potassium white micas as a barometric indicator in low-grade metamorphism of pelitic schists. *Contrib. Mineral. Petrol.* **45**, 143–152.

Siedlecka, A. and Siedlecki, S. (1972). Lithostratigraphical correlation and sedimentology of the late Precambrian of Varanger Peninsula and neighbouring areas of east Finnmark, north Norway. *International Geological Congress* **24**(6), 249–258.

Siedlecki, S. (1980). Geologiske kart over Norge, berggrunnskart VADSØ—M 1:250,000. Norges Geologiske Undersøkelse.

Siedlecki, S. and Levell, B. K. (1978). Lithostratigraphy of the late Precambrian Løkvikfjell Group of Varanger Peninsula, east Finnmark, north Norway. *Norges Geol. Unders.* **242**, 73–85.

Spray, J. G. and Williams, G. D. (1980). The sub-ophiolite rocks of the Ballantrae Igneous Complex, S. W. Scotland. *J. Geol. Soc. Lond.* **137**, 359–368.

St-Onge, M. R. (1981). Normal and inverted metamorphic isograds and their relation to syntectonic Proterozoic batholiths in the Wopmay Orogen, Northwest Territories, Canada. *Tectonophysics* **76**, 295–316.

Sturt, B. A., Pringle, I. R. and Ramsay, D. M. (1978). The Finnmarkian phase of the Caledonian orogeny. *J. Geol. Soc. Lond.* **135**, 597–610.

Teisseyre, J. H. (1972). Geological investigations in the area between Kjolnes and Trollfjorden (Varanger Peninsula). *Norges Geol. Unders.* **278**, 81–92.

Thompson, A. B. and England, P. C. (1984). Pressure-temperature-time paths of regional metamorphism II. Their inference and interpretation using mineral assemblages in metamorphic pelites. *J. Petrol.* **25**, 929–955.

Townsend, C. (1986). Thrust tectonics within the Caledonides of northern Norway. Unpublished Ph.D. thesis, University of Wales.

Townsend, C. (1987). Thrust transport directions and thrust sheet restoration in the Caledonides of Finnmark. *J. Struct. Geol.* **9**, 345–352.

Townsend, C., Roberts, D., Rice, A. H. N. and Gayer, R. A. (1986). The Gaissa Nappe, Finnmark, North Norway: an example of a deeply eroded external imbricate zone within the Scandinavian Caledonides. *J. Struct. Geol.* **8**, 431–440.

Townsend, C., Mackay, A. and Rice, A. H. N. (1988). Stratigraphy and structure of the southwest part of the Gaissa Thrust Belt and adjoining areas, *this volume.*

Watkins, K. P. (1985). Geothermometry and geobarometry of inverted metamorphic zones in the west Central Scottish Dalradian. *J. Geol. Soc. Lond.* **142**, 157–165.

Williams, G. D., Chapman, T. J. and Milton, N. J. (1984). Generation and modification of finite strain patterns by progressive thrust faulting in the Laksefjord Nappe, Finnmark. *Tectonophysics* **107**, 177–186.

Worthing, M. A. (1986). Variation in garnet and plagioclase compositions in pelitic blastomylonites across a zone of declining metamorphic grade on Seiland, North Norway. *Norges Geol. Unders. Bull.* **406**, 67–82.

Zwaan, K. B. and Roberts, D. (1978). Tectonostratigraphic succession and development in the Finnmarkian nappe sequence. *Norges Geol. Unders.* **343**, 53–71.

16 Metamorphic evolution of the Caledonian nappes of north central Scandinavia

Andrew J. Barker

Department of Geology, The University, Southampton, SO9 5NH, UK

The tectonometamorphic evolution of Caledonian nappes from Tysfjord to Tromsø is reviewed, and peak metamorphic conditions are quantified. Excluding the Seve Nappe and Tromsø Nappe Complex, which locally recorded pre-Scandian eclogite facies metamorphism, evidence to date suggests only a Scandian (mid to late Silurian) orogenic history for other nappes of the region. A similar D1 to D4 structural evolution is recorded in these nappes with peak metamorphism largely syn-D2 followed by a later period of chlorite grade retrogression concentrated near to thrust zones. External thrust slices record sub-greenschist facies metamorphism, whereas the highest allochthonous units record kyanite–sillimanite grade conditions characteristic of deep burial in association with continental collision. A combination of geothermobarometry and petrographic observation has allowed assessment of P–T–t loops for individual nappes, but the full detail of these paths requires further evaluation.

INTRODUCTION

The area under consideration extends northwards from Tysfjord to Tromsø, and eastwards to the Caledonian frontal region of Torneträsk (Fig. 1). It is beyond the scope of this paper to present a detailed nappe-by-nappe analysis through the region. The intention is to provide a general introduction and description of the tectonostratigraphy of the region, and subsequently outline the sequence of structural and metamorphic events recognized in each nappe. The interrelationships between deformation and metamorphism have been described, and P–T conditions have been quantified by comparison of geothermobarometric work of the present studies with that of other authors.

REGIONAL SETTING

Following the terminology introduced by Kulling (1972) and Gee (1975), the Caledonian nappes of a large part of Scandinavia have for simplicity been categorized as Lower, Middle, Upper and Uppermost Allochthon.

For ease of understanding, Table I gives the present interpretation of how local nappe sequences fit into this scheme, and how different units correlate with each other. Figure 2 shows a generalized cross-section from Lofoten to Torneträsk (also see Fig. 1), to give an overall impression of the regional structure.

The Lower Allochthon consists of basement-cover imbricates dominated by mylonitized Proterozoic granitoid rocks and Vendian to Cambrian clastic sediments (Kulling, 1964, 1972; Føyn and Glaessner, 1979; Ahlberg, 1980, 1985). The Lower Allochthon is succeeded by mylonitic "hardschists" of the Middle Allochthon (Abisko Nappe, Kulling, 1960, 1964), which pass westwards into phyllonitic units (Storfjell Group and (lower) Rombak Group; Gustavson, 1966, 1974a; Fossbakken Nappe, Barker, 1986; Målselv Nappe, Andresen *et al.*, 1985). In eastern areas the "hardschists" are overlain by the Seve nappes comprising distinctive augen gneisses, and metasediments intruded by mafic dykes (Kulling, 1964). The Seve shows many similarities with sequences of the underlying nappes, especially the Särv Nappe (Strömberg, 1955, 1961) farther south in Jämtland. For this reason it has been grouped as the upper part of the Middle Allochthon. The Seve nappes are absent from western parts of the study area, where units equivalent to the Lower Køli nappes comprised largely of pelitic schist and marble rest with thrust contact above phyllonites of the Middle Allochthon. The Lower Køli nappes of the region include units such as the Høgtind Nappe (Barker, 1986; formerly (upper) Rombak Group of Gustavson, 1974a) and the Senja Nappe (Bergh and Andresen, 1985). These units are succeeded by a variety of nappes including amphibolites and high-grade gneisses, of the Narvik Nappe Complex (Barker, 1986; =Narvik Group of Gustavson, 1966), in turn overlain by nappes dominated by marble and pelitic

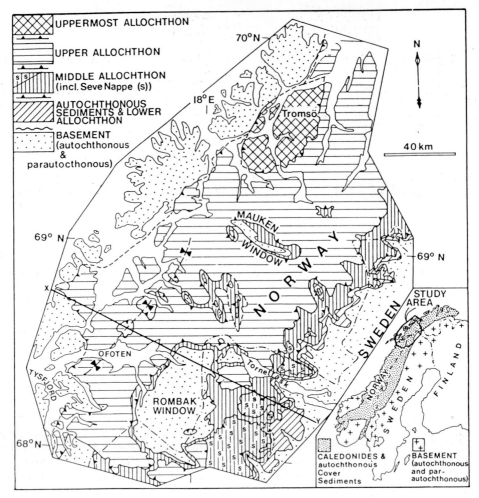

Fig. 1. Location of study area, and simplified geology of the region. Based on Gustavson (1974b), Fareth (1981), Kulling (1964), Lindstrom *et al*. (1985), Björklund (1985), Hodges (1985) and the author's own field observations. (Cross-section X-Y is shown in Fig. 2.)

Table I. Nappe correlations in the region from Tysfjord to Tromsø. (All boundaries are thrusts, except that at the base of the Dividal Group, which is an unconformity above Precambrian crystalline basement).

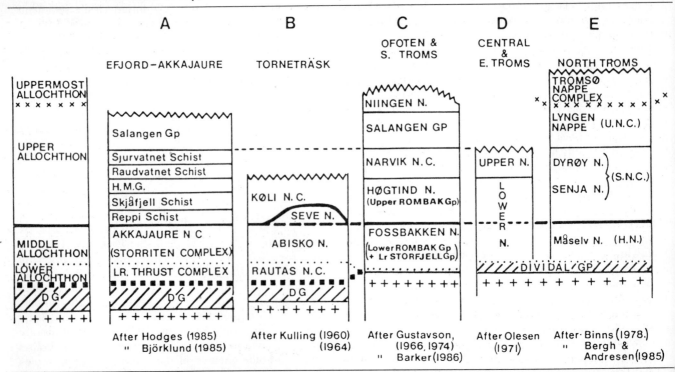

Abbreviations: D.G. = Dividal Group; H.M.G. = Hjertevatnet Migmatitic Gneiss; H.N. = Helligskogen Nappe; N = Nappe; N.C. = Nappe Complex; S.N.C. = Skibotn Nappe Complex; U.N.C. = Ullsfjord Nappe Complex.

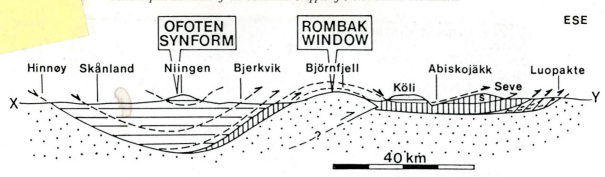

Fig. 2. Generalized WNW–ESE cross-section from Lofoten to Torneträsk. For line of section and key to ornamentation, see Fig. 1. Vertical scale is of necessity exaggerated about 5 ×, in order to show structural relationships in the more or less flat-lying eastern part of the section.

schist. Such nappes include the Småtinden and Gratangen Nappes (Barker and Anderson, 1988; = Salangen Group of Gustavson, 1966), Ullsfjord Nappe Complex (Binns, 1978) and Lyngen Nappe (Bergh and Andresen, 1985). The Niingen Nappe (Barker, 1986; = Niingen Group of Gustavson, 1966) of SW Troms, and Tromsø Nappe-Complex (Fareth, 1981; Andresen *et al.*, 1985) of NW Troms form the highest tectonic units of the region. The Tromsø Nappe Complex is distinctly different from immediately underlying units and as well as containing high-grade gneisses it also contains retrogressed eclogites (see Fig. 9). The Niingen Nappe, although metamorphosed to mid-amphibolite facies is not distinctly different from some underlying units and consequently is placed within the "Upper" rather than "Uppermost" Allochthon. Many of the higher nappes contain variable-sized synorogenic "granitoid" bodies (see Gustavson, 1966, 1969, 1974a,b).

STRUCTURAL–METAMORPHIC HISTORY

Two important orogenic events have contributed to the structural–metamorphic development of the Scandinavian Caledonides. The earliest of these (ca. 540–490 Ma) is termed the Finnmarkian phase (Sturt *et al.*, 1978), and the later (ca. 440–410 Ma) is termed the Scandian phase (Gee, 1975). The Finnmarkian phase dominantly affects rocks of northernmost Scandinavia (Finnmark), but it has been shown farther south in Nordland (Boyle *et al.*, 1985; Boyle, 1987; Billett, 1987) that a pre-Scandian (possibly Finnmarkian) structural–metamorphic history is present in some of the rocks at Sulitjelma. Additionally, Dallmeyer and Gee (1986) record a significant pre-Scandian event (491 ± 8 Ma) causing eclogite facies metamorphism in rocks from the Seve Nappe Complex of Norrbotten, prior to upper amphibolite facies Scandian metamorphism (Williams and Claesson, 1987). Thorough dating has yet to be undertaken in Troms and north Norrbotten, but dates by Krogh *et al.* (1982) and Dallmeyer *et al.* (1987) suggest a similar pattern of dominant Scandian metamorphism but with an earlier ("Finnmarkian"?) history in a few nappes. Attempts to determine the nature and extent of possible Finnmarkian elements in the Troms region by field and thin-section studies have so far proved fruitless. This is due to the intensely pervasive S2 schistosity that obscures earlier structures.

The S2 fabric is common to all nappes, and is demonstrably Scandian because mid Ordovician to Silurian fossils are recorded in some of the units that it effects (e.g., Sagelvatn Group, Olaussen, 1976; Bjørlykke and Olaus-

sen, 1981; Skibotn Nappe Complex, Binns and Gayer, 1980; Balsfjord Supergroup, Binns and Matthews, 1981). Because of the limited knowledge of pre-S2 structures, it is speculative to say whether such fabrics are "Finnmarkian" or "Scandian". Clearly, in units containing Ordovician–Silurian fossils, S1 fabrics represent the earliest part of the Scandian history. However, in units such as the unfossiliferous Narvik Nappe Complex, Niingen Nappe and Tromsø Nappe Complex S1 structures may include early Scandian as well as pre-Scandian elements. Indeed, this might be expected in view of nappe correlations into central Nordland, where Boyle *et al.* (1985) and others have demonstrated the presence of a pre-Scandian history. Since conclusive evidence for Finnmarkian structures has yet to be established, the present study, with some reservations, finds no grounds for considering S1 fabrics of the Troms region as anything other than early Scandian; appreciating that any porphyroclasts contained within this fabric may represent relicts from pre-Scandian events.

Binns and Matthews (1981), Steltenpohl and Bartley (1984), Bergh and Andresen (1985), Hodges (1985) and Barker (1986) have, with some differences, generally established a similar tectonometamorphic history for the various nappes in the region from Tysfjord to Tromsø. To summarize the evolution of so many nappes and such a large area will always lead to oversimplifications. However, there appear to be certain metamorphic and deformation events that are common throughout. That is not to say their development was contemporaneous in all units, since there would undoubtedly be some diachroneity from hinterland to foreland.

D1

Knowledge of D1 structures in the region is extremely limited, the clearest evidence for this event being represented by inclusion trails in garnet porphyroblasts. In most cases this inclusion fabric is straight, but in other examples a curved trail towards the rim suggests synchronous development, and relative rotation between porphyroblast and developing schistosity. The S1 fabric is defined by quartz, opaques, epidote and less commonly mica. Preservation of chloritoid porphyroblasts postdating S1 (Fig. 3) suggests that for the Hogtind Nappe at least, S1 developed under greenschist facies conditions. The oblique chloritoids probably record an early stage in the development of S2 that ultimately completely transposed S1.

D2 and MP1

D1 progressed into D2 as Scandian orogenesis intensified, and S2 developed by the progressive transposition of S1.

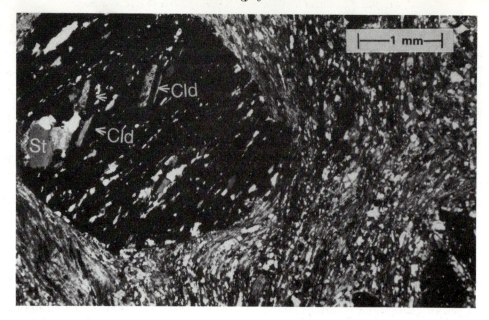

Fig. 3. Chloritoid (Cld) inclusions preserved in garnet due to shielding from *P–T* conditions reached in matrix, overprint an earlier fabric (S1). Staurolite (St) and kyanite (but *not* chloritoid) occur as small prophyroblasts in the matrix of the rock. The regional schistosity (S$_2$) continued development after garnet growth had ceased, to give discordant relationships between inclusion fabric and matrix schistosity. Scale bar = 1 mm.

Main porphyroblast growth occurred largely during the waning stages of D1 and onset of D2 (Fig. 8). However, porphyroblastesis had largely terminated before the regional S2 schistosity was completely developed. Consequently, the included fabrics (Si) usually show discordance with external S2 fabrics (Fig. 3). The pervasive S2 schistosity is axial planar to F2 folds, and envelops garnet, staurolite and kyanite porphyroblasts. Peak metamorphic grade attained during MP1 varies from nappe to nappe (compare Figs. 9 and 1), and will be discussed in detail in a later section.

S2 and F2 developed under conditions of intense ductile shear as the deforming crust was uplifted and transported southeastwards during the collision of Baltica and Laurentia. However distinct decollement horizons associated with this stage in the tectonometamorphic evolution have not been recognized. It is probably most realistic to envisage very broad shear zones/slide zones, and intense internal strain distributed throughout the rock mass (Fig. 4(*b*)). Only at a late stage in D2 and into D3 did the strain focus into discrete (narrow) thrust zones along which major translations subsequently occurred (Fig. 4(*c*)).

D3

D3 structures evolved progressively from early recumbent isoclines to later more upright open folds (Barker and Anderson, 1988). The recumbent folds are often well-developed near to thrusts. They often display an S3 axial planar cleavage of variable intensity. S–C mylonites and extensional crenulations are well-developed at major thrust boundaries. This is especially true at lower levels in the nappe pile, where phyllonites are extensively developed. The formation of F3 and S3 structures was not coeval, but varied according to when a given thrust operated. Intense shear associated with thrust movement has led to the development of a composite S2–S3 fabric at major thrust boundaries.

Since main thrust displacements were post-D2, not only were F2 fold structures truncated but, because peak metamorphism occurred early in D2, isograds were also truncated (Fig. 4). This has given rise to general meta-

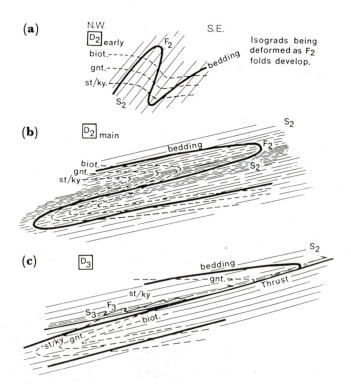

Fig. 4. Schematic diagram to illustrate how isograds associated with peak metamorphism became progressively folded as D2 intensified (*a, b*), and then truncated (*c*) as discrete thrust planes developed during D$_3$. Note that higher grade rocks are emplaced southeastwards over lower grade rocks, and that inverted isograds are preserved below the thrust (e.g., Høgtind Nappe).

morphic inversion as deeper-level higher-grade rocks were emplaced onto lower-grade units as thrusts propagated south-eastwards towards the Baltic foreland. Complication of this simple picture of metamorphic grade variations has arisen owing to late D3, and D4 out-of-sequence thrust movements. Barker (1986) describes this in relation to the Kvernmo Nappe, and Anderson (this volume) recognizes the Grønfjellet Thrust marking the base of the Narvik Nappe Complex as a late out-of-sequence thrust, activated in response to D4 basement imbrication.

D4

D4 witnessed the development of upright open fold structures ranging from crenulations to broad regional folds. Axes dominantly trend NNE–SSW, and the upright nature suggest little or no simple shear. Structures such as the Ofoten Synform (Figs. 1 and 2), fold major thrust boundaries and it seems likely that they formed in response to basement imbrication and ramp development (Fig. 2). Basement rocks are exposed in a structural high from Mauken to Rombak (Fig. 1) and indeed farther south to Nasafjell along the same line. This gives a strong suggestion of basement duplication, which Bax (1986) and Anderson (this volume) have been able to demonstrate in basement areas of Torneträsk and S. Troms respectively. A second less well-developed WNW–ESE set of F4 structures may reflect deformation above lateral ramps, but this has still to be proved.

RETROGRESSION AND LATE-STAGE PORPHYROBLAST GROWTH

Throughout the area, intense retrogression of higher-grade assemblages near to thrusts is well-reported (Gustavson, 1966; Olesen, 1971; Barker, 1986). Initial alteration of porphyroblasts may have occurred in association with thrust movement, but complete pseudomorphing of garnet to random aggregates of chlorite (Fig. 5) must have occurred when shearing had ceased. The "soft" aggregate of chlorites would certainly have been smeared-out into the regional fabric if this were not so. Since the garnet to chlorite reaction requires large amounts of water to proceed, it demonstrates that thrust zones acted as major fluid pathways well after shearing terminated.

Several workers (Bergh and Andresen, 1985; Barker, 1986) have reported post-S2 rim growth (Fig. 6(a),(b)) on certain garnets, but as yet no obvious relationships with particular lithologies or specific levels in the nappe pile have been realized. If one were to model garnet rim developments in terms of down-heating from an overthrust "hot" slab, all garnets beneath a given thrust should exhibit such rims. Since this is not the case, an alternative explanation must be sought.

As there is no particular pattern to the distribution of garnets with post-S2 rims, it seems most plausible that bulk rock chemistry and local changes in fluid chemistry during metamorphism may be responsible. Despite suitable P–T conditions, garnet growth may be prematurely halted in certain rocks owing to exhaustion of reactants. However, if other reactions and/or external fluid flux change the local fluid chemistry, it is possible for renewed garnet growth at a later stage. This would give texturally distinctive rim growths such as those observed.

In a single thin (200 m) unit (Småtinden Nappe), Barker and Anderson (1988) have observed post-S2 and generally post-F3 garnets (Fig. 7(a), (b)). Similar relationships have also been recorded by Bergh and Andresen (1985) within the Balsfjord Group of the Lyngen Nappe, but such relationships are not recorded in other units. Garnets of the thin Småtinden nappe are abundant and typically of 5–10 mm diameter, which makes it unlikely that bulk rock chemistry inhibited growth. As an alternative, it is possible that this nappe acted as a broad D2 shear zone, such that intense strain rates gave strong dissolution and removal of ions. As argued by Bell (1985) and Bell *et al.* (1986), this would inhibit porphyroblast nucleation and growth. Only when shear had focused into discrete thrust planes (i.e. D3) could porphyroblasts develop in this unit, and consequently such porphyroblasts show a combination of post-S2 and post-F3 relationships. It is notable that these syn-/post-D3 garnets (Småtinden Nappe) yield significantly lower temperatures than units with typical syn-D2 garnets (Table II).

VARIATIONS IN METAMORPHIC GRADE OF NAPPES

Figure 9 shows a map of peak metamorphic grade for Caledonian nappes from Tysfjord to Tromsø. Parts of east Troms and Sweden are inadequately reported in the literature, and thus the accuracy of the map in these areas is less well constrained. Nevertheless, the broad picture presented should give a reasonable impression of grade variations within the region. With the exception of part of the Lyngen Nappe, and the thin Gratangseidet and Øse Nappes, all the Upper Allochthon (Køli Nappe equivalents and higher) attained at least garnet grade and

Fig. 5. Garnet pseudomorphed to an aggregate of chlorite (near to Høgtind Thrust). Scale bar = 1 m.

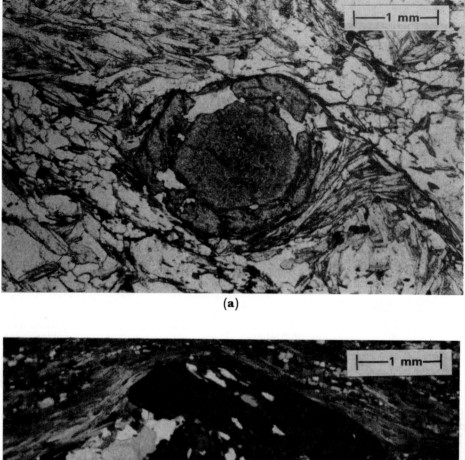

(a)

(b)

Fig. 6. Late (post-S$_2$) rim growths on garnet. (*a*) Almost complete post-S2 rim developed around older core of garnet (Høgtind Nappe). Scale bar = 1 mm. (*b*) Detail of partially developed post-S2 rim on garnet (Grønfjellet Nappe), showing discordance between core (S1) and rim (S2) fabrics. Note the parallelism between S2 inclusion fabric of rim and that of the matrix. Scale bar = 1 mm.

generally reached staurolite/kyanite grade, with silli-manite locally present (Bartley, 1981; Andresen, *et al.*, 1985).

In north Norrbotten and southeastern Troms the Seve Nappe exhibits amphibolite facies metamorphism, with widespread garnet (Kulling, 1964; Björklund, 1985; Lindstrom *et al.* 1985), and local kyanite (L. Page, personal communication, 1987). In southern Norrbotten, retro-gressed eclogite lenses/boudins occur within the amphi-bolite facies (kyanite grade) psammitic and pelitic schists of the Seve Nappe (Andréasson *et al.*, 1985). The eclogitic assemblages are recorded in the cores of large boudins representing dismembered metadolerite sheets (Andréasson *et al.*, 1985). The margins of these boudins are retrogressed to garnet amphibolite, and other pods are entirely garnet amphibolite. The chemistry of the Seve eclogites/metadolerites is tholeiitic, and they are inter-preted as higher grade equivalents of mafic dykes found in

the underlying Särv nappe (Andréasson *et al.*, 1985). The eclogite assemblages record pre-Scandian ages (491 ± 8 Ma, Dallmeyer and Gee, 1986) and have been retrogressed to amphibolite facies during Scandian metamorphism.

With the exception of the Seve nappe, the Middle Allochthon (Abisko Nappe, Kulling, 1960, 1964; Ak-kajaure Nappe Complex, Björklund, 1985; Målselv Nappe, Andresen *et al.*, 1985; and Fossbakken Nappe, Barker, 1986) only attained chlorite–biotite (greenschist facies) conditions. These units are well-exposed in the frontal regions of east Troms and Torneträsk, as well as around the Rombak window, and smaller windows in central and south Troms (Figs. 1 and 9). The Lower Allochthon, essentially only represented in the east (Kul-ling, 1960, 1964), consists of basement-cover imbricates metamorphosed to anchizone or at maximum low-green-schist facies conditions (Anderson, this volume).

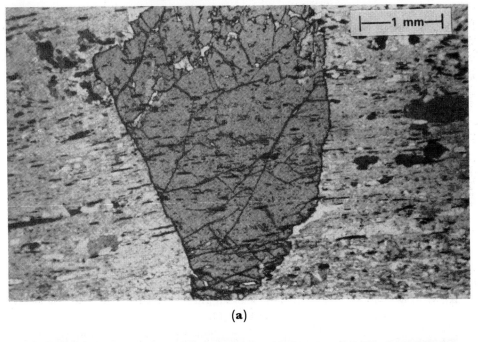

(a)

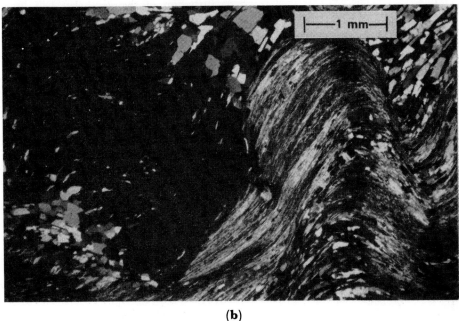

(b)

Fig. 7. (*a*) Garnet porphyroblast (Småtinden Nappe) developed entirely post-S2 regional schistosity. Scale bar = 1.5 mm. (*b*) Garnet porphyroblast (Småtinden Nappe) overgrowing F3 (or F4?) crenulation. Scale bar = 1 mm.

Within some individual nappes of the Upper Allochthon, metamorphic inversion is recognized. For example, within the Høgtind Nappe (Barker, 1986), which is laterally equivalent to the "Lower Nappe" (Olesen, 1971); the lower part of the nappe is at garnet grade whilst the higher levels of the nappe are at staurolite–kyanite grade. The nature of F2 fold vergence corroborates these metamorphic observations to demonstrate that the Høgtind Nappe represents the inverted limb of a thrust-truncated F2 anticline (see Fig. 4, and Barker, 1986). It is notable that farther south in central Nordland, a major F2 anticline at Sulitjelma similarly causes inversion of metamorphic isograds (Boyle *et al.*, 1979, 1985; Mason, 1984; Boyle, 1987). Since this occurs at a structurally identical level in the tectonostratigraphy, and since there are strong lithological similarities between the two areas, it seems highly probable that this metamorphic inversion is of regional extent.

QUANTITATIVE ASSESSMENT OF *P–T* CONDITIONS

Using calibrations based largely on Ferry and Spear (1978) and adopting the approach of Ghent *et al.* (1979), Ghent and Stout (1981) and Hodges and Royden (1984), several recent papers (Bergh and Andresen, 1985; Crowley and Spear, 1987; Steltenpohl and Bartley, 1987; Barker and Anderson, 1988) have assessed *P–T* conditions within nappes of the Upper Allochthon. Table II documents this information, and accepting the probable large error bars (±1 kb) on *P* estimates, it is clear that for the vast majority of nappes maximum equilibrium conditions were of the order $T = 500°–650°C$ and $P = 6–9$ kb. It is encouraging that there is general agreement between different workers, and that the estimates obtained are consistent with rock mineralogy. For example, the *P–T* conditions obtained are consistent with the fact that kyanite is the stable

(a)

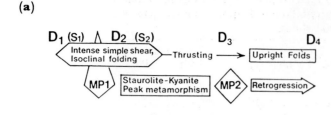

(b)

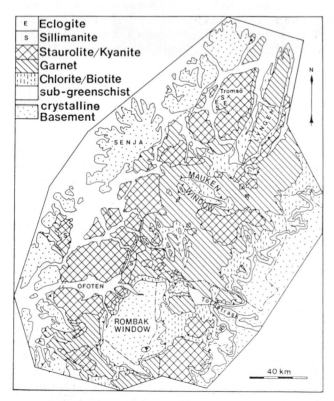

Fig. 8. (*a*) Simplified summary diagram illustrating relative timing of key deformation and metamorphic events recorded in nappes from Tysfjord to Tromsø. (*b*) Timing of mineral growth/prophyroblastesis relative to deformation events.

Al_2SiO_5 polymorph throughout the region, andalusite being entirely absent and sillimanite uncommon. There are a few units that yield P–T conditions outside the above range. The Tromsø N.C. for example gives $T = 700°C$ $P = 17$ kb for the early eclogite facies event, with

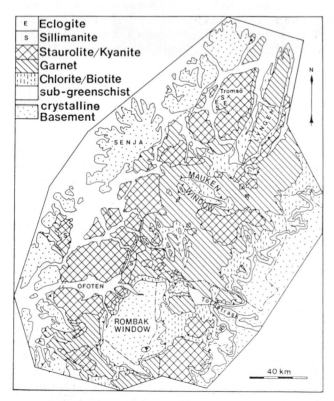

Fig. 9. Map of peak metamorphic grade in the Caledonides of northcentral Scandinavia (Tysfjord to Tromsø).

$T = 660$–$700°C/P = 9$ kb for later metamorphism (Krogh *et al.* (unpub. data), referred to by Andresen *et al.* (1985)). At the other end of the range, the Lyngen and Småtinden nappes consistently give low T estimates, the former yielding $T = 400$–$445°C$ (Bergh and Andresen, 1985) whilst the latter gives $T = 430$–$500°C$ (Barker and Anderson, 1988). P estimates for these units are uncertain.

P–T–t Paths

During initial continental collision there is rapid burial of crustal material and accompanying P increase (England and Thompson, 1984). This crustal thickening and burial is typically followed by a period of heating as the crust attempts to revert to the normal geotherm. Rapid temperature rises occur in the first few tens of millions of years after thickening; thereafter, rates slow. It is argued (England and Thompson, 1984) that erosion rarely commences prior to 20 Ma after burial and as such has a small influence on peak temperatures attained by a given rock. Onset of thrusting gives relatively rapid uplift and decompression, but thermal equilibration following perturbation of the geotherm will be very much slower (Oxburgh and Turcotte, 1974). Sleep (1979) showed that it is possible for large-scale folds to develop within a timespan of 1 Ma, and that isograds could be folded while hot. If uplift was fairly rapid (e.g., by thrusting), lower levels would not heat up significantly because of the slow nature of conductive heat transfer in rocks. Consequently, it is possible to preserve isograds, such as those seen in the Troms and Sulitjelma regions.

Almost regardless of the rocks' position in the thickened pile, England and Thompson (1984) note that rocks spend a high proportion (of the order 30–40%) of their P–T–t path within 50 to 100°C of the maximum temperature experienced. However, P conditions change considerably over this period, and P recorded at peak T conditions is considerably lower than maximum P (generally by several kilobars). With this in mind, it is reasonable to expect that rocks of the study area were buried slightly more deeply than P estimates of Table II would indicate. This is shown schematically by the upper parts of the P–T loops of Figs. 10 and 11.

On the basis of their modelling, Thompson and England (1984) concluded that the association most diagnostic of continental collision metamorphism is that of the Kyanite–sillimanite facies series. This is exactly the situation found in the Scandian nappes from Tysfjord to Tromsø, North Norway. The eclogite facies metamorphism recorded in parts of the Tromsø Nappe Complex, and in the Seve Nappe Complex farther south, is an unrelated pre-Scandian event, now overprinted by amphibolite facies Scandian metamorphism.

By examination of included phases in porphyroblasts (especially garnet), coupled with geothermobarometry it has been possible to gain more detailed insight into the P–T–t paths that rocks of the Caledonian nappes of north-central Scandinavia have experienced. Certain rocks from the Høgtind Nappe contain garnets with abundant chloritoid inclusions (Fig. 3), whilst staurolite and kyanite occur throughout the matrix. This gives convincing petrographic evidence that staurolite formed at the expense of chloritoid by reactions such as chloritoid + muscovite + quartz $\rightleftharpoons$ staurolie + biotite + H_2O (Tilley, 1925) or chloritoid + chlorite $\rightleftharpoons$ staurolite + biotite + H_2O (Albee, 1972). Since a few staurolite grains are included at the garnet margins, it shows that staurolite development commenced before garnet growth was com-

Table II. *P–T* estimates for various Caledonian Nappes in the Tysfjord–Tromsø region (all are from the Upper Allochthon, except Tromsø Nappe Complex, which is Uppermost Allochthon).

Senja N.	[a]$T = 520°C$	Bergh and
Dyroy N.	[a]$PT = 530°C/7\,kb$ (core) to $660°C/10\,kb$ (rim)	Andresen (1985)
Lyngen N.	[a]$T = 440–445°C$	
Marko N.	From $PT = 528°C/6.6\,kb$ (mean $= 567° \pm 32°C/$	Crowley and
	to PT $620°C/8.8\,kb$ $8.0 \pm 0.9\,kb$)	Spear (1987)
Langvatn N.	$T = 500–550°C$ (max. $570°C$)	
	$P = 7.5–9.2\,kb$	
Niingen Gp.	Cooled from $600°C/8.5\,kb$	Steltenpohl and
+	to $450°C/7.0\,kb$	Bartley (1987)
Salangen Gp.		
Niingen N.	$T = 520–600°C$ $P = 5.3–6.3\,kb$	Barker and
[d]Gratangen N.	$T = 470–560°C$ $P = 6.9–8.2\,kb$	Anderson (1988)
[d]Småtinden N.	$T = 430–500°C$ –	
[b]Kvernmo N.	$T = 610–680°C$ $P = 6.2–7.3\,kb$	
[b]Grønfjellet N.	$T = 530–620°C$ $P = 8.2–8.9\,kb$	
[c]Høgtind N.	$T = 480–580°C$ $P = 7.7–9.4\,kb$	
Tromsø	Early event $T = 700°C$ $P = 17\,kb$	Krogh *et al.*
N.C.	Later event $T = 660–700°C$ $P \simeq 9\,kb$	(unpub. data referred to by Andresen *et al.* 1985)

[a] Uncorrected Ferry and Spear calculations, therefore *T* estimate is probably low.
[b] Comparable to part of Narvik Group of Gustavson (1966).
[c] Comparable to part of Rombak Group of Gustavson (1966).
[d] Comparable to part of Salangen Group of Gustavson (1966).

plete. The chloritoid to staurolite transformation is a prograde reaction, which at moderate pressures (5–8 kb) occurs at 520–540°C (Fig. 10). This indicates that in the examples considered the garnet core developed at temperatures below this value, whereas the rim formed at higher temperatures (Fig. 10). On the basis of the curves of Richardson (1968) and Salje (1986) (Fig. 10), the presence of staurolite + kyanite shows that peak metamorphic conditions in excess of 4.5 kb/530°C were experienced by the upper part of the Høgtind Nappe.

In zoned garnets it might be expected that core-to-rim profiles record the prograde history. Alternatively, maximum *T* might be reached at some time midway through growth such that the final increments of growth occurred during the cooling stages. However, the fact that high-grade assemblages are generally preserved in rocks suggests that reaction rates during decompression are usually slow. Consequently, rims will normally represent conditions at or soon after peak temperatures (Spear *et al*, 1984). Exceptions to this rule occur where retrogression has been intense and the garnet margins have been resorbed. In such cases "rim-matrix" garnet–biotite pairs yield lower *T* compared to "core-inclusion" garnet–biotite pairs. Rather than representing the prograde history, such cases record part of the uplift path. A retrogressed kyanite gneiss from the Niingen Nappe (Fig. 10) illustrates such a path.

The examples described demonstrate that knowledge of both core and rim conditions of zoned garnets is essential for accurate interpretation of the *P–T–t* history. Bearing in mind the modelling of England and Thompson (1984), and *P–T* information obtained from rocks of the region, the types of *P–T–t* loop followed by nappes of north central Scandinavia during the Scandian Orogeny are schematically shown in Fig. 11. Paths C and B represent those most likely for the Lower and Middle Allochthon

respectively. A′, A″ and A‴ represents *P–T–t* paths for rocks of the Upper Allochthon, with A′ being for the most deeply buried rocks.

CONCLUSIONS

With the exception of the Seve Nappe and Tromsø Nappe Complex, which locally record a pre-Scandian eclogite facies metamorphism, evidence to date suggests only a Scandian (mid to late Silurian) orogenic history for other nappes of the region.

A common, though not necessarily coeval, tectono-metamorphic sequence is recorded in these Scandian nappes. D1 represented by S1 inclusion fabrics in garnet is succeeded by intense D2 ductile shearing with an associated pervasive regional schistosity axial planar to F2 folds. Peak metamorphism and porphyroblastesis largely occurred early in D2 and as major F2 structures developed, some regional metamorphic isograd became inverted. It was not until D3 that discrete thrust surfaces developed and the major nappes of the region were emplaced. F2 folds and regional metamorphic isograds were truncated at this stage and higher-grade nappes were emplaced southeastwards onto lower-grade units. A period of greenschist facies retrograde metamorphism occurred post-thrusting, and is concentrated in the vicinity of thrust zones.

Peak metamorphic grade attained during Scandian metamorphism varies from sub-greenschist facies in the imbricated Lower Allochthon of Sweden, to upper amphibolite facies in the highest allochthonous units of Troms. Quantitative estimates of peak metamorphic conditions in nappes of the Upper and Uppermost Allochthon indicate that peak temperatures of the order of 530–650°C were attained in association with pressures of 6–9 kb. Modelling

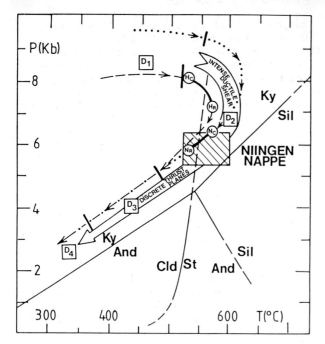

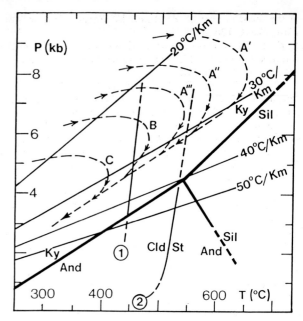

Fig. 10. Projected *P–T* loops for the Niingen Nappe and upper part of the Høgtind Nappe based on matrix assemblages, included mineral phases in garnet, garnet–biolite geothermometry (core and rim), and geobarometry (rim/matrix conditions only), using data from Barker and Anderson (1988) and unpublished results. Largely based on samples 27/80 (Høgtind Nappe) and 74/85 (Niingen Nappe), which have favourable included mineralogies.

Hc, Hr = Høgtind Nappe garnet core (c) and rim (r) respectively.

Nc, Nr = Niingen Nappe garnet core (c) and rim (r) respectively.

––– = projected *P–T* loop for Høgtind Nappe.

··· = projected *P–T* loop for Niingen Nappe.

D1, D2, D3 and D4 represent successive deformation events. These are considered part of a progressive sequence, though bars on the *P–T* loop show approximately where one event ends and the next starts. Kyanite (Ky)–andalusite (And)–sillimanite (Sill) stability fields after Salje (1986); chloritoid (Cld)⇌ staurolite (St) reaction curve of Richardson (1968)

Fig. 11. Projected *P–T* loops for various nappes of the Caledonides of northcentral Scandinavia. Based on petrographic/mineralogic observations, and *P–T* estimates of Table II.

A′, A″ and A‴ = projected *P–T* path for nappes of the Upper Allochthon. (A′ representing nappes attaining highest *P–T* conditions and A‴ the lowest *P–T* conditions).

B = projected *P–T* path for nappes of Middle Allochthon (excluding Seve Nappe).

C = projected *P–T* path for nappes of Lower Allochthon.

Ky-And-Sill stability fields; after Salje (1986). (1) Stilpnomelane + phengite ⇌ biotite + chlorite + epidote + H_2O (Nitsch, 1970). (2) Chloritoid ⇌ staurolite reaction curve (Richardson, 1968).

by England and Thompson (1984) demonstrates that these pressure values are likely to be minimum estimates, since maximum burial is attained before peak temperatures have been experienced.

Acknowledgements

Particular thanks are extended to Rod Gayer, Mark Anderson, Richard Binns, Magne Gustavson and Arild Andresen, who by way of discussions have contributed much to my understanding of the Troms region. Special thanks also go to Maurits Lindstrom, Ebbe Zachrisson, David Gee, Gerhard Bax, Mats Johnsson, Benno Kathol, Laurence Page and Dan Zetterberg, who have given me an excellent insight into the geology of the Torneträsk-Kebnekaise area of Sweden during my recent fieldwork. Additional thanks to Roger Mason and Hugh Rice for their critical comments on an early draft of this paper.

REFERENCES

Ahlberg, P. (1980). Early Cambrian trilobites from northern Scandinavia. *Norsk Geol. Tidsskr.* **60**, 153–159.

Ahlberg, P. (1985). Lower Cambrian trilobites faunas from the Scandinavian Caledonides–a review. In: Gee, D. G. and Sturt, B. A. (Eds.) *The Caledonide Orogen—Scandinavia and Related Areas*, pp. 339–346. Wiley, Chichester.

Albee, A. L. (1972). Metamorphism of pelitic schists: reaction relations of chloritoid and staurolite. *Bull. Geol. Soc. Amer.* **83**, 3249–3268.

Anderson, M. W. (1988). Basement evolution during Caledonian Orogenesis, Troms, N. Norway, *this volume.*

Andréasson, P.-G., Gee, D. G. and Sukotji, S. (1985). Seve eclogites in the Norrbotten Caledonides, Sweden. In: Gee, D. G. and Sturt, B. A. (eds.) *The Caledonide Orogen—Scandinavia and Related Areas*, pp. 887–901. Wiley, Chichester.

Andresen, A., Fareth, E., Bergh, S., Kristensen, S. E. and Krogh, E. (1985). Review of Caledonian lithotectonic units in Troms, north Norway. In: Gee, D. G. and Sturt, B. A. (eds.) *The Caledonide Orogen—Scandinavia and Related Areas*, pp. 569–578. Wiley, New York.

Barker, A. J. (1986). The geology between Salangsdalen and Gratangenfjord, Troms, Norway. *Norges Geol. Unders. Bull.* **405**, 41–56.

Barker, A. J. and Anderson, M. W. (1988). The Caledonian structural-metamorphic evolution of the Troms region, Norway. In: Daly, J. S., Cliff, R. and Yardley, B. W. D. (eds.). *The Evolution of Metamorphic Belts. Geol. Soc. Lond. Spec. Publ. No.*

Bartley, J. M. (1981). Lithostratigraphy of the Storvann Group, east Hinnøy, north Norway, and its regional implications. *Norges Geol. Unders.* **370**, 11–24.

Bax, G. (1986). Basement involved Caledonian nappe tectonics in the Swedish part of the Rombak-Sjangeli Window. *Geol. Fören. Stockh.* **108**, 3.

Bell, T. H. (1985). Deformation partitioning and porphyroblast rotation in metamorphic rocks: a radical reinterpretation. *J. Metam. Geol.* **3**, 109–118.

Bell, T. H., Fleming, P. D. and Rubenach, M. J. (1986). Porphyroblast nucleation, growth and dissolution in regional metamorphic rocks as a function of deformation partitioning during foliation development. *J. Metam. Geol.* **4**, 37–67.

Bergh, S. G. and Andresen, A. (1985). Tectonometamorphic evolution of the allochthonous Caledonian rocks between Malangen and Balsfjord, Troms, north Norway. *Norges Geol. Unders. Bull.* **401**, 1–34.

Billett, M. F. (1987). The geology of the northern Sulitjelma area and its relationship to the Sulitjelma Ophiolite. *Norsk Geol. Tiddskr.* **67**, 71–83.

Binns, R. E. (1978). Caledonian nappe correlations and orogenic history in Scandinavia north of lat. 67°N. *Bull. Geol. Soc. Amer.* **89**, 1475–1490.

Binns, R. E. and Gayer, R. A. (1980). Silurian or Upper Ordovician fossils at Guolasjav'ri, Troms, Norway. *Nature* **284**, 53–55.

Binns, R. E. and Matthews, D. W. (1981). Stratigraphy and Structure of the Ordovician–Silurian Balsfjord Supergroup, Troms, north Norway. *Norges Geol. Unders.* **365**, 39–54.

Björklund, L. (1985). The Middle and Lower Allochthons in the Akkajaure–Tysfjord area, northern Scandinavian Caledonides. In: Gee, D. G. and Sturt, B. A. (eds.) *The Caledonide Orogen—Scandinavia and Related Areas*, pp. 515–528. Wiley, Chichester.

Bjørlykke, A. and Olaussen, S. T. (1981). Silurian sediments, volcanics and mineral deposits in the Sagelvvatn area, Troms, North Norway. *Norges Geol. Unders.* **365**, 1–38.

Boyle, A. P. (1987). A model for stratigraphic and metamorphic inversions at Sulitjelma, central Scandes. *Geol. Mag.* **124**, 451–466.

Boyle, A. P., Griffiths, A. J. and Mason, R. (1979). Stratigraphical inversion in the Sulitjelma area, Central Scandinavian Caledonides. *Geol. Mag.* **116**, 393–402.

Boyle, A. P., Hansen, T. S. and Mason, R. (1985). A new tectonic perspective of the Sulitjelma region. In: Gee, D. G. and Sturt, B. A. (eds.) *The Caledonide Orogen—Scandinavia and Related Areas*, pp. 529–542. Wiley, Chichester.

Crowley, P. D. and Spear, F. S. (1987). The *P–T* evolution of the Middle Køli Nappe Complex, Scandinavian Caledonides (68°N) and its tectonic implications. *Contrib. Mineral. Petrol.* **95**, 512–522.

Dallmeyer, R. D. and Gee, D. G. (1986). $^{40}Ar/^{39}Ar$ mineral dates from retrogressed eclogites within the Baltoscandian miogeocline: Implications for a polyphase Caledonian orogenic evolution. *Bull. Geol. Soc. Amer.* **97**, 26–34.

Dallmeyer, R. D., Clark, A., Cumbert, J., Hames, W. E. and McKinney, J. (1987) Polyphase Caledonian tectonothermal evolution of the Western Gneiss Terrane, Senja, Troms, Norway. *The Caledonian and Related Geology of Scandinavia.* Abstracts, p. 12. University College, Cardiff.

England, P. C. and Thompson, A. B. (1984). Pressure-temperature-time paths of regional metamorphism I. Heat transfer during the evolution of thickened continental crust. *J. Petrol.* **25**, 894–928.

Fareth, E. (1981). Tromsø 1:250,000 Berggrunskart, foreløpig utgave. Norges Geologiske Undersøkelse.

Ferry, J. M. and Spear, F. S. (1978). Experimental calibration of the partitioning of Fe and Mg between biotite and garnet. *Contrib. Mineral. Petrol.* **66**, 113–117.

Føyn, S. and Glaessner, M. F. (1979). Platysolenites, other animal fossils, and Precambrian–Cambrian transition in Norway. *Norsk Geol. Tidsskr.* **59**, 25–46.

Gee, D. G. (1975). A tectonic model for the central part of the Scandinavian Caledonides. *Amer. J. Sci.*, **275A**, 468–515.

Ghent, E. D. and Stout, M. Z. (1981). Geobarometry and geothermometry of plagioclase–biotite–garnet–muscovite assemblages. *Contrib. Mineral. Petrol.* **76**, 92–97.

Ghent, E. D., Robbins, D. B. and Stout M. Z. (1979). Geothermometry, geobarometry, and fluid compositions of meta-

morphosed calc-silicates and pelites, Mica Creek, British Columbia. *Amer. Mineral.*, **64**, 874–885.

Gustavson, M. (1966). The Caledonian mountain chain of the Southern Troms and Ofoten areas; Part I: Basement rocks and Caledonian metasediments. *Norges Geol. Unders.* **239**, 162p.

Gustavson, M. (1969). The Caledonian mountain chain of the Southern Troms and Ofoten areas; Part II: Caledonian rocks of igneous origin. *Norges Geol. Unders.* **261**, 162p.

Gustavson, M. (1974a). Beskrivelse til det berggrunnsgeologiske gradteigskart N9-1:100,000. *Norges Geol. Unders.* **308**, 34p.

Gustavson, M. (1974b). Geologiske kart over Norge, 1:250,000, Narvik. Norges Geologiske Undersøkelse.

Hodges, K. V. (1985). Tectonic stratigraphy and structural evolution of the Efjord-Sitasjaure area, northern Scandinavian Caledonides. *Norges Geol. Unders.* **399**, 41–60.

Hodges, K. V. and Roydon, R. L. (1984). Geologic thermobarometry of retrograded metamorphic rocks: an indication of the uplift trajectory of a portion of the northern Norwegian Caledonides. *J. Geophys. Res.* **89**, 7077–7090.

Krogh, E. J., Andresen, A., Bryhni, I. and Kristensen, S. E. (1982). Tectonic setting, age and petrography of eclogites within the uppermost tectonic unit of the Scandinavian Caledonides, Tromsø area, northern Norway. *Terra cognita* **2**, 316.

Kulling, O. (1960). On the Caledonides of Swedish Lapland. In: Kulling, O. and Geijer, P. (eds.). Guide to Excursions Nos A25 and C20, International Geological Congress, 21 Session Norden 1960, pp. 18–39, Sveriges Geologiska Undersökning, Stockholm.

Kulling, O. (1964). Oversikt over Norra Norrbottenfjallens Kaledonberggrund. *Sveri. Geol. Unders.* **Ba 19**.

Kulling, O. (1972). In: Strand, T. and Kulling, O. (eds.) *Scandinavian Caledonides*. Wiley, Chichester.

Lindstrom, M., Bax, G., Dinger, M., Dworatzek, M., Editmann, W., Fricke, A., Kathol, B., Klinge, H., von Pape, P. and Stumpf, U. (1985). Geology of a part of the Torneträsk of the Caledonian front, northern Sweden. In: Gee, D. G. and Sturt, B. A. (eds.) *The Caledonide Orogen—Scandinavia and Related Areas*, pp. 509–513. Wiley, Chichester.

Mason, R. (1984). Inverted isograds at Sulitjelma, Norway: the result of shear-zone deformation. *J. Metam. Geol.* **2**, 77–82.

Nitsch, K-H. (1970). Experimentelle bestimmung der Oberen Stabilitatsgrenze von stilpnamelan (Experimental determination of the upper stability limit of stilpnamelane). *Abstr. Fortschr. Mineral.* **47**, 48–49.

Olaussen, S. (1976). Palaeozoic fossils from Troms, Norway. *Norsk Geol. Tidsskr.* **56**, 457–459.

Olesen, N. Ø. (1971). The relative chronology of fold phases, metamorphism, and thrust movements in the Caledonides of Troms, North Norway. *Norsk Geol. Tidsskr.* **51**, 355–377.

Oxburgh, E. R. and Turcotte, D. L. (1974). Thermal gradients and regional metamorphism in overthrust terrains with special reference to the eastern Alps. *Schweiz. Mineral. Petrogr. Mitt.* **54**, 641–662.

Richardson, S. W. (1968). Staurolite stability in a part of the system Fe-Al-Si-O-H. *J. Petrol.* **9**, 468–488.

Salje, E. (1986). Heat capacities and entropies of andalusite and sillimanite: The influence of fibrolitization on the phase diagram of Al2Si05 polymorphs. *Amer. Mineral.* **71**, 1366–1371.

Sleep, N. H. (1979). A thermal constraint on the duration of folding with reference to Acadian geology, New England. *J. Geol.* **87**, 583–589.

Spear, F. S., Selverstone, J., Hickmott, D., Crowley, P. and Hodges, K. V. (1984). *P–T* paths from garnet zoning: A new technique for deciphering tectonic processes in crystalline terranes. *Geology* **12**, 87–90.

Steltenpohl, M. G. and Bartley, J. M. (1984). Kyanite grade metamorphism in the Evenes and Bogen Groups, Ofoten, North Norway. *Norsk Geol. Tidsskr.* **64**, 21–26.

Steltenpohl, M. G. and Bartley, J. M. (1987). Thermobarometric profile through the Caledonian nappe stack of Western Ofoten, North Norway. *Contrib. Mineral. Petrol.* **96**, 93–103.

Strömberg, A. (1955). Zum gebirgsbau de skanden im mittleren Harjedalen. *Bull. Geol. Inst. Univ. Uppsala*, **35**, 199–243.

Strömberg, A. (1961). On the tectonics of the Caledonides in the Southwestern part of the County of Jämtland, Sweden. *Bull. Geol. Inst. Univ. Uppsala* **39**, 92pp.

Sturt, B. A., Pringle, I. R. and Roberts, D. M. (1978). The Finnmarkian phase of the Caledonian orogeny. *J. Geol. Soc. Lond.* **135**, 597–610.

Thompson, A. B. and England, P. C. (1984). Pressure–temperature–time paths of regional metamorphism II. Their inference and interpretation using mineral assemblages in metamorphic rocks. *J. Petrol.* **25**, 929–955.

Tilley, C. E. (1925). A preliminary survey of metamorphic zones in the southern Highlands of Scotland. *Q. J. Geol. Soc. Lond.* **81**, 100–112.

Williams, I. S. and Claesson, S. (1987). Isotopic evidence for the Precambrian provenance and Caledonian metamorphism of high grade paragneisses from the Seve Nappes, Scandinavian Caledonides II. Ion microprobe zircon U–Th–Pb. *Contrib. Mineral. Petrol.* **97**, 205–217.

17 High-pressure ultramafic rocks from the Allochthonous Nappes of the Swedish Caledonides

H. L. M. van Roermund,

State University of Utrecht, Institute of Earth Sciences,
P.O. Box 80.021, 3508 TA, Utrecht, The Netherlands

Some new occurrences of high-pressure ultramafic rocks are reported from high-pressure low-temperature thrust sheets in the lower part of the Upper Allochthon (Seve Nappe Complex), central-north Scandinavian Caledonides. Two distinct types have been recognized.

(1) Garnet peridotites (and interlayered metabasic rocks) in Jämtland, Sweden, reveal a retrograde 5-stage tectonometamorphic evolutionary history. A coarse-grained 4-phase mineral assemblage I (grt 1+Ol1+cpx 1+opx 1) transforms by strain-induced recrystallization into a 6-phase mineral assemblage 2 (grt 2+ol2+cpx2+opx2+spl+amph 1), while a porphyroclastic or equigranular tabular microstructure is formed. Stage 3 develops during subsequent decompression when two types of kelyphites around garnet are formed, e.g. an intimate intergrowth of spinel and pyroxenes is overgrown by coarser-grained spinel–orthopyroxene–amphibole assemblages. In metabasic rocks, omphacite is replaced by cpx–plag symplectites. Later-stage metamorphic readjustments result in amphibolite (4) and greenschist (5) facies mineral assemblages.

(2) High-pressure ultramafic rocks in Norrbotten, Sweden, are characterized by prograde metamorphic mineral assemblages overgrowing inferred magmatic minerals. In peridotites the grt–ol assemblage is absent, but intercalated metabasic rocks contain the high-pressure assemblage garnet-2 pyroxenes. Subsequent reworking during amphibolite and greenschist facies resembles that of type I.

General aspects of the origin and tectonic setting of both ultramafic rock types are discussed in the context of the Caledonian orogeny. Type 1 is interpreted as representing Baltoscandian subcontinental upper mantle material brought into the crust by deep-level ductile continental imbrication of the Baltoscandian plate during the Caledonian eclogite-forming high-pressure metamorphic event. Type 2 is interpreted as representing lower-pressure protoliths, possibly of oceanic affinity, that have undergone prograde high-pressure metamorphism during the Caledonian orogeny.

INTRODUCTION

High-pressure ultramafic rocks and associated country rock ecologites have considerable significance for the information they may provide on the nature of crust–mantle interaction during orogenesis (Carswell and Gibb, 1980; Medaris, 1984). Their usefulness has been particularly enhanced by recent work on the stability field of garnet peridotites (O'Neill, 1981) and by studies of P-T-dependent solid solutions among garnet, pyroxene, olivine and spinel (e.g., Mori and Green, 1978; Ellis and Green, 1979; O'Neill and Wood, 1979; Harley, 1984a,b). High-pressure ultramafic rocks in the Scandinavian Caledonides are thus far only described from the "autochthonous" basement gneisses along the west coast of Norway (western Gneiss Region). The purpose of the present contribution is to document some newly discovered occurrences of high-pressure ultramafic rocks in

the allochthonous units of the Seve Nappes, Jämtland and Norrbotten, Sweden, to assess their P-T histories and to discuss the results in terms of a petrogenetic model.

Ultramafic rocks in the Scandinavian Caledonides

The Scandinavian Caledonides are characterized by a sequence of thin, large-scale nappes (Törnebohm, 1896; Asklund, 1938; Gee, 1975) that have been thrust from west to east onto the autochthonous rocks of the Baltoscandian platform during the Caledonian orogeny. The allochthonous units can be described in terms of four major tectonic units, i.e. Lower, Middle, Upper and Uppermost Allochthons (Strand and Kulling, 1972; Gee and Zachrisson, 1979; Roberts and Gee, 1985). Ultramafic rocks occur at two distinct structural levels: in the (autochthonous?) basement gneisses of the Baltoscandian platform (western

Gneiss Region (WGR)) and in the metamorphosed supra-crustal rocks and associated basement slivers of the (Middle?) Upper and Uppermost Allochthon. Areal distribution maps of ultramafic rocks in the Scandinavian Caledonides are given by Qvale and Stigh (1985), who also subdivided the ultramafic rocks in the Scandinavian Caledonides into five distinct groups. Among these five groups, Alpine-type peridotites, including ophiolitic members and solitary ultramafic bodies, are by far the most widespread category. Among this latter group, the solitary ultramafic bodies have been subdivided into the following two groups.

Group 1 contains prograde metamorphic mineral parageneses, superimposed locally on primary magmatic textures. Magmatic textures are exclusively found in rocks of Cambrian–Silurian age; metamorphic textures are found in medium- to high-grade rocks of varying age, in both the autochthonous basement gneisses and in the far-travelled nappes of the (Middle?) Upper and Uppermost Allochthons.

Group 2 contains retrograde metamorphic mineral parageneses superimposed on a primary high-grade metamorphic mineral assemblage: olivine + orthopyroxene ± garnet ± clinopyroxene ± amphibole. They are typically associated with country rock eclogites (and locally anorthosites). Thus far, group 2 has been restricted in its spatial distribution to the high-grade "basement" gneisses along the west coast of Norway (WGR). This paper extends its distribution to include the high-grade rocks of the Seve Nappes (Upper Allochthon) in Jämtland, Sweden.

The metamorphic grade indicated by the metamorphic mineral assemblages of group 1 Alpine-type peridotites is consistent with the metamorphic grade of the country

rocks and generally no problems exists concerning their petrogenetic origin. This is in marked contrast with that of group 2, for which the following two alternative petrogenetic models have been proposed.

(i) the garnet peridotites formed by the prograde metamorphism of lower-pressure protoliths (serpentinites and/or spinel/plagioclase lherzolites) (Griffin and Qvale, 1985; Griffin et al., 1985);

(ii) the garnet-bearing mineral assemblage is directly inherited from the subcontinental mantle during orogenesis (Medaris, 1984; Carswell, 1986).

This subdivision of Alpine-type peridotites into "ophiolite- and root-zone type" peridotites dates back to Den Tex (1969).

The present study areas, in the Swedish counties of Jämtland (key area I) and Norrbotten (key area II) can be found by superposition of the metamorphic map of the Scandinavian Caledonides (Bryhni and Andreasson, 1985) onto the areal distribution map of Alpine-type peridotites of Qvale and Stigh (1985). Geologically, both key areas belong to the Seve Nappe Complex (Upper Allochthon). Geographical coordinates of both key areas are given below.

REGIONAL GEOLOGY

The Seve Nappe Complex, extending for ca. 800 km along the strike of the mountain belt, is composed of medium- and high-grade metapelites and metapsammites enclosing subordinate marbles, together with extensive metabasites and occasional solitary ultramafites. Biostratigraphic evidence is lacking, but geochronological data suggests that at least some of the gneisses and metabasites are derived from Middle Proterozoic protoliths (Reymer et al.,

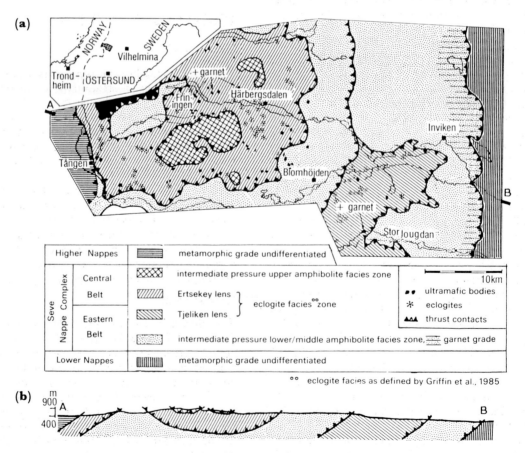

Fig. 1. Metamorphic facies map (Fig. 1(a)) and related structural profile (Fig. 1(b)) through the Tången-Inviken area, Jämtland (key area 1).

Table I. Vertical hypothetical cross-section through the Seve Nappe Complex (parallel to the strike of the orogen) showing the tectonostratigraphic correlation between both key areas in Jämtland and Norrbotten. Lens terminology as in text.

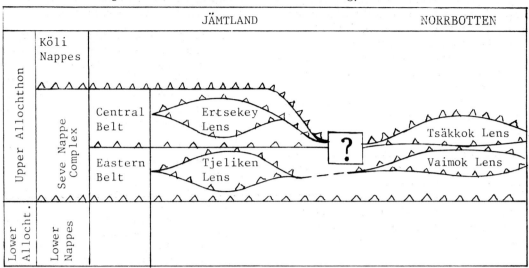

1980; Claesson, 1982, 1987; Mearns and Van Roermund, 1985). In both key areas, the Seve Nappe Complex consists of a pile of alternating thrust sheets comprising units that have undergone either high-pressure/low-(high) temperature (HP/LT) or the classical intermediate-pressure type of metamorphism. This is illustrated by the accompanying metamorphic facies map (Fig. 1(a)) and related structural profile (Fig. 1(b)) of key area I. For a more detailed discussion of the tectonostratigraphy, structure and metamorphism of both key areas, the reader is referred to Zachrisson (1969), Zwart (1974), Williams and Zwart (1977), Van Roermund and Bakker (1984), Stephens and Van Roermund (1984) and Andréasson *et al.* (1985). For the nomenclature and the relative tectonostratigraphic correlation of the HP/LT tectonic lenses, the reader is referred to Table I, and for references to Van Roermund (1982, 1985), Van Roermund and Bakker (1984), Stephens and Van Roermund (1984) and Andréasson *et al.* (1985).

In the HP/LT thrust sheets in both key areas, metabasic rocks of appropriate composition have been transformed into eclogites. P–T estimates of eclogite formation are ca. 780°C/18 kb (Ertsekey lens), ca. 575°C/15 kb (Tjeliken lens), 550–700°C/12–15 kb (Vaimok lens) and 500–600°C/12–15 kb (Tsåkkok lens).

Isolated lenticular bodies of ultramafic rocks occur in most units of the Seve Nappe Complex (Calon, 1979; Stigh, 1980). Their spatial distribution in key area I is indicated in Fig. 1(a). In key area I ultramafic rocks with prograde metamorphic mineral parageneses (group 1) are restricted to those tectonic units that are characterized by an intermediate-pressure metamorphic facies series, whilst ultramafic rocks with retrograde mineral assemblages (group 2) superimposed on the high-grade metamorphic mineral assemblage olivine + orthopyroxene ± garnet ± clinopyroxere ± amphibole ± spinel are confined to the HP/LT thrust-sheets. In terms of bulk rock chemistry (following Jackson and Thayer, 1972), group 2 Alpine-type peridotites are dunitic, harzburgitic and lherzolitic in composition. This paper is restricted to the (aluminous) lherzolitic subtype, but examples of the dunitic/harzburgitic subtype are found, among others at Ruotats (Du Rietz, 1935). Among the (aluminous) lherzolites there are distinct differences in the relative bulk chemistry, e.g. Ertsekey lens occurrences are closer to

(aluminous) dunites/harzburgites than Tjeliken lens group 2 peridoties.

In the Ertsekey lens, the garnet peridotites studied occur around Lake Friningen (Fig. 1; 60°70′ N; 14°15′ E; see also Du Rietz, 1935; plates 1 and 2), whilst in the Tjeliken lens occurrences are found along the northern shore of Lake Stor Jougdan (Fig. 1; 64°60′ N; 14°70′, E). Less detailed information is so far available from key area II, but high-pressure ultramafic rocks are present in the Vaimok lens at the southwest side of Lake Sartaure (66°55′N; 17°E).

FIELD OBSERVATIONS

Key area I

The three garnet-bearing peridotites in the Ertsekey lens occur close to the structural base of an eclogite-bearing migmatitic ky–sil–K feldspar gneiss (ky→sil→ky (Van Roermund and Bakker, 1984)). The direct contact with the enclosing country rocks is not exposed. The largest body, with dimensions of 30 × 300 m and situated directly east of Lake Friningen, has been studied in some more detail (Fig. 1).

Outcrops of garnet-bearing peridotites in the Tjeliken lens are very small, only a few metres across, but occur over a wide area, indicating the existence of an ultramafic body extending several hundred metres along-strike (Fig. 1). The immediate country rock is not exposed, but from regional field mapping it is interpreted to be the eclogite-bearing quartzofeldspathic gneiss (Fig. 1).

Key area II

The ultramafic rocks in the Tsåkkok lens are all of the dunitic/harzburgitic subtype and are therefore not discussed further in this paper. Some ultramafic rocks of the Vaimok lens are lherzolitic in composition but are not garnet peridotites. Garnet–pyroxene mineral assemblages are, however, widespread in the intercalated metabasic layers and in this way testify to the high-pressure metamorphic nature of the rocks. The country rocks, although not directly exposed, are unmigmatized eclogite-bearing garnet–kyanite gneisses.

Owing to variations in the relative amounts of the major

Fig. 2. (*a*) Compositional layering (S$_1$) in garnet peridotite from the Ertsekey lens. (*b*) Porphyroclastic S$_2$ (∥ to S$_1$) foliation in garnet peridotite from the Ertsekey lens. (*c*) S$_3$ shear zones cross-cutting the S$_2$ porphyroclastic microstructure in garnet periodotite from the Ertsekey lens. (*d*) Late-stage talc vein. Vaimok lens.

rock-forming minerals, all high-pressure ultramafic bodies contain a conspicuous compositional layering (S$_1$) (Fig. 2($a + b$)). In addition, up to 30 cm thick layers of garnet pyroxenites, websterites or garnet amphibolites occur (Fig. 2(a)), these are especially common in the Vaimok lens occurrences (up to 50 vol.%). A prominent feature of the garnet-bearing ultramafic rocks of the Ertsekey lens is a porphyroclastic microstructure. The elongated coarse-grained porphyroclasts of olivine I, clinopyroxene I, orthopyroxene I and/or garnet I (now replaced by ortho-pyroxene–spinel symplectites) define a foliation (S$_2$; Figs. 2(b) and 3(a)) and are mantled by an optically strain-free tabular to equidimensional microstructure of olivine II, opx II, cpx II, amph I, spinel I, grt II (Figs. 3(a) and 6(a)). This typical porphyroclastic microstructure is lacking (except some thick garnet pyroxenite layers) in garnet-bearing peridotites of the Tjeliken lens. This is interpreted to be due to complete strain-induced recrystal-lization. Here the microstructure of the olivine-bearing assemblage can best be described as equigranular tabular (Nicolas and Poirier, 1976, plat 24; Fig. 3(c)) with a few distinct lenticular domains of garnet, spinel, amphi-bole, opx. (Fig. 3(d)). Cigar-shaped garnets II, with aspect ratios 1 : 5, form the cores of these lenticular domains with amphibole II and/or orthopyroxene III, both with

dispersed spinel inclusions, as a kelyphitic rim around it (Fig. 3(d)).

In ultramafic rocks of the Vaimok lens, the porphyro-clastic microstructure is completely absent. Here the S$_2$ foliation consists of a fine- to medium-grained equi-granular tabular microstructure of ol, opx, amph, spinel. Intercalated metabasic layers are coarser grained and contain well-developed porphyroblasts of garnet, ortho-pyroxene, clinopyroxene and/or amphibole overgrowing lower-grade foliations (Fig. 4(a)).

At all occurrences, S$_1$ has been folded into asymmetrical tight to isoclinal folds with a prominent axial plane foliation (S$_2$). In general, S$_2$ is parallel to S$_1$ but locally small but consistent deviations of up to 20^0 occur. S$_2$ is the most conspicuous structure in all bodies; its original mineral assemblage can, however, be replaced mimetically by later-stage mineral parageneses.

In addition, all the high-pressure ultramafic rocks contain irregular anastomozing shear zones (S$_3$) that transect both earlier foliations (Fig. 2(c)) and are particul-arly well-developed along the margins of the ultramafic bodies. S$_3$ foliation planes are connected with lower-grade mineral assemblages (low/middle amphibolite facies). Subvertical fractures, accentuated by coarse-grained asbestos layers, testify to a third deformation event within

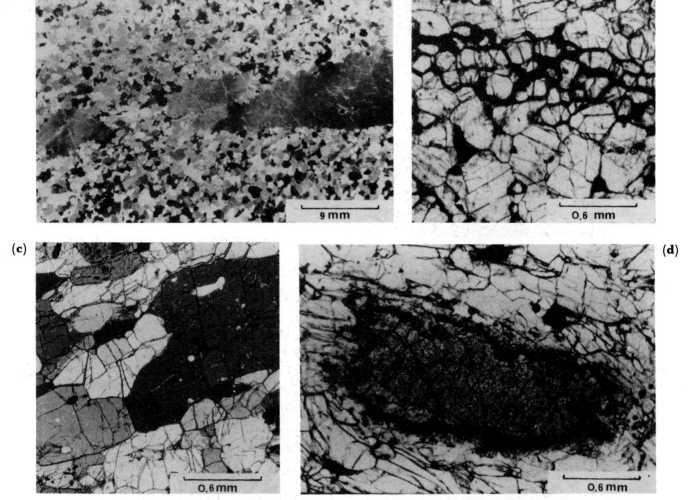

Fig. 3. (*a*) Porphyroclastic microstructure in dunite from the Ertsekey lens. (*b*) Recrystallized (M₂) garnets in garnet websterite from the Ertsekey lens. (*c*) Equigranular, tabular microstructure in Tjeliken lens garnet peridotite. (*d*) Lenticular domains of garnet surrounded by a shell of orthopyroxenite, spinel and amphibole in garnet peridotite from the Tjeliken lens.

the brittle regime (Fig. 2(*d*)). Table II summarizes the main structural elements in the ultramafic bodies.

PETROGRAPHIC OBSERVATIONS

Field observations and optical microscopy indicate significant mineralogical and textural variations that reflect the complex tectonometamorphic evolution of the high-pressure ultramafic and intercalated metabasic rocks. Five important stages of metamorphic mineral growth have been recognized and are subsequently referred to here as M_1, M_2, M_3, M_4 and M_5. M_1 is related to S_1; it is only present in Jämtland. Intense penetrative deformation (S_2) marks the onset of M_2. M_3 post-dates a period of decompression as is exemplified by the development of amphibole/orthopyroxene–spinel symplectites secondary after garnet in olivine-bearing rocks. M_4 is related to the localized development of S_3 shear-zones. Furthermore, M_4 can be discriminated from M_3 on the basis of the observed mineral paragenesis (upper amphibolite versus lower/middle amphibolite facies respectively), indicating a regime of declining *P–T* conditions. M_5 is related to local serpentinization of the bodies as well as to subvertical fractures (S_4) developed during greenschist facies, as is evidenced by their vein assemblages. A paragenetic diagram of the high-pressure ultramafic and metabasic rocks is given in Table III.

MINERAL CHEMISTRY

Electron microprobe (EMP) analyses have been obtained with an ARL–EMX electron microprobe fitted with a Link energy-dispersive system, correction program ZAF–4 (Statham, 1976), at the Mineralogisk-Geologisk Museum at Oslo (Norway). Selected EMP analyses of relict-garnet lherzolite (Ertsekey lens), garnet pyroxenites (Ertsekey lens), garnet websterites (Ertsekey lens) and garnet peridotites (Tjeliken lens) are illustrated in Table IV. Selected EMP analyses from comparable garnet pyroxene layers of all three ultramafic rock types are presented in Table V. In the following, only the most important findings will be summarized.

Garnet

In Jämtland, garnets in garnet pyroxenites (Ertsekey lens, grt I + II; Tjeliken lens, grt II) and garnet-bearing peridotites (Tjeliken lens, grt II) are uniformly pyrope-rich (57–67% pyr) and have grossularite contents of 12.0–19.0 mol.%, they have low values of MnO (0.21–0.80 wt.%) but contain significant amounts of Cr_2O_3 (<1.59 wt.%; Fig. 5(*a*)). Garnets from internal eclogites (Ertsekey lens, grt I + II) are Cr-free and contain a grossularite content of 26–28 mol.%. The corresponding pyrope values are 40–45 mol.%, while MnO values are low (0.20–0.39 wt.%; Fig. 5(*a*)). Garnets from interlayered

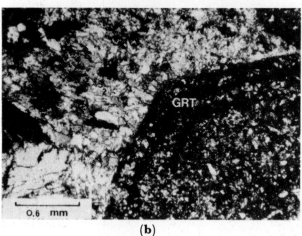

Fig. 4. (*a*) Orthopyroxene porphyroblast overgrowing S_2 foliation (Vaimok lens). (*b*) Texturally zoned garnet porphyroblast in garnet websterite (Vaimok lens).

Table II. Synoptic table showing the major structural elements of the high-pressure ultramafic rocks in Jämtland and Norrbotten.

	Jämtland *Ertsekey + Tjeliken lens*	*Norrbotten* *Vaimok lens*
S_1: Compositional layering	Metamorphic minerals (secondary?)	Magmatic minerals (primary)
S_2: Main foliation	Porphyroclastic to equigranular/ tabular microstructure	Porphyroblastic to equigranular/ tabular microstructure
S_3: Shearzones	Subordinate (common at the margins of the bodies)	Subordinate (common at the margin of the body)
S_4: Subvertical fractures	Talc/asbestos veins	Talc/asbestos veins

metabasic rocks in the ultramafites of the Vaimok lens contain significant amounts of Cr_2O_3 (<1.89 wt.%) but are less pyrope-rich (31–54 mol.%) and have grossularite contents between 14 and 25 mol.% (Fig. 5(*a*)).

EMP analyses revealed two types of garnets:

(i) chemically homogeneous garnets;
(ii) compositionally zoned ganets.

Compositionally zoned garnets I + II from metabasic rocks in ultramafic bodies from Jämtland show a marked retrograde zoning profile, i.e. from core to rim increasing iron and decreasing magnesium, typically restricted to the outer 200 μm of the garnet grains (Table V). In addition, garnets from metabasic rocks intercalated with ultramafites of the Vaimok lens as well as two fully recrystallized M_2 garnets from garnet-bearing peridotites of the Tjeliken lens reveal a "prograde" zoning pattern (Fig. 5(*b*),(*c*)). The prograde-zoned M_2 garnets in garnet peridotites (Tjeliken lens) also reveal two texturally distinct phases of garnet growth: the inner core is more densely crowded with tiny spinel inclusions than the surrounding outer rim (Fig. 6(*b*)). Differences in chemistry in spinel and other inclusions in these two garnet growth zones have not been detected. The other inclusions in these garnets are orthopyroxenes, clinopyroxenes and amphiboles.

Orthopyroxene

Orthopyroxene has low calcium-contents (Wo 0.5 mol.%); its enstatite component varies from 89.5–95.0 mol.%. M_2 and M_3 orthopyroxenes in contact with garnet are compositionally zoned with increasing Al_2O_3 contents from core to rim. M_2 (with M_3 rims) orthopyroxenes in contact with garnets range from 0.80 to 2.70 wt.% Al_2O_3 with the most common core values between 1.2 and 1.5 for the Tjeliken lens (Fig. 5(*e*)), (Table V). The range 0.7–1.30 wt.% Al_2O_3 with the most common core values between 0.95 and 1.05 for the Ertsekey lens (Fig. 5(*d*)), (Table 5). The most common values for the Vaimok lens range between 1.14 and 1.40 wt.% Al_2O_3 (Table V).

Frequently, M_2 garnets of garnet-bearing peridotites in the Tjeliken lens are more or less replaced by M_3 mineral assemblages containing orthopyroxene (opx 3). Such opx 3 has characteristically higher Al_2O_3 wt.% than opx 2, it ranges between 1.5 and 5.0 with most common values between 1.7 and 2.0 wt.%.

PETROGRAPHY AND MINERAL-CHEMISTRY SUMMARY

From a petrogenetic point of view, the most important petrographic and mineral-chemistry observations are:

(1) M_2 garnets in contact with olivine are replaced by kelyphites containing spinel-pyroxene and/or amphiboles (Fig. 3(*d*)).
(2) Retrograde zoning of M_1 garnets in garnet-pyroxenites in Jämtland (Ertsekey lens, Fig. 5(*a*)).
(3) Prograde compositional zoning in some garnets from metabasic layers in Vaimok ultramafites (Fig. 5(*a*), (*c*)).
(4) M_2 garnets in Tjeliken garnet-bearing peridotites are riddled with spinel inclusions; some of these garnets reveal a prograde compositional zoning profile (Figs. 5(*b*), 6(*b*)).
(5) Garnet and orthopyroxene in metabasic layers of Vaimok ultramafites show a porphyroblastic tex-

Table III. Synoptic table showing the metamorphic mineral evolution of the high-pressure ultramafic/basic rocks.

	Jämtland		Norrbotten	Rock type
	Ertsekey	Tjeliken		
M_0	?	?	?	
M_1	$ol_1 + cpx_1 + opx_1 + grt_1$	$ol_1 + cpx_1 + opx_1 + grt_1$	chlorite peridotites	meta-peridotites
	$grt_1 + opx_1 + cpx_1$	$grt_1 + opx_1 + cpx_1$	amphibolites	meta-basic rocks
M_2	$ol_2 + cpx_2 + opx_2 + grt_2$ $(\pm amph_1) + spinel_1$	$ol_2 + cpx_2 + opx_2 + grt_2$ $+ spinel_1 \,(\pm amph_1)$	$ol + opx + amph$ spinel	meta-peridotites
	$grt_2 + opx_2 + cpx_2$ $\pm amph_1 + phlog_1$	$grt_2 + opx_2 + cpx_2$ $\pm amph_1 \pm phlog_1$	$grt + opx + cpx$ $\pm amph$	meta-basic rocks
M_3	$ol + grt \rightarrow spinel_1 + pyr_3$ $ol + grt \rightarrow amph_2 + opx_3 + spinel_2$		no characteristic phases developed	meta-peridotites
	omph. pyr. $\rightarrow cpx_3 + plag$			meta-basic rocks
M_4	chlorite peridotites		chlorite peridotites	meta-peridotites
	amphibolites		amphibolites	meta-basic rocks
M_5	$serp + talc + amph + chl$		$serp + talc + amph + chl$	meta-peridotites
	$+ chlorite$		$+ chlorite$	meta-basic rocks

ture overgrowing a fine-grained S_2 foliation (Figs. 4(a), (b)).

(6) Orthopyroxene 2 in contact with garnet 2 in Tjeliken lens ultrabasic rocks has higher Al_2O_3 wt.% than identical orthopyroxene 2 in ultrabasic bodies of the Ertsekey lens (Fig. 5(d), (e); Table V).

These results are consistent with a prograde metamorphic origin for the high-pressure ultramafic rocks in Norrbotten. In contrast, the high-pressure ultramafic rocks in Jämtland reveal a retrograde metamorphic history, with some indications that during M_2 temperature and/or pressure might have been prograde resulting in the prograde nature of some M_2 garnets. In addition, the different P–T conditions during M_2 for the various ultramafic rocks are indicated by their mineral chemistry.

PRESSURE–TEMPERATURE ESTIMATES

A number of experimentally calibrated mineralogical geothermometers/barometers based on element-partitioning relationships are available. For garnet peridotites, equilibrium temperatures are most commonly estimated from the pyroxene miscibility gap (Wells, 1977) and/or the partitioning of iron and magnesium among garnet, clinopyroxene, orthopyroxene and olivine (e.g. Mori and Green, 1978; Ellis and Green, 1979; O'Neill and Wood, 1979; Kawasaki, 1979), whilst equilibrium pressures can be obtained from the solubility of $MgAl_2SiO_6$ in orthopyroxenes (Wood, 1974; Perkins and Newton, 1980; Harley, 1984a). For a critical evaluation of most of these methods, the reader is referred to Carswell and Gibb (1980), Griffin *et al.* (1985) and Carswell *et al.* (1985).

In the present study, estimates of equilibrium temperatures have been obtained using the iron–magnesium partitioning between M_2 garnets and M_2 orthopyroxenes (Harley, 1984b). Equilibrium pressures have been estimated from Al_2O_3 contents of M_2 orthopyroxenes using the recent experimental work of Harley (1984a).

In all calculations, iron has been taken as total iron,

owing to the difficulty in estimating Fe^{2+}/Fe^{3+} by charge balance from EMP analyses and to the low Fe^{3+} contents that are to be expected in the various phases (except amphibole) present. As noted earlier, garnet and orthopyroxene are chemically zoned. In applying geothermometers/barometers to compositionally zoned minerals, one is faced with the problem of deciding what portions of grains, if any, represent equilibrium compositions. Extensive discussion of this problem exists in the literature, e.g. Krogh (1977), Medaris (1984); Carswell *et al.* (1985).

In the present case, in fully recrystallized garnet pyroxenites from Jämtland, cores of M_2 orthopyroxenes have been combined with core values of unzoned recrystallized M_2 garnets. In the latter case, careful selection took place to avoid interference with superimposed retrograde trends. In garnet pyroxenites from Norrbotten, most data are obtained from chemically unzoned garnets that are in textural equilibrium with relatively unzoned orthopyroxenes (same method). The results of average temperature and pressure estimates are given in Fig. 7, while selected mineral-chemical data are presented in Table V. In addition, it should be noted that the initial preliminary mineral-chemical data of M_1 metamorphic minerals from the ultramafic/basic rocks of Jämtland (Tjeliken and Ertsekey lens) indicate initially higher P–T conditions during M_1.

DISCUSSION

To evaluate the petrogenetic significance of the present study, it is necessary to compare the results from the ultramafic rocks with similar studies done on country-rock eclogites enclosed in the surrounding gneisses (Van Roermund and Bakker, 1984; Stephens and Van Roermund, 1984; Andréasson *et al.*, 1985; Van Roermund, 1985). The eclogites are interpreted as having been formed from lower-pressure protoliths during "westwards subduction" of the Baltic plate (Krogh, 1977; Van Roermund, 1985). Recent $^{40}Ar/^{39}Ar$ mineral ages of amphiboles secondary after garnets from retro-eclogites in the Vaimok

Table IV
(a) Relict-garnet lherzolite (Ertsekey lens)

	Clinopyroxene I+II					Orthopyroxene I+II (away from garnet)						Olivine I+II					Spinel II
	Clast I Core	Clast I Rim	Clast I Rim	Clast II Rim	Clast II Core	Clast I Core	Clast I Rim	Clast II Rim	Clast II Core	Rexr.	In sympl. after garnet	Clast I Core	Clast I Rim	Clast II Core	Clast II Rim	Rexr.	In sympl. after garnet
SiO_2	52.30	52.47	52.45	53.79	53.48	56.24	55.96	56.91	56.87	57.25	56.89	41.32	41.34	41.00	41.37	41.38	0.00
Al_2O_3	4.66	4.52	3.22	3.44	3.05	2.99	1.78	2.08	1.23	1.26	1.50	0.18	0.00	0.00	0.00	0.00	35.15
TiO_2	0.39	0.45	0.22	0.30	0.29	0.14	0.09	0.01	0.20	0.03	0.06	0.06	0.00	0.00	0.00	0.00	0.00
MgO	15.64	15.75	16.21	15.70	16.72	35.37	35.35	35.37	36.27	35.84	34.87	51.59	51.41	51.63	51.62	52.01	15.25
FeO	1.42	1.41	1.45	1.32	1.30	4.49	4.62	4.27	4.67	4.60	6.19	6.95	7.03	6.86	7.07	6.26	16.15
MnO	0.07	0.16	0.05	0.00	0.11	0.19	0.13	0.08	0.10	0.03	0.24	0.09	0.10	0.10	0.05	0.11	0.00
CaO	21.09	21.54	22.05	22.17	22.35	0.20	0.18	0.19	0.23	0.17	0.16	0.00	0.00	0.00	0.00	0.00	0.00
Na_2O	1.85	1.85	1.40	1.63	1.23	0.24	0.22	0.15	0.07	0.20	0.16	0.00	0.00	0.00	0.00	0.00	0.00
K_2O	0.21	0.34	0.23	0.03	0.05	0.16	0.26	0.00	0.00	0.00	0.00	0.00	0.00	0.00	0.00	0.00	0.00
Cr_2O_3	1.98	1.67	1.56	1.94	1.34	0.67	0.61	0.26	0.16	0.23	0.35	0.02	0.00	0.05	0.00	0.00	33.23
	99.61	100.16	98.84	100.32	99.32	100.69	99.20	99.32	99.80	99.61	100.42	100.21	99.88	99.64	100.11	99.76	99.78
Si	1.903	1.903	1.927	1.941	1.937	1.918	1.939	1.956	1.952	1.966	1.955	0.997	1.001	0.995	0.999	0.999	0.000
Al	0.200	0.193	0.139	0.146	0.130	0.120	0.073	0.085	0.050	0.051	0.061	0.005	0.000	0.000	0.000	0.000	1.203
Ti	0.011	0.012	0.006	0.008	0.008	0.004	0.002	0.000	0.005	0.001	0.002	0.001	0.000	0.000	0.000	0.000	0.000
Mg	0.848	0.851	0.888	0.844	0.903	1.798	1.826	1.812	1.856	1.835	1.786	1.855	1.855	1.867	1.858	1.872	0.660
FeII	0.043	0.043	0.045	0.040	0.039	0.128	0.134	0.123	0.134	0.132	0.178	0.140	0.142	0.139	0.143	0.126	0.392
Mn	0.002	0.005	0.002	0.000	0.003	0.005	0.004	0.002	0.003	0.001	0.007	0.002	0.002	0.002	0.001	0.002	0.000
Ca	0.822	0.837	0.868	0.857	0.867	0.007	0.007	0.007	0.008	0.006	0.006	0.000	0.000	0.000	0.000	0.000	0.000
Na	0.131	0.130	0.100	0.114	0.086	0.016	0.015	0.010	0.005	0.013	0.011	0.000	0.000	0.000	0.000	0.000	0.000
K	0.010	0.016	0.011	0.001	0.002	0.007	0.011	0.000	0.000	0.000	0.000	0.000	0.000	0.000	0.000	0.000	0.000
Cr	0.057	0.048	0.045	0.055	0.038	0.018	0.017	0.007	0.004	0.006	0.010	0.000	0.000	0.000	0.000	0.000	0.763
	2.028	2.037	2.030	2.008	2.015	2.021	2.027	2.003	2.018	2.011	2.014	2.000	1.999	2.010	2.002	2.001	3.018

(b) Garnet pyroxenite (Ertsekey lens)

| | Clinopyroxene | | | Garnet | | | |
| | Inclusion in garnet | | | | | | |
	Core	Rim	Matrix	Core	Rim	Core	Rim
					Internal eclogite		
SiO_2	53.91	53.67	53.26	40.89	40.42	39.73	40.35
Al_2O_3	9.21	9.32	9.00	23.91	23.59	23.06	23.50
TiO_2	0.25	0.23	0.30	0.00	0.00	0.00	0.00
MgO	11.97	12.16	12.23	12.09	12.06	11.95	11.09
FeO	1.87	2.04	1.89	13.66	13.84	14.06	13.85
MnO	0.00	0.00	0.00	0.39	0.35	0.34	0.21
CaO	19.24	19.49	19.01	10.29	9.88	10.16	10.28
Na_2O	3.64	3.41	3.51	0.00	0.00	0.00	0.00
	100.09	100.32	99.20	101.23	100.14	99.30	99.28
Si	1.925	1.915	1.920	2.977	2.976	2.962	2.997
Al	0.388	0.392	0.382	2.051	2.047	2.026	2.058
Ti	0.007	0.006	0.008	0.000	0.000	0.000	0.000
Mg	0.637	0.647	0.657	1.312	1.324	1.328	1.228
FeII	0.056	0.061	0.057	0.832	0.852	0.877	0.860
Mn	0.000	0.000	0.000	0.024	0.022	0.021	0.013
Ca	0.736	0.745	0.734	0.803	0.779	0.812	0.818
Na	0.252	0.236	0.245	0.000	0.000	0.000	0.000
	2.000	2.001	2.004	4.998	5.000	5.025	4.974

(c) **Garnet websterite (Ertsekey lens)**

	Garnet clinopyroxenite				Garnet websterite										
	Clinopyroxene II		Garnet II		Clinopyroxene II		Garnet II		Orthopyroxene II			Olivine II	Phlogopite I	Spinel I	
	Core	Rim	Core	Rim	Core	Rim	Core	Rim	Core	Rim	Rim	Rim	Rim	Core	Rim
SiO_2	54.33	54.54	41.30	41.05	54.90	54.90	41.30	41.40	56.57	57.61	56.50	41.61	38.48	0.00	0.00
Al_2O_3	2.82	2.77	23.92	24.01	0.98	1.22	23.37	23.74	0.85	1.11	1.48	0.00	18.89	27.47	30.42
TiO_2	0.23	0.19	0.09	0.04	0.14	0.17	0.01	0.03	0.05	0.02	0.05	0.00	1.44	0.00	0.00
MgO	16.60	16.89	16.54	16.41	18.01	17.84	18.60	18.30	34.38	34.76	34.94	50.73	23.44	12.44	12.24
FeO	1.78	1.86	10.87	11.60	1.54	1.53	9.65	10.69	6.62	6.71	6.27	8.23	3.00	17.73	18.16
MnO	0.00	0.01	0.45	0.53	0.00	0.20	0.38	0.45	0.10	0.06	0.05	0.06	0.05	0.00	0.00
CaO	23.79	23.62	7.42	6.76	24.75	24.02	5.18	5.08	0.27	0.23	0.10	0.00	0.00	0.00	0.00
Na_2O	1.02	1.04	0.02	0.18	0.45	0.57	0.06	0.09	0.10	0.04	0.09	0.00	0.00	0.00	0.00
K_2O	0.00	0.00	0.00	0.00	0.00	0.00	0.00	0.00	0.00	0.00	0.00	0.00	9.39	0.00	0.00
Cr_2O_3	0.15	0.13	0.01	0.09	0.05	0.12	1.01	0.90	0.38	0.22	0.00	0.00	0.00	41.40	39.79
	100.72	101.05	100.62	100.67	100.82	100.57	99.56	100.68	99.32	100.76	99.48	100.63	94.69	99.04	100.61
Si	1.953	1.954	2.965	2.954	1.974	1.976	2.970	2.957	1.969	1.973	1.957	1.004	5.427	0.000	0.000
Al	0.120	0.117	2.024	2.037	0.041	0.052	1.981	1.998	0.034	0.044	0.060	0.000	3.139	0.988	1.067
Ti	0.006	0.005	0.005	0.002	0.004	0.005	0.001	0.002	0.001	0.001	0.001	0.000	0.153	0.000	0.000
Mg	0.890	0.902	1.770	1.760	0.965	0.957	1.994	1.948	1.783	1.774	1.804	1.825	4.927	0.566	0.543
FeII	0.054	0.056	0.653	0.698	0.046	0.046	0.580	0.639	0.193	0.192	0.182	0.166	0.354	0.453	0.452
Mn	0.000	0.000	0.027	0.032	0.000	0.006	0.023	0.027	0.003	0.002	0.001	0.001	0.006	0.000	0.000
Ca	0.916	0.907	0.571	0.521	0.953	0.926	0.399	0.389	0.010	0.008	0.004	0.000	0.000	0.000	0.000
Na	0.071	0.072	0.003	0.025	0.031	0.040	0.008	0.012	0.007	0.003	0.006	0.000	0.000	0.000	0.000
K	0.000	0.000	0.000	0.000	0.000	0.000	0.000	0.000	0.000	0.000	0.000	0.000	1.689	0.000	0.000
Cr	0.004	0.004	0.001	0.005	0.001	0.003	0.057	0.051	0.010	0.006	0.000	0.000	0.000	0.999	0.936
	2.014	2.017	5.053	5.081	2.017	2.011	5.044	5.066	2.011	2.003	2.015	1.992	7.695	3.006	2.998

Sample 84-54

(d) Garnet peridotite (Tjeliken lens)

	Garnet		Olivine II	Orthopyroxene II+III			Amphibole			Garnet Inclus.	Spinel Matrix Inclus.	Amphibole Kelyphite I	Amphibole Kelyphite II
	Rim II	Core		Inclus.	Core	Rim	Inclus.	Core	Rim				
SiO_2	41.54	40.35	40.50	54.28	57.30	56.24	44.74	44.18	44.60	00.00	00.00	00.00	00.00
Al_2O_3	23.65	24.26	00.00	4.27	1.12	1.52	13.61	13.04	13.26	39.39	26.39	38.51	50.73
TiO_2	0.08	0.06	0.00	0.00	0.00	0.10	0.44	0.55	0.50	00.00	00.00	00.00	00.00
MgO	17.46	18.53	49.30	34.16	34.96	34.44	18.64	18.27	18.09	15.55	10.16	14.25	17.45
FeO	10.61	9.29	8.54	5.00	5.70	5.87	2.94	3.02	2.95	14.27	21.00	16.56	13.04
MnO	0.78	0.58	0.13	0.16	0.15	0.21	0.00	0.11	0.09	00.00	00.00	00.00	00.00
CaO	5.63	4.92	0.00	0.25	0.25	0.23	12.01	12.25	12.60	00.00	00.00	00.00	00.00
Na_2O	0.00	0.00	0.00	0.00	0.00	0.00	2.42	2.14	2.46	00.00	00.00	00.00	00.00
K_2O	0.00	0.00	0.00	0.00	0.00	0.00	0.93	1.04	0.47	00.00	00.00	00.00	00.00
Gr_2O_3	0.88	1.57	0.00	0.77	0.33	0.29	1.73	1.84	0.95	30.60	41.37	30.20	17.68
NiO	0.09	0.15	0.31	0.24	0.08	0.00	0.21	0.21	0.26	00.00	00.00	00.00	00.00
	100.72	99.71	98.78	99.13	99.89	98.90	97.67	96.65	96.23	98.81	98.92	99.52	99.90
Si	2.967	2.901	1.000	1.893	1.973	1.960	6.335	6.339	6.394	0.000	0.000	0.000	0.000
Al	1.992	2.054	0.000	0.174	0.045	0.063	2.279	2.208	2.240	1.312	0.970	1.308	1.622
Ti	0.017	0.004	0.000	0.000	0.000	0.002	0.051	0.060	0.052	0.000	0.000	0.000	0.000
Mg	1.859	1.989	1.813	1.767	1.793	1.788	3.928	3.907	3.869	0.658	0.472	0.609	0.705
Fe II	0.635	0.558	0.177	0.145	0.163	0.172	0.349	0.362	0.353	0.339	0.547	0.400	0.295
Mn	0.047	0.035	0.003	0.005	0.004	0.006	0.000	0.016	0.008	0.000	0.000	0.000	0.000
Ca	0.429	0.381	0.000	0.009	0.008	0.008	1.820	1.880	1.939	0.000	0.000	0.000	0.000
Na	0.000	0.000	0.000	0.015	0.000	0.000	0.332	0.586	0.345	0.000	0.000	0.000	0.000
K	0.000	0.000	0.000	0.000	0.000	0.000	0.085	0.190	0.043	0.000	0.000	0.000	0.000
Cr	0.043	0.086	0.000	0.021	0.008	0.004	0.187	0.207	0.103	0.687	1.018	0.685	0.378
Ni	0.004	0.009	0.006	0.007	0.002	0.000	0.025	0.026	0.026	0.000	0.000	0.000	0.000
	7.993	8.021	2.999	4.636	3.996	4.003	15.392	15.781	15.372	2.996	3.007	3.002	3.000

Table V. EMP analyses and calculated structural formulae for selected M_2 garnets and M_2 orthopyroxenes from garnet pyroxenites.

| | Sample 84–068 (metabasite) Tjeliken lens | | | | Sample 73388 (metabasite) Ertsekey lens | | | | Sample Nor-1B (metabasite) Vaimok lens | | | |
| | Garnet II | | Orthopyroxene II | | Garnet II | | Orthopyroxene II | | Orthopyroxene | | Garnet | |
	Rim	Core	Rim	Core	Core	Rim	Core	Rim	Core	Rim	Core	Rim
SiO_2	41.21	41.59	56.92	58.01	41.30	41.40	56.57	57.61	55.77	55.84	40.49	40.58
TiO_2	0.03	0.08	—	—	0.01	0.03	0.05	0.02	—	—	—	—
Al_2O_3	23.35	23.58	1.76	1.34	23.37	23.74	0.85	1.11	1.15	1.26	22.68	22.68
FeO	10.36	9.54	5.78	5.90	9.65	10.69	6.62	6.71	10.53	10.05	15.25	15.17
MnO	0.71	0.51	0.17	0.08	0.38	0.45	0.10	0.06	—	0.10	0.50	0.67
MgO	17.84	18.29	35.36	34.94	18.60	18.30	34.38	34.76	31.62	31.99	13.54	13.41
CaO	5.69	5.75	0.17	0.21	5.18	5.08	0.27	0.23	0.20	0.25	7.04	7.48
Na_2O	—	—	—	—	0.06	0.09	0.10	0.04	—	—	—	—
K_2O	—	—	—	—	—	—	—	—	—	—	—	—
Cr_2O_3	1.29	1.22	0.14	0.12	1.01	0.90	0.38	0.22	0.28	0.34	1.43	1.15
NiO	0.06	0.09	0.11	—	—	—	—	—	0.11	0.13	—	0.03
Total	100.54	100.65	100.41	100.60	99.56	100.68	99.32	100.76	99.66	99.96	100.93	101.17
Si	2.956	2.972	1.951	1.980	2.970	2.957	1.969	1.973	1.967	1.961	2.969	2.971
Ti	0.001	0.004	—	—	0.001	0.002	0.001	0.001	—	—	—	—
Al	1.975	1.978	0.070	0.053	1.981	1.998	0.034	0.044	0.048	0.052	1.960	1.957
Fe	0.621	0.570	0.165	0.168	0.580	0.639	0.193	0.192	0.311	0.295	0.935	0.929
Mn	0.043	—	0.004	0.002	0.023	0.027	0.003	0.002	—	0.003	0.031	0.042
Mg	1.907	1.948	1.807	1.777	1.994	1.948	1.783	1.774	1.662	1.674	1.480	1.463
Ca	0.438	0.440	0.006	0.008	0.399	0.389	0.010	0.008	0.008	0.009	0.553	0.587
Na	—	—	—	—	0.008	0.012	0.007	0.003	—	—	—	—
K	—	—	—	—	—	—	—	—	—	—	—	—
Cr	0.037	0.069	—	—	0.057	0.051	0.010	0.006	0.008	0.009	0.083	0.067
Ni	0.003	0.005	0.004	0.004	—	—	—	—	0.003	0.004	—	—

lens (Norrbotten) indicate Finnmarkian rather than Scandian times for this subduction event (Dallmeyer and Gee, 1986). A Caledonian age for the high-pressure metamorphism in Jämtland has also been inferred from Sm–Nd eclogitic garnet–whole-rock ages (Mearns and Van Roermund, 1985) as well as from U–Pb ages from zircons out of the enclosing gneisses (Claesson, 1987). Sm–Nd dates from clinopyroxene–garnet pairs from eclogites elsewhere in the Scandinavian Caledonides, however, give early Scandian rather than Finnmarkian ages (Griffin and Brueckner, 1980; Mearns and Lappin, 1982).

$P-T$ estimates for the formation of the country-rock eclogites in the Seve nappes are in reasonable agreement with $P-T$ estimates during M_2 in the present work except for the Tjeliken lens ultramafic rocks (Van Roermund, 1985). M_2 temperature estimates from the Tjeliken lens ultramafic rocks are consistently 100–150°C above those from the surrounding country-rock eclogites; pressure estimates, however, do correspond. If the formulation and application of the geothermometers and barometers is correct then these results indicate that the Tjeliken garnet peridotites have been recrystallized during M_2 at much higher temperatures than the surrounding country-rock eclogites, which implies tectonic introduction of the Tjeliken garnet–peridotites during the Caledonian orogeny. Tectonic introduction (from the subcontinental mantle) has to be inferred also for the Jämtland garnet–peridotites on the basis of their petrographic and mineral-chemical data. This is in marked contrast with the high-pressure ultramafic rocks in the Vaimok lens, which reveal a prograde metamorphic evolution subsequently followed by a retrograde evolutionary trend. The latter are therefore interpreted to resemble a dismembered group 1 Alpine-type peridotite body of the ophiolite subtype (Den Tex, 1969) that has been brought down during the Caledonian orogeny to deeper crustal

levels than similar ultramafic bodies elsewhere in the Caledonides.

The garnet peridotites in Jämtland, however, resemble the rare European occurrences of garnet lherzolites enclosed within (i) high-grade basement gneisses of the Western Gneiss Region (Norway), (ii) the Lepontine Belt of the Central Alps and (iii) the basement gneisses of the Bohemian massif. Their petrogenetic significance has recently been discussed by, for example, Carswell and Gibb (1980), Medaris (1984), and Carswell (1986). In their petrogenetic models the garnet lherzolites have been interpreted to resemble subcontinental upper mantle material introduced into the crust by deep-level ductile "imbrication" of the continental crust during orogenesis.

Current tectonic models for the Central Caledonides (Dallmeyer et al., 1985; Dallmeyer and Gee, 1986) reveal westward-directed subduction of the Baltic plate during early Caledonian times followed by a final collision event between the Baltic and Laurentian plates in the Scandian. The metamorphic grade of the overlying eugeoclinal sediments of the Köli Nappes (Upper Allochthon) in Jämtland is too low to explain the high-pressure metamorphism in the underlying Seve Nappes by the burden of the overriding Laurentian plate. Consequently, the rock burden that produced the high-pressure metamorphism in the Seve Nappes must have been due to internal ductile imbrication of the outermost part of the Baltoscandian platform that most likely pre-dated the final collision event. In this model, the Seve garnet peridotites represent subcontinental Baltoscandian upper mantle material brought into the crust during the eclogite-forming Caledonian event. This implies deep-level ductile continental imbrication within the Baltoscandian plate during "early" Caledonian times followed subsequently by late-stage easterward-directed regional thrusting during the Scandian collision event in order to bring the

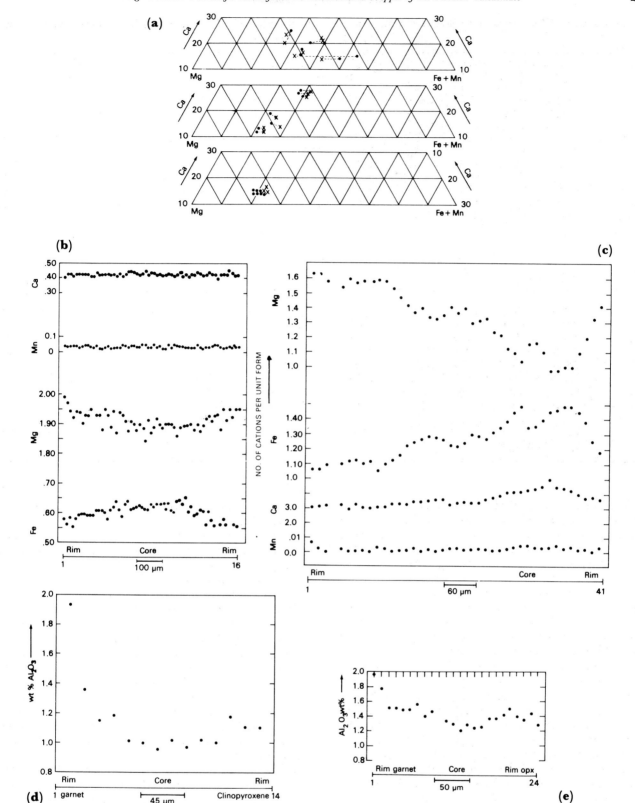

Fig. 5. (*a*) Truncated triangular garnet end-member diagram showing the compositional variations of garnets. *Top* from metabasic layers in Vaimok lens; *middle* from metabasic layers in Ertsekey lens; *bottom* from garnet periodotites and garnet pyroxenites in Tjeliken lens. Symbols: dots = cores, crosses = rims. (*b*) Microprobe stepscan profile across M_2 garnet from garnet periodotite: Tjeliken lens. (*c*) Microprobe stepscan profile across M_2 garnet from metabasic layer; Vaimok lens. (*d*) Microprobe stepscan profile across orthopyroxene II from garnet pyroxenite; Ertsekey lens. (*e*) Microprobe stepscan profile across orthopyroxene II from garnet pyroxenite; Tjeliken lens.

high-pressure metamorphic rocks (including ultramafics) of the Seve Nappes into their present high-level tectonic position. In this respect, the high-pressure ultramafic rocks of the Seve Nappes play a crucial role in unravelling the complex petrogenetic evolution of the Central Caledonides.

Acknowledgements

This study would have been impossible without the help of Bill Griffin, Mike Stephens and Ebbe Zachrisson. I owe them my special thanks. Also, the work has benefited from discussions with Tony Carswell.

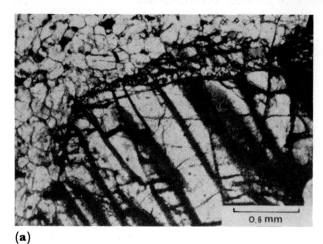

(a)

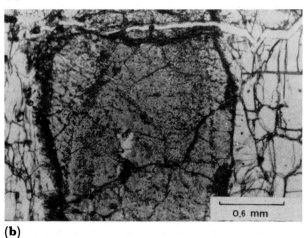

(b)

Fig. 6. (*a*) M₁ garnet phorphyroblast surrounded by a rim of dynamically recrystallized M₂ garnets.Garnet–websterite; Ertsekey lens. (*b*) Texturally zoned M₂ garnet riddled with spinel inclusions. This garnet has been used for the EW microprobe stepscan profile indicated in Fig. 5(*b*). Garnet peridotite; Tjeliken lens.

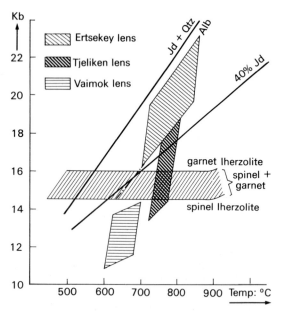

Fig. 7. Estimated *PT* conditions during M₂ (boxes encompasses the range of determined values including error estimates).

REFERENCES

Andréasson, P.-G., Gee, D. G. and Sukotjo, S. (1985). Seve eclogites in the Norrbotten Caledonides, Sweden. In: Gee, D. G. and Sturt, B. A. (eds.) *The Caledonide Orogen—Scandinavia and Related Areas*, pp. 887–902. Wiley, Chichester.

Asklund, B. (1938). Hauptzuge der Tektonik und Stratigraphie der mitteleren Kaledoniden in Schweden. *Sver. Geol. Unders. Ser. C* **417**, 99 pp.

Bryhni, I. and Andreasson, P. G. (1985). Metamorphism in the Scandinavian Caledonides. In: Gee, P. G. and Sturt, B. A. (eds.) *The Caledonide Orogen—Scandinavia and Related Areas*, pp. 763–781. Wiley, Chichester.

Calon, T. J. (1979). A study of the Alpine-type peridotites in the Seve–Köli Nappe Complex, Central Swedish Caledonides, with special reference to the Kittelfjäll peridotite. Ph.D. thesis, University of Leiden, The Netherlands, 236 pp. (Available on request.)

Carswell, D. A. (1986). The metamorphic evolution of Mg–Cr type Norwegian garnet peridotites. *Lithos* **19**, 279–297.

Carswell, D. A. and Gibb, F. G. F. (1980). The equilibration conditions and petrogenesis of European crustal garnet lherzolites. *Lithos* **13**, 19–29.

Carswell, D. A., Krogh, E. J. and Griffin, W. L. (1985). Norwegian orthopyroxene eclogites: calculated equilibration conditions and petrogenetic implications. In: Gee, D. G. and Sturt, B. A. (eds.). *The Caledonide Orogen—Scandinavia and Related Areas*, pp. 323–342. Wiley, Chichester.

Claesson, S. (1982). Caledonian metamorphism of Proterozoic Seve rocks on Mt Åreskutan, southern Swedish Caledonides. *Geol. Fören. Stockh. Forh.* **103**, 291–304.

Claesson, S. (1987). Isotopic evidence for the Precambrian provenance and Caledonian metamorphism of high grade paragneisses from the Seve Nappes, Scandinavian Caledonides. *Contrib. Mineral. Petrol.* **97**, 196–204.

Den Tex, E. (1969). Origin of ultramafic rocks, their tectonic setting and history. *Tectonophysics* **7**, 457–488.

Dallmeyer, R. D. and Gee, D. G. (1986). ⁴⁰Ar/³⁹Ar mineral dates from retrogressed eclogites within the Baltoscandian miogeocline: implications for a polyphase Caledonian orogenic evolution. *Geol. Soc. Amer. Bull.* **97**, 26–34.

Dallmeyer, R. D., Gee, D. G. and Beckholmen, M. (1985). ⁴⁰Ar–³⁹Ar mineral age record of early Caledonian tectonothermal activity in the Baltoscandian miogeocline, Central Scandinavia. *Amer. J. Sci.* **285**, 532–568.

Du Rietz, T., (1935). Peridotites, serpentinites and soapstones of Northern Sweden. *Geol. Fören. Stockh. För.* **57**(2), 401, 133–157.

Ellis, D. J. and Green, D. H. (1979). An experimental study of the effect of Ca upon garnet–clinopyroxene Fe–Mg exchange equilibria. *Contrib. Mineral. Petrol.* **71**, 13–22.

Gee, D. G. (1975). A geotraverse through the Scandinavian Caledonides—Östersund to Trondheim. *Sver. Geol. Unders., Ser. C,* **717**, 66 pp.

Gee, D. G. and Zachrisson, E. (1979). The Caledonides in Sweden. *Sver. Geol. Unders., Ser. C,* **769**, 48 pp.

Griffin, W. L. and Brueckner, H. K. (1980). Caledonian Sm–Nd ages and a crustal origin for Norwegian eclogites. *Nature* **285**, 319–321.

Griffin, W. L. and Qvale, H. (1985). Superferrian eclogites and the crustal origin of garnet peridotites, Almklovdalen, Norway. In: Gee, D. G. and Sturt, B. A. (eds.) *The Caledonide Orogen—Scandinavia and Related Areas*, pp. 803–812. Wiley, Chichester.

Griffin, W. L. *et al.* (1985). High pressure metamorphism in the Scandinavian Caledonides. In: Gee, D. G. and Sturt, B. A. (eds.) *The Caledonide Orogen—Scandinavia and Related Areas*, pp. 783–802. Wiley, Chichester.

Harley, S. (1984a). Solubility of alumina in orthopyroxene coexisting with garnet in FeO–MgO–Al₂O₃–SiO₂ and CaO–FeO–MgO–Al₂O₃–SiO₂. *J. Petrol.* **25**, 665–696.

Harley, S. (1984b). An experimental study for the partitioning of Fe and Mg between garnet and orthopyroxene. *Contrib. Mineral. Petrol.*, **86**, 359–373.

Jackson, E. P. and Thayer, T. P. (1972). Some criteria for distinguishing between stratiform concentric and Alpine-type peridotite-gabbro complexes. *Int. Geol. Congr. 24th Montreal,* Sect. 2, pp. 289–296.

Kawasaki, T. (1979). The thermodynamic analysis on the Fe–Mg exchange equilibrium between olivine and garnet: an application to the estimation of P–T relations of ultramafic rocks. *J. Jpn. Assoc. Mineral. Petrol. Econ. Geol.* **74**, 395–405.

Krogh, E. J. (1977). Evidence for a Precambrian continent–continent collision in western Norway. *Nature* **267**, 17–19.

Mearns, E. W. and Lappin, M. A. (1982). A Sm–Nd isotopic study of internal and external eclogites, garnet Unerzolite and gray gneiss from Almklovdalen, western Norway. *Terra Cognita* **2**, 324.

Mearns, E. W. and Van Roermund, H. L. M. (1985). On the age of protolith and eclogite formation in metabasic rocks from the Seve Nappes. Central Scandinavian Caledonides. *Terra Cognita* **5**, (4), 434.

Medaris, L. G. (1984). A geothermobarometric investigation of garnet peridotites in the Western Gneiss Region of Norway. *Contrib. Mineral. Petrol.* **87**, 72–86.

Mori, T. and Green, D. H. (1978). Laboratory duplication of phase equilibria observed in natural garnet lherzolites. *J. Geol.* **86**, 83–97.

Nicolas, A. and Poirier, J. P. (1976). *Crystalline Plasticity and Solid State Flow in Metamorphic Rocks.* Wiley, New York.

O'Neill, H. St. C., (1981). The transition between spinel lherzolite and garnet lherzolite, and its use as a geobarometer. *Contrib. Mineral. Petrol.* **77**, 185–194.

O'Neill, N. St. C. and Wood, B. J. (1979). An experimental study of Fe–Mg partitioning between garnet and olivine and its calibration as a geothermometer. *Contrib. Mineral. Petrol.* **70**, 59–70.

Perkins, D. and Newton, R. C. (1980). The composition of coexisting pyroxene and garnet in the system CaO–MgO–Al_2O_3–SiO_2 at 900°–1100°C and high pressures. *Contrib. Mineral. Petrol.* **75**, 291–300.

Qvale, H. and Stigh, J. (1985). Ultramafic rocks in the Scandinavian Caledonides. In: Gee, D. G. and Sturt, B. A. (eds.) *The Caledonide Orogen—Scandinavia and Related Areas*, pp. 693–715. Wiley, Chichester.

Reymer, A. P. S., Boelrijk, N. A. I. M., Hebeda, E. H., Priem, H. N., Verduren, E. A. Th. and Verschuren, R. H. (1980). A note on Rb–Sr whole-rock ages in the Seve Nappe of the Central Scandinavian Caledonides. *Norsk Geol. Tidsskr.* **60**, 139–147.

Roberts, D. and Gee, D. G. (1985). An introduction to the structure of the Scandinavian Caledonides. In: Gee, D. G. and Sturt, B. A. (eds.) *The Caledonide Orogen—Scandinavia and Related areas*, pp. 55–68. Wiley, Chichester.

Statham, P. J. (1976). A comparative study of techniques for quantitative analysis of the X-ray spectra obtained with a Si(Li) detector. *X-ray Spectros.* **5**, 16–28.

Stephens, M. B. and Van Roermund, H. L. M. (1984). Occurrence of glaucophane and crossite in eclogites of the Seve Nappes, southern Norrbotten Caledonides, Sweden. *Norsk. Geol. Tidskr.* **69**, 155–163.

Stigh, J. (1980). Detrital serpentinites of the Caledonian allochthon in Scandinavia. In: Wones, D. R. (ed.), *The Caledonides in the USA.* Virginia Polytechnic Inst. and State University, Dept. Geol. Sci. Mem. **2**, 149–156.

Strand, T. and Kulling, O. (1972). *The Scandinavian Caledonides.* Wiley Intrascience, London.

Törnebohm, A. E. (1896). Grunddragen af det centrala Skandinaviens bergbyggnad. Kongl. *Svenska. Vetensk. Akad. Handl.* **28**, 212 pp.

Van Roermund, H. L. M. (1982). On eclogites from the Seve Nappe, Jämtland, Central Scandinavian Caledonides. Ph.D. thesis, University of Utrecht, The Netherlands, 99 pp. (Available on request.)

Van Roermund, H. L. M. (1985). Eclogites of the Seve Nappe, Central Scandinavian Caledonides. In: Gee, D. G. and Sturt, B. A. (eds.) *The Caledonide Orogen—Scandinavia and Related Areas*, pp. 873–886. Wiley, Chichester.

Van Roermund, H. L. M. and Bakker, E. (1984). Structure and metamorphism of the Tången–Inviken area, Seve Nappe, Central Scandinavian Caledonides. *Geol. Fören. Stockh. Förh.* **105**, 301–319.

Wells, P. R. A. (1977). Pyroxene thermometry in simple and complex systems. *Contrib. Mineral. Petrol.* **62**, pp. 129–139.

Williams, P. F. and Zwart, H. J. (1977). A model for the development of the Seve–Köli Caledonian Nappe Complex. In: S. K. Saxena and Bhattacharji (eds.). *Energetics of Geological Processes*, pp. 169–187. Springer Verlag, New York.

Wood, B. J. (1974). Solubility of alumina in orthopyroxene coexisting with garnet. *Contrib. Mineral. Petrol.* **46**, 1–15.

Zachrisson, E. (1969). Caledonian geology in northern Jämtland southern Västerbotten. *Sver. Geol. Unders. Ser. C.*, **644**, 33 pp.

Zwart, H. J. (1974). Structure and metamorphism in the Seve-Köli Nappe Complex and its implications concerning the formation of metamorphic nappes. *Cent. Soc. Geol. Belg. Geologie des domaines crystalline*, pp. 129–144.

18 Status of the supracrustal rocks in the Western Gneiss Region, S Norway

Inge Bryhni

Mineralogisk-geologisk Museum, Sars Gate 1, 0562 Oslo 5, Norway

A large number of zones of presumed supracrustal rocks of uncertain position, interfolded with more granitic and migmatitic gneisses, are indicated on regional geological maps of the Western Gneiss Region, S Norway. For each individual zone, one of two origins is possible: either it is part of the Caledonised Precambrian basement, possibly as a Precambrian cover sequence; or it is a Lower Paleozoic cover and/or a Caledonian nappe sequence that has been folded into the basement and subsequently metamorphosed at high pressure.

Various associations with different histories can be recognized:

(1) supracrustal rocks deformed and metamorphosed together with adjacent Precambrian gneisses and granites but unaffected by Caledonian movements;

(2) anorthosite-, amphibolite- and ultramafite-bearing gneisses with large volumes of feldspathic quartzite and mica schist forming tectonic successions comparable with the Lower and Middle Allochthons of the Scandinavian Caledonides;

(3) banded micaceous gneisses, amphibolites and marbles that may represent either Precambrian rocks rejuvenated during the Caledonian orogeny or a Caledonian depositional/tectonic cover sequence.

INTRODUCTION

The Western Gneiss Region of South Norway is a large area made up of mainly Precambrian gneissic and granitic rocks that occur between the Faltungsgraben (Central Trough) and the coast (Fig. 1). It may be subdivided into a southern Bergen–Kristiansund area and a northern Fosen–Namsos area, of which mainly the former will be considered here.

Minor amounts of obviously metamorphosed supracrustals are present in the Western Gneiss Region as quartzites, feldspathic quartzites, mica schists, garnetiferous and micaceous gneisses, kyanite- and/or sillimanite-bearing gneisses, banded amphibolites, calc-silicate rocks and marbles (Bryhni and Sturt, 1985). Hernes (1965) also considered that the more homogeneous gneisses in the region had formed from original supracrustal rocks. He suggested a tectonostratigraphy comprising:

(I) *Upper Tingvoll Group* with heterogeneous gneisses, augen gneisses, quartzites and amphibolites;

(II) *Lower Tingvoll Group* with homogeneous gneisses, augen gneisses and quartzites;

(III) *Raudsand Group* with homogeneous gneisses and amphibolites; and

(IV) *Frei Group* with heterogeneous gneisses, augen gneisses, mica schists, marbles and frequent inclusions of eclogite, amphibolite, dolerite and ultramafite.

Råheim (1972) inferred that equivalents of the Rausand and Lower Tingvoll groups (his Kristiansund and Sandvik groups) were originally volcanites and sediments.

Bryhni (1966) and Bryhni and Grimstad (1970), suggested that the southern part of the region might contain nappes similar to those of the Faltungsgraben, but in a more deformed and metamorphosed state. For all such units, the term Fjordane Complex was tentatively proposed (Table I). They distinguished between an infrastructure of granites and relatively homogeneous gneisses and a more heterogeneous suprastructure containing supracrustals with interlayers of metamorphic plutonic rocks, now mainly gneisses, meta-anorthosites and amphibolites. Similar relations were found by Brueckner (1977) and others in the Sotaseter–Grotli–Tafjord area.

At the eastern margin of the Scandinavian Caledonides, there is a regular succession of thrust sheets termed Lower, Middle and Upper Allochthon. A similar tectonostratigraphy is now also tentatively recognized on regional-scale maps by Bryhni and Sturt (1985) and Gee *et al.* (1985) for parts of the Western Gneiss Region.

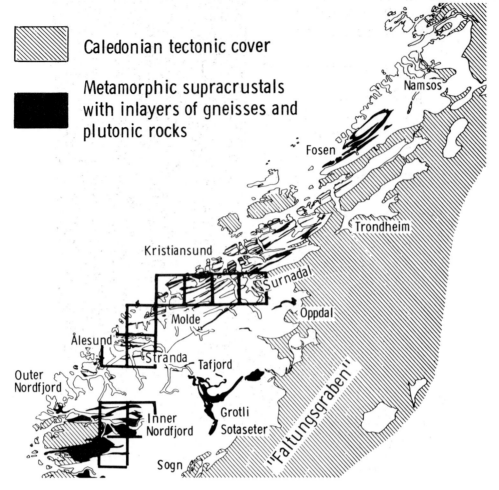

Fig. 1. Western Gneiss Region of S Norway. Squares indicate areas in the southern part of the region where new preliminary geological quadrangle maps at the scale 1:50 000 have been published or are in preparation.

A new programme of mapping of the mainly gneisses in the region is in progress (Fig. 1). Special consideration was given to a group of banded amphibolites and biotite–garnet-rich gneisses associated with calcite marbles, included in the Frei Group of Hernes. Similar sequences

Table I. Tentative geological correlation between the outer Fjordane part of the Western Gneiss Region and the Faltungsgraben in inner Sogn. (From Bryhni, 1966, Fig. 2.)

Western Norway	Faltungsgraben
Fjordane Complex	
Cambro–Silurian schists with granodiorite and granite	Cambro-Silurian schists
Feldspathic quartzite, dolomite, marble and feldspathic quartz–mica schist	Valdres sparagmite
Layered quartz–biotite–plagioclase gneiss, augen gneiss, granitic gneiss and amphibolite with inclusions of ultramafite, eclogite or gabbro and rocks of the Anorthosite kindred	Jotun thrust masses
Jostedal Complex	
Migmatite, granite Two-feldspar gneiss Augen gneiss in the west	Precambrian (pre-Eocambrian)

have in the Fosen–Namsos area been considered to vary "from far-travelled allochthonous units containing rocks of Late Precambrian to Early Paleozoic age, to parauto-chthonous elements of older Precambrian rocks" (Roberts, 1986; with reference to Tucker, 1986).

A minor occurrence within the latter area has been described by Johansson *et al.* (1987) as rootless Paleozoic granodioritic dykes (>430 Ma) as evidence in favour of a Caledonian tectonic cover, possibly corresponding to the Gula nappe (Upper Allochthon).

DESCRIPTION OF INDIVIDUAL AREAS

Outer Nordfjord area

The outer Nordfjord area comprises micaceous gneisses with mica schists and pelitic gneisses interlayered with banded gneisses supposed to be metamorphosed supra-crustal rocks (Bryhni, 1966). Figure 2 shows the geology of an area with mainly layered light and dark grey mica-rich gneisses from Allmenningen, which is typical of the outer Nordfjord part of the Western Gneiss Region. The main mineral assemblage, biotite–plagioclase–quartz, is sup-plemented by variable amounts of microcline, garnet or white mica. Apatite, sphene, zircon, rutile, black iron minerals, epidote (often with orthite cores) and chlorite (often as integrowths with biotite) are present as accessory or secondary minerals. Quartzitic interlayers are infre-quent at Allmenningen, but there are significant inter-calations elsewhere (see Bryhni, 1966).

Numerous inclusions of dunite, eclogite and anorthosite

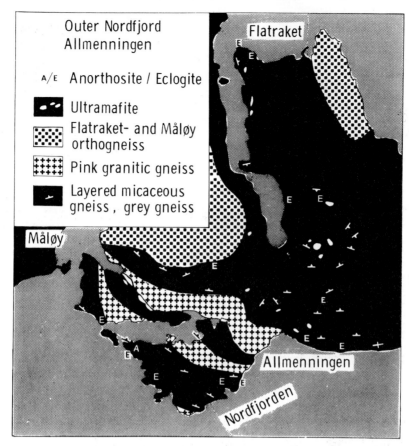

Fig. 2. Part of outer Nordfjord, showing locations of small inclusions of anorthosite, eclogite and ultramafite in layered micaceous gneiss.

occur in the layered micaceous gneisses, which alternate with a pink, granitic gneiss. A large body of granulitic, originally plutonic rock, the Flatraket–Måløy orthogneiss of syenitic to monzonitic/syenogabbroic composition, occurs above the banded mica-rich gneisses. The body, which is transected by mafic dykes and contains mantled, large orbs of alkali feldspar, is locally transformed into a coarse augen gneiss.

Quartzite and garnet–mica schists from adjacent areas have yielded Rb–Sr whole-rock dates of ca. 1500, 1100 and 400 Ma, interpreted as Svecofennian sediments (ca. 1760 Ma) affected by open chemical behaviour during Sveconorwegian and/or Caledonian events (Lappin *et al.*, 1979). However, the local association with feldspathic quartzites (especially on the south side of Nordfjord), anorthositic, mafic and ultramafic inclusions and "dyked" plutonic rocks with large, alkali feldspar augen, may rather relate to certain rocks of the Lower and Middle Allochthon in the marginal parts of the Caledonides, similar to those discussed below.

Sunnfjord–Inner Nordfjord area

Granitic gneisses and granites in the east are here overlain by large volumes of feldspathic quartzite and mica schist with gneissified or mylonitized, originally plutonic, rocks (Bryhni and Grimstad, 1970). The association differs from that of outer Nordfjord in containing much more feldspathic quartzite- and anorthosite-bearing gneisses as interfolded sheets (Fig. 3).

Sliding tectonics and recumbent folding make the tectonostratigraphic relations uncertain, but the area is essentially composed of two locally repeated units.

(1) *Eikefjord Group* with mainly metamorphosed anorthosite, amphibolite, ultramafite and mylonitized

orthogneisses of syenitic to monzonitic and granitic composition. The orthogneisses sometimes contain large, mantled alkali-feldspar megacrysts, and relicts with granulite-facies mineral assemblages, and show transitions from coarse-grained, massive granitoidal rocks via augen gneisses to fine-grained, dark, finely layered or laminated mylonitic gneisses.

(2) *Lykkjebø Group* with mainly feldspathic quartzite and feldspathic mica schist, amphibolite, micaceous gneisses and dolomitic, and locally also calcitic marble.

The rocks of Sunnfjord–inner Nordfjord are lithologically similar to those of the Oppdal area, described by Krill (1985, 1987; Krill and Sigmond, 1986). Inclusions of amphibolite and metagabbro are present in the Lykkjebø Group, but extensive developments of rocks correlatable with the Saetra- or Ottfjäll-type mafic dyke swarms have not been found. Thus, it is most likely that the Lykkjebø and Eikefjord units correspond to the Åmotsdal and Risberget units of Oppdal respectively, and have accordingly been related to the Lower and Middle Allochthon on regional-scale maps by Bryhni and Sturt (1985) and Gee *et al.* (1986).

Ålesund area

Presumed supracrustal rocks alternate with migmatitic gneisses as zones up to 2 km thick in the area to the south and southeast of Ålesund (Fig. 4), and can be described under the term Langevåg Group.

At Langevåg (UTM coordinates 550 270), this group consists of distinctly layered amphibolites together with garnet–biotite gneisses, calc-silicate rocks and calcite marbles. The layered amphibolite is usually a very irregular rock with layers or lenticles of garnet amphibo-

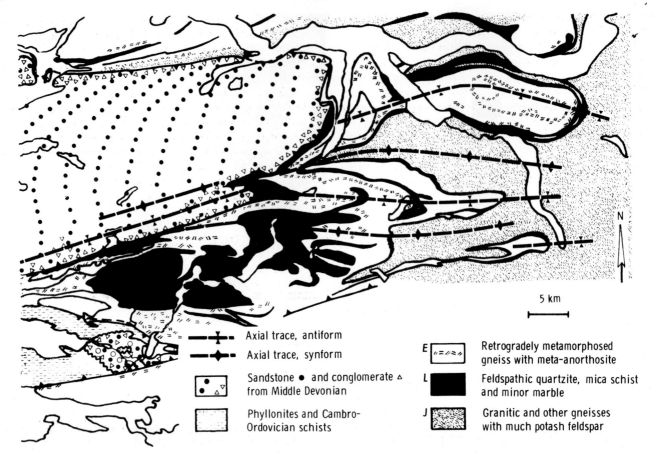

Fig. 3. Geology of parts of the Sunnfjord–inner Nordfjord area, showing interfolded sequences of supracrustals and original plutonic rocks above a gneissic/granitic basement to the east. "E" and "L" indicate the Eikefjord Group and Lykkjebø Group respectively, while "J" indicates the Jostedal Complex.

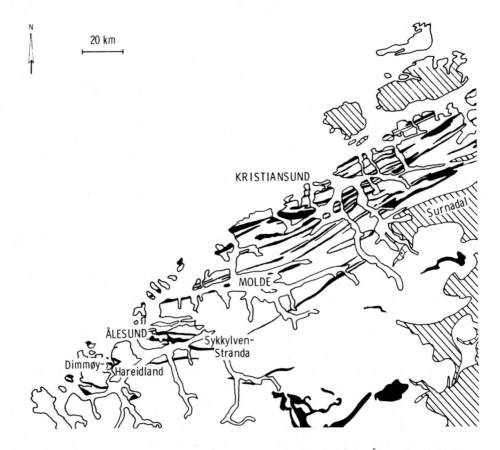

Fig. 4. General position of presumed supracrustal rocks (black) in the Ålesund and Molde areas.

lite, garnetite, hornblendite, light garnet–quartz–plagioclase gneiss, calc-silicate rocks and marble.

The garnet–biotite gneisses occur as narrow layers in amphibolite or as units up to several hundred metres thick. They often contain abundant quartz and plagioclase and may be termed garnet–biotite–quartz–plagioclase gneiss, with the constituent minerals listed in order of increasing abundance. Kyanite or sillimanite mainly occur as inclusions in biotite aggregates.

The calc-silicate rocks and silicate marbles contain diopside, tremolite, hornblende, epidote, zoisite, phlogopite, quartz and plagioclase as characteristic minerals besides calcite.

The unit contains inclusions of a websteritic ultramafite (cpx–opx) and is cut by tonalitic dykes not found in the adjacent gneisses as well as granites and pegmatites that also may be present in the gneisses. Eclogitic inclusions are not common close to Ålesund, but an exceptionally large body ("the world's largest eclogite" according to Mysen and Heier, 1972) appears to be enclosed by metamorphosed supracrustal rocks at Dimmøy–Hareidland west-southwest of Ålesund.

Molde area

Recent geological mapping on the Molde peninsula has revealed the existence of several slices of presumed supracrustal rocks interlayered with migmatitic gneisses, augen gneisses and gneissified granites. These are lithologically similar to those of the Langevåg Group described above and also to the Ertvågøy Group of Askvik and Rokoengen (1985). At least seven such zones occur on the Molde peninsula, with thickness up to more than 2 km, striking ENE–WSW (Bryhni *et al.*, 1986; Reksten, 1986; Bjørnstad, 1987; Kullerud, 1987). The strike is parallel to a zone of schists along Moldefjord that has been related to a the Trondheim Nappe (Upper Allochton) as a deeply downfolded structure with Lower Paleozoic rocks, e.g. an extension of the Surnadal synform (Hernes, 1955). Recently, Krill and Sigmond (1986) have suggested that the tectonostratigraphy from Surnadal continues with the Støren, Surna–Blåhø and Saetra–Risberget nappes (Upper and Middle Allochthon) along Moldefjord all the way to the west coast.

The zones across the Molde peninsula are composed of interlayered amphibolite, garnet amphibolite, hornblende–garnet or biotite–garnet gneiss and calcite marble, locally with strongly retrogressed eclogitic rocks. The composition is dominated by amphibole, plagioclase, quartz and garnet, with variable amounts of biotite, epidote and clinopyroxene. The biotite–garnet gneiss is interlayered with locally concentrated garnet–amphibole, calcite–garnet, clinopyroxene and garnet–quartz. A typical rock is kyanite-bearing with 30–40% garnet, much biotite and quartz and more than 20% plagioclase. Sillimanite is occasionally present along shear zones that transect the main foliation (Reksten, 1986). Migmatization is extensive in places, usually characterized with plagioclase–pegmatitic pods, and the zones are transected by numerous dykes of two-feldspar pegmatites. White layers of quartz–plagioclase or pink garnet–quartz–plagioclase rocks occur in minor amounts, and a few ultramafic bodies (dunites, garnet peridotites) are preferentially found in the severely sheared contact zones.

There appears to be a metamorphic facies contrast between the schists along rocks of Moldefjord and the surrounding eclogite-bearing supracrustals and gneisses, but this may be explained by localized retrograde alterations within the zone. There are, in fact, relics of garnet

amphibolites in the hornblende schists, and the garnet–biotite gneisses or schists especially are lithologically very similar to those that occur in the zones of supracrustals related to the Ertvågøy Group rocks. Carswell and Harvey (1985) noted the presence of relict kyanite in the Moldefjord schists and suggested that they had suffered the same high metamorphic pressures as those experienced by the other supracrustals. To this can be added that kyanite is quite common as inclusions in biotite in the Moldefjord schists and that rutile is present both in these schists and in the similar-looking supracrustals on the mainland, as are coarse white marbles except when locally transformed to grey mylonites.

Thus, the Moldefjord schists are not unique in the area, and probably had the same early history as the zones of supracrustal rocks on the mainland. The present differences may be related mainly to a late Caledonian phase of deformation, whereby the rocks along Moldefjord became more penetratively retrogressed than the other zones of supracrustal rocks.

DISCUSSION

Tectonics

There appears to be a structural contrast between the southern and the northern part of the Western Gneiss Region. The southern part has a regional fabric with EW to WNW–ESE lineation parallel to the axes of regional folds (F2 of Bryhni and Grimstad, 1970, see Fig. 3). This is also the major strike of foliations and lithologic boundaries at the eastern margin of the region, where it is cut by the NE–SW trend of the Faltungsgraben. The most recent maps of the southeastern margin (Lutro, 1987) show ca. 1000 Ma old gneisses and granites with the EW to WNW–ESE trend to be unconformably overlain by autochthonous sediments of probable Eocambrian–Lower Paleozoic age. Thus, the EW to ESE–WNW structural grain in the basement is either of late Precambrian age (Sveconorwegian) or produced during Caledonian reactivation of Precambrian structures along the same directions. Parallel Caledonian reactivation of older structures is likely, since the EW to ESE–WNW linear structures also occur in the cover of Lower Paleozoic rocks in the coastal parts of Sunnfjord.

There remains the alternative, that the EW to WNW–ESE linear fabric was produced during an early Paleozoic, pre-Scandian phase of the Caledonian orogeny. This, however, is unlikely from the evidence at present available.

In the north, around Molde and towards the Fosen–Namsos area, there is a distinct ENE–WSW trend, parallel to the coast and other major topographic features. Here, there is strong evidence that the structural grain was imposed during the Scandian phase (late Silurian/early Devonian) of the Caledonian orogeny. If the schists in the extension of the Caledonian Surnadal synform at Moldefjord really are the more schistose equivalents of the supracrustals occurring adjacent to them, then there may be significant intercalations in this part of the Western Gneiss Region of tectonic units related to the Middle and Upper Allochton.

Muret (1960) interpreted the Western Gneiss Region as composed of very large recumbent fold nappes of Pennine type. His idea that Precambrian basal gneisses were enveloped by supracrustals of Cambro–Ordovician age must still be considered as a plausible working hypothesis for explaining interlayering of the supracrustals with the migmatitic gneisses. Some of the metamorphosed supracrustal rocks (e.g., the Seve equivalents) may, however, be of Precambrian rather than of Lower Paleozoic age.

Tectonostratigraphy, age and correlations

The rocks of Sunnfjord–inner Nordfjord resemble those at Oppdal, where the Åmotsdal and Risberget tectonic units may correspond to the mainly psammitic Lykkjebø Group and the anorthosite- and orthogneiss-bearing Eikefjord Group respectively.

Lithologically similar rocks occur above a gneissic basement at Sotaseter–Grotli–Tafjord (Fig. 1), where M. M. Shouls (in Strand and Kulling, 1972) first noted that there were two separate thrust sheets folded into the basement, probably representing a continuation of nappes of the Faltungsgraben.

Farther to the northwest, where the eclogites indicate progressively higher *P–T* conditions (Griffin *et al.*, 1985), the supracrustals are more similar to the Blåhø and the Surna units of the Oppdal area. In places, they are associated with coarse augen gneisses where the feldspars show "rapakivi-like" structures similar to those of the Risbekken unit of Oppdal. Thus, on lithological grounds, some of the rocks here may correspond to the Upper and Middle Allochthon.

Certain of the supracrustals in the Gneiss Region are, however, definitely of Precambrian age. The geological map of Norway (Sigmond *et al.*, 1984) locates zones of Precambrian metasandstone in the gneisses south of Sognefjorden and an orthoquartzite is unconformably overlain by Lower Paleozoic phyllites north of Vik, Sogn (Bryhni *et al.*, 1979). Even more convincingly, the southeast marginal zone of the Gneiss Region in Sogn contains a unit of micaschist, banded mica gneiss, quartzite, meta-arkose and also possible mafic volcanites unconformably overlain by a low-grade autochthonous quartzite of presumed lower Paleozoic age (Lutro, 1987).

It is much more difficult to establish the age of the supracrustals in more northwestern parts of the region, where Caledonian folding and metamorphism may have affected both basement and Caledonian cover. Here, there is as yet no direct field evidence of age differences between the supracrustals and their surrounding migmatitic gneisses. All contacts are concordant, and usually strongly deformed to the extent that the original relations are obscured. Nor has any structural contrast been observed. Both units contain migmatitized rocks, although true granitic neosomers are most common in the migmatitic gneisses presumed to represent a basement.

On the Molde peninsula, Carswell and Harvey (1985) interpreted the paragneisses as being deposited and first metamorphosed 1700–1800 Ma ago, intruded by 1500 Ma monzonites/granodiorites, and thereafter metamorphosed during the Sveconorwegian and Caledonian cycles. They infer that "it is feasible that the Paragneiss Complex ... may contain both Svecofennian basement and Caledonian (cover) lithostratigraphic units."

A radiometric age difference between rocks of the basement (about 1600 Ma) and the supracrustals (900–1100 Ma) was indicated by Aas (1986) in the Sykkylven–Stranda area. She demonstrated that the coupling of the supracrustals to their basement had already taken place in Late Proterozoic (Sveconorwegian) and that they were subsequently folded isoclinally and metamorphosed together in the amphibolite facies. The apparent Sveconorvegian Rb–Sr dates indicated that the supracrustals were Proterozoic and *unrelated* to the rocks of likely Lower Paleozoic age (Blåhø Group) in the Oppdal area.

Studies by Krill (1985, 1987; Krill and Sigmond, 1986) from the latter area indicate that one of the nappe sheets there, the Saetra unit, contains premetamorphic, predeformation dykes dated to 745 ± 37 Ma, whilst the overlying nappe contains supracrustal rocks (e.g., Blåhø gneiss) presumably younger than 583 ± 69 Ma. This implies that emplacement of the nappes and their early recumbent folding with the basement took place late during the peak of the Caledonian cycle (ca. 425 Ma). The tectonostratigraphy is involved in later Caledonian folding in the Surnadal synform (Krill, 1985, 1987), possibly as far west as to the Moldefjord area (Krill and Sigmond, 1986). Thus, some of the intercalations of supracrustals within the Western Gneiss Region might, by analogy, be formed in a similar way by Caledonian infolding of a cover and/or a pile of nappes. This could apply also to the frequent interlayers of supracrustals of the Ertevågøy Group or Langevåg unit, which are lithologically similar to the, admittedly more schistose, rocks of Moldefjorden.

Relation of eclogites, ultramafites, dolerites and later dykes to the supracrustals

The eclogites of western Norway are most frequently enclosed in migmatitic or granitic gneisses, but also occur in rocks of presumed supracrustal origin. Their less frequent occurrence in the supracrustals is probably related to the "wetter" environment in the latter, where the flux of fluids generated during progressive metamorphism is likely to reduce the possibilities for inclosed mafic rocks to transform into eclogites. Mafic rocks that did transform became frequently retrogressed to granulites and amphibolites, where the eclogitic parentage is only revealed by inclusions of omphacite in garnet.

Ultramafites are very common in the coastal parts of the Western Gneiss Region and do not appear to be preferentially associated with any of the two major rock units. The few known occurrences of pyroxenites are confined to the supracrustals. Many of the dunitic or peridotitic bodies are located in the contact zone between the supracrustals and the migmatitic gneisses, and could have been emplaced along zones of concentrated strain during the Caledonian orogeny.

Gabbros ("dolerites") are very common in the gneisses near Ålesund and further along the coast to the Fosen–Namsos area. To date, they have only been found in the migmatitic or granitic rocks. This, if sustained, indicates that the supracrustals were emplaced upon the migmatitic or granitic rocks after emplacement of the gabbros; thereby representing a cover/basement couple that was subsequently folded and metamorphosed.

Recorded radiometric dates for the dolerites range from 1517 ± 60 Ma (Rb–Sr rock isochron: Tørudbakken, 1982) and 1289 ± 48 Ma, 1258 ± 56 Ma, 1198 ± 56 Ma and 926 ± 70 Ma (Sm–Nd whole-rock mineral isochrons: Mearns, 1984; Mørk and Mearns, 1986) to as late as about 425 Ma ($^{39}Ar/^{40}Ar$ plateau age: Lynch, 1976, cited in Carswell and Harvey, 1985). Although there may have been various episodes of intrusion (Mørk and Mearns, 1986), it is more likely that the younger dates reflect more or less complete equilibration during the Scandian phase of the Caledonian orogeny. This is in accord with the close relationship between the dated rocks and the coronites and eclogites indicative of a penetrative Caledonian neometamorphism of the gabbros. Accordingly, it is most likely that the Western Gneiss Region was intruded by an extensive suite of plutonic rocks (1600–1500 Ma) in which the gabbros were emplaced relatively late. The plutonites, together with earlier metamorphic rocks (e.g. migmatites), provided a basement for the supracrustals. In the southern part of the region, in the area of regional WNW–ESE regional grain, there were also substantial additions to the crust by plutonic activity in Sveconorwegian times at about 1000 Ma.

The tonalitic, granitic and pegmatitic dykes in the supracrustals of the Ålesund area (e.g., at Magerholm, UTM 712 250) are as yet undated, but might provide important information on the timing of accretion of the supracrustals to the surrounding migmatitic gneisses.

Wilson model

In a standard Wilson model of rifting, drifting and collision, the cover of supracrustals would be caught together with oceanic crust between the colliding plates and interlayered with their "basement" by obduction, imbrication and folding. Ultramafites from the upper mantle might be incorporated in the supracrustals and their migmatitic basement during such a process, and dry, mafic rocks transformed into eclogites in the overthickened crust after collision (see Cutberth *et al.* (1983) for a more detailed model and Gebauer *et al.* (1985) for an alternative comprising also a Sveconorwegian orogeny). In Fig. 5, the collision phase with overthickened crust is outlined for a simple Wilson model where the supracrustals and oceanic crust become squeezed, folded and imbricated between the colliding plates. After tectonic and erosional stripping of the upper plate, the supracrustals might form a cover intimately interfolded and imbricated with the lower crustal block. Such a model might apply for cover sequences like those of the Surnadal synform close to the Trondheim area (Krill, 1985) and similar occurrences.

One of the problems in making plate tectonic reconstructions for the Western Gneiss Region is the uncertain

age of the supracrustals. In view of this, the term Fjordane Complex can still be used as a provisional designation for any sequence of supracrustals, with or without sheets of crystalline rocks, that tectonically overlies more homogeneous, migmatitic and granitic gneisses.

CONCLUSIONS

(1) Certain of the supracrustal rocks at the southeast margin of the Western Gneiss Region are of undoubted Precambrian age and are only negligibly overprinted during the Caledonian orogeny.

(2) Layered micaceous gneisses with frequent inclusions of ultramafite, eclogite and anorthosite in outer Nordfjord were strongly affected by the Caledonian orogeny. It is uncertain whether they have formed from Precambrian protoliths or rather represent a strongly transposed nappe sequence originally corresponding to that of the marginal part of the orogen.

(3) Nappe sequences corresponding to the Lower and Middle Allochthon of the marginal part of the orogen can be identified in the mainly psammitic rocks (Lykkjebø Group) and the anorthosite-bearing orthogneisses (Eikefjord Group) of Sunnfjord–inner Nordfjord, Tafjord–Grotli–Sotaseter and the Oppdal district.

(4) In the central part of the Western Gneiss Region, the most conspicuous supracrustal rocks are a suite of banded amphibolites, garnet–biotite gneisses and marbles, termed the Ertevågøy Group. This suite of rocks occurs as more or less parallel zones in migmatitic gneisses and granitoids dated to 1700–1500 Ma and was strongly modified together with adjacent migmatitic and granitic gneisses during the high-pressure Scandian orogenic event (ca. 425 Ma) and later. There is an unsatisfactory coverage of radiometric dating for the supracrustals, although Sr and Nd isotope data indicate that at least one of the zones is 900–1100 Ma old and that it was attached to the adjacent migmatitic gneisses already in the Late Proterozoic (Aas, 1986). For other occurrences, especially in the Oppdal, Surnadal and the Fosen–Namsos areas, there is some evidence that the supracrustals were emplaced during the Caledonian orogeny (Krill, 1985, 1987; Johannson *et al.*, 1987).

(5) Ultramafic and eclogitic rocks occur in both supracrustals and migmatitic gneisses. However, many of the basic rocks in the supracrustals never reached the eclogite stage during the high-pressure metamorphism. Those that did, were to a large extent later retrograded to garnet amphibolites.

(6) Gabbro (dolerite) is apparently restricted to the migmatitic gneisses. This may indicate that the supracrustals occur above a basement with a different history, at least up to the time of gabbro intrusion around 1500 Ma or perhaps even as late as 1200 or 900 Ma ago. The two units were, however, accreted when they picked up inclusions of ultramafites from the upper mantle during the Caledonian orogeny.

(7) A separate history of the supracrustal rocks is indicated by dykes of tonalitic to granodioritic composition, not found in the adjacent "basement" gneisses. Accretion of the "basement" and "cover" probably occurred subsequent to the emplacement of the dykes.

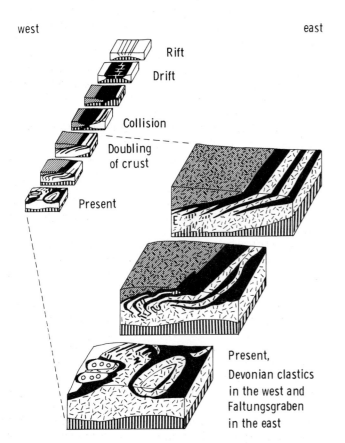

west

east

Rift

Drift

Collision

Doubling
of crust

Present

Present,
Devonian clastics
in the west and
Faltungsgraben
in the east

Fig. 5. Possible relation of supracrustal rock and oceanic crust (black) between two colliding plates in a Wilson model provided that the supracrustals really represent infolded Caledonian cover. Upper left inset redrawn from Bryhni (1983).

Acknowledgements

Field work was financed by Møre og Romsdal County via Norges Geologiske Undersøkelse, for which I express my gratitude. I also thank Professor B. A. Sturt for help.

Discussion and pleasant company in the field with H. Austrheim, B. Bjørnstad, L. Kullerud, K. Reksten, A. G. Aas and others is gratefully acknowledged.

REFERENCES

Aas, A. G. (1986). Tektonometamorf utvikling for ortogneiser og suprakrustale enheter i Sykkylven-Stranda området, Sunnmøre. Cand. Scient. thesis, University of Oslo, 149 pp.

Askvik, H. and Rokoengen, K. (1985). Geologisk kart over Norge. berggrunnskart KRISTIANSUND–M. 1:250 000. *Norges Geol. Unders.*

Bjørnstad, B. (1987). Geologisk utviking av proterozoiske gneiser i ytre Romsdal, Vest–Norge. Cand. Scient. thesis, University of Oslo, 131 pp.

Brueckner, H. (1977). A structural, stratigraphic and petrologic study of anorthosites, eclogites, and ultramafic rocks and their country rocks, Tafjord area, western south Norway. *Norges Geol. Unders.* **332**, 1–53.

Bryhni, I. (1966). Reconnaissance studies of gneisses, ultrabasites, eclogites and anorthosites in outer Nordfjord, western Norway, *Norges Geol. Unders.* **241**, 68pp.

Bryhni, I. (1983). Regional overview of metamorphism in the Scandinavian Caledonides. In: *Regional Trends in the Geology of the Appalachian–Hercynian–Mauritanide Orogen*, pp. 193–204. Reidel, New York.

Bryhni, I. and Grimstad, E. (1970). Supracrustal and infracrustal rocks in the Gneiss Region of the Caledonides west of Breimsvatn. *Norges Geol. Undes.* **266**, 105–140.

Bryhni, I. and Sturt, B. A. (1985). Caledonides of southwestern Norway. In: Gee, D. G. and Sturt, B. A. (eds.) *The Caledonian Orogen—Scandinavia and Related Areas*, pp. 89–108. Wiley, Chichester.

Bryhni, I. Brastad, K. and Jacobsen, V. W. (1979). Berggrunns-geologisk kart LEIKANGER 1317 II 1:50 000. Foreløpig utgave. Norges Geologiske Undersøkelse.

Bryhni, I., Austrheim, H. Bjørnstad, B. Kullerud, L. and Reksten, K. (1986). Berggrunnsgeologiske kart EIDE 1320 4 og TINGVOLL 1320 1, foreløpige utgaver. Norges Geologiske Undersøkelse.

Carswell, D. A. and Harvey, M. A. (1985). The intrusive history and tectonometamorphic evolution of the Basal Gneiss Complex in the Moldefjord area, west Norway. In: Gee, D. G. and Sturt, B. A. (eds.) *The Caledonide Orogen–Scandinavia and Related Areas*, pp. 843–858. Wiley, Chichester.

Cutberth, S. Harvey, M. A. and Carswell, D. A. (1983). A tectonic model for the metamorphic evolution of the Basal Gneiss Complex, western South Norway. *J. Metam. Geol.* **1**, 63–90.

Gebauer, D., Lappin, M.A., Grunenfelder, M. and Wyttenbach, A. (1985). The age and origin of some Norwegian eclogites: a U–Pb zircon and REE study. *Chem. Geol.* **52**, 227–247.

Gee, D. G., Kumpulainen, R., Roberts, D., Stephens, M. B., Thon, A. and Zachrisson, E. (compilers) (1985). Scandinavian Caledonides Tectonostratigraphic map. In: Gee, D. G. and Sturt, B. A. (eds.) *The Caledonide Orogen—Scandinavia and Related Areas*. Wiley, Chichester.

Griffin, W. L., Austrheim, H., Brastad, K., Bryhni, I., Krill, A. G., Krogh, E. J., Mørk, M. B. E., Qvale, H. and Tørudbakken, B. (1985). High-pressure metamorphism in the Scandinavian Caledonides. In: Gee, D. G. and Sturt, B. A. (eds.) *The Caledonide Orogen—Scandinavia and Related Areas*, pp. 783–801. Wiley, Chichester.

Hernes, I. (1955). Trondhjemsskifrene ved Molde. *Norsk Geol. Tidsskr.* **34**, 123–137.

Hernes, I. (1965). Die Kaledonische Schictenfolge in Mittelnorwegen. *N. Jb. Geol. Palaent. Mh.* **2**, 69–84.

Johansson, L., Andreasson, P.-G. and Schøberg, H. (1987). An occurrence of the Gula nappe in the Western Gneiss Region, central Scandinavian Caledonides. *Norsk Geol. Tidsskr.* **67**, 85–92.

Krill, A. G. (1985). Relationship between the Western Gneiss Region and the Trondheim Region: Stockwerk–Tectonics reconsidered. In: Gee, D. G. and Sturt B. A. (eds.) *The Caledonide Orogen—Scandinavia and Related Areas*, pp. 475–484. Wiley, Chichester.

Krill, A. G. (1987). Berggrunnskart STANGVIK 1420 4, 1:50 000, foreløpig utgave, Norges Geologiske Undersøkelse.

Krill, A. G. and Sigmond, E. M. O. (1986). Surnadalens dekkelagfølge og dens fortsettelse mot vest. *Geologinytt*, **21**, 35.

Kullerud, L. (1987). Rb–Sr studie av gneiser i ytre Romsdal, Vest Norge. Cand. Scient. thesis. University of Oslo, 173 pp.

Lappin, M. A., Pidgeon, R. T. and Breemen, O. van (1979). Geochronology of basal gneisses and mangerite syenites of Stadlandet, west Norway. *Norsk. Geol. Tidsskr.* **59**, 161–181.

Lutro, O. (1987). Mørkrisdalen 1418 II. Description of the geological map M. 1 50 000. *Norges Geol. Unders.*, in prep.

Mearns, E. W. (1984). *See* Mørk, M. B. E. and Mearns, E. W. 1986.

Muret, G. (1960). Partie de la culmination du Romsdal, Chaine Caledonienne, Norvege. *Int. Geol. Congr. Norden 1960*, **19**, 28–32.

Mysen and Heier (1972). Petrogenesis of eclogites in high-grade metamorphic gneisses, exemplified by the Hareidlandet eclogite, western Norway. *Contrib. Mineral. Petrol.* **36**, 73–94.

Mørk, M. B. E. and Mearns, E. W. (1986). Sm–Nd isotopic systemaics of a gabbro–eclogite transition. *Lithos* **19**, 255–267.

Reksten, K. (1986). En petrologisk studie av Eide-området, Romsdalshalvøya. Cand. Scient. thesis, University of Oslo, 222 pp.

Roberts, D. (1986). Structural-photogeological and general tectonic feaures of the Fosen–Namsos Western Gneiss Region of Central Norway. *Norges Geol., Unders.* **407**, 13–25.

Råheim, A. (1972). Petrology of high grade metamorphic rocks of the Kristiansund area. *Norges Geol. Unders.* **57**, 317–328.

Sigmond, E. M. O., Gustavson, M. and Roberts, D. (1984). Berggrunnskart over Norge, 1:1000 000. Norges Geologiske Undersøkelse.

Strand, T. and Kulling, O. (eds.) (1972). *Scandinavian Caledonides.* Wiley, Chichester.

Tucker, R. D. (1986). Geology of the Hemnefjord–Orkanger area, South–central Norway. *Norges Geol. Unders.* **404**, 1–24.

Tørudbakken, B. (1982). En geologisk undersøkelse av nordre Trollheimen ved Surnadal med hovedvekt på sammenhengen mellom strukturer, metamorfose og aldersforhold. Cand. real. thesis. University of Oslo, 259 pp.

Part V

Palaeontology and Biostratigraphy

19 Stratigraphy and faunas of the Parautochthon and Lower Allochthon of southern Norway

David L. Bruton,[1] *David A. T. Harper*[2] *and John E. Repetski*[3]

[1]Paleontologisk Museum, Sars Gate 1, 0562 Oslo 5, Norway
[2]Department of Geology, University College, Galway, Ireland
[3]U.S. Geological Survey, U.S. National Museum, Washington D.C. 20560, USA

Parautochthonous successions in Rogaland and on Hardangervidda together with allochthonous sequences in east Jotunheimen contain brachiopods, trilobites, graptolites and condonts ranging in age from Early Cambrian to Middle Ordovician. Fossils from four intervals, late Middle Cambrian, early and late Tremadoc and late Arenig–early Llanvirn, permit correlation of strata and events across the entire Baltoscandian platform. Where developed, the upper Middle Cambrian is uniform in both fauna and facies. Although presumed lower Tremadoc strata on Hardangervidda preserve a siliciclastic facies, elsewhere facies and faunas comparable with the Tremadoc Ceratopyge Shale, Ceratopyge Limestone and Dictyonema Shale are present. The last is developed throughout the Parautochthon and Lower Allochthon. During the Arenig and early Llanvirn, siliciclastic sediments were deposited on Hardangervidda, whilst broadly coeval turbidites within the Lower Allochthon indicate the initiation of tectonism in this part of externalides.

Despite the rarity of fossils in the deformed rocks of the Scandinavian Caledonides, palaeontological information is of critical importance to the understanding of this complex part of the orogen. Bruton (1986) has recently classified the better-documented fossiliferous successions within the Scandinavian Caledonides on the basis of the tectonic position of each together with the age and biogeography of their faunas. Currently, considerable attention has been focused on the fossiliferous rocks of the Upper Allochthon in the Trondheim Region (e.g., Neuman and Bruton, 1974, Bruton and Bockelie, 1980, 1982), where faunal data have provided important constraints on the development of models for the internal parts of the orogen.

In contrast, the faunas of the locally fossiliferous successions in the Parautochthon and Lower Allochthon have received relatively little attention. Nevertheless, study of such parts of the tectonostratigraphy of the Caledonides affords an opportunity to examine events and processes within this part of the externalides of the orogen. We briefly review the existing palaeontological data for this part of the Scandes and illustrate and discuss new faunal information from the Cambrian of the Parautochthon and the Cambrian and Ordovician of the Lower Allochthon in southern Norway (see Fig. 1 for locality information).

GEOLOGICAL SETTING

The proven Cambrian and Ordovician rocks of the Parautochthon and Lower Allochthon of southern Norway were deposited in a variety of environments on and adjacent to the western edge of the Baltoscandian platform (Størmer, 1967). To the east and southeast, the rocks of the Oslo Region occupy an intermediate position between the developing Caledonides and the more stable Baltoscandian platform. Carbonate sources in the east contrast with emergent siliciclastic sources to the west within the Caledonides and both promoted an asymmetry of facies distribution on both local and regional scales.

The Cambrian rocks of this part of the Scandinavian Caledonides have featured in reviews by Martinsson (1974), Bergström (1980) and Bergström and Gee (1985), whilst Bruton *et al.* (1985) and Bruton and Harper (1988) have reviewed the Ordovician successions of the region. Both groups of studies have emphasized, to varying degrees, the relationships of the successions within the Caledonides with those coeval sequences developed elsewhere on the platform whilst Bruton and Harper (1988) have attempted to relate events manifest within the platform successions with those occurring to the west in the developing Caledonides.

The Caledonide Geology of Scandinavia © R. A. Gayer (Graham & Trotman), 1989, pp. 231–241.

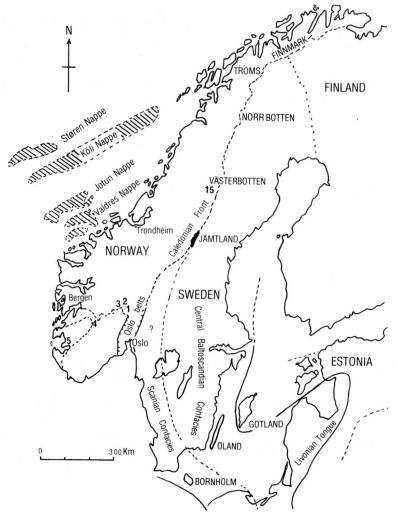

Fig. 1. Map of Scandinavia and adjacent Baltic area indicating localities discussed in text. Presumed pre-thrust positions of four of the allochthons, prior to the Scandian event, are superimposed as are the Baltoscandian confacies belts. Numbered localities are as follows: 1, Aurdal; 2, Synfjell; 3, Mellane; 4, Hardangervidda; 5, Ritland. Localities 1–3 are in the Lower Allochthon whilst 4 and 5 are in the Parautochthon.

PARAUTOCHTHON

Cambrian and Ordovician rocks cropping out within the Parautochthon of southern Norway have been reviewed briefly by Brynhi and Sturt (1985). They were deposited towards the western edge of the Baltoscandian platform and have suffered limited translation. The most complete succession, on Hardangervidda, is illustrated in Fig. 2, column 4.

Cambrian

On Hardangervidda, Lower Cambrian rocks have been documented from both Finse and Ustaoset. At Finse (Goldschmidt, 1912), conglomerates with *Torellella laevigata* (Linnarsson, 1871) overlie the Precambrian basement. The succession at Ustaoset is dominated by conglomerates, shelly sandstones and limestones (Goldschmidt, 1925) that overlie the Precambrian and contain *Torellella laevigata* and *Strenuaeva linnarssoni* (Kiær, 1917) (=? *Proampyx linnarssoni*; see Ahlberg, 1985, p. 341) (Størmer, 1925).

Rocks of Middle Cambrian age crop out along the west-facing scarp of Austmannshovud, east of Ritland farm, not west as stated by Henningsmoen (1952, p. 14), in the Hjelmeland district, Josenfjord, northwest of Stavanger. Here, black, fossiliferous, rusty-weathering

shales, with a basal sandstone containing an ichnofauna including *Cruziana* crop out. Henningsmoen (1952) documented a trilobite fauna, with *Peronopsis fallax* (Linnarsson, 1869), *Paradoxides oelandicus* Sjögren, 1872 and *Ptychoparia anderseni* Henningsmoen, 1952, together with the brachiopods *Lingulella* spp., *Acrothele* (*Redlichella*) *granulata* Linnarsson, 1876, *Acrotreta socialis* Seebach, 1865 and an orthoid; hyoliths, sponge spicules and various problematica are associates. Henningsmoen (1952) concluded an early Middle Cambrian (*P. oelandicus* Zone) age for the fauna and, since rocks of this age are apparently absent in the Oslo Region, Västergötland and Scania, he envisaged a shallow-water marine environment, for the Ritland fauna, seaward of a landmass to the east.

The shales are at least 120 m thick; the highest 20 m are exposed in a vertical cliff. Approximately half way up the cliff, a bed of limestone concretions is developed. The limestone, collected from loose blocks at the base of the cliff, has yielded an abundant trilobite fauna (Fig. 3) including, *Peronopsis* cf. *quadrata* (Tullberg, 1880), *P. fallax*, *Ptychagnostus* sp., *Solenopleura* cf. *bucculenta* Grönwall, 1902, *Solenopleura* spp., *Anomocaroides* cf. *limbatus* Angelin, 1854 and *Parasolenopleura* sp.

An abundant inarticulate brachiopod fauna (Fig. 3), recovered from etched limestone residues, is dominated by "*Acrotreta*" cf. *socialis* together with *Dictyonina* cf. *ornatella*

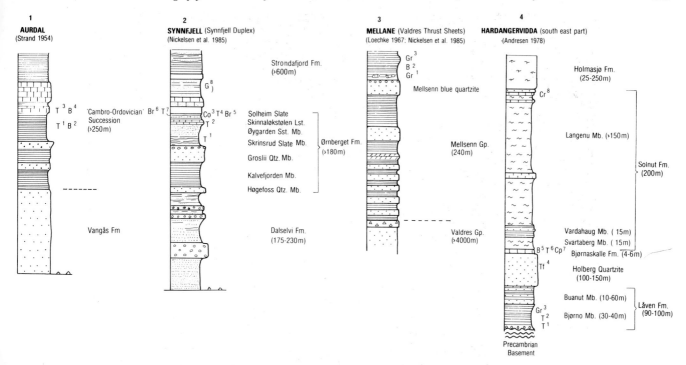

Fig. 2. Schematic representations of the stratigraphy in each of the thrust sheets of East Jotunheimen area and that of the Parautochthon on Hardangervidda. Faunal symbols are: B, brachiopod; Co, conodont; Cp, cephalopod; Cr, crinoid ossicle; Gr, graptolite; T, trilobite; Tf, trace fossil. Superscripts are as follows: Aurdal: 1, 2—Middle and Upper Cambrian trilobites and brachiopods; 3, 4—Late Tremadoc (Ceratopyge Limestone) trilobites and brachiopods. Synfjell: 1, 2—Middle Cambrian trilobites; 3–5—Tremadoc (Dictyonema Shale, Ceratopyge Limestone) conodonts, trilobites and brachiopods; 6, 7—Arenig–Llanvirn shelly fauna; 8—Arenig–Llanvirn graptolites. Mellane: 1, 2—Tremadoc (Dictyonema Shale) graptolites and brachiopods (?Ceratopyge Shale); 3—Arenig–Llanvirn graptolites. Hardangervidda: 1—Lower Cambrian faunas; 2—Middle Cambrian trilobites; 3—Tremadoc (Dictyonema Shale) graptolites; 4—tube-like traces; 5–7—Arenig–Llanvirn shelly fauna; 8—Middle Ordovician echinoderm.

(Linnarsson, 1876), *Acrothele* cf. *granulata* and rare lingulids and obolids. The residues also contain hesslandonid ostracods, abundant phosphatized echinoderm plates, cystoid stem columnals, sponge spicules and numerous problematica. The fauna is currently being monographed by the first two authors.

The entire fauna and associated facies suggest a correlation with the upper Middle Cambrian Andrarum Limestone (*Solenopleura* [= *Jincella*] *brachymetopa* Zone; see Martinsson, 1974, Fig. 4), particularly its development on the island of Bornholm (Berg–Madsen, 1985). The Andrarum Limestone and associated facies (Fig. 6(a)) probably had a wide distribution across the submerged parts of the Baltoscandian platform.

The complete extent of the Cambrian succession at Ritland is not known with certainty. Nevertheless, blocks of limestone containing abundant specimens of *Hyolithes* sp. have been collected by Dr A. G. Doré whilst specimens of *Eccaparadoxides* sp. have been recovered recently from the lower scree slopes confirming the presence of the lower Middle Cambrian (Henningsmoen, 1952). This sequence is underlain by a spectacular polymict, immature breccia containing angular blocks of gneiss several metres wide. Doré (personal communication) considers the breccia to be a highly immature deposit and not simply a basal conglomerate associated with the lower Cambrian transgression. Rather deposition by alluvial fans or debris flows in a tectonically active environment is envisaged. Spjeldnaes (1985, p. 139) has suggested depositional processes involving rifting or faulting.

Bruton *et al.* (1985) have recently identified strata correlated with the uppermost zone of the Middle Cambrian (*Lejopyge laevigata* Zone) overlying a basal conglomerate near Nasatjønn on Hardangervidda.

Ordovician

The presence of Tremadoc rocks in the Parautochthon has been demonstrated by the occurrence of *Dictyonema flabelliforme* across the region (Størmer, 1941), further confirming the uniformity of facies across large parts of Baltoscandia. (We follow Legrand (1985) in retaining the name *Dictyonema* in preference to *Rhabdinopora* (Erdtmann, 1982).) There is as yet no evidence for younger Tremadoc strata in the Parautochthon.

In the vicinity of Hermodsholtjønn, near Haukelifjell on Hardangervidda, a shelly fauna from parautochthonous metasedimentary rocks was described by Bruton *et al.* (1985). Brachiopods from the Bjørnaskalle Formation, in the middle part of the succession, indicate a correlation with the uppermost Arenig–lowest Llanvirn interval. The brachiopods, together with associated, as yet undescribed trilobites and cephalopods are typically Baltic and suggest an origin within that province. The fauna is located within a calcareous, sandy horizon; elsewhere the succession is dominated by siliciclastic material. Although the fauna suggests a correlation with the Orthoceras Limestone and related units developed elsewhere on the Baltoscandian platform, the facies of the Bjørnaskalle Formation is in marked contrast to those carbonate-dominated units. To the north and west, deeper-water facies are developed (Bruton and Harper, 1988) whilst, significantly, to the northeast some 200 km along the strike of the Caledonian Front, intra-Iapetus faunas are present within the lower part of the Upper Allochthon (Bruton and Harper, 1981).

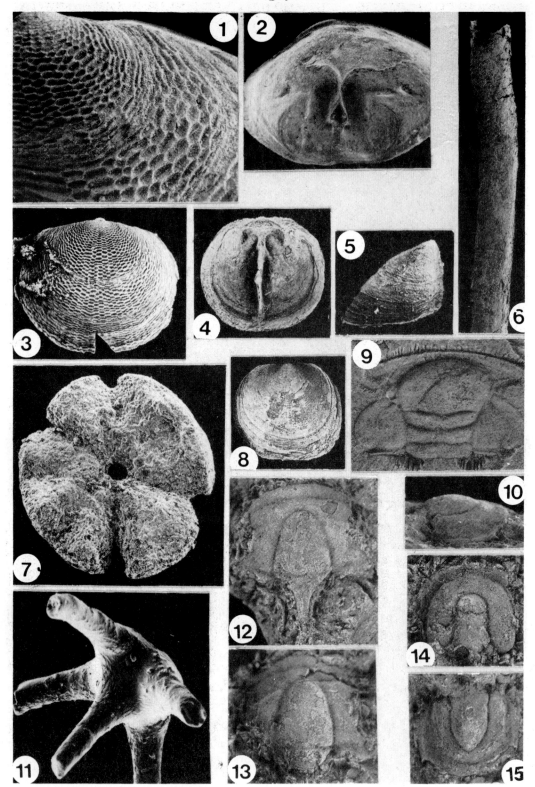

Fig. 3. All material from west-facing scarp of Austmannshovud, east of Ritland Farm (Lyngvatnet Sheet 1313 III, GR 533682), Hjelmeland district, Josenfjord, northwest of Stavanger, Part (9) from loose scree material of? early Middle Cambrian age; the remainder from limestone blocks of late Middle Cambrian age (*Solenopleura* [= *Jincella*] *brachymetopa* Zone). Specimens on parts 1–8, 11 recovered after etching with acetic acid.

1,3: *Dictyonina* cf. *ornatella* (Linnarsson, 1876). **1**: Detail of distinctive pitted ornament on posterolateral part of pedicle valve, × 30, P.M.O. 117.041. **3**: Pedicle valve exterior, × 7, P.M.O. 117.042.

2,4,5,8: "*Acrotreta*" cf. *socialis* (Seebach, 1865). **2**: Interior of pedicle valve, × 20, P.M.O. 117.043. **4**: Interior of brachial valve, × 8, P.M.O. 117.044. **5**: Lateral view of pedicle valve exterior, × 9, P.M.O. 117.045. **8**: Exterior of brachial valve, × 9, P.M.O. 117.046.

6: Phosphatic tube, × 48, P.M.O. 117.047.

7: Phosphatized cystoid stem ossicle, × 144, P.M.O. 117.048.

9: *Eccaparadoxides* sp. Dorsal view of cranidium, × 2, P.M.O. 117.049. Coll. O.T.B. Nokling, 1981.

10,14,15: *Peronopsis* cf. *quadrata* (Linnarsson, 1869) **10,15**: Lateral and dorsal views of pygidium, × 6, P.M.O. 117.050. **14**: Dorsal view of cephalon, × 6, P.M.O. 117.051.

11: Sponge spicule, × 120, P.M.O. 117.064.

12: *Parasolenopleura* sp. Dorsal view of cranidium. × 4, P.M.O. 117.052.

13: *Solenopleura* sp. Dorsal view of cranidium, × 4, P.M.O. 117.053.

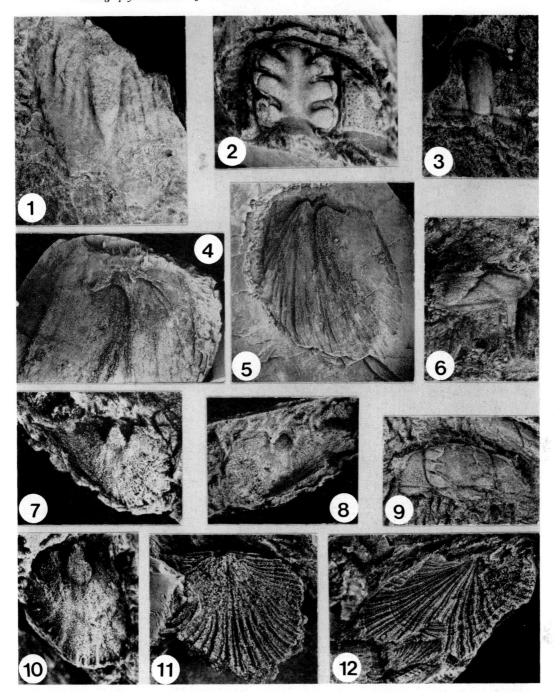

Fig. 4. Specimens in parts 1,3,6 from Tremadoc Ørnberget Formation (Solheim Slate, Ceratopyge Limestone equivalent), southwest of Groslii and south Øygarden, Valdres (Fullsenn Sheet 1717 III, GR 173703).

1: *Dikelokephalina dicraeura* (Angelin, 1854). Dorsal view of pygidium, × 5, P.M.O. 117.054. Coll. D.L.B., 1973.

2: *Parapilekia* sp. Dorsal view of cephalon, × 5, P.M.O. 117.055. Coll. D.L.B. 1973.

3,6: *Ceratopyge forficula* (Sars, 1835) **3**: Dorsal view of cranidium, × 5, P.M.O. 117.056. **6**: Dorsal view of incomplete pygidium, × 5, P.M.O. 117.057. Coll. G. Henningsmoen, 1969.

4,5: Taffiid brachioped gen. et sp. indet. Latex cast and internal mould of brachial valve, × 2, P.M.O. 117.058. Tremadoc Mellsenn Group (?Ceratopyge Shale equivalent), Mellenesæter Authors' Coll., 1985.

7,8,10–12: *Archaeorthis* cf. *christianiae* (Kjerulf, 1865). Ørnberget Formation (Solheim Slate, "Orthoceras Limestone" equivalent), Groslii. Authors' Coll., 1984. **7,8**: Internal mould and latex cast of pedicle valve, × 6, P.M.O. 117.059. **10**: Internal mould of pedicle valve, × 6, P.M.O. 117.060. **11**: Latex cast of external mould of brachial valve, × 6. **12**: External mould of brachial valve exterior, × 6, P.M.O. 117.062.

9: *Pliomera* sp. Dorsal view of cranidium, × 5, P.M.O. 117.063. Coll. D.L.B., 1973.

All material shown in Figs. 3 and 4 housed in the Paleontologisk Museum, Oslo (P.M.O.).

LOWER ALLOCHTHON

The stratigraphy of the Lower Allochthon in East Jotunheimen has been reviewed in detail by Nickelsen *et al.* (1985). Four major tectonostratigraphic units have been recognized: in ascending order, the Aurdal, Synfjell and Valdres thrust sheets and the Strondafjord Formation.

The stratigraphy and facies of the Aurdal Thrust Sheet are fundamentally different from those of the other three. Carbonate facies are dominant and invite close comparison with those of the successions of the Oslo Region. We review some of the existing palaeontological information from these tectonic units and document new data from the Synfjell and Valdres thrust sheets.

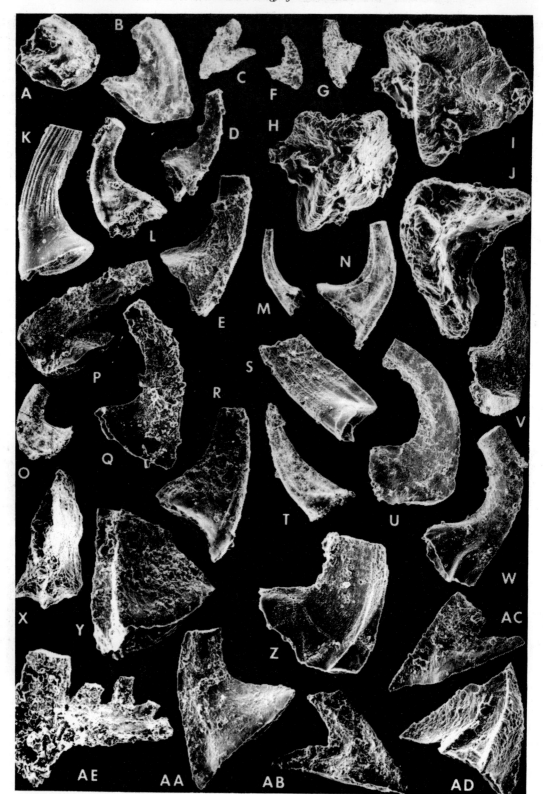

Fig. 5. Late Tremadoc conodonts from the Ørnberget Formation (Solheim Slate, Ceratopyge Limestone equivalent), southwest of Groslii and south of Øygarden, Valdres (Fulsenn Sheet 1717 III, Gr 173703). SEM photomicrographs of carbon-coated specimens. Illustrated specimens are reposited in conodont type collections of Department of Paleobiology, U.S. National Museum (U.S.N.M.), Washington, D.C., U.S.A.; non-illustrated conodonts are housed in U.S.G.S. conodont collections, under Cambro-Ordovician sample numbers 10650-CO and 10651-CO.

A, B: *Drepanoistodus* aff. *D. forceps* (Lindström). Inner lateral views of oistodontiform, × 77, and non-geniculate (drepanodontiform) coniform, × 50, elements, respectively, U.S.N.M. 424953 and 424954, loc. U.S.G.S. 10650-CO.

C–G: *Acodus* sp. Lateral views of oistodontiform (**C**, × 65), ramiform (**D, E**, × 75 and × 150), drepanodontiform (**F**, × 55), and acondontiform (**G**, × 55) elements. **D** and **E**, U.S.N.M. 424955 and U.S.N.M. 424956 from loc. U.S.G.S. 10650-CO. **C, F**, and **G**: U.S.N.M. 424957, 424958 and 424959 from loc. U.S.G.S. 10651-CO.

H–J: Genus and species indeterminate. Stelliscaphate platform elements. **H, I**: Upper and oblique-upper views, × 162 and × 170, U.S.N.M. 424960. **J**: upper view, U.S.N.M. 424961, both from loc. U.S.G.S. 10650-CO.

K: "*Scolopodus*" *peselephantis* Lindström. Lateral view, × 150, U.S.N.M. 424962 from loc. U.S.G.S. 10650-CO.

L: *Oneotodus variabilis* Lindström. Inner lateral view, × 70, U.S.N.M. 424963, loc. U.S.G.S. 10651-CO.

M: *Scolopodus* sp. Lateral view, × 50, U.S.N.M. 424964, loc. U.S.G.S. 10651-CO.

N: *Protopanderodus?* sp. C. Inner lateral view, × 50, U.S.N.M. 424965, loc. U.S.G.S. 10650-CO.

O–Q: *Paroistodus numarcuatus* (Lindström). Inner lateral views of oistodontiform (**P**, × 68; U.S.N.M. 424966) and drepanodontiform

Cambrian

The Skindsrud Slate of the Ørnberget Formation, in the Synfjell Thrust Sheet, has yielded the trilobite *Paradoxides* together with hyoliths (Bruton, unpublished data), indicative of a Middle Cambrian age for this part of the succession (Nickelsen *et al.*, 1985, Fig. 5). Within the Skinnaløkstølen Limestone Member of the Ørnberget Formation, a fauna of agnostids and large trilobites was correlated with those of the Middle Cambrian by Henningsmoen (in Nickelsen *et al.*, 1985, p. 373). There is as yet no palaeontological evidence to suggest the presence of Lower or Upper Cambrian strata in this part of the Lower Allochthon.

Ordovician

Bruton and Harper (1988) have reviewed the palaeontological information, to date, derived from the Ordovician rocks of the Aurdal, Synfjell and Valdres thrust sheets and have summarized the current internal stratigraphy of each unit (Fig. 2, columns 1–3, modified from Nickelsen *et al.* (1985)). The lowermost unit in the tectonostratigraphy, the Aurdal Thrust Sheet, has a facies development with some similarities to that of the Oslo Region; the upper two units, however, differ in the sparse development of carbonates.

The middle unit, the Synfjell Thrust Sheet, contains about 35 m of poorly exposed and tectonized grey slate, assigned by Nickelsen *et al.* (1985, p. 373) to the Solheim Slate Member of the Ørnberget Formation. Several metres above the underlying Skinnaløkstølen Limestone, light-grey weathering shales contain *Dictyonema*. The shales are overlain by a thin bed of limestone considered by Henningsmoen (in Nickelsen *et al.* 1985, p. 373) to be a lateral equivalent of the Ceratopyge Limestone of the Oslo Region; whilst Repetsky (*sic*) reported a conodont fauna of latest *Paltodus deltifer* Zone (late Tremadoc) age from acid residues (in Nickelsen *et al.* 1985, p. 373). A correlation of the lower part of the Solheim Slate with both the lower and upper Tremadoc may be suggested.

A significant amount of new data has been assembled from intensive collection of the Solheim Slate at Groslii sæter. The presence of the dendroid graptolite *Dictyonema* has been confirmed in the light-grey weathering shales above the Skinnaløkstølen Limestone. From beds immediately above the limestone, a brachiopod fauna (Fig. 4), dominated by abundant *Archaeorthis* cf. *christianiae* (Kjerulf, 1865) together with an indeterminate orthide, a taffiid and syntrophid, has been collected from horizons within a locally calcareous slate.

The overlying limestone unit contains a trilobite fauna (Fig. 4) dominated by *Ceratopyge forficula* (Sars, 1835) and species of *Parapilekia*, *Euloma*, *Niobe* and *Shumardia*. Brachiopods are sparse, but nevertheless a species of *Lingulella* and *Archaeorthis* cf. *christianiae* together with three indeterminate orthidines are present. Conodonts (Fig. 5) recovered from acid residues include those listed in Table I. The Colour Alteration Index (CAI) of the conodonts is 5–5.5, indicative of a minimum finite temperature of 300–400°C.; on the basis of these data, Nickelsen et al. (1985, p. 377) suggested an overburden of some 10–13 km, assuming a geothermal gradient of 30°C/km.

The conodont fauna clearly established a late Tremadoc age and equivalency with the Ceratopyge Limestone elsewhere on the Baltoscandian platform. Lindström (1955) documented several of the above species from the Ceratopyge Limestone in south central Sweden, and most of the Valdres species, including *Paltodus d. deltifer*, *Oneotodus variabilis*, and *"Scolopodus" peselephantis*, occur in the Ceratopyge Limestone at Bjerkåsholmen, Asker, on the Oslofjord (Repetski, unpublished material). Viira (1974) reported this fauna, including the critical (*"peracutus"*) element of the index subspecies *P. d. deltifer*, in the upper part of the Varangu Member ($A_{III}V$) of the upper Tremadoc of North Estonia, and she tentatively traced it into the subsurface of W Latvia. She included the Varangu Member in her *Scandodus varanguensis* Biozone. Szaniawski (1980), on the basis of study of Tremadoc faunas from the Holy Cross Mountains, Poland, and examination of other Baltoscandian faunas, synonymized *Scandodus varanguensis* Viira with the apparatus of *Paltodus deltifer pristinus* (Viira). He also subdivided the *Paltodus deltifer* Zone of Lindström (1971) into a lower *P. d. pristinus* Subzone and upper *P. d. deltifer* Subzone. The latter is characterized by the appearance of the nominal subspecies and *Paroistodus numarcuatus*. Two more recent reports also record the indices of the *P. d. deltifer* Zone in the uppermost part of the Tremadoc in the eastern Baltoscandian platform region (Dubinina, 1983; Borovko *et al.*, 1983).

Acodus sp. (Fig. 5, C–G) is quite similar to, and indeed might be conspecific with, the apparatus identified as *Acodus?* sp. by Szaniawski (1980). However, inadequate preservational quality of these small elements does not warrant assignment except to agree with Szaniawski (1980) that these do conform to the *Acodus* apparatus plan. Of additional note in the Valdres fauna is the recovery of two elements of a platform taxon (Fig. 5, H–J), which are, unfortunately, also too poorly preserved for further systematic analysis at this time.

A locally abundant brachiopod fauna higher in the succession contains *Archaeorthis* aff. *christianiae*, *"Orthambonites"* cf. *novitas* Öpik, 1939, *Ranorthis* cf. *norwegica* Öpik, 1939 and *Oslogonites* cf. *costellatus* Öpik, 1939; all the named species, with the exception of *A. christianiae*, occur with varying degrees of abundance in the middle part of the Orthoceras Limestone, the Expansus (or

---

(**O, Q**, × 47 and × 125, U.S.N.M. 424967 and 424968, respectively) elements. **O** from loc. U.S.G.S. 10651-CO; **P** and **Q** from loc. U.S.G.S. 10650-CO.

R: *Protopanderodus?* sp. B. Inner lateral view, × 100, U.S.N.M. 424969, loc. U.S.G.S. 10650-CO.

S, T: *Protopanderodus?* sp. A. Inner lateral views; **S**: × 150, U.S.N.M. 424970. **T**: × 50, U.S.N.M. 424971, from locs. U.S.G.S. 10650-CO and 10651-CO, respectively.

U–W: *Drepandous arcuatus* Pander. Inner lateral views, respectively, of sculponeaform, × 88, graciliform, × 75, and arcuatiform, × 75, elements, U.S.N.M. 424972, 424973, and 424974. All from loc. U.S.G.S. 10650-CO.

X–AD: *Paltodus deltifer deltifer* (Lindström). **X,Y**: Upper and anterolateral views, × 200 and × 150, of suberectiform ("peracutus") elements, U.S.N.M. 424975 and 424976. **Z, AD**: Lateral views, × 150 and × 175, of multicostate coniform elements, U.S.N.M. 424977 and 424978. **AA**: Inner lateral view of drepanodontiform element, × 88, U.S.N.M. 424979. **AB, AC**: Inner lateral views, × 75 and × 88, of oistodontiform elements, U.S.N.M. 424980 and 424981. All from loc. U.S.G.S. 10650-CO.

AE: *Cordylodus* sp. Inner lateral view of compressed element, × 115, U.S.N.M. 424982, loc. U.S.G.S. 10651-CO.

Table I. Conodonts from the Ørnberget Formation

	Sample	
	10650–CO; (375 g)	10651–CO; (1205 g)
Acodus sp.		
Acondontiform (P) elements	12	8
Drepanodontiform + ramiform (S) elements	46	16
Oistodontiform (M) elements	25	3
Cordylodus sp.		
Rounded elements		1
Compressed elements		1
Drepanodus arcuatus Pander		
Arcuatiform elements	67	37
Pipaform and sculponeaform elements	30	22
Graciliform elements	3	3
Drepanoistodus aff. *D. forceps* (Lindström)		
Drepanodontiform elements	5	2
Oistodontiform elements	12	1
Oneotodus variabilis Lindström	10	10
Paltodus deltifer deltifer (Lindström)		
Drepanodontiform elements	8	11
Multicostate coniform elements	20	20
Suberectiform ("*peracutus*" elements	4	5
Oistodontiform elements	29	56
Paroistodus numarcuatus (Lindström)		
Drepanodontiform elements	84	42
Oistodontiform elements	21	15
Protopanderodus? sp. A	23	3
P.? sp. B	2	1
P.? sp. C	1	
Scolopodus sp.		2
"*Scolopodus*" *peselephantis* Lindström	18	5
Gen et sp. indet. stelliscaphate platform elements	2	
Indet. scandodontiform elements	1	
Indet. oistodontiform elements	7	
Indet. coniform elements	42	5
Phosphannulus universalis Müller, Nogami and Lenz	1	
Undet. phosphatic brachiopod valves	62	3

Asaphus) Shale, of the Oslo Region (Öpik, 1939). The fauna was recovered from a slate.

Graptolites indicating "3b" to "4a" (in terms of the Norwegian etager) horizons have been noted from the Solheim Slate (Strand, 1938). These horizons are considered by Nickelsen *et al.* (1985, p. 374) to be correlatives of the "Graptolitskifer" in Vestre Gausdal (Bjørlykke, 1893). The graptolite fauna from these beds at Bratland in Vestre Gausdal has been revised in detail by Williams (1984), who concluded (p. 10), that all the taxa may be found within the middle Arenig–lower Llanvirn interval (*D. nitidus* or *I. gibberulus* Subzone to the "*D. bifidus*" Zone of the British sequence).

The upper tectonic unit, the Valdres Thrust Sheet, contains the Mellsenn Group, which ranges in age from late Precambrian to probable Llanvirn. At Mellane the group is overturned. The Upper (Blue Quartz) Quartzite is overlain by 4 m of black slate containing specimens of *Dictyonema* (Peach, in Nickelsen *et al.*, 1985). Immediately above the black slates, 5 to 8 m of dark-grey slate with limestone nodules have yielded orthide brachiopods (Loeschke, 1967; Peach, personal communication). Bjørlykke (1905) reported graptolites of "3b" to "4a" age from grey slates higher in the group. Detailed collection of the section at Mellenesaeter has confirmed and partly expanded the faunal information already available. Rare and strained brachiopods were obtained from the

limestone nodules and dark-grey slate. The fauna contains *Ceratopyge forficula* together with *Archaeorthis* cf. *christianiae* and an indeterminate taffiid brachiopod and is probably comparable to that from the supposed correlatives of the Ceratopyge Shale within the Solheim Slate of the Synfjell Thrust Sheet.

ANALYSIS

The new and existing palaeontological information available for the Parautochthon and Lower Allochthon of the southern parts of Norway permits a preliminary analysis of the development and evolution of facies patterns in this part of the Scandinavian Caledonides. Moreover, such patterns may be related to both local and regional tectonic events within the orogen.

The development of the Lower Cambrian in the Parautochthon is sporadic and where present is associated with a variety of coarse siliciclastic facies. Although Strand (1954) has presented palaeontological evidence indicating the presence of Lower Cambrian strata within the Aurdal Thrust Sheet of the Lower Allochthon, there are as yet no comparable fossil data from the higher tectonostratigraphical units in the East Jotunheimen area; nevertheless, the apparent palaeontological hiatus may be represented by, to date, unfossiliferous arenaceous and conglomeratic

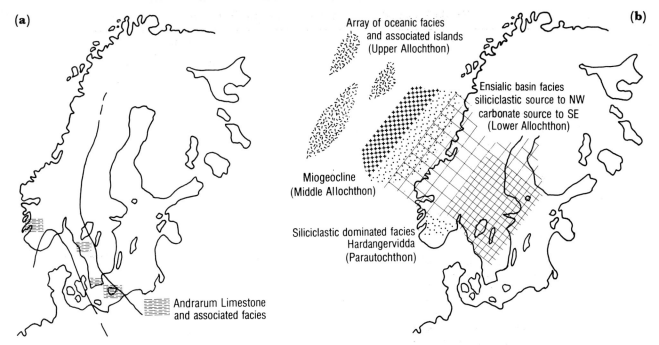

Fig. 6. (*a*) Distribution of upper Middle Cambrian Andrarum Limestone and associated facies throughout Baltoscandia (palaeogeography modified from Martinsson 1974, Fig. 15). (*b*) Hypothetical distribution of late Arenig – early Llanvirn magnafacies across southern part of Scandinavian Caledonides.

facies. In contrast, the higher Middle Cambrian is fairly uniform where developed on the Baltoscandian platform (Fig. 6(*a*)). There is as yet no faunal evidence for Upper Cambrian strata throughout the Parautochthon and Lower Allochthon in southern Norway. The highest alum shale deposits, containing *Dictyonema flabelliforme*, have a wide distribution (Størmer, 1941, and herein) across the Autochthon, Parautochthon and Lower Allochthon, suggesting an interval of uniform and pervasive sedimentation

Facies variants of the Ceratopyge Shale and Ceratopyge Limestone have been recognized in all three tectonic units within the Lower Allochthon. The brachiopod fauna of these formations is dominated by small orthidines similar in composition and diversity to coeval faunas in the Oslo Region (Harper, 1986). The Tremadoc articulate brachiopods of the Baltoscandian area are relatively poorly known. The species are usually referred to genera that have been hitherto documented in most detail from the North American Province (see, for example, Tjernvik, 1956, and Henningsmoen, 1964). Whether these genera had such a cosmopolitan distribution during the Tremadoc or merely reflect parallel evolution of these primitive orthide stocks requires further investigation of the Baltic faunas.

In the Parautochthon on Hardangervidda, the Tremadoc horizons above the Dictyonema shale are presumably represented by a succession of sandstones and shales.

The post-Tremadoc rocks of the Parautochthon and Lower Allochthon display the greatest mutual contrasts and differences, with coeval successions both in the Oslo Region and eastwards on the Baltoscandian platform. On Hardangervidda, the lithofacies and ichnofauna of the Holberg Quartzite indicate shallow-water, siliciclastic-dominated environments during much of the Arenig; sandy carbonates, the Bjørnaskalle Formation, correlate with the Orthoceras Limestone of the Oslo Region. Northwards and westwards, deeper-water facies are developed (Andresen, 1978) and transitions from shelf environments to those of a miogeosyncline have been suggested (Andresen and Færseth, 1982).

Within the Lower Allochthon of East Jotunheimen, the systematic development of various siliciclastic turbidite facies may be related to the early tectonic evolution of this part of the Baltic margin (Nickelsen *et al.*, 1985). Distal turbidites are developed within the Solheim Slate of the Synfjell Thrust Sheet at levels above the Tremadoc, although sandy beds are present within correlatives of the Ceratopyge Limestone and Ceratopyge Shale. Higher in the Solheim Slate more proximal turbidites are present, whilst within the higher Valdres Thrust Sheet proximal facies are developed at significantly lower horizons. The diachronous development of these various turbidite facies has been related to the progressive approach, from the northwest, of the higher basement nappes of the Middle Allochthon (e.g., the Jotun Nappe) preceded by an apron of siliciclastic sediment (Nickelsen *et al.*, 1985). Reappraisal of the biostratigraphy of this part of the Lower Allochthon suggests initiation of this event within the Arenig.

Nystuen (1981) has suggested that parts of the Lower Allochthon may have been translated some 200–400 km southeastwards. The detailed palinspastic reconstruction of Hossack *et al.* (1985) indicates a minimum translation of 290 km for the Jotun Nappe of the Middle Allochthon; moreover, restoration of the thrust sheets of the Lower Allochthon require the existence of a shelf some 400 km wide (Nickelsen *et al.*, 1985). This extensive shelf was bounded to the northwest by the miogeocline (Fig. 6(*b*)) whose post-Tremadoc advance southeastwards ultimately disrupted and destroyed this external part of the orogen. Seaward of the miogeocline, a variety of oceanic facies are now located within the Upper Allochthon. The associated, often diverse, faunas have been assigned to intra-Iapetus island complexes (Neuman, 1984; Bruton and Harper, 1985) within the internal parts of the Caledonides; recent isotopic studies of the hydrography of the Ordovician Iapetus Ocean (Keto and Jacobsen, 1987) are consistent with the existence of such island habitats.

Acknowledgements

We are grateful to Dr. R. P. Nickelsen, Bucknell University, U.S.A. for placing his Synfjell fossil collection at our disposal and for guiding Bruton in the field. Dr. A. G. Doré (Conoco, U.K. Ltd.) and Mr B. T. G. Wandås (Saga Petroleum A/S), kindly provided locality and fossil data from Ritland, Rogaland. The SEM photographs were taken with the help of Mrs. Eva Braaten, Biologisk Institut, Electronmikroskopi Laboratoriet, University of Oslo. Field work in Jotunheimen by Bruton and Harper and in Rogaland by Bruton was financed by Norges Almenvitenskapelige Forskningsråd. Bruton wishes to thank Mr. Frank Nikolaisen for field assistance in Rogaland and numerous discussions about the fauna. Lucy Kellehan helped with the typing. Repetski thanks R. T. Lierman and R. C. Orndorff for technical assistance and publishes with permission of the Director of the U.S. Geological Survey.

REFERENCES

Ahlberg, P. (1985). Lower Cambrian trilobite faunas from the Scandinavian Caledonides—a review. In: Gee, D. G. and Sturt, B. A. (eds.) *The Caledonide Orogen—Scandinavia and Related Areas*, pp. 339–346. Wiley, Chichester.

Andresen, A. (1974). New fossil finds from the Cambro-Silurian metasediments on Hardangervidda. *Norges Geol. Unders.* **304**, 55–60.

Andresen, A. (1978). Lithostratigraphy of the autochthonous/parautochthonous Lower Palaeozoic metasediments on Hardangervidda, South Norway. *Norges Geol. Unders.* **338**, 59–69.

Andresen, A. and Færseth, R. (1982). An evolutionary model for the southwest Norwegian Caledonides. *Amer. J. Sci.* **282**, 756–782.

Angelin, N. P. (1854). Palaeontologica Scandinavica I. *Crustacea formationis transitionis*. Fasc. 2, pp. 21–92. Lund.

Berg-Madsen, V. (1985). Middle Cambrian biostratigraphy, fauna and facies in southern Baltoscandia. Acta Universitatis Upsaliensis. *Abstracts of Uppsala Dissertations Faculty Science* **781**, 37pp.

Bergström, J. (1980). The Caledonian margin of the Fennoscandian shield during the Cambrian. In: Wones, D. R. (ed.) *The Caledonides in the U.S.A.* Dept. Geol. Sci. Va. Poly. Inst. Univ. Mem. 2, pp. 9–13.

Bergström, J. and Gee, D. G. (1985). The Cambrian in Scandinavia. In: Gee, D. G. and Sturt, B. A. (eds.) *The Caledonide Orogen—Scandinavia and Related Areas*, pp. 247–270. Wiley, Chichester.

Bjørlykke, K. O. (1893). Gausdal. *Norges Geol. Unders.* **13**, 1–36.

Bjørlykke, K. O. (1905). Det centrale Norges fjellbygning. *Norges Geol. Unders.* **39**, 1–595.

Borovko, N. G., Popov, L. E. and Sergeeva, S. P. (1983). New data on the fauna characteristic and on the range of the Upper Tremadocian deposits in the eastern part of the Baltic-Ladoga klint. *Dok. Akad. Nauk. USSR* **273**, 404–407 [In Russian].

Bruton, D. L. (1986). Recognition and significance of fossil-bearing terranes in the Scandinavian Caledonides. *Geol. Fören. Stockh. Förh.* **108**, 272–273.

Bruton, D. L. and Bockelie, J. F. (1980). Geology and palaeontology of the Hølonda area, western Norway—a fragment of North America? In: Wones, D. R. (ed.) *The Caledonides in the U.S.A.* Dept. Geol. Sci. Va. Poly. Inst. State Univ. Mem., 2, pp. 41–47.

Bruton, D. L. and Bockelie, J. F. (1982). The Løkken–Hølonda–Støren areas. In: Bruton, D. L. and Williams, S. H. (eds.) *Field excursion guide, IV. Int. Symp. Ordovician System. Palaeont. Contrib. Univ. Oslo*, **279**, 77–91.

Bruton, D. L. and Harper, D. A. T. (1981). Brachiopods and trilobites of the early Ordovician serpentine Otta Conglomerate, south central Norway. *Norsk Geol. Tidsskr.* **61**, 153–181.

Bruton, D. L. and Harper, D. A. T. (1985). Early Ordovician (Arenig–Llanvirn) faunas from oceanic islands in the Appalachian–Caledonide orogen. In: Gee, D. G. and Sturt, B. A. (eds.) *The Caledonide Orogen—Scandinavia and Related Areas*, pp. 359–368. Wiley, Chichester.

Bruton, D. L. and Harper, D. A. T. (1988). Arenig–Llandovery stratigraphy and faunas across the Scandinavian Caledonides. In: Harris, A. L. and Fettes, D. J. (eds.) *The Caledonian—Appalachian Orogen. Geol. Soc. Lond. Spec. Publ.* pp. 477–498.

Bruton, D. L., Harper, D. A. T., Gunby, I. and Naterstad, J. (1985a). Cambrian and Ordovican fossils from the Hardangervidda Group, Haukelifjell, Southern Norway, *Norsk Geol. Tidsskr.* **64**, 313–324.

Bruton, D. L., Lindström, M. and Owen, A. W. (1985b). The Ordovician of Scandinavia. In: Gee, D. G. and Sturt, B. A. (eds.) *The Caledonide Orogen—Scandinavia and Related Areas*, pp. 273–282, Wiley, Chichester.

Brynhi, I. and Sturt, B. A. (1985). Caledonides of southwestern Norway. In: Gee, D. G. and Sturt, B. A. (eds.) *The Caledonide Orogen—Scandinavia and Related Areas*, pp. 89–107. Wiley, Chichester.

Dubinina, S. (1983). Conodont stratigraphy of the lower Ordovician and the lower part of the Middle Ordovician in North-West Latvia. *Eesti NSV Tead. Akad. Toim.* **32**, 45–52 [In Russian].

Erdtmann, B.-D. (1982). A reorganisation and proposed phylogenetic classification of planktic Tremadoc (early Ordovician) dendroid graptolites. *Norsk Geol. Tidsskr.* **62**, 121–144.

Goldschmidt, V. M. (1912). Ein kambrisches konglomerat bei Finse und dessen Metamorphose. *Vid. Selsk. Skr. M.N.Kl.*, 18.

Goldschmidt, V. M. (1925). Ueber fossilfuhrende unterkambrische Basalablagerungen bei Ustaoset. *Fennia* **45**, 3–11.

Grönwall, K. A. (1902). Bornholms Paradoxideslag og deres fauna. *Danmarks Geol. Unders.* 2 raekke Nr. 13, i–xii, 230 pp., 4pls.

Harper, D. A. T. (1986). Distributional trends within the Ordovician brachiopod faunas of the Oslo Region, south Norway. In: Racheboeuf, P. R. and Emig, C. C. (eds.), *Les Brachiopodes fossiles et actuels. Biostratigraphie du Paléozoique* **4**, 465–475.

Henningsmoen, G. (1952). Early Middle Cambrian fauna from Rogaland. *Norsk Geol. Tidsskr.* **30**, 13–32.

Henningsmoen, G. (1964). Liste over fossiler i Norge. Under Ordovicium (nr. 3). Fossil-Nytt 3.

Hossack, J. R., Garton, M. R. and Nickelsen, R. P. (1985). The geological section from the foreland up to the Jotun thrust sheet in the Valdres area, south Norway. In: Gee, D. G. and Sturt, B. A. (eds.) *The Caledonide Orogen—Scandinavia and Related Areas*, pp. 443–456, Wiley, Chichester.

Keto, L. S. and Jacobsen, S. B. (1987). Nd and Sr isotopic variations of Early Paleozoic Oceans. *Earth Planet. Sci. Lett.* **84**, 27–41.

Kiær, J. (1917). The Lower Cambrian *Holmia* fauna at Tømten in Norway. *Norsk Vid.-Akad. Skr. M. N. Kl.* 1916: 10, 140 pp.

Kjerulf, T. 1865. Veiviser ved geologiske excursioner i Christiania omegna. Univ. Programm 2. Halvaat 1865, Christiania, 43 p.

Legrand, P. (1985). *Dictyonema – Rhabdinopora. Norsk Geol. Tidsskr.* **65**, 224–225.

Lindström, M. (1955). Conodonts from the lowermost Ordovician strata of south-central Sweden. *Geol. Fören. Stockh. Förh.* **76**, 517–604.

Lindström, M. (1971). Lower Ordovician conodonts of Europe. In: Sweet, W. C. and Bergström, S. M. (eds.) *Symposium on Conodont Biostratigraphy*, pp. 21–61, Geol. Soc. Amer. Mem. 127.

Linnarsson, J. G. O. (1869). Om Vestergötlands Cambriska och Siluriska aflagringar. *Kongl. Sv. Vet.-Akad. Handl.* **8**(2), 1–89.

Linnarsson, J. G. O. (1871). Geognostika och palaeontologiska iakttagelser öfver Eophytonsandstenen i Vestergötland. *Kongl. Sv. Vet.-Akad. Handl.* **9**(7).

Linnarsson, J. G. O. (1876). Brachiopoda of the Paradoxides beds of Sweden. *Kongl. Sv. Vet.-Akad. Handl.* **3**(12), 1–34.

Loeschke, J. (1967). Zur petrographie des Valdres Sparagmites zwischen Bitihorn und Langsuen/Valdres (Sud Norwegen). *Norges Geol. Unders.* **243**, 67–98.

Martinsson, A. (1974). The Cambrian of Norden. In: Holland, C. H. (ed.) *Cambrian of the British Isles, Norden, and Spitsbergen*, pp. 185–283, Wiley, Chichester.

Nickelsen, R. P., Hossack, J. R., Garton, M. and Repetsky [*sic*], J. (1985). Late Precambrian to Ordovician stratigraphy and correlation in the Valdres and Synfjell thrust sheets of the Valdres area, southern Norwegian Caledonides, with some comments on sedimentation. In: Gee, D. G. and Sturt, B. A. (eds.), *The Caledonide Orogen—Scandinavia and Related Areas*, pp. 369–379. Wiley, Chichester.

Neuman, R. B. (1984). Geology and paleobiology of islands in the Ordovician Iapetus Ocean: Review and implications. *Bull. Geol. Soc. Amer.* **95**, 1188–1201.

Neuman, R. B. and Bruton, D. L. (1974). Early Middle Ordovician fossils from the Hølonda area, Trondheim Region, Norway. *Norsk Geol. Tidsskr.* **54**, 69–115.

Nystuen, J. P. (1981). The late Precambrian "Sparagmites" of southern Norway: a major Caledonian Allochthon—the Øsen-Røa Nappe Complex. *Amer. J. Sci.* **281**, 69–94.

Öpik, A. A. (1939). Brachiopoden und ostrakoden aus dem Expansusschiefer Norwegens. *Norsk Geol. Tidsskr.* **19**, 117–142, pls. 1–6.

Sars, M. (1835). Ueber einige neue oder unvollstandig bekannte Trilobiten. *Okens Isis. Jg.* 1835, 28, cols. 33–343.

Seebach, K. von (1865). Beitrage zur Geologie der Insel Bornholm. *Zeit Deutsch Geol. Gesell.* **17**, 338–347.

Sjögren, A. (1872). Om några försteningar i Ölands kambriska lager. *Geol. Fören. Stockh. Förhandl.* **1**, 67–80, pl. 5.

Spjeldnæs, N. (1985). Biostratigraphy of the Scandinavian Caledonides. In: Gee, D. G. and Sturt, B. A. (eds.) *The Caledonide Orogen—Scandinavia and Related Areas*, pp. 317–329. Wiley, Chichester.

Størmer, L. (1925). On a Lower Cambrian fauna at Ustaoset. *Fennia* **45**, 12–22.

Størmer, L. (1941). Dictyonema shales outside the Oslo Region. *Norsk Geol. Tidsskr.* **20**, 161–169.

Størmer, L. (1967). Some aspects of Caledonian geosyncline and foreland west of the Baltic Shield. *Q. J. Geol. Soc. Lond.* **123**, 183–214.

Strand, T. (1938). Nordre Etnedal. Beskrivelse til det geologiske gradteigskart. *Norges Geol. Unders.* **180**, 54 pp.

Strand, T. (1954). Aurdal. Beskrivelse til det geologiske gradteigskart. *Norges Geol. Unders.* **185**, 1–71. [In Norwegian with English Summary.]

Szaniawski, H. (1980). Conodonts from the Tremadocian chalcedony beds, Holy Cross Mountains (Poland). *Acta Palaeont. Pol.* **25**, 101–121, pls. 15–18.

Tjernvik, T. E. (1956). On the Early Ordovician of Sweden: stratigraphy and fauna. *Bull. Geol. Inst. Uppsala* **36**, 107–284.

Tullberg, S. A. (1880). Om *Agnostus*-arterna i de kambriska aflagringarne vid Andrarum. *Sveriges Geol. Unders. C*, **42**.

Viira, V. (1974) Ordovician conodonts of the east Baltic. *Eesti NSV Tead. Akad. Geol. Inst. 'Valgus'* 140 pp., pls. 1–13. [In Russian with English summary.]

Williams, S. H. (1984). Lower Ordovician graptolites from Gausdal, central southern Norway: a reassessment of the fauna. *Norges Geol. Unders.* **395**, 1–24.

Part VI

Devonian Geology

20 Microtectonic evidence of Devonian extensional westward shearing in Southwest Norway

Alain Chauvet and Michel Séranne

Laboratoire de Tectonique, C.N.R.S. UA266 U.S.T.L., Place E. Bataillon
34060 Montpellier Cédex, France

A microtectonic investigation of both the Devonian Basins of Western Norway and the adjacent part of the Western Gneiss Region provides evidence of extensional tectonics. The Devonian Basins are bounded by a long-lived, low-angle, hangingwall down to the west, extensional detachment that controlled the syntectonic sedimentation and induced the deformation of the basin-fill. On the basis of shear criteria analysis in the basement, it is shown that the detachment corresponds to the last stage in the evolution of a wide extensional down to the west 2 km thick shear zone. It is argued that these Devonian extensional tectonics overprinted and retrogressed previous Caledonian compressive structures, thus providing an efficient process for the thinning of the previously orogenically thickened crust.

INTRODUCTION

The Caledonian Orogen in southern Scandinavia is characterized by a sequence of large-scale nappes that were thrust from west to east onto the autochthonous rocks of the Baltoscandian platform during Middle Ordovician to Upper Silurian times (Sturt *et al.*, 1975; Sturt and Thon, 1978; Roberts and Sturt, 1980; Gee, 1982). This Mid-Paleozoic orogenesis resulted in a thickened crust with an associated high-grade and eclogitic metamorphism (Bryhni and Brastad, 1980; Cuthbert *et al.*, 1983).

The relation of the Devonian Basins of western Norway with the Caledonian Orogen is not clearly defined. Recently the extensional origin of the basin has been demonstrated (Hossack, 1984; Séranne and Séguret, 1985a) and was designated as a post-Caledonian extensional event (Norton, 1986; McClay *et al.*, 1986; Séranne and Séguret, 1987). Moreover, the occurrence of eclogitic rocks as young as 400–420 Ma and yielding pressures of 12–18 kbar and temperatures of 500–800°C (Griffin and Brueckner, 1980; Griffin *et al.*, 1985) close to the Devonian paleosurface provides a very strong constraint for a late event allowing the rapid uplift of the high-pressure rocks between the Caledonian Orogeny and Devonian basin formation (Norton, 1986). However, it must be stressed that if the eclogites formed during the Caledonian Orogeny, their exhumation is not only the result of extensional faulting, as the thickness of the pre-extensional orogenic crust would be too great (40–60 km for the formation of high-pressure rocks added to the present 30 km beneath the eclogite outcrops). Thrusting subsequent to high-pressure metamorphism (Cuthbert *et al.*, 1983) and prior to extension is though to have played a role in the ascent of the eclogites in the crust.

The microtectonic investigation of the Devonian basin-fill and the basement of these basins reveals the existence of extensional structures linked to the activity of a large-scale low-angle extensional shear zone that controlled the deposition of the sediments (Chauvet *et al.*, 1987).

The aim of this study is to present and discuss additional evidence of Devonian extensional westward shearing with the help of a microtectonic-based method involving analysis of ductile shear criteria and brittle structures and to give a chronological evolution of the post-Caledonian extension.

This paper first summarizes a structural analysis of the basins and then studies the evolution of the deformation in the pre-Devonian basement. Evidence will be given for a single extensional event that associates the evolution of the two zones. Finally, the relationship of these structures will be discussed with respect to the Caledonian Orogeny.

GEOLOGICAL SETTING

Large-scale Caledonian nappes emplaced by eastward-directed thrusting are presently preserved east of the Western Gneiss Region. The four Devonian basins crop out on the western margin of the gneiss-window. From north to south, they are: Hornelen, Hasteinen, Kvamshesten and Solund basins (Fig. 1).

The basement consists of "supracrustal" rocks of uncertain age folded with Precambrian gneisses and highly deformed Paleozoic rocks (Bryhni, 1966, 1987).

The Devonian sediments are bounded by tectonic contacts except on the western boundary where they unconformably overlie slightly deformed mangerites-syenites, Ordovician–Silurian metasediments and ophiolitic rocks of Ordovician age (Brekke and Solberg, 1988). These units may represent outliers of Caledonian nappes that were thrust over the Western Gneisses and are related to the metasedimentary cover and the Jotun Nappe (Skjerlie, 1969, 1974).

STRUCTURAL ANALYSIS

Devonian basins

Although some differences can be noted, the four basins show the same general organization. The infill consists of coarse clastic sediments: conglomerates developed along the active marginal faults interfingering with fluvio-lacustrine sandstones (Steel *et al.*, 1977). In the Solund Basin, the basin-fill consists mostly of coarse conglomerates of questionable sedimentological interpretation (Nilsen, 1968; Steel *et al.*, 1985). The stratification dips at an average angle of 25° east in the axes of the basins (Fig. 1 and 2). In the Hornelen Basin where bedding has a similar orientation to the Solund Basin, a 25 km thick

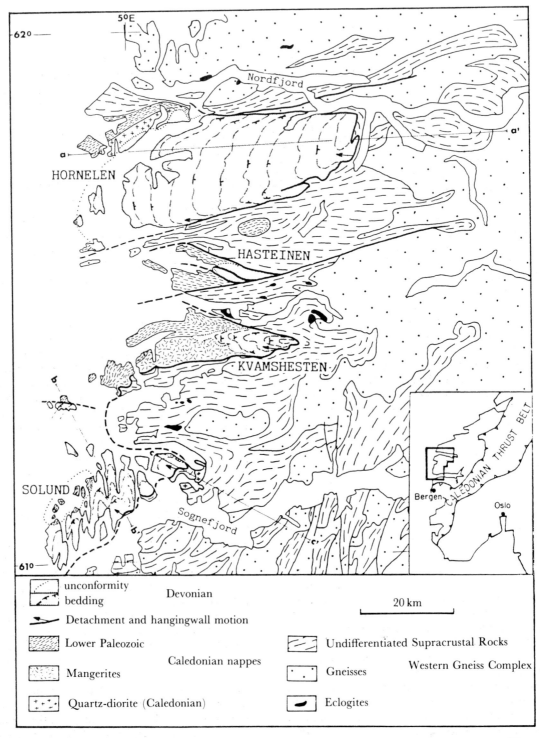

Fig. 1. Geological map of the study area. (Adapted from Kildal (1970) and Sigmond *et al.*, (1983).

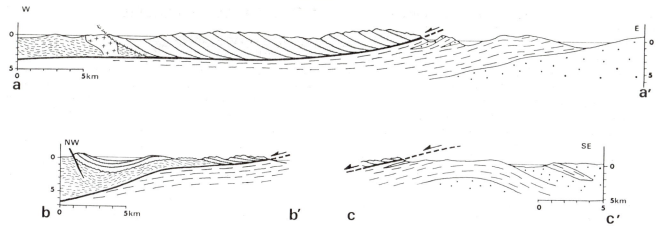

Fig. 2. Geological sections through the study area. Location on Fig. 1. Legend as Fig. 1.

stratigraphic sequence has been measured (Bryhni, 1964). Such a thickness does not represent the true depth of the basins as there is no significant change in strain or in metamorphic grade between the oldest and the youngest strata. This considerable sediment thickness constitutes the main constraint on the mechanism of the basin formation. Various mechanisms of depocentre migration have been proposed (Bryhni, 1964; Steel and Gloppen, 1980; Bryhni and Sturt, 1985) by westward-dipping retrogressive high-angle normal faults. Those models take into account the discrepancy between stratal thickness and actual basin depth (which does not exceed few kilometres) but they fail to explain (i) the lack of transverse faults into the basins and (ii) the occurrence of the low-angle normal detachment that bounds the basins to the east.

The eastern margin of the basins is a low-angle dip-slip detachment that dips 25° west in the Hornelen Basin and is shallowly dipping (locally subhorizontal) in the Kvamshesten and Solund Basins. It consist of highly sheared rocks involving both Devonian sediments and basement. The microtectonic analysis demonstrates east–west lineations and dip-slip slickensides parallel to the maximum dip direction. Towards the north and south, the normal detachment progressively changes to lateral ramps that exhibit oblique slickensides. In each basin, the northern lateral ramp yields a dextral offset, whereas the southern one is sinistral (Fig. 1). These shear-senses are supported by microstructures such as slickensides, and by large-scale Riedel faults associated with the lateral ramps. The lateral ramps and westward-dipping basal detachment do not cut each other and are, therefore, assumed to link. The normal displacement on the westward-dipping detachment is transferred to strike-slip along the east–west trending lateral ramps.

The basin-fill displays various progressive unconformities that are well exposed within the steeply dipping formations along the northern margins of Hornelen and Kvamshesten Basins (Bryhni and Skjerlie, 1975). In Hornelen, the thickness variation of the sedimentary sequences between the marginal conglomerates and the axis of the basin (ratio from 1 to 3 respectively) gives evidence for a syntectonic progressive unconformity. These facts demonstrate the syntectonic character of the sedimentation by movements along the detachment.

It has been proposed that the four basins were bounded by a single westward-sliding detachment where the north and south lateral ramps correspond to east–west trending culminations of the basement (Hossack, 1984; Séranne and Séguret, 1985a, 1987). These two studies concluded that the horizontal displacement along the detachment is about

50 km during Middle Devonian sedimentation but the cumulative displacement may be greater (Norton, 1986).

Some recent paleomagnetic studies suggested that folding of the Devonian series could be due to a late tectonic event as young as Jurassic in age (Torsvik *et al.*, 1986, 1987). However, this deformation affected the basin-fill while sediments were not fully lithified (Séranne and Séguret, 1985b; 1987), which supports a synsedimentary deformation.

Microtectonic study of the basin-fill

The sediments of the Hornelen and Kvamshesten Basins are not seen to be affected by a penetrative deformation, because they represent shallow erosion levels (Seranne and Séguret, 1987). The Hasteinen basin-fill underwent a severe deformation that is linked to the development of a sinistral ENE–WSW strike-slip system but which is outside the topic of this study.

The Solund Basin displays the deep structures of the basins because erosion reached deeped levels than in Hornelen. The conglomerates that represent most of the infill are deformed. Strain consists of reorientation of the long axes of the clasts in a SE–NW direction that is normal to bedding strike. The preferred orientation is associated with vertical and NE–SW striking tension-gashes (i.e., normal to the clasts preferred orientation) (Fig. 3). The tension-gashes only affect the pebbles and die out in the matrix, thus indicating a rheological contrast between the clasts and the whole conglomerate. The angular relationship between bedding-strike, clast orientation and tension-gashes described above has also been locally observed in the Hornelen basin.

In the Solund Basin, the matrix possesses a flat-lying cleavage that does not affect the clasts. The schistosity contains a NW–SE stretching lineation which is also identified on the surfaces of the clasts. Pebbles and cobbles display horizontal and NW–SE oriented tectonic pressure-shadows. They consist of preferred cementation of the sandy matrix by crystallization of quartz, chlorite and epidote at the end of the clasts. The surfaces of the cobbles bear slickensides identical to those described on fault planes by Petit (1987). They indicate that the clasts were able to move against each other within the conglomerate.

It has been suggested that in the Solund Basin, clasts were reoriented by strain toward the NW–SE direction by body-rotation within a non-fully lithified conglomerate (Séranne and Séguret, 1985b, 1987). If the conglomerate originally had sedimentary fabric (Nilsen, 1968), the

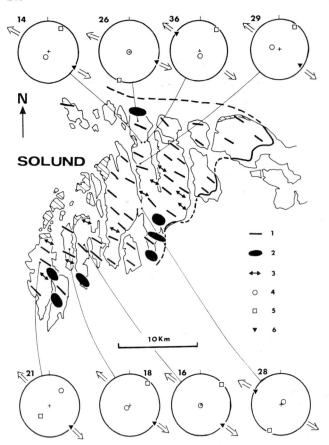

Fig. 3. Structural map of Solund Basin: 1, orientation of the long axes of clasts by body rotation within the conglomerate (Data from Nilsen, 1968); 2, ductile deformation of the clasts close to the basal detachment; 3, extension determined by tension-gashes; 4, 5 and 6: respectively the maximum, intermediate and minimum stresses computed from groups of striated microfaults (Etchecopar *et al.*, 1981). Extension (arrows) is given by the orientation of the minimum stress in the Schmidt stereonet, lower hemisphere. Number of microfaults is given for each microtectonic station.

(1) The tectonic regime was an east–west extension during the formation of the Devonian basins.

(2) This extension is compatible with the strain axes determined from the ductile deformation and the tension gashes.

(3) The basin-fill deformation is contemporaneous with the basin formation.

(4) Deformation increases toward the basal detachment.

(5) Eastward shearing within the basin-fill is antithetic to the major westward shearing along the basal detachment.

Microstructure evolution in the pre-Devonian basement

The detachment displays the highest strain rate in the area. It defines a heterogeneous shear zone that involved Devonian sediments (in the hangingwall) and basement rocks (in the footwall). Except for local variations due to varying original lithologies, the fault-rocks associated with the detachment show a north–south change from cataclastic rocks in Hornelen to mylonitic rocks in Solund. This has been related to a deepening level of erosion towards the south (Séranne and Séguret, 1987).

The substratum of the four basins consists of Paleozoic and Precambrian rocks that show a similar deformation pattern. They exhibit a regional foliation plane that tends to parallel the Devonian detachment. The general foliation results from polyphase tectonics and includes isoclinal folding of an early foliation (Santarelli, 1977).

A stretching lineation is clearly developed throughout the study area. It is formed either by elongated feldspars, or by reoriented micas and amphiboles. The first type defines a penetrative lineation, whereas the second is carried by foliation planes only. The latter postdates the former. They both tend towards parallelism with an E–W to NW–SE direction when one goes towards the detachment (Fig. 4), whereas they have a inconsistent orientation away from the detachment.

Numerous microstructures characteristic of a westward non-coaxial deformation can be observed in the study area. Some asymmetric feldspars and rolling structures can be seen in the rocks showing an elongated feldspar lineation, in sections parallel to the lineation. Figure 5(a) displays a typical pattern of rotated feldspar where the shear-sense is indicated by the embayments of matrix close to the porphyroclast (Passchier and Simpson, 1986). In Fig. 5(b) the sense of vorticity is shown by the sigmoidal aspect of the deformed porphyroclasts (Passchier and Simpson, 1986). Porphyroclasts are also observed associated with pressures shadows of distinct composition. The asymmetry of pressure-shadows is diagnostic of rotation of the clasts in a non-coaxial progressive deformation (Etchecopar and Malavieille, 1987). These shear criteria clearly indicate a westward displacement of the hangingwall.

C–S and C′–S fabrics are the most frequent structures (Fig. 5(c)). The angular relation between the schistosity and the centimetre-scale shear planes defines the sense of shearing (Berthé *et al.*, 1979; Simpson and Schmid, 1983). They consistently show a westward sense of shearing. In thin section, the shear bands are characterized by some crystallization of biotite, quartz, chlorite, albite, and epidote assemblages that diagnose a retrogressive metamorphism associated with the deformation (Fig. 5(d)). C–S and C′–S microstructures are clearly associated with the mica and amphibole lineation. Frequently, the shear planes deform the sigmoidal porphyroclasts and the pressure shadows described above.

effects of the tectonic-induced rotation would have been greatly enhanced.

The stratification planes sometimes show sheared sandy intervals and grooved or slickensided surfaces that are evidence of dip-slip movements parallel or close to the dip-direction. The constant attitude of the long axes of the clasts with respect to bedding-dip (i.e., preferred orientation parallel to structural dip), which is consistent throughout the Solund Basin and reported within the steeply dipping bedding planes of the other basins, especially in northern Hornelen (Bryhni, 1978) and Hasteinen, argues for a reorientation of the clasts by eastward dip-slip shearing along the bedding planes.

Towards the tectonic basal detachment, the deformation increases and the pebbles are ductilely stretched (Fig. 3). The ductile strain recorded by these clasts yields a stretching direction parallel to the preferred orientation observed within the less-deformed area.

The conglomerates in the Solund Basin are also affected by metre-scale microfaults. A computer-aided method (Etchecopar *et al.*, 1981; Etchecopar, 1984) applied to the slickensides allowed us to determine the orientation of the stress tensor that was responsible for the development of these microfaults (Fig. 3). The minimum (extensional) stress (σ_3) is consistently oriented NW–SE and horizontal, whereas the maximum stress (σ_1) is vertical.

Microtectonic analysis of the basin fill led to the following results.

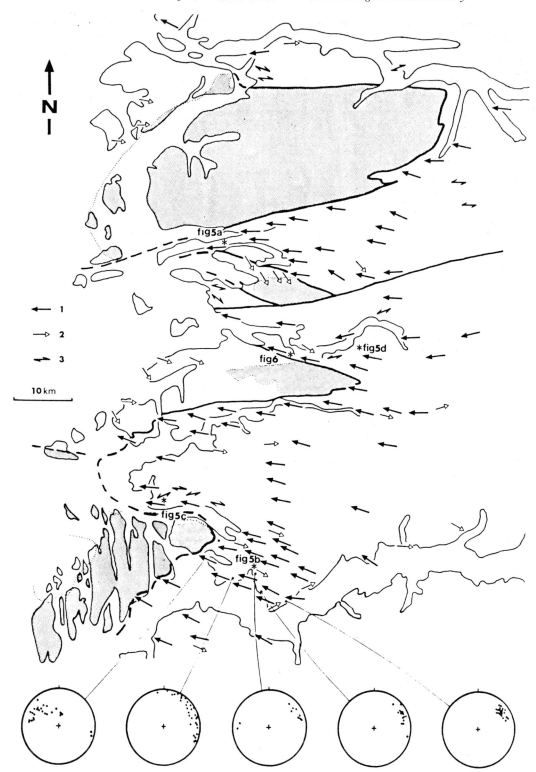

Fig. 4. Stretching lineation and shearing vergence in the basin basement: 1, westward shear criteria; 2, eastward shear criteria; 3, shear criteria defined in vertical foliation. Folds axes orientation plotted on Schmidt stereonet (lower hemisphere), tending to parallel the stretching lineation towards the detachment.

The plastic deformation of quartzitic rocks also supports a westward shearing. The obliquity of the quartz $\langle c \rangle$ axes maxima give evidence of a westward shearing (Bouchez and Pécher, 1981) (Fig. 6). The distribution of the maxima favours a basal and rhombohedric glide of $\langle a \rangle$ type characteristic of a low-grade metamorphic facies (Tullis *et al.*, 1973) consistent with the low-grade assemblages that characterize the shear bands. Numerous examples of quartz elongated grain shape favour a late tectonic imprint of the quartz texture (Brunel, 1980; Simpson and Schmid, 1983).

The different shear criteria are initiated during a progressive shearing event beginning in relatively hot con-

ditions (early feldspar lineation and related deformed porphyroclasts) and continuing at lower temperatures (retrogressive lineation contemporaneous with the shear bands).

All these structures are associated with a metre-scale folding that shows curved axes and subhorizontal axial planes. These folds have an asymmetric geometry and they post-date the previous isoclinal ones. Regionally, the fold axes tend to lie parallel to the stretching lineation towards the basin detachment (Fig. 4) (Norton, 1986, 1987). They are characteristic of "*a*" folds generated in progressive deformation by simple shear (Williams, 1978; Cobbold

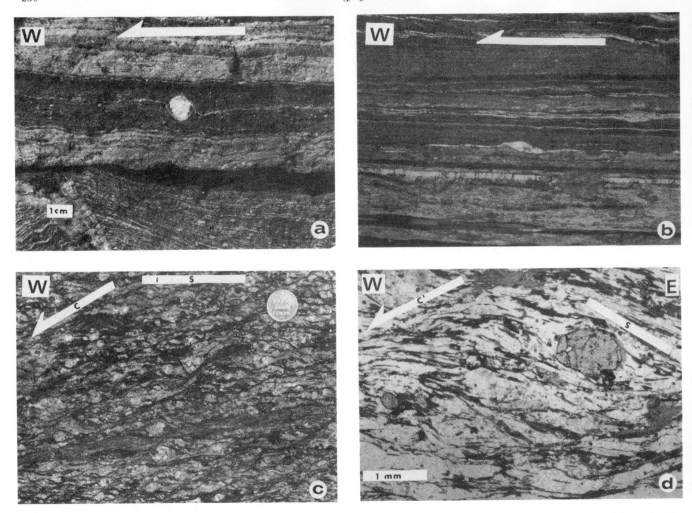

Fig. 5. Microstructures providing shear criteria in the basement; location on Fig. 4. (*a*) Feldspar porphyroclast with rolling tail; (*b*) asymmetric feldspar prophyroclast associated with drag folds toward the west; (*c*) shear bands in paragneiss (C–S structures); (*d*) retromorphic character of the metamorphic conditions marked by the crystallization of chlorite and calcite in shear bands.

and Quinquis, 1980; Malavieille *et al.*, 1984). These structures are diagnostic of heterogeneous shear zones whose related strain culminates in the detachment.

The shear criteria demonstrate a regionally westward vergence in the study area (Fig. 4). Close to the basin,

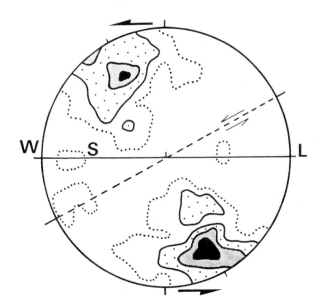

Fig. 6. Typical example of quartz ⟨*c*⟩ axis preferred orientation diagram measured in quartz-rich phyllonite; 110 measurements; contours are 1, 3, 6, max. 9%; the calculated best shearing plane is indicated. Location on Fig. 4.

where the foliation dips west, they indicate a normal fault displacement. Nevertheless, farther east, where the foliation dips east, ductile shear criteria also yield a westward vergence and thus a reverse fault displacement. This indicates a broad folding of the foliation in an anticlinal structure that post-dates the early ductile westward shearing (Fig. 2).

Eastward shear-sense criteria are very scarce. The only examples are found below the Devonian unconformity, in the hangingwall of the detachment and more scarcely in preserved nuclei of the substratum (Fig. 4).

When one goes towards the east, the westward shear criteria decrease in density outside a 30 km wide zone around the basin. The central part of the Gneiss Complex displays only the polyphasal Precambrian tectonic events characterized by migmatitic structures and inconsistent rotational shear criteria. Additional westward rotational deformation evidence only occurs in some thin heterogeneous zones.

The ductile fabric is overprinted by a brittle event that is well expressed in the substratum of the Devonian Basins. Tension-gashes filled by quartz, chlorite, epidote and calcite, and normal or strike-slip microfaults are the main brittle structures. Measurements of striae and slickensides (with crystallization of quartz, chlorite and epidote) in groups of microfaults provide the orientation of the stress tensors (Etchecopar, 1984) throughout the area. They define a direction of extension similar to that of the basin-fill. The similarities of stress axes and tectonic structures, together with the identical crystallization on microfault planes, observed within both basin and sub-stratum, suggest that brittle tectonics affected both zones during a single event occurring contemporaneously with sediment deposition, i.e. Middle Devonian time (Séranne and Séguret, 1987). Pseudotachylytes and cataclastic rocks associated with the brittle structures confirm more superficial conditions for this deformation (Sibson, 1977).

On the basis of the microtectonic observations in the basement, we suggest the following sequence of deformational events:

(1) Development of the early westward shear criteria in a deep part of the crust (ductile shear criteria such as rolling and asymmetric feldspars).

(2) Decrease of the *P–T* conditions and occurrence of less-ductile structures such as shear-bands and asymmetric folds. These structures also indicate a westward transport.

(3) Brittle deformation indicating more superficial conditions of extensional deformation.

INTERPRETATION

The geometric and structural data gathered from the four basins fit with the development of extensional sedimentary basins. The deformation of the basin-fill is consistent with a heterogeneous shearing that displays an increase of strain towards the basal detachment. The constant orientation of strain axes with respect to bedding planes throughout the basins indicates that the eastward dip–slip shearing of the basin-fill is controlled by the tilted bedding.

During basin formation, the syntectonic sediments were deposited along the low-angle detachment. The depocentre remained located along the footwall cut-off of the active detachment but the sediments were continuously carried away towards the west on top of the westward-moving hangingwall (Fig. 7). The enormous stratigraphic thickness of sediments is therefore explained (Séranne and Séguret, 1985a, 1987).

The deformation observed within the basin-fill results from the development of the basins. The dip–slip shearing parallel to stratification planes are antithetic to the westward-sliding detachment (Fig. 7). It acted like the normal faults antithetic to a main fault, such as invoked by Gibbs (1984) in extensional basins. Within the Devonian infill, this deformation accommodated the westward translation of the hangingwall.

The inferred mechanisms of basin formation require that extension was taken up below the basin by a westward normal simple shear (Wernicke, 1985). This may be consistent with the ductile deformation observed in the basement. The existence of ductile deformation in the Western Gneiss has always been related to Caledonian or previous compressive deformation (Dietler *et al.*, 1985). However, the observation of shear criteria allows us to question this assumption and to suggest an alternative. Since the Caledonian thrust transport direction is eastward, it is likely that the westward shear criteria belong to a distinct tectonic event. As the westward shearing is localized at the vicinity of the basins and represents the last imprint on the Western Gneisses, we assume that this ductile event post-dates the Caledonian structures that are preserved east of the study area.

We interpret the extensional tectonics in a model of a crustal-scale low-angle shear zone. The long-lasting east–west orientation of the displacement (initially marked by ductile stretching lineation in the basement, then marked by brittle extension axes both in the basement and in the basins) argues for the continuity of the ductile and brittle events. It demonstrates the extensional character of the ductile westward shearing evidenced in the study.

The evolution of the deformation is presently observed in the footwall (Fig. 8). First, the rocks suffered a ductile deformation expressed in a kilometre-wide shear-zone that occurred at a deep structural level. Since its age is later than that of the Caledonian compressive tectonics that terminated during Upper Silurian time (Hossack and Cooper, 1986), the initiation of extensional tectonics consequently began during the Early Devonian (Fig. 8(*b*)).

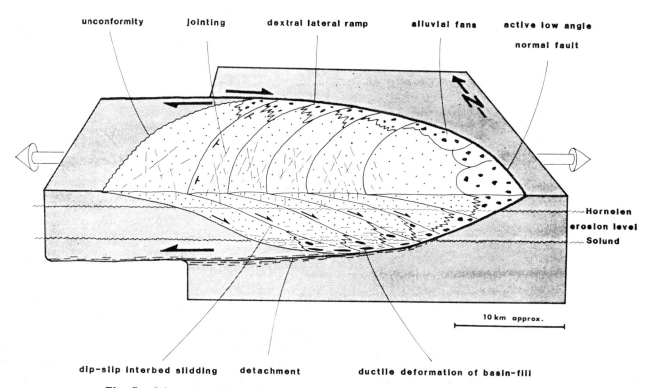

Fig. 7. Schematic of the mechanism of basins formation and related infill deformation.

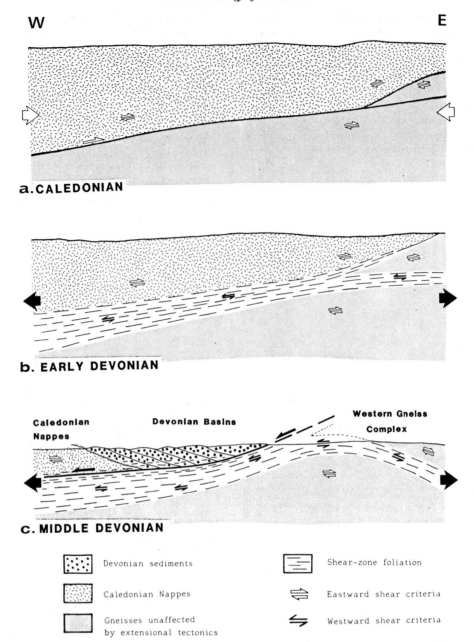

Fig. 8. Tentative interpretation of the evolution of the extensional shear zone (without scale). (*a*) Situation at the end of Caledonian orogeny, the eastward-thrusting developed eastward shear criteria within the Caledonian nappes and within the underlying gneissic basement. (*b*) Initiation of extensional westward shearing along the pre-existing discontinuities, development of ductile westward-directed structures in the deep-seated parts of the shear-zone. (*c*) Situation at middle Devonian time; the extension induced the ascent of the footwall rocks which thus yielded brittle/ductile to brittle structures; the displacement was taken up by the detachment that bounds the basins and controlled the sedimentation, whereas the uplift of the Gneiss Complex induced a broad folding and the locking of the east limb of the shear zone.

Secondly, the uplift of the footwall brought the rocks to lower *P–T* conditions in which brittle/ductile to brittle structures developed (Fig. 8(*c*)). Uplift also induced the broad folding of the foliation and therefore of the shear zone. This may explain the early westward shear criteria in reverse fault sense presently cropping out in the east of the zone. During this stage, ductile strain was restricted to a thin zone localized at the detachment, whereas the brittle deformation affected the substratum and the Devonian sediments. The brittle event was contemporaneous with Middle Devonian syntectonic sedimentation controlled by the detachment.

DISCUSSION

The reactivation of pre-existing thrust planes as normal faults has been proposed in the North Sea (Brewer and Smythe, 1984) and in the Basin and Range Province from deep seismic reflection profile interpretation (Allmendinger *et al.*, 1983) and surface geology (Malavieille, 1987). In Norway, the basal contact of the Devonian Basins has been proposed as a reworked Caledonian thrust plane (Hossack, 1984; Séranne and Séguret, 1987). This is supported by the low-angle dip of the shear zone. Analogue modelling (Faugére and Brun, 1984; Vendeville *et al.*, 1987) has shown that normal faults initiate at low angle only when they activate a pre-existing low-angle heterogeneity. Furthermore, our study demonstrates that the shear zone does not cross-cut the foliation at a high angle. It is likely that the foliation anticline structure in the Western Gneiss Region resulted as an uplift-generated dome rather than a large-scale drag related to the normal shear zone, as proposed by Norton (1986). This supports the assumption that the fault resulted from the reactivation of the previous structures

(Fig. 8). On the other hand, one would expect to find eastward Caledonian-related shear criteria overprinted by the westward extensional ones. However, the extensional process may have overprinted the pre-Devonian shear criteria except in the hangingwall and in isolated nuclei within the footwall.

Although the age of the normal shear zone initiation is not yet precisely defined, the extensional tectonics described in this paper provide an efficient mechanism that allowed the thinning of the Caledonian thick orogenic crust, down to a normal thickness, by the end of Middle Devonian times.

Acknowledgements

This work was supported by ELF Aquitaine Norge S.A. and by the ATP "Plis-Faille" and "Hercynien". This paper owes a lot to the advice and the constant help of M. Brunel and M. Séguret. We also thank J. Malavieille for useful comments and M. Mattauer for stimulating discussions.

Contribution CNRS-INSU-DBT N° 10. Theme "Croissance des Continents".

REFERENCES

Allmendinger, R. W., Sharp, J. W., Vontish, D., Serpa, L., Brown, L., Kaufman, S., Oliver, J. and Smith, R. B. (1983). Cenozoic and Mesozoic structure of the Eastern Basin and Range Province, Utah, from COCORP seismic-reflection data. *Geology* **11**, 532–536.

Berthé, D., Choukroune, P. and Jegouzo, P. (1979). Orthogneiss, mylonite and non-coaxial deformation of granites: the example of the South Armorican Shear Zone. *J. Struct. Geol.* **1**, 31–42.

Bouchez, J. L. and Pecher, A. (1981). The Himalayan Main Central Thrust pile and its quartz-rich tectonites in Central Nepal. *Tectonophysics* **78**, 23–50.

Brewer, J. A. and Smythe, D. K. (1984). MOIST and the continuity of crustal reflector geometry along the Caledonian-Appalachian orogen. *J. Geol. Soc. Lond.* **141**, 105–120.

Brekke, H. and Solberg, P. O. (1988). The geology of Atloy. *Norges Geol. Unders.*

Brunel, M. (1980). Quartz C-Axis fabrics in shear-zone mylonites evidence for a major imprint due to late strain increments. *Tectonophysics* 64, T33–T34.

Bryhni, I. (1964). Migrating basins on the Old Red Continent. *Nature* **202**, 384–385.

Bryhni, I. (1966). Reconnaissance studies of Gneisses, Ultrabasites, Eclogites and Anorthosites in Outer Nordfjord, Western Norway. *Norges Geol. Unders.* **241**, 67 pages.

Bryhni, I. (1978). Flood deposit in the Hornelen Basin, west Norway (Old Red Sandstone). *Norsk Geol. Tidsskr.* **58**, 273–300.

Bryhni, I. (1987). Status of the Supracrustal Rocks in the Gneiss Region of S. Norway. The Caledonian and Related Geology of Scandinavia Meeting, Cardiff, 22–23 Sept. 1987.

Bryhni, I. and Brastad, K. (1980). Caledonian regional metamorphism in Norway, *J. Geol. Soc. Lond.*, **137**, 251–259.

Bryhni, I. and Skjerlie, F. J. (1975). Syndepositional tectonism in the Kvamshesten district (Old Red Sandstone), Western Norway. *Geol. Mag.* **112**, 593–600.

Bryhni, I. and Sturt, B. A. (1985). Caledonides of Southwestern Norway. In: Gee, D. G. and Sturt, B. A. (eds.) *The Caledonide Orogen—Scandinavia and Related Areas*, pp. 89–107. Wiley, Chichester.

Chauvet, A., Séranne, M., Brunel, M. and Séguret, M. (1987). Devonian Basins of W. Norway and the stretching of an overthickened crust. *Terra Cognita* 7, 202.

Cobbold, P. R. and Quinquis, H. (1980). Development of sheath folds in shear regimes. *J. Struct. Geol.* **2**, (1/2), 119–126.

Cuthbert, S. J., Harvey, M. A. and Carswell, D. A. (1983).

A tectonic model for the metamorphic evolution of the Basal Gneiss Complex, Western South Norway. *J. Metam. Geol.* **1**, 63–90.

Dietler, T. N., Koestler, A. G. and Milnes, A. G. (1985). A preliminary structural profile through the Western Gneiss Complex, Sognefjord, Western Norway, *Norsk Geol. Tidsskr.* **65**, 233–235.

Etchecopar, A. (1984). Etude des états de contrainte en tectonique cassante et simulations de déformations plastiques (approche mathématique). Thèse d'état, Université de Montpellier (France). 269 pp.

Etchecopar, A. and Malavieille, J. (1987). Computer models of pressure shadows: a method of strain measurement and shear-sens determination. *J. Struct. Geol.* **9**, (5/6), 667–677.

Etchecopar, A., Vasseur, G. and Daignieres, M. (1981). An inverse problem in microtectonics for the determination of stress tensors from fault striation analysis. *J. Struct. Geol.* **3**, (1), 51–65.

Faugère, E. and Brun, J. P. (1984). Modélisation expérimentale de la distension continentale. *C.R. Acad. Sci. Paris*, t. **299**, II, n° 7, 365–370.

Gee, G. D. (1982). The Scandinavian Caledonides. *Terra Cognita* **2**, 89–96.

Gibbs, A. D. (1984). Structural evolution of extensional basin margins. *J. Geol. Soc. Lond.* **41**, pp. 609–620.

Griffin, W. L., Austrheim, H., Brastad, K., Bryhni, I., Krill, A. G., Krogh, E. J., Mork, M. B. E., Qvale, H. and Torudbakken, B. (1985). High pressure metamorphism in the Scandinavian Caledonides. In: Gee, D. G. and Sturt, B. A. (eds.) *The Caledonide Orogen—Scandinavia and Related Areas*, pp. 783–801. Wiley, Chichester.

Griffin, W. L. and Brueckner, H. K. (1980). Caledonian Sm–Nd Ages and a crustal origin for Norwegian eclogites. *Nature* **285**, 319–321.

Hossack, J. R. (1984). The geometry of Listric Growth faults in the Devonian basins of Sunnfjord, W Norway. *J. Geol. Soc. Lond.* **141**, 629–637.

Hossack, J. R. and Cooper, M. A. (1986). Collision tectonics in the Scandinavian Caledonides. In: Coward, M. P. and Ries, A. C. (eds.) *Collision Tectonics. Geol. Soc. Lond. Spec. Publ.*, **19**, 287–304.

Kildal, E. S. (1970). Geologisk kart over Norge. 1:250.000 Maloy. Norges Geologiske Undersøkelse.

McClay, K. R., Norton, M. G., Coney, P. and Davis, G. H. (1986). Collapse of the Caledonian Orogen and the Old Red Sandstone. *Nature* **323**, 147–149.

Malavieille, J., Lacassin, R. and Mattauer, M. (1984). Signification tectonique des linéations d'allongement dans les Alpes occidentales. *Bull. Soc. Géol. Fr.* (7), XXVI, n° 5, 895–906.

Malavieille, J. (1987). Kinematics of compressional and extensional ductile shearing deformation in a metamorphic core complex of the Northeastern Basin and Range. *J. Struct. Geol.* **9**, (5/6), 541–554.

Nilsen, T. H. (1968). The relationship of sedimentation to tectonics in the Solund area of Southwestern Norway. *Norges Geol. Unders.* **259**, 108 p.

Norton, M. G. (1986). Late Caledonide Extension in Western Norway: a Response to Extreme Crustal Thickening. *Tectonics* **5**, (2), 195–204.

Norton, M. G. (1987). The Nordfjord-Sogn Detachment, W. Norway. *Norsk Geol. Tidsskr.* **67**, 93–106.

Passchier, C. W. and Simpson, C. (1986). Porphyroclast systems as kinematic indicators. *J. Struct. Geol.* **8**, (8), 831–843.

Petit, J. P. (1987). Criteria for the sense of movement on fault surfaces in brittle rocks. *J. Struct. Geol.* **9**, (5/6), 597–608.

Roberts, D. and Sturt, B. A. (1980). Caledonian deformation in Norway. *J. Geol. Soc. Lond.* **137**, 241–250.

Santarelli, N. (1977). Le soubassement gneissique des internides calédoniennes scandinaves. Limite de la calédonisation. *Rev. Geogr. Phys. Geol. Dyn.* **XIX**, (5), 405–420.

Séranne, M. and Séguret, M. (1985a). Etirement Ductile et Cisaillement Basal dans les Bassins Dévoniens de l'Ouest Norvège. *C. R. Acad. Sci. Paris*, t. 300, Série II, 373–378.

Séranne, M. and Séguret, M. (1985b). Ductile Deformation of Soft Conglomerates in Western Norway Devonian Basins. *Geol. Soc. London Meeting*, London 15–17 April 1985, p. 30.

Séranne, M. and Séguret, M. (1987). The Devonian Basins of Western Norway: Tectonics and Kinematics of an extending crust. In: Coward, M. P., Dewey, J. F. and Hancock, P. L. (eds.) *Continental Extensional Tectonics. Geol. Soc. Lond. Spec. Publ.* **28**, 537–548.

Sibson, R. H. (1977). Fault Rocks and Fault mechanisms. *J. Geol. Soc. Lond.* **133**, 191–213.

Sigmond, E. M., Gustavson, M. and Roberts, D. (1984). Berggrunnskart over Norge 1/1 000 000. Norges Geolog Undersøkelse.

Simpson, C. and Schmid, S. H. (1983). An evolution of criteria to deduce the sense of movements in sheared rocks. *Geol. Soc. Amer. Bull.* **94**, 1281–1288.

Skjerlie, F. J. (1969). The pre-Devonian rocks in the Askvoll-Gaular area and adjacent districts. Western Norway. *Norges Geol. Unders.* **258**, 325–359.

Skjerlie, F. J. (1974). The Lower Palaeozoic Sequence of the Stavfjord District, Sunnfjord. *Norges Geol. Unders.* **302**, 1–32.

Steel, R. and Gloppen, T. G. (1980). Late Caledonian (Devonian) basin formation, Western Norway: signs of strike-slip Tectonics during infilling. *Spec. Publ. Int. Assoc. Sediment.* **4**, 79–103.

Steel, R., Siedlecka, A., Roberts, D. (1985). The Old Red Sandstone basins of Norway and their deformation: a review. In: Gee, D. G. and Sturt, B. A. (eds.) *The Caledonide Orogen—Scandinavia and Related Areas*, pp. 293–315. Wiley, Chichester.

Steel, R. J., Maehle, S., Nilsen, H., Roe, S. L. and Spinnanger, A. (1977). Coarsening upward cycles in the alluvium of Hornelen Basin (Devonian) Norway: Sedimentary response to tectonic events. *Geol. Soc. Amer. Bull.* **88**, 1124–1134.

Sturt, B. A. and Thon, A. (1978). Caledonides of Southern Norway. *Geol. Surv. Can.* Paper 78. **13**.

Sturt, B. A., Pringle, I. R. and Roberts, D. (1975). Caledonian Nappe Sequence of Finmark, Norther Norway and the timing of orogenic deformation and Metamorphism. *Geol. Soc. Amer. Bull.* **86**, 710–718.

Torsvik, T. H., Sturt, B. A., Ramsay, D. M., Kisch, H. J. and Bering, D. (1986). The tectonic implications of Solundian (Upper Devonian) magnetization of the Devonian rocks of Kvamshesten, Western Norway. *Earth Planet. Sci. Lett.* **80**, 337–347.

Torsvik, T. H., Sturt, B. A., Ramsay, D. M. and Vetti, V. (1987). The tectonic magnetic signature of the Old Red Sandstone and pre-Devonian strata in the Hasteinen Area, Western Norway, and implications for the later stage of the Caledonian Orogeny. *Tectonics* **6**, (3), 305–322.

Tullis, J. A., Christie, J. M. and Griggs, D. T. (1973). Microstructures and preferred orientations of experimentally deformed quartzites. *Geol. Soc. Amer. Bull.* **84**, 297–314.

Vendeville, B., Cobbold, P. R., Davy, P., Brun, J. P. and Choukroune, P. (1987). Physical models of extensional tectonics at various scales. In: Coward, M. P., Dewey, J. F. and Hancock, P. L. (eds.) *Continental Extensional Tectonics. Geol. Soc. Lond. Spec. Publ.* **28**, 95–108.

Wernicke, B. (1985). Uniform-sense normal simple shear of the continental lithosphere. *Can. J. Earth Sci.* **22**, 108–125.

Williams, G. D. (1978). Rotation of contemporary folds into the *x* direction during overthrust processes in Laksefjord, Finnmark. *Tectonophysics* **48**, 29–40.

Part VII

East Greenland Caledonian Geology

21 The Late Proterozoic sedimentary record of East Greenland: its place in understanding the evolution of the Caledonide Orogen

M. J. Hambrey

Department of Earth Sciences, University of Cambridge, Downing Street, Cambridge CB2 3EQ, UK

Late Proterozoic sediments are well preserved in a variety of depositional settings in the Caledonide Orogen. Vendian sequences in particular have a variety of lithofacies (especially tillites) that permit correlation throughout the orogen. Palaeogeographical reconstructions based on facies analysis have been assembled for the East Greenland sequence. Here, continental basement is capped by tillites southwest of the central East Greenland fjord region, whilst the main development of Vendian and bounding strata accumulated in an ensialic setting over the fjord region itself.

Two levels of tillite-bearing strata are matched by similar occurrences in Svalbard and northern Norway (type area of Varanger glaciation). Other parts of Scandinavia are represented by only one level of tillite (the younger of the two), whereas Scotland and Ireland have a major tillite complex that could belong to either glacial epoch.

The East Greenland Vendian succession most closely matches that of NE Svalbard, and both were probably deposited in a contiguous basin with a common source area. Western Svalbard and the other localities belonged to different basins. The relative positions of the different Late Proterozoic basins have major implications for strike-slip movements within the Arctic Caledonides.

INTRODUCTION

Throughout the Caledonide Orogen and associated fold belts (Fig. 1), there are several well-preserved Late Proterozoic sedimentary sequences, reflecting a wide variety of depositional environments. A common characteristic of many of these sequences is the presence of Varanger tillites, which are indicative of a major climatic event that can be documented throughout the whole North Atlantic region and many other parts of the world. These tillites provide an excellent means of correlation, whereas other methods, such as biostratigraphy and geochronology, are insufficiently refined to permit correlation with the same precision as for Phanerozoic rocks.

The East Greenland Late Proterozoic sequence has been compared with those of Scandinavia, Scotland and Svalbard, largely on the basis of tillites. As long ago as the 1930s, Kulling (1934) noted the similarity between East Greenland and Nordaustlandet (Svalbard), a correspondence developed further by Harland (1965, 1985) and Hambrey (1983). Yet, now lying between these two areas are the Late Proterozoic strata of western Svalbard, which

although containing tillite, in most respects (notably tectonic setting, composition and style of sedimentation) differ from the other areas. The implications of these contrasts, and of the strike-slip nature of major faults in Svalbard, led Harland and others to develop the hypothesis of large-scale (1000 km+) sinistral displacements in the Arctic Caledonides (Harland, 1969; Harland and Gayer, 1972; Harland and Wright, 1979; Harland *et al.*, 1984). However, these ideas have been disregarded in many recent pre-Caledonian reconstructions of the Arctic–North Atlantic region, even though evidence for major sinistral strike-slip has been forthcoming from the other end of the Caledonide Orogen in Newfoundland and the Appalachians (e.g., Morris, 1976; Kent and Opdyke, 1978; Van der Voo, 1980). On the other hand, the evidence from the British Caledonides has been conflicting, Van der Voo and Scotese (1981) proposing large-scale strike-slip along the Great Glen Fault, whilst Winchester (1973), Smith and Watson (1983) and Soper and Hutton (1984) preferred much smaller displacements on this fault and in the British Isles generally.

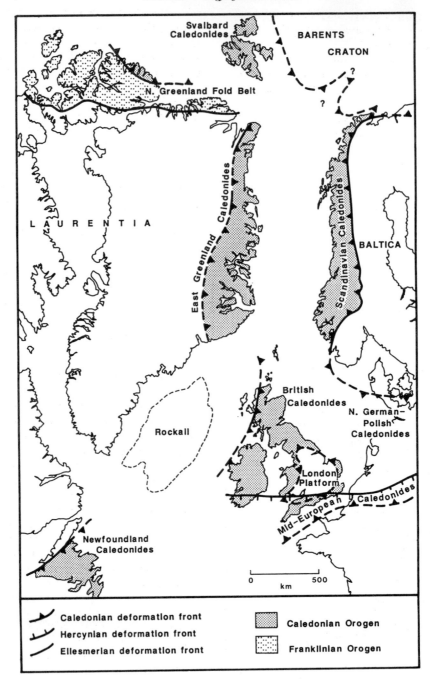

Fig. 1. Caledonian and related fold belts in the Arctic–North Atlantic region according to their distribution prior to Cenozoic sea-floor spreading(configuration after Le Pichon *et al.*, 1977), indicating principal areas discussed in this paper.

In order to test the hypothesis of major strike-slip in the Arctic Caledonides, and to provide constraints on the configuration of land and sea in Late Proterozoic time, Cambridge expeditions have been making systematic stratigraphic, sedimentological, structural and metamorphic studies in both East Greenland and Svalbard. The essential data from Svalbard have been presented elsewhere (Hambrey, 1982; Fairchild and Hambrey, 1984; Fairchild and Spiro, 1987; Hambrey *et al.*, in prep.). The data from East Greenland are summarized in this and the following papers in this volume, and in Hambrey and Spencer (1987); these papers provide a palaeogeographical framework for studies of basin evolution and subsequent tectonic evolution of the Arctic Caledonides, in particular the significance of strike-slip and the timing and location of the opening of the Iapetus Ocean (Hambrey and Harland, in prep.).

This paper provides the background and outlines the regional setting for the more detailed studies of the East Greenland Caledonides that follow, providing a brief comparison between the so-called foreland tillite localities and the main geosynclinal succession to the east.

EAST GREENLAND "FORELAND" SUCCESSIONS

The main development of Late Proterozoic, and in particular Vendian sediments in East Greenland is in the fjord region between 71°30′ N and 74°30′ N, where they form the relatively unmetamorphosed part of the Caledonian Fold Belt. The Vendian succession, represented essentially by the Tillite Group with two main tillite-bearing formations, is up to 1 km thick and rests on ca. 13 km of supposedly Late Riphean (?Sturtian) clastic and carbonate sediments of the Eleonore Bay Group. It is in turn overlain by some 3 km of Cambro–Ordovician rocks, dominated by carbonates (Henriksen and Higgins, 1976; Henriksen, 1985) (Fig. 2, Table I).

Lying 100–200 km to the west and southwest of this region are three recorded areas of tillite resting directly on basement (Haller, 1971) (Fig. 2). Of these, the Gåseland (Gnejssø) "tillites" (Wenk, 1961; Phillips and Friderichsen, 1981) are so deformed that few indications of their mode of deposition are preserved. Indeed, there is little positive evidence to suggest that they are tillites (and are best described as psephites), nor that their age is Vendian (Manby and Hambrey, this volume).

About 22 km to the north, in Paul Stern Land, Phillips and Friderichsen (1981) described an only slightly deformed tillite, and this shows strong similarities with the upper of the two tillite formations in the Tillite Group of the fjord region (Moncrieff, this volume). The thickness of the Paul Stern Land tillite ranges from 0 to 43 m. About 180 km farther north at Tillit Nunatak in Charcot Land, Henriksen (1981) described several hundreds of metres of tillite resting on basement. Our work has shown

this estimate to be excessive, and reflects previously unrecognized tectonic complications (Manby and Hambrey, this volume). The original thickness of the exposed succession could not have exceeded 50 m, but the top is missing as a result of erosion. The Charcot Land tillite has no obvious similarities with any of the other tillites in East Greenland, although its glacial origin is not in doubt (Moncrieff, this volume). The rock is strongly cleaved and shows growth of metamorphic chlorite, so any similarity may have been disguised.

Each of the above localities has been regarded as a window through Caledonian thrust sheets that comprise high-grade schists of Middle Proterozoic age, with the tillites resting on autochthonous or parautochthonous basement (Henriksen and Higgins, 1976; Henriksen, 1985). However, there is evidence in both Gåseland and Charcot Land of large-scale translation to the west or northwest of both psephites and tillites, and of the basement directly beneath, and to a lesser extent in Paul Stern Land. Thus true foreland must lie further to the west than previously envisaged, and the tillites may have originated several tens of kilometres to the east (Manby and Hambrey, this volume).

EAST GREENLAND FJORD REGION

The 17 km thick Late Proterozoic to Early Palaeozoic succession of East Greenland (Table I) rests on a metamorphic complex that has yielded Proterozoic as well as Caledonian ages. The Vendian Period in East Greenland, as defined on biostratigraphic grounds (Vidal, 1976, 1979), is represented by the entire Tillite Group. The bulk of it belongs to the Varanger Epoch, embracing the two main glacial stages defined in Finnmark, northern Norway, namely the Smalfjord (earliest) and Mortensnes epochs; there is no positive evidence that the Ediacaran Epoch is present, and there is a hiatus at the top of the group, so most of it may be unrepresented (Harland *et al.*, 1982, modified 1982 time-scale).

The Tillite Group consists of five formations: the Ulvesø (at the base), Arena, Storeelv, Canyon and Spiral Creek. These have been redefined and described by Hambrey and Spencer (1987) following several decades of sporadic investigations by Danish expeditions (reviewed by Haller, 1971; Henriksen and Higgins, 1976; Higgins, 1981). Although thicknesses of individual rock units vary substantially, most can be traced throughout the fjord region.

The most distinctive rocks are the two levels of diamictite, interpreted as tillite (in the Ulvesø and Storeelv formations). The two formations compositionally are quite different and, although both contain intrabasinal rocks from the Eleonore Bay Group, only the Storeelv Formations contains significant amounts of far-travelled exotic material (granites, gneisses and volcanic rocks).

Overall, the dominant aspect is of deposition from glaciers in a near-shore setting, the variety of facies reflecting shifts in the grounding line across the sites of deposition. There are no indications of deposition of supraglacial debris, suggesting that ice cover on the adjacent land was continuous. The non-glacial units were deposited in offshore to shallow and marginal marine settings.

Palaeocurrent information comes from glacially abraded pavements and diamictite (lodgement till) clast fabrics in the Ulvesø and Storeelv formations, together with cross-laminations, wave ripples, gutter casts and elongate shrinkage cracks in the non-glacial units. Although these data give quite variable directions, they

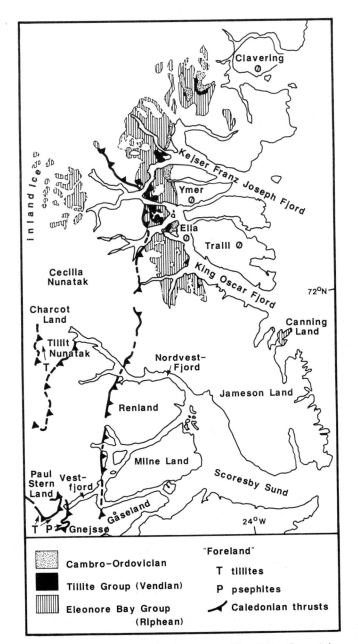

Fig. 2. Distribution of Vendian strata (the Tillite Group) in East Greenland (after Higgins and Phillips 1979).

Table I. Late Proterozoic to Early Palaeozoic succession that overlies the metamorphic complex of central East Greenland. Age key: L = Late, M = Middle, E = Early, Ed = Ediacara, Var = Varanger. (After Henriksen and Higgins, 1976; Henriksen, 1985; Hambrey and Spencer, 1987).

Age		Group	Formation			Lithology	Typical thickness (m)
Ordovician	L		Heimbjerge			Limestone	320
	M		Narwhale Sound			Dolostone and dolomitic limestone	462
	E		Cape Weber			Limestone and dolomitic limestone	1040–1165
			Antiklinalbugt			Limestone	212–270
						—— unconformity ——	
Cambrian	M		Dolomite Point			Dolostone	371–421
			Hyolithus Creek			Dolostone	210–215
	E		Ella Island			Arenaceous limestone	80–102
			Bastion			Glauconitic shale, sandstone and limestone	137–152
			Kløftelv			Quartzite and sandstone	70–75
						—— unconformity ——	
Vendian	Ed.?		Spiral Creek			Sandstone, mudstone and limestone breccia	25–55
			Canyon			Dolostone, mudstone, marl and limestone	250–300
	Var.	Tillite	Storeelv			Diamictite, sandstone and conglomerate	up to 200
			Arena			Sandstone and dolomitic mudstone	150–310
			Ulvesø			Diamictite, sandstone, conglomerate and mudstone	up to 320
Sturtian?		Upper Eleonore Bay	Limestone–Dolomite "series"	Bed nos.	19/20	Thin-bedded dolostone, limestone and shale/limestone	210/80
					18	Limestone and stromatolitic dolostone	ca. 400
					17	Dolostone breccia	60–70
					16	Bituminous and oolitic limestone	80–200
					15	Limestone with breccia and dolostone	50–80
					14	Limestone and breccia and dolostone	140–300
			Multicoloured "series"		7–13	Varicoloured sandstone, shale, limestone, dolostone, breccia and mudstone	1000
Late Riphean			Quartzite "series"		1–6	Quartzite, sandstone, psammite and pelite	2000
		Lower Eleonore Bay	U. Argillaceous–Arenaceous "series"			Quartzite, slate and shale	1200–1400
			Calc–Argillaceous "series"			Siliceous limestone, shale and thin-bedded quartzite	150
			L. Arenaceous–Argillaceous "series"			Quartzite and greywacke	8000

indicate a fairly consistent transport direction towards the north and northeast (Moncrieff and Hambrey, 1988).

The palaeocurrent information conforms with palaeogeographical considerations based on facies in the diamictites and their relationship to the underlying rocks (Moncrieff, this volume). Throughout most of the fjord region, the Ulvesø Formation rests conformably on both Beds nos. 19 and 20 of the Limestone–Dolomite "series", with which it may be partly a lateral equivalent (Hambrey and Spencer, 1987). In the south of the fjord region, the Ulvesø Formation rests on a weathered surface of rocks characteristic of Bed no. 18. In Canning Land to the southeast, the lower part of the Tillite Group is missing

and the Storeelv Formation rests unconformably on a weathered palaeosurface on Bed. 18. In some "foreland" localities, the Storeelv Formation rests directly on basement. The available evidence therefore points to a source area for the sediment in the SW–SE sector, with increasing stratigraphic depth of erosion in this direction.

Although continuity of deposition is indicated in most of the fjord region from Eleonore Bay Group to Tillite Group time, the incorporation of Eleonore Bay Group (especially Limestone–Dolomite "series") clasts in the tillite implies substantial uplift and erosion in the source area. Evidence of tectonic instability in the main depositional basin, however, is limited to subaqueous debris flows in Bed no.

19 (Herrington and Fairchild, this volume), although marked facies variations in the Tillite Group suggest uneven subsidence and uplift.

COMPARISON WITH OTHER AREAS

The Vendian successions in East Greenland show a transition from deposition in a terrestrial environment on continental basement in the southwest to deposition in a subsiding ensialic-type basin to the northeast, with a NW–SE-trending coastline. Two glacial horizons are present in the ensialic basin, but only the later one was deposited in the cratonic areas.

The two glacial horizons of East Greenland can be correlated biostratigraphically and lithostratigraphically with the Smalfjord and Mortensnes formations of Finnmark (the type area of Varanger glaciation). However, the Finnmark tillites were deposited in an epicontinental setting, above a dominantly continental clastic succession (e.g., Edwards, 1984). In southern Norway, a single tillite horizon (the Moelv Formation) is of continental aspect, developed within an aulacogen in basement rocks (e.g., Nystuen, 1976, 1982); it is correlated with the upper tillite of Finnmark.

The Scottish Port Askaig Formation of the Dalradian Supergroup is far thicker than any of the above sequences but, although comprising as many as 17 glacial horizons (Spencer, 1971), it cannot be correlated specifically with either of the two glacial horizons of Finnmark or East Greenland. The character of the Dalradian depositional basin as ensialic (Harris *et al.*, 1978; Anderton, 1985) suggests similarities with the East Greenland succession. Similar glacial facies, with at least partial marine aspect (Eyles and Eyles, 1983), reinforce the broader similarities with East Greenland, discussed by Higgins and Phillips (1979). However, compositionally and in terms of detailed stratigraphy, the areas are distinct and cannot have belonged to the same depositional basin.

The comparisons with NE Svalbard, but not western Svalbard (discussed by Harland and Wright, 1979; Hambrey, 1983; Harland, 1985), is much closer than with other areas. Since the recognition of two glacial horizons in NE Svalbard (Hambrey, 1982), the two areas can be matched almost bed-by-bed, suggesting that they were part of a contiguous sedimentary basin that was then divided by the strike-slip movements referred to by Harland (e.g., 1985).

REFERENCES

Anderton, R. (1985). Sedimentation and tectonics in the Scottish Dalradian. *Scott. J. Geol.* **21**, 407–436.

Edwards, M. B. (1984). Sedimentology of the Upper Proterozoic glacial record, Vestertana Group, Finnmark, North Norway. *Norges Geol. Unders.* **394**, 76pp.

Eyles, C. H. and Eyles, N. (1983). Glaciomarine model for the upper Precambrian diamictites of the Port Askaig Formation, Scotland. *Geology* **11**, 692–696.

Fairchild, I. J. (1983). Effects of glacial transport and neomorphism on Precambrian dolomite crystal sizes. *Nature* **304**, 714–716.

Fairchild, I. J. (1988). Dolomitic stromatolite-bearing units with storm deposits from the Vendian successions of East Greenland and Scotland: a case of facies equivalence, *this volume*.

Fairchild, I. J. and Hambrey, M. J. (1984). The Vendian succession of northeastern Spitsbergen: petrogenesis of a dolomite-tillite association. *Precambrian Res.* **26**, 111–167.

Fairchild, I. J. and Spiro, B. (1987). Petrological and iso-topic implications of some contrasting Late Precambrian carbonates, NE Spitsbergen. *Sedimentology* **34**, 973–989.

Haller, J. (1971). *Geology of the East Greenland Caledonides.* Wiley, Chichester.

Hambrey, M. J. (1982). Late Precambrian diamictites of northeastern Svalbard. *Geol. Mag.* **119**, 527–551.

Hambrey, M. J. (1983). Correlation of Late Proterozoic tillites in the North Atlantic region and Europe. *Geol. Mag.* **120**, 209–232.

Hambrey, M. J. and Spencer, A. M. (1987). Late Precambrian glaciation of central East Greenland. *Meddel. Grønland. Geoscience* **19**, 50pp.

Hambrey, M. J., Harland, W. B. and Waddams, P. Vendian geology of Svalbard. *Norsk Polarinst. Skrifter*, in press.

Harland, W. B. (1965). The tectonic evolution of the Arctic-North Atlantic region. *Phil. Trans. Soc. Lond.* **258A**, 59–75.

Harland, W. B. (1969). Contribution of Spitsbergen to under-standing of tectonic evolution of North Atlantic region. *Amer. Assoc. Petrol. Geol. Mem.* **12**, 817–851.

Harland, W. B. (1978). A reconsideration of Late Precambrian stratigraphy of southern Spitsbergen. *Polarforschung* **48**, 44–61.

Harland, W. B. (1985). Caledonide Svalbard. In: Gee, D. G. and Sturt, B. A. (eds.) *The Caledonide Orogen—Scandinavia and Related Areas*, pp. 999–1016. Wiley, Chichester.

Harland, W. B. and Gayer, R. A. (1972). The Arctic Caledonides and earlier oceans. *Geol. Mag.* **109**, 289–314.

Harland, W. B. and Wright, N. J. R. (1979). Alternative hypothesis for the pre-Carboniferous evolution of Svalbard. *Norsk Polarinst. Skr.* **179**, 89–118.

Harland, W. B., Cox, A. V., Llewellyn, P. G., Pickon, C. A. G., Smith, A. G. and Walters, R. (1982). *A geological time scale.* Cambridge University Press, Cambridge, 131 pp.

Harland, W. B., Gaskell, B. A., Heafford, A. P., Lind, E. K. and Perkins, P. J. (1984). Outline of Arctic post-Silurian continental displacements. In: Spencer, A. M. (ed.) *Petroleum Geology of the North European Margin*, pp. 137–148. Norwegian Petroleum Society, Graham & Trotman, London.

Harris, A. L., Baldwin, C. T., Bradbury, H. J., Johnson, H. D. and Smith, R. A. (1978). Ensialic basin sedimentation: the Dalradian Super-group. In: Bowes, D. R. and Leake, B. E. (eds.) *Crustal Evolution in Northwestern Britain and Adjacent Regions. Geol. J. Spec. Issue* **10**, 115–138. Seel House Press, Liverpool.

Henriksen, N. (1981). The Charcot Land tillite, Scoresby Sund, East Greenland. In: Hambrey, M. J. and Harland, W. B. (eds.), *Earth's pre-Pleistocene Glacial Record*, pp. 776–777. Cambridge University Press, Cambridge.

Henriksen, N. (1985). The Caledonides of central East Greenland 70°–76° N. In: Gee, D. G. and Sturt, B. A. (eds.) *The Caledonide Orogen—Scandinavia and Related Areas*, pp. 1095–1113. Wiley, Chichester.

Henriksen, N. and Higgins, A. K. (1976). East Greenland Caledonian Fold Belt. In: Escher, A. and Watt, W. S. (eds.) *Geology of Greenland*, pp. 182–246. Grønlands Geologiske Undersøgelse, Copenhagen.

Herrington, P. M. and Fairchild, I. J. (1988). Carbonate shelf and slope facies evolution prior to Vendian glaciation, central East Greenland, *this volume*.

Higgins, A. K. (1981). The Late Precambrian Tillite Group of the Kong Oscars Fjord and Kejser Franz Josephs Fjord region of East Greenland. In: Hambrey, M. J. and Harland, W. B. (eds.) *Earth's pre-Pleistocene Glacial Record*, pp. 778–781. Cambridge University Press, Cambridge.

Higgins, A. K. and Phillips, W. E. A. (1979). East Greenland Caledonides—a continuation of the British Caledonides. In: *The Caledonides of the British Isles reviewed. Geol. Soc. Lond. Spec. Publ.* **8**, 19–32.

Kent, D. V. and Opdyke, N. D. (1978). Paleomagnetism of the Devonian Catskill redbeds: evidence for motion of the coastal New England–Canadian Maritime region relative to cratonic North America. *J. Geophys. Res.* **83**, 4441–4450.

Kulling, O. (1934). Scientific results of the Swedish–Norwegian Arctic Expedition in the summer of 1931. XI "The Hecla Hoek Formation' around Hinlopenstretet. *Geografiska Annaler* **16**, 161–254.

Le Pichon, X., Sibuet, J.-C. and Francheteau, J. (1977). The fit of continents around the North Atlantic Ocean. *Tectonophysics* **38**, 169–209.

Manby, G. M. and Hambrey, M. J. (1988). Structural setting of Late Proterozoic tillites in East Greenland, *this volume*.

Moncrieff, A. C. M. (1988). Lat Proterozoic palaeogeography of East Greenland: evidence from the Tillite Group and related rocks, *this volume*.

Moncrieff, A. C. M. and Hambrey, M. J. (1988). Late Precambrian glacially related grooved and striated surfaces in the Tillite Group of central East Greenland. *Palaeogeog. Palaeoclimatol. Palaeoecol.* **65**, 183–200.

Morris, W. A. (1976). Transcurrent motion determined paleomagnetically in the northern Appalachians and Caledonides and the Acadian Orogeny. *Can. J. Earth Sci.* **13**, 1236–1243.

Nystuen, J.-P. (1976). Facies and sedimentation of the Late Precambrian Moelv Tillite in the eastern part of the Sparagmite Region, southern Norway. *Norges Geol. Unders.* **329**, 70pp.

Nystuen, J.-P. (1982). Late Proterozoic basin evolution on the Baltoscandian Craton: the Hedmark Group, southern Norway. *Norges Geol. Unders.* **375**, 74pp.

Phillips, W. E. A. and Friderichsen, J. D. (1981). The Late Precambrian Gaaseland tillite, Scoresby Sund, East Greenland. In: Hambrey, M. J. and Harland, W. B. (eds.) *Earth's pre-Pleistocene Glacial Record*, pp. 773–775. Cambridge University Press, Cambridge.

Smith, D. I. and Watson, J. W. (1983). Scale and timing of movements on the Great Glen Fault, Scotland. *Geology* **11**, 523–526.

Soper, N. J. and Hutton, D. H. W. (1984). Late Caledonian sinistral displacements in Britain: implications for a three-plate collision model. *Tectonics* **3**, 781–794.

Spencer, A. M. (1971). Late pre-Cambrian glaciation in Scotland. *Geol. Soc. Lond. Mem. No. 6*.

Van der Voo, R. (1980). Reply concerning "A paleomagnetic pole position from the folded Upper Catskill redbeds, and its tectonic implications". *Geology* **8**, 259–260.

Van der Voo, R. and Scotese, C. R. (1981). Paleomagnetic evidence for a large (2000 km) sinistral offset along the Great Glen fault during Carboniferous time. *Geology* **9**, 583–589.

Vidal, G. (1976). Late Precambrian acritarchs from the Eleonore Bay Group and Tillite Group in East Greenland. A preliminary report. *Grønlands Geol. Undersøgelse. Rapp. Nr. 78*, 19pp.

Vidal, G. (1979). Acritarchs from the Upper Proterozoic and Lower Cambrian of East Greenland. *Grønland Geol. Undersøgelse. Bull.* **134**, 55pp.

Wenk, E. (1961). On the crystalline basement and the basal part of the pre-Cambrian Eleonore Bay Group in the south-western part of Scoresby Sund. *Meddelelser om Grønland* **168** (1).

Winchester, J. A. (1973). Pattern of regional metamorphism suggests a sinistral displacement of 160 km along the Great Glen Fault. *Nature (Phys. Sci.)* **246**, 81–84.

22 Carbonate shelf and slope facies evolution prior to Vendian glaciation, central East Greenland

Paul M. Herrington and Ian J. Fairchild

School of Earth Sciences, University of Birmingham,
P.O. Box 363, Birmingham B15 2TT, UK

Two sections through 1 km of Upper Riphean carbonates have been studied in the fjord zone of central East Greenland at Ella Ø and just west of Kap Weber, which lies 65 km north. A close stratigraphic resemblance exists that highlights some confusions in previous literature about correlations in the fjord zone.

The lower 700 m lies within Bed group 18 of the standard lithostratigraphy and can be divided into 16 correlatable members. The lower 12 members compare closely in lithology and thickness between the two sections and are dominated by microsparry and pisolitic limestones with subordinate dolostones that are commonly stromatolitic. Members 4 to 10 define a 100 m thick section in which stromatolitic dolostones are particularly abundant. Deposition on a stable carbonate shelf is inferred. Members 13 and 15 are variably shaly carbonates with an intervening dolostone unit (member 14) that has sufferred collapse brecciation, probably shortly after sedimentation. Member 14 is thicker and higher in the section at Kap Weber. Member 16 is a marker horizon of interlaminated calcite and dolomite that parallels the layer-cake stratigraphy defined by members 1 to 12 and the members of Bed group 19.

Bed group 19 (ca. 200 m) can be divided into seven members that are correiatable between the two sections. The basal member contains peritidal carbonates and is abruptly overlain by dolomitic shales interpreted as slope-apron environment deposits. Extensive early diagenetic formation of ferroan carbonates was related to bacterial activity. The original sediments were rich in kaolinite and iron oxides, derived from a humid land area. Overall, there is a shallowing upwards trend and this is demonstrated both by an upward development of carbonaceous laminae within background sediments, interpreted as microbial mats, and also by an upward increase in the development of submarine lithification of carbonate-rich turbidites. Most turbidites are centimetre-scale, muddy, and sourced from within the slope environment. Carbonate-rich turbidites are decimetre-scale, spasmodically distributed and sourced from a carbonate shelf. Slumping of lithified carbonate-rich turbidites is conspicious. The sudden development of slope deposits is attributed to a phase of basin extension.

In the northern section, a carbonate shelf limestone (Bed group 20) succeeds slope deposits and is followed by subaerial Vendian glacial deposits. In contrast, glaciomarine deposits succeed slope carbonates in the southern section.

In the continuation of the sedimentary basin in NE Spitsbergen, slope sediments are not developed and the sedimentation rate is correspondingly lower.

INTRODUCTION

The magnificent Upper Proterozoic sections of the fjord zone of central East Greenland are currently the focus of several investigations embracing lithostratigraphy, biostratigraphy, chemostratigraphy, palaeomagnetism, sedimentology and palaeoecology. Of these, a detailed lithostratigraphic framework is necessary to underpin the other studies. The thick succession of carbonate shelf sediments below Vendian glacial units is a distinctive feature of both East Greenland and NE Svalbard sections, which are reckoned as part of the same depositional basin (Kulling, 1934; Harland and Gayer, 1972; Hambrey, 1983; Knoll *et al.*, 1986a). Detailed studies of the upper kilometre of these carbonates, probably Upper Riphean in age (Vidal, 1985), have been carried out in two sections in East Greenland. This has led to an improved perception of the degree of lateral continuity of sedimentation within the basin. Whilst there is apparently greater lateral continuity of a number of thin lithofacies units than was previously supposed, syndepositional tectonic instability has been recognized in the upper part of the section by the presence of marine slope facies. This points to a previously un-recognized period of basin extension.

LITHOSTRATIGRAPHY

The Upper Eleonore Bay Group (Katz, 1961), comprising approximately the uppermost 6.5–7 km of the

Proterozoic succession, has traditionally been divided into 20 numbered "Bed groups" (Teichert, 1933; Eha, 1953). Their distribution in the fjord zone of central East Greenland was studied in a series of mapping projects covering (from south to north) North Scoresby Land (Fränkl, 1953b), Lyell Land (Sommer, 1957), Ymer Ø, Ella Ø and Suess Land (Eha, 1953; also Schaub, 1950, 1955 on Ella Ø), Andrée Land (Fränk, 1953a) and

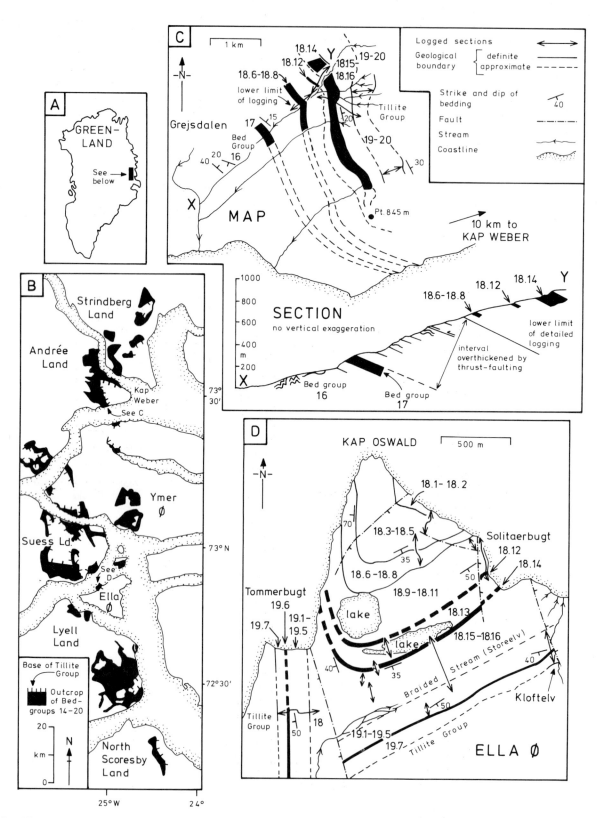

Fig. 1. (A) Location of fjord zone of central east Greenland. (B) Distribution of Limestone–Dolomite "series" (Bed groups 14–20) in the fjord zone (after Hambrey and Spencer, 1987), and location of Kap Weber and Ella Ø sections. (C) Map, derived from an aerial photograph, of studied sections near Kap Weber and cross-section along line X–Y marked on map. The cross-section was produced by sighting angles with an Abney level and measuring altitude with an altimeter. (D) Map, derived from aerial photographs, of the Kap Oswald district of Ella Ø.

Strindberg Land (Katz, 1952). Because of this body of work, an attempt by Katz (1961) to alter the "Bed group" nomenclature to conform with modern stratigraphic practice was not accepted by Haller (1971) and Henriksen and Higgins (1976), pending integrated regional studies.

We are here concerned with the upper part ("Bed groups" 18 to 20) of the Limestone–Dolomite "series" (Bed groups 14–20, Haller, 1971). Bed groups 18 and 19 were thought to show considerable lateral variation in thickness and internal subdivisions (Henriksen and Higgins, 1976), but it appears that this is partly due to inaccurate estimates of thickness in some cases, but also to miscorrelations, perhaps not suprising given that the regional mappers were not able to study other parts of the fjord region. Our work has demonstrated a close similarity of two sections 65 km apart that form the basis of our account. However, unequivocal demonstration of the relationship of our sections to the standard Bed group scheme owes much to the work of A. H. Knoll and K. Swett (personal communication, and Green *et al.*, 1987), who have logged one complete and two partial sections through the whole Limestone–Dolomite "series".

In our work, the sections have been divided into the smallest units that can be correlated. These units are designated as informal members of the Bed groups, bearing in mind that any future revision of the stratigraphy is likely to give formation status to Bed groups. Members defined here have simply been numbered sequentially from bottom up, e.g. Bed group 18, member 1 or simply member 18.1. The numbering system is used here as a convenient expedient for description. The system for Bed group 18 will need to be altered in future work, since we did not measure a complete section through to the bottom of the Bed group.

The localities studied (Fig. 1) are 10 km west of Kap Weber (around 73°30′ N, 24°50′ W) in Andrée Land, and on Ella Ø (72°52′ N, 25°7′ W). Sections through Bed group 18 are shown in Figs. 3 and 4, and of Bed group 19 in Fig. 5 in sufficient detail to demonstrate the correlations made between the two areas. In several cases, these sections are composites of individual transects that overlap stratigraphically: the precise locations are shown on Fig. 1.

Bed group 18

This unit consists predominantly of dark-grey weathering pisolitic, arenitic and microsparry limestones with subordinate pale-yellow to cream weathering dolostone

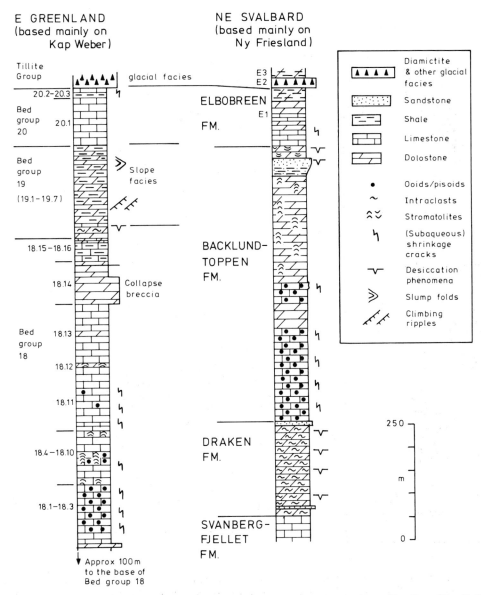

Fig. 2. Summary lithostratigraphy of the uppermost 1 km of carbonates underlying the lower Vendian tillite. E Greenland column is from data of this paper; NE Svalbard column is a combination of data from Wilson (1961), Fairchild and Hambrey (1984) and unpublished data of Fairchild.

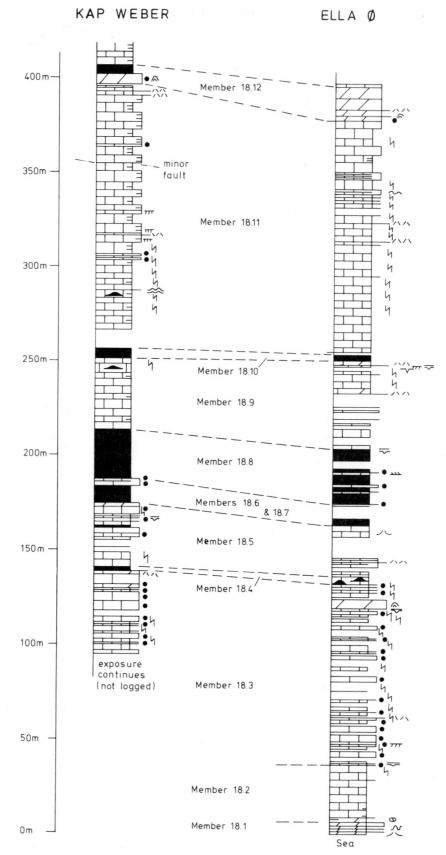

Fig. 3. Comparative sections of the lower logged portion of Bed group 18 at Kap Weber and Ella Ø. Key on Fig. 4.

members that commonly contain stromatolites. The sequence has been divided into 16 members (Figs. 3 and 4), the lowest two members only being studied on Ella Ø. The lowest beds recorded at Kap Weber correspond to the top 45 m of member 18.3 as defined on Ella Ø. Only the upper 600 m (Kap Weber) to 700 m (Ella Ø) of Bed group 18 were recorded in this study. The distinctively compact

pale-weathering, non-stromatolitic dolostone that comprises Bed group 17 in the Kap Weber area (Fränkl, 1953a) is ca. 700 m below the base of the logged interval at Kap Weber (Fig. 1C). Knoll (personal communication, 1986) indicated that only 100–120 m of lower Bed group 18 should be present below the base of our logged interval on the basis of his experience nearby to the west in Grejsdalen;

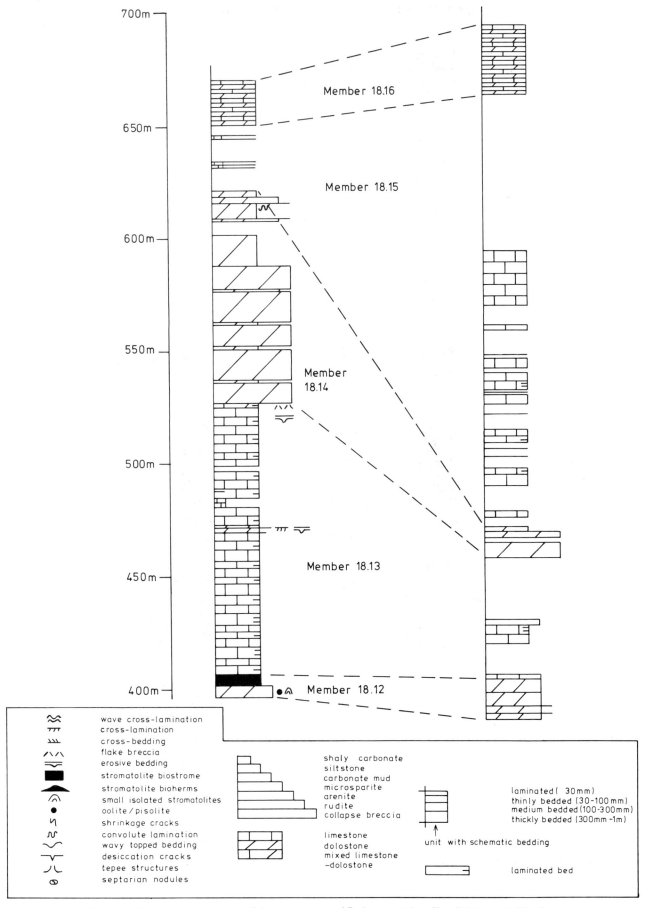

Fig. 4. Comparative sections of the upper part of Bed group 18 at Kap Weber and Ella Ø.

overthickening by thrusting at Kap Weber can be inferred. Some specific thrust planes were observed below our logged section but no repetitions were observed within the logged interval. Fränkl's (1953a) estimate of a 400 m thickness of Bed group 18 (to the top of member 18.14) is clearly a considerable underestimate. Also, Bed groups 14 to 17 are not represented on Ella Ø (Knoll, personal communication, 1986; Green *et al.*, 1987) contrary to previous interpretations (Eha, 1953; Schaub, 1955; Haller, 1971).

Members 18.1 to 18.3 (ca. 135 m) form a major coarsening upwards sequence of interbedded dark-grey pisolitic and microsparry limestones. In members 18.4 to 18.10 (ca. 120 m), similar but thinner coarsening up cycles (ca. 2 m to 40 m thick) alternate with yellow weathering dolomitic stromatolite biostromes and bioherms (ca. 2 m to 50 m thick). Members 18.11 to 18.12 form a second major (ca. 145 m) coarsening upwards sequence of calcarenites and microsparites that culminates in a yellow weathering ca. 15 m thick stromatolitic arenitic dolostone.

Marked thickness variations occur within the overlying members of Bed group 18 between Kap Weber and Ella Ø (Fig. 4), although a similar sequence of lithologies occurs. At Kap Weber, Member 18.13 forms a thick (ca. 130 m) monotonous sequence of alternating dark-grey weathering shaly limestones and limestone microsparites. The equivalent interval on Ella Ø is up to 61 m thick, although only 8 m of dark-grey weathering microsparite, overlain by 2.5 m of black chert, are exposed. Member 18.14 at Kap Weber is ca. 95 m thick and consists predominantly of cream weathering dolomite microspar. The basal 65 m have been brecciated and cemented by fibrous dolomite and are overlain by massive dolomicrosparites and subordinate intraclastic dolorudites and dolarenites. On Ella Ø the member is thinner (45 m max.), although only an isolated 10 m section is exposed, all of breccia lithology. Member 18.15 has a maximum thickness of 29 m at Kap Weber, where it is poorly exposed, and

consists of dark-grey interbedded silty limestone microsparite and calcareous shale. In contrast, on Ella Ø, the member has a maximum thickness of 195 m, although only the basal 125 m are exposed, and is composed of mid-grey weathering limestone microsparites. Member 18.16 is 20 m thick at Kap Weber and has a minimum thickness of 26 m on Ella Ø (150 m max.). At both localities, the member is composed of interlaminated limestone and dolostone microsparite. The top of this marker horizon is used to define the Bed group 18–19 boundary in accordance with Schaub (1950, 1955), Eha (1953) and Hambrey and Spencer (1987), rather than the top of member 18.14 (e.g., Fränkl, 1953a). This is because member 18.14 cuts across the "layer-cake" inferred from the correlation of the other members and, in any case, owes part of its distinctive character to diagenetic dissolution. Unfortunately, some regional descriptions (Fränkl, 1953b; Sommer, 1957) are insufficiently detailed to distinguish between shaly carbonates of Bed groups 18 and 19 as now defined.

It should be noted that Katz (1952, 1961), Eha (1953), Fränkl (1953a) and Sommer (1957) reported that the light-coloured stromatolitic dolostone horizons sandwiched between dark limestones within Bed group 18 (Members 18.4, 18.6–8, 18.10, 18.12) vary locally in thickness and stratigraphic position. At least in the central part of the fjord zone between Kap Weber and Ella Ø, it is clear that this standard view (Haller, 1971; Henriksen and Higgins, 1976) is incorrect.

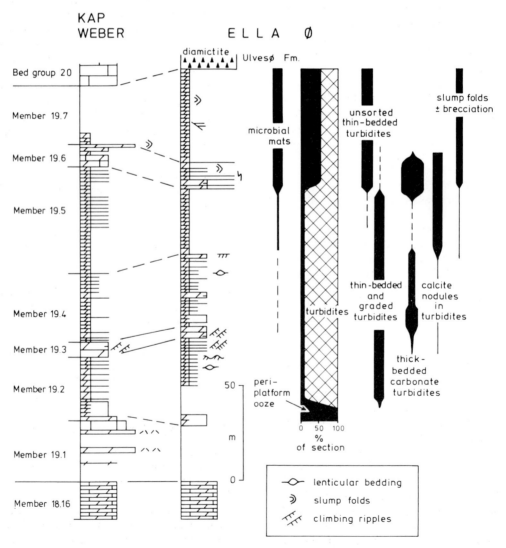

Fig. 5. Comparative sections of Bed group 19 at Kap Weber and Ella Ø with sedimentological data and interpretations of the latter. Local variations in the position and thickness of individual sediment gravity flows occur on Ella Ø. Key as on Fig. 4 with additions given.

Bed group 19

The complete thickness was logged on both Ella Ø and at Kap Weber, totalling 220 and 210 m respectively, and has been divided into seven members (Fig. 5). Bed group 19 consists predominantly of green-grey or red weathering dolomitic shales. Subordinate buff weathering dolomicrite units commonly contain grey weathering calcite microspar bands that are locally disrupted to form breccias. The Bed group is made distinctive by this variety of weathering colours.

Member 19.1 is only exposed at Kap Weber where it is ca. 30 m thick. Its lower half is unexposed and is overlain by mixed calcite and dolomite arenites and intraclastic breccias, replaced 2 km southeast by non-rudaceous limestones and dolostones. Member 19.2 is ca. 40–50 m thick and composed of 5 m of buff weathering dolomicrite overlain by dolomitic shales with thin dolomicrite beds. Member 19.3 is a ca. 8 m thick dolomicrite unit in which beds commonly contain distinctive climbing ripple horizons. Members 19.4 to 19.5 are again composed largely of dolomitic shales with isolated dolomicrite units that commonly contain calcite microspar-rich bands. At Kap Weber, the members are thicker (ca. 100 m) than those on Ella Ø (ca. 70 m). Member 19.6 is composed of concretionary limestones within dolomicrite beds and variably exhibits slump folds and disruption to form breccias containing angular, tabular clasts of calcite microspar in a dolomicrite matrix. There are extreme local variations in the development of breccias. On Ella Ø, member 19.7 is ca. 50 m thick and composed of organic-rich dolomitic shale. At Kap Weber only the basal 6 m are exposed and consist of dark-grey weathering dolomitic shale. The member has a maximum thickness of 32 m and the boundary between Bed group 19 and 20 is unexposed.

On Ella Ø, Bed group 19 is directly overlain by the Ulvesø Formation of the Tillite Group. At Kloftelv (Fig. 1) the contact appears sharp and conformable, whereas at Tommerbugt underlying Bed group 19 mudstones appear to have been injected into the Ulvesø Formation by soft sediment deformation, perhaps due to rapid deposition of the overlying tillite. Uppermost Bed group 19 shale must have been unlithified at the time of Ulvesø Formation deposition and the contact is therefore probably conformable.

Detailed correlation of members within Bed group 19 appears to be possible within the fjord zone and is perhaps only hampered by inadequate descriptions of sections. However, care is needed when correlating the breccia lithology of member 19.6, which is laterally discontinuous and can occur over a wide stratigraphic interval.

Bed group 20

Bed group 20 only occurs north of Ella Ø and is therefore only described in this study from Kap Weber. Member 20.1 (?105 m) is a dark-grey weathering microsparite. Member 20.2 (16 m) is composed of mid-grey weathering calcareous shale. Member 20.3 (7 m) comprises an interlaminated limestone and dolomite microsparite with subaqueous shrinkage cracks, grading up into partially dolomitized calcarenites. The contact with the Tillite Group is sharp and of probable tectonic origin, since local shear planes and minor folds are common. Members 20.2 and 20.3 are only locally present and the overall thickness of the Bed group varies greatly because of this deformation. Its true thickness is difficult to assess. Bed group 20 does not appear to be laterally equivalent to the breccia units of Bed group 19 member 6, as suggested by Schaub (1955) and

Katz (1961), as breccia units are present in the same stratigraphic position within Bed group 19 on Ella Ø and at Kap Weber, underlying Bed group 20 in the latter area. This is in agreement with the observations of Hambrey and Spencer (1987) at other localities in the fjord zone.

SEDIMENTOLOGY

Detailed investigations incorporating petrographic work have been carried out on Bed group 19 that add significantly to the field data pointing to a marine slope environment. Bed groups 18 and 20 have as yet been studied less intensively in the laboratory, but field evidence points to a carbonate shelf setting.

Bed group 18 and member 19.1

Much of the carbonate has been recrystallized to yield microsparry textures in which original fabrics have been lost (cf. Tucker, 1983, 1985). Oolitic and pisolitic fabrics are variably preserved, a shallow marine origin being demonstrated by the discovery of endolithic cyanobacteria identical to modern forms from the Bahama Banks (Knoll et al., 1986b). The existence of stromatolite-bearing, coarsening-upwards cycles is consistent with repeated shoaling that is a common motif in shelf carbonates (James, 1979).

Isolate to poorly linked shrinkage cracks filled by pure microspar are present within numerous horizons of microsparry limestone. The cracks are thought to originate by subaqueous shrinkage and their fillings represent recrystallized aragonite cement (Fairchild and Spiro, 1987). Such cementation also points to shallow marine conditions.

Subaerial exposure is indicated by tepee structures at some horizons, and peritidal deposition is suggested by horizons with reworked intraclasts.

Members 18.13 to 18.16 contain fewer sedimentary structures. Members 18.13 and 18.15 display limestone-shaley limestone alternations that are typical of offshore shelf carbonates (Ricken, 1986). In contrast, member 18.14 is composed of pure dolostone that shows all transitions from undisturbed, through dolostone cut by sets of subhorizontal fractures through to disorganized breccia. Downward sequences of this type are indicative of collapse brecciation (Middleton, 1961). Fractures are filled by thick crusts of fibrous dolomite and there is evidence of erosion of one collapsed unit prior to resumption of deposition. The collapse was presumably related to dissolution of underlying evaporites or soluble carbonate. The environmental setting of member 18.16 with its thin laminae of dolomite-calcite alternations is not known.

Most of member 19.1 is unexposed, but peritidal conditions for at least some of the member can be inferred from the abundance of intraclastic lithologies and presence of tepee structures at the member top.

Bed group 19: members 19.2 to 19.7

The bulk of the sediments are shaly carbonates now composed of quartz silt, berthierine after kaolinite and microcrystalline iron-rich carbonate (ankerite, siderite or ferroan dolomite). Carbonate phases are believed to be an early diagenetic product of organic matter oxidation via methanogenesis and iron reduction (cf. Baker and Burns, 1985) and are thought to have replaced an earlier detrital carbonate phase. Iron within carbonates was probably originally present as detrital iron oxide and disseminated iron oxides are still abundant within certain horizons. An association between iron oxides and kaolinite in the original sediment may have been produced by weathering of a lateritic source area.

A sequence of structures is variably developed within these shales, consistent with their interpretation as an accumulation of fine-grained turbidites superimposed upon peri-platform ooze. Sediments within the upper part of Bed group 19 were modified by slumping, to form debrites (debris flow deposits) and turbidites.

The vast majority of turbidites are up to a few centimetres thick and are either graded, with well developed sedimentary structures (lower members) or massive and unsorted (upper members) (Fig. 5). The sequence of structures in graded turbidites is similar to those described by van der Lingen (1969) for silt turbidites and is closely linked to C, D and E divisions of the "Bouma sequence" (Fig. 6). Massive turbidites contain no sedimentary structures or sorting and can be thought of as the deposits of high-density turbidity currents or dilute debris flows (Pickering *et al.*, 1986).

Background sediment between turbidites displays fine irregular lamination picked out by subtle variation in clay, carbonate and quartz silt content. The lowermost few metres of member 19.2 are composed of a relatively thick accumulation of carbonate-rich, quartz-silt-poor background sediment thought to have formed during an initial period of sediment starvation following the formation of the slope environment.

The transition upwards from graded to massive turbidites is marked by a sharp increase in the abundance of background sediment, which becomes subequal in abundance to turbidites (Fig. 5). This increase is associated with development of anastomosing organic-rich laminae, similar to examples in the literature thought to represent microbial mats (Horodyski, 1980; Schieber, 1986). Massive unsorted turbidites contain tabular clasts or "flakes" of organic-rich material, up to 10 mm in diameter, thought to have formed via slumping of such organic-rich sediment.

Thus, a lower unit of structured turbidites is overlain by a unit dominated by the development of microbial mats, slumping and the formation of structureless debrites/turbidites.

Superimposed on these sediments, especially in members 19.3 and 19.6, are isolated beds and units of beds of more thickly bedded (up to 30 cm thick) dolomicrite-rich turbidites, some of which contain climbing ripples (Fig. 5). Since overlying slope sediments have a high siliciclastic content and could not therefore have formed the source of these beds, it is assumed that, in contrast to thinly bedded tubidites, thick-bedded carbonate-rich turbidites were sourced from outside the slope environment.

Within thick-bedded turbidites calcite-microspar forms nodules, bands and complete beds, increasing in abundance and development towards the top of the sequence (Fig. 5). Early lithification due to this development of calcite-microspar, and subsequent slumping, led to the formation of sedimentary breccias associated with member 19.6.

Development of subaqueous shrinkage cracks and microbial mats within upper members indicates a shallowing-up or prograding sequence. Such cracks are only described from shallow-water sediments (e.g., Fairchild and Spiro, 1987; Horodyski, 1976; Donovan and Foster, 1972), whereas Brock (1976) limits the occurrence of cyanobacterial mats to 50 m water depth as far as light is concerned. It is suggested that the development of microbial mats has formed a major control, principally by binding sediment, on the development of the carbonate slope.

Limited palaeocurrent data from climbing-ripple horizons within member 19.3 indicate these turbidites were flowing to the NNE, oblique to slump fold vergence directions to the SE measured from member 19.6. However, since Kap Weber lies NNE of Ella Ø, turbidite facies, in particular member 19.3, must be laterally continuous over tens of kilometres in a downflow direction. This indicates turbidity currents moved along the axis of a linear basin fed at frequent intervals by turbidity currents moving towards the basin axis. This is consistent with models of turbidity currents flowing parallel to the axes of elongate basins reported in the literature (e.g., MacDonald and Tanner, 1983; Walker, 1978; Woodcock, 1976).

A slope-apron environment (Mullins and Cook, 1985) for Bed group 19 is indicated by the following.

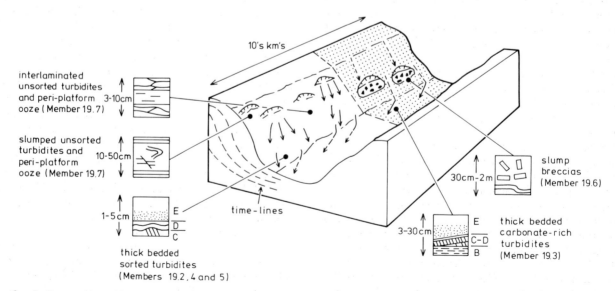

Fig. 6. Sedimentological model for the deposition of Bed group 19 in a basin (half-graben) with axial flow of turbidity currents. The basin is inferred to have been created by normal faulting, but there is insufficient information to determine the polarity of this faulting. C, D and E refer to the Bouma divisions of turbidites.

(1) There is a good correlation of members between Ella Ø and Kap Weber.

(2) There are no channel deposits in our studied sections or in several other studied by Hambrey and Spencer (1987).

(3) Slump folding and discordant bedding are present (Fig. 5).

(4) Bed group 19 is directly overlain by shelf sediments of Bed group 20. Background sediments within Bed group 19 were probably derived from these shelf sediments and can therefore be termed peri-platform ooze (Schlager and James, 1978).

Figure 6 is a schematic representation of the basin geometry and sedimentary processes operating during the formation of Bed group 19. The rear of the block diagram represents times, perhaps related to sea-level fluctuations, when thick carbonate-rich turbidites, sourced primarily from outside the slope environment, were being deposited. Turbidity currents that flowed into the axis of the basin were deflected to flow parallel to the basin axis. Deposition within the slope environment, early lithification and subsequent slumping led to the formation of slump breccias. The front of the diagram represents the more normal sedimentation of thin-bedded turbidites. Increased accumulation of peri-platform ooze owing to trapping by microbial mats at higher levels led to slope instability and slumping. Slump masses evolved by progressive dilution into debris flows/high-density turbidity currents (depositing unsorted, thin-bedded turbidites) and low-density turbidity currents (depositing thin-bedded, sorted turbidites).

The abrupt change upward within Bed group 19 from inferred peritidal to slope-apron sediments points to tectonically induced subsidence and tilting. In this otherwise stable geological setting, extensional tectonics is indicated. Unlike more extreme examples of this phenomenon (e.g., Bernoulli and Jenkyns, 1974), this rifting phase is abortive, since shallow-water conditions become re-established.

Bed group 20

Member 20.1 consists of recrystallized, organic-rich limestones, similar to many limestones in Bed group 18, with recognizable allochems in places. High strontium contents (up to 2700 ppm) indicate an originally aragonitic composition. Member 20.3 contains wave-generated crosslamination, subaqueous shrinkage cracks and sorted and rounded intraclasts pointing to a shallow marine setting. The member can be thought of as part of a carbonate shelf prograding onto the slope-apron deposits of Bed group 19. Carbonate shelf conditions evidently could not be established in the Ella Ø region where Bed group 20 is absent. It is likely therefore that the upper part of member 19.7 on Ella Ø is time-equivalent to Bed group 20.

It is significant that, whereas on Ella Ø Bed group 19 passes up into glaciomarine tillite of the Ulvesø Formation, at Kap Weber Bed group 20 passes up into a sandstone thought largely to be aeolian in origin (A. C. M. Moncrieff, personal communication), reflecting the differing positions relative to sea-level of the two areas. The basal diamictites on Ella Ø are extremely calcareous clasts floating in a recrystallized microsparry matrix: it seems likely that much aragonitic debris was incorporated, derived from a shelf area. Given the evidence for flow of ice to the north (Moncrieff, this volume), this shelf may not have been the one represented by Bed group 20 deposits, but a separate shelf to the south.

COMPARISON WITH NE SVALBARD

Given the overall similarity of the late Proterozoic geological history of NE Svalbard and East Greenland mentioned earlier, it is instructive to compare the pre-glacial carbonate sections (Fig. 2). It is clear that the development of slope facies in Greenland has no counterpart in NE Svalbard below the Elbobreen Formation. The facies consistently exhibit shallow-water structures such as desiccation phenomena, prolific stromatolites or ooids/pisoids. In the basal member of the Elbobreen Formation, randomly oriented subaqueous shrinkage cracks in the basal part, and intraclastic and stromatolitic dolostones in the upper part, attest to continuing shelf deposition. However, one locality (Polarisbreen: Fairchild and Hambrey, 1984) has a central unit consisting of uniform carbonates, lacking evidence of sediment gravity flows but with a single chert bed exhibiting slump folds overturned to the southeast. Even so, there was no development of slope facies on the scale of East Greenland. This can be better understood by comparing the whole upper Proterozoic sections in the two areas.

Both in East Greenland and NE Svalbard (NE Spitsbergen and Nordaustlandet) there is a Riphean to Ordovician succession that was involved in Caledonian deformation, but which in the upper part has escaped metamorphism and is only locally noticeably deformed. Below Cambrian rocks, there are 6.5–7 km of unmetamorphosed sediments (Upper Eleonore Bay Group) in East Greenland (Haller, 1971; Knoll *et al.*, 1986a) and 6.5 km of comparable Middle to Upper Hecla Hoek sediments in NE Spitsbergen (Wilson, 1961; Swett and Knoll, 1985; Knoll *et al.*, 1986a). In each case a dominantly terrigenous succession gives way upwards to a carbonate-dominated section. Knoll *et al.* (1986a) have proposed a correlation scheme tied by a series of lithostratigraphic markers for the unmetamorphosed late Proterozoic sections as a whole. This is backed by carbon isotope analyses of carbonates and organic carbon that provides the first attempt at a chemostratigraphy for the sequences. Comparison of intervals between pairs of lithostratigraphic markers given in Knoll *et al.*, (1986a) would indicate a complex picture of basin development with sedimentation rates in East Greenland ranging from 50 to 300% of those in NE Spitsbergen. However, allowing for draughting errors in Knoll *et al.* (1986a) and a revised correlation of the base of the Draken Formation (Fig. 2) with the base of Bed group 15 (Knoll personal communication, 1987), a much simpler picture emerges. Taking the base of Bed group 18 to correspond with the base of the Backlundtoppen Formation, the underlying 3–3.5 km of sediments can be divided into seven units with similar thicknesses in the two areas. A lithological similarity persists upwards for at least 250–300 m with a predominance of pisolitic limestones in the basal Backlundtoppen Formation and Bed group 18 (Fig. 2). The rough equation of the upper Backlundtoppen Formation and member E1 of the Elbobreen Formation of NE Svalbard with members 18.11 to 18.16 plus Bed groups 19 and 20 of east Greenland indicates a subsidence rate about twice as great for the latter. This fits well with the evidence for accelerated basin subsidence in the form of slope facies only in the Greenland section.

An improvement in our understanding of the timing of these events requires more biostratigraphical data. The position of the Riphean–Vendian boundary is still unknown in the East Greenland area (Knoll, personal communication, 1987). The absence of lower Vendian fossils in both regions (Vidal, 1985) is puzzling given that no important hiatuses are evident from the sedimentological studies.

CONCLUDING DISCUSSION

The Eleonore Bay Group, Tillite Group and Cambro–Ordovician of East Greenland and the Riphean to Ordovician Hecla Hoek of NE Svalbard were deposited in a basin that does not seem to have been adjacent to a continental margin, accepting the cratonic nature of the Barents Sea area east of Svalbard (Harland and Gayer, 1971; Roberts and Gale, 1978). The southern end of the basin may have trended into the Iapetus ocean (Harland and Gayer, 1971) when it opened in the Cambrian (Anderton, 1982). On this basis, the Greenland basin may therefore be classifiable as a failed rift rather than an intracratonic basin. Evidence for a rifting phase may be found in the volcanic deposits of the Lower Eleonore Bay Group (Haller, 1971) and the Lower Hecla Hoek (Harland *et al.*, 1966), although the continuity of sedimentation with overlying deposits is unclear. McKenzie's (1978) model of basin development predicts a gradually decreasing thermal subsidence, following initial rapid subsidence on rifting. However, the sedimentary basin seems to have been active for several hundred million years, which is too long for a single rifting event. Also, this paper evidences tectonic instability at a much later stage of basin evolution.

Members 18.1 to 18.12 show both lateral continuity and evidence of deposition as carbonate shelf sediments. Members 18.13 to 18.16 exhibit lateral variations. Although the geometry of these variations needs further investigation, they might indicate the onset of syndepositional faulting leading to partitioning of environments in the basin. More certain evidence of tectonic activity, presumably related to renewed basin extension is the rapid drowning and tilting of the shelf that can be inferred around the 19.1–19.2 boundary.

Subsequently, deposition of Bed group 19 slope sediments probably occurred within a restricted elongate basin. A carbonate shelf found in the north of the fjord zone kept pace with subsidence and (at Kap Weber) was capped by subaerial deposits within the Ulvesø Formation. In the south of the fjord zone, the slope environment persisted into Tillite Group times. Marked facies variations within the Tillite Group may also be related to differential uplift and subsidence (Hambrey, 1988, this volume).

In summary, this paper has demonstrated the extent of lateral facies continuity within the Upper Riphean shelf carbonates in the central fjord zone and revealed an important episode of inferred basin extension as evidenced by slope-related carbonate deposition.

Acknowledgements

Fieldwork was enabled by N.E.R.C. grant GR3/5438 awarded to Mr W. B. Harland (Cambridge University). We are grateful to Mike Hambrey (Cambridge), for his leadership of the expedition, knowledge and support, and to Andy Moncrieff for companionship in the field. PMH acknowledges an N.E.R.C. studentship. Figures were draughted by Carl Burness.

REFERENCES

Anderton, R. (1982). Dalradian deposition and the later Precambrian–Cambrian history of the N Atlantic region: a review of the early history of the Iapetus ocean. *J. Geol. Soc. Lond.* **139**, 421–31.

Anderton, R. (1985). Sedimentation and tectonics in the Scottish Dalradian. *Scott. J. Geol.* **21**, 407–36.

Baker, P. A. and Burns, S. J. (1985). Occurrence and formation of dolomite in organic-rich continental margin sediments. *Amer. Assoc. Petrol. Geol. Bull.* **69**, 1917–1930.

Bernoulli, D. and Jenkyns, H. C. (1974). Alpine, Mediterranean and central Atlantic Mesozoic facies in relation to the early evolution of the Tethys. In: Dott, Jr, R. H. and Shaver, R. H. (eds.) *Modern and Ancient Geosynclinal Sedimentation*, pp. 129–160. Spec. Publ. Soc. Econ. Paleont: Minerol., Tulsa, **19**.

Brock, T. D. (1976). Environmental microbiology of living stromatolites. In: Walter, M. R. (ed.) *Stromatolites*, pp. 141–148. Elsevier, Amsterdam.

Donovan, R. N. and Foster, R. J. (1972). Subaqueous shrinkage cracks from the Caithness flagstone series (Middle Devonian) of North-East Scotland. *J. Sediment. Petrol.* **42**, 309–317.

Eha, S. (1953). The Pre-Devonian sediments on Ymers Ø, Suess Land, and Ella Ø (East Greenland) and their tectonics. *Meddr. Grønland* **111**(2), 105pp.

Fairchild, I. J. (1988). Dolomitic stromatolite-bearing units with storm deposits from the Vendian of East Greenland and Scotland: a case of facies equivalence, *this volume*.

Fairchild, I. J. and Hambrey, M. J. (1984). The Vendian of NE Spitsbergen: petrogenesis of a dolomite-tillite association. *Precambrian Res.*, **26**, 111–167.

Fairchild, I. J. and Spiro, B. (1987). Petrological and isotopic implications of some contrasting Late Precambrian carbonates, NE Spitsbergen. *Sedimentology* **34**, 973–989.

Fränkl, E. (1953a). Geologische Untersuchungen in Ost-Andrées Land (NE-Grønland). *Meddr. Grønland* **113**(4), 160pp.

Fränkl, E. (1953b.) Die Geologische Karte von Nord-Scoresby Land (NE-Grønland). *Meddr. Grønland* **113**(6), 56pp.

Green, J., Knoll, A. H., Golubic, S. and Swett, K. (1987). Paleobiology of distinctive benthic microfossils from the Upper Proterozoic Limestone-Dolomite "Series", central East Greenland. *Amer. J. Bot.* **74**, 928–40.

Haller, J. (1971). *Geology of the East Greenland Caledonides.* Interscience, New York, 413pp.

Hambrey, M. J. (1983). Correlation of Late Proterozoic tillites in the North Atlantic region and Europe. *Geol. Mag.* **120**, 209–232.

Hambrey, M. J. (1988). The Late Proterozoic sedimentary record of East Greenland: its place in understanding the evolution in the Caledonide Orogen, *this volume*.

Hambrey, M. J. and Spencer, A. M. (1987). Late Precambrian glaciation of central east Greenland. *Meddr. Grønland Geoscience* **19**, 50pp.

Harland, W. B. and Gayer, R. (1971). The Arctic Caledonides and earlier oceans. *Geol. Mag.* **109**, 289–314.

Harland, W. B., Wallis, R. H. and Gayer, R. A. (1966). A revision of the Lower Hecla Hoek succession in central north Spitsbergen and correlation elsewhere. *Geol. Mag.* **103**, 70–97.

Henriksen, N. and Higgins, A. K. (1976). East Greenland Caledonian Fold Belt. In: Escher, A. and Watt, W. S. (eds.) *Geology of Greenland*, pp. 183–245. Geological Survey of Greenland, Copenhagen.

Horodyski, R. J. (1976). Stromatolites of the Upper Siyeh Limestone (Middle Proterozoic), Belt Supergroup, Glacier National Park, Montana. *Precambrian Res.* **3**, 517–536.

Horodyski, R. J. (1980). Middle Proterozoic shale facies microbiota from the lower Belt Supergroup, Little Belt Mts., Montana. *J. Paleont.* **54**, 649–663.

James, N. P. (1979). Shallowing-upward sequences in carbonates. In: Walker R. G. (ed.) *Facies Models*, pp. 126–136. Geoscience Canada Reprint Series 1.

Katz, H. R. (1952). Zur Geologie von Strindbergs Land (NE Grønland). *Meddr. Grønland* **111**(1), 150pp.

Katz, H. R. (1961). Late Precambrian to Cambrian stratigraphy in East Greenland. In: Raasch, G. O. (ed.) *Geology of the Arctic*, vol. 1, pp. 299–328. Toronto University Press, Toronto.

Knoll, A. H., Hayes, J. M., Kaufman, A. J., Swett, K. and Lambert, I. B. (1986a). Secular variation in carbon isotopic ratios from Upper Proterozoic successions of Svalbard and East Greenland. *Nature* **321**, 832–8.

Knoll, A. H., Golubic, S., Green, S. and Swett, K. (1986b). Organically preserved microbial endoliths from the late Proterozoic of East Greenland. *Nature* **321**, 856–7.

Kulling, O. (1934). The "Hecla Hoek Formation" round Hinlopenstretet. *Geogr. Annaler. Stock. Arg.* **XVI**, Häft 4, 161–254.

MacDonald, D. I. M. and Tanner, P. W. G. (1983). Sediment dispersal patterns in part of a deformed Mesozoic back arc basin on South Georgia, South Atlantic. *J. Sediment. Petrol.* **53**, 83–104.

McKenzie, D. (1978). Some remarks on the development of sedimentary basins. *Earth Planet Sci. Lett.* **40**, 25–32.

Middleton, G. V. (1961). Evaporite solution breccias from the Mississippian of southwest Montana. *J. Sediment. Petrol.* **31**, 189–195.

Moncrieff, A. C. M. (1988). The Tillite Group and related rocks of East Greenland: implications for late Proterozoic palaeography, *this volume*.

Mullins, H. T. and Cook, H. E. (1986). Carbonate apron models: Alternatives to the submarine fan model for palaeoenvironmental analysis and hydrocarbon exploration. *Sediment. Geol.* **48**, 37–79.

Pickering, K. T., Stow, D. A. V., Watson, M. P, and Hiscott, R. N. (1986). Deep-water facies processes and models: a review and classification scheme for modern and ancient sediments. *Earth-Sci. Rev.* **23**, 75–174.

Ricken, W. (1986). Diagenetic Bedding. *Lecture Notes in Earth Sciences* **6**. Springer, Berlin.

Roberts, D. and Gale, G. H. (1978). The Caledonian–Appalachian Iapetus ocean. In: D. H. Tarling (ed.) *Evolution of the Earth's Crust*, pp. 255–342. Academic Press, London.

Schaub, H. P. (1950). On the Precambrian to Cambrian sedimentation in NE-Greenland. *Meddr. Grønland* **114**(10), 50pp.

Schaub, H. P. (1955). Tectonics and morphology of Kap Oswald (NE-Greenland). *Meddr. Grønland* **103**(10), 33pp.

Schieber, J. (1986). The possible role of benthic microbial mats during the formation of carbonaceous shales in shallow Mid-Proterozoic basins. *Sedimentology* **33**, 521–536.

Schlager, W. and James, N. P. (1978). Low-magnesian calcite limestones forming at the deep-sea floor, Tongue of the Ocean, Bahamas. *Sedimentology* **25**, 675–702.

Sommer, M. (1957). Geologie von Lyells Land (NE-Grønland). *Meddr. Grønland* **155**, 157pp.

Swett, K. and Knoll, A. (1985). Stromatolitic bioherms and microphytolites from the Late Proterozoic Draken Conglomerate Formation, Spitsbergen. *Precambrian Res.* **28**, 327–347.

Teichert, C. (1933). Untersuchungen zum Bau des kaledonischen Gebirges in Ostgronland. *Meddr. Grønland* **95**(1), 121pp.

Tucker, M. E. (1983). Sedimentation of organic-rich limestones in the Late Precambrian of southern Norway. *Precambrian Res.* **22**, 295–315.

Tucker, M. E. (1985). Calcitized aragonite ooids and cements from the Late Precambrian Biri Formation of Southern Norway. *Sediment. Geol.* **43**, 67–84.

Van der Lingen, G. J. (1969). The turbidite problem. *N. Z. J. Geol. Geophys.* **12**, 7–50.

Vidal, G. (1985). Biostratigraphic correlation of the Upper Proterozoic and Lower Cambrian of the Fennoscandinavian Shield and the Caledonides of East Greenland and Svalbard. In: Gee, D. G. and Sturt, B. A. (eds.) *The Caledonide Orogen—Scandinavia and Related Areas* pp. 331–338. Wiley, Chichester.

Walker, R. G. (1978). Deep water sandstone facies and ancient submarine fans: models for exploration for stratigraphic traps. *Bull. Amer. Assoc. Petrol Geol.* **62**, 932–966.

Wilson, C. B. (1961). The Upper Middle Hecla Hoek rocks of Ny Friesland, Spitsbergen. *Geol. Mag.* **98**, 98–116.

Woodcock, N. H. (1976). Ludlow Series slumps and turbidites and the form of the Montgomery Trough, Powys, Wales. *Proc. Geol. Assoc.* **87**, 169–182.

23 Dolomitic stromatolite-bearing units with storm deposits from the Vendian of East Greenland and Scotland: a case of facies equivalence

Ian J. Fairchild

School of Earth Sciences, University of Birmingham,
PO Box 363, Birmingham B15 2TT, UK

Comparison of Vendian rocks in Scotland and East Greenland reveals an occurrence of a congruent group of dolomitic facies at one stratigraphic position in each succession. Member 3 (0–200 m thick) of the Dalradian Bonahaven Formation of Islay, Scotland, contains three sedimentary facies. The layered facies is represented by mudrocks with thin storm sand layers, subaqueous shrinkage cracks, anhydrite pseudomorphs and rare desiccation features. The sandstone facies contains tidal cross-stratification and desiccated dolostone layers. The stromatolite facies consists of biostromes and bioherms of varied morphology, but with only rare examples of regular columnar types.

The upper part of the Canyon Formation on Ella Ø, central East Greenland, displays a major shallowing-upwards sequence from offshore mudrocks through storm sediments (with dolomitic crusts and intraclasts) to a dolostone unit (the stromatolitic member, 25 m thick) with increasing evidence of emergence upwards, followed by playa lake sediments. The stromatolitic member is closely similar to member 3 of the Bonahaven Formation: the latter differs mainly in its near-random facies stacking and greater siliciclastic content.

The two occurrences are in different sedimentary basins and, although both are late Vendian in age, the Greenland example is likely to be the younger by up to several tens of millions of years. The facies congruence reflects partly the similar age, partly a similar climate, but especially a similar environment. This was an evaporative lagoon with only local tidal activity and low wave energy, except during infrequent severe storms. The energy of these storms was largely dissipated in the open shoreface at the lagoon margin.

The combination of environmental parameters is not found in standard facies models, yet two examples are illustrated here. Others ought to be represented in the stratigraphic record.

INTRODUCTION

Shallow marine sedimentary facies vary enormously in character in response to climate, proximity to sources of sediment supply, physical parameters (e.g., tide, wave and storm activity) and organic activity at different times in Earth history. It is unusual to find examples of closely similar facies associations deposited in different sedimentary basins at different times because of this vast potential for variation. An example is described here of two such equivalent facies associations (mixed carbonate–siliciclastic coastal deposits of Vendian age) and an attempt is made to elucidate the controls on their similarity

and minor differences. By implication, this kind of study helps to clarify the reasons for the variation of other Late Proterozoic shallow marine facies.

One facies association is drawn from the metasediments of the Dalradian Supergroup of Scotland, and the other from the Tillite Group of central East Greenland. A late Proterozoic reconstruction for the northeast Atlantic region (Fig. 1(a)) illustrates that the two areas are likely to have been part of separate depositional basins whose preserved portions were not less than 1000 km apart. Central East Greenland and NE Spitsbergen display a similar late Proterozoic to Ordovician record of sedimentation with evidence that their common deposi-

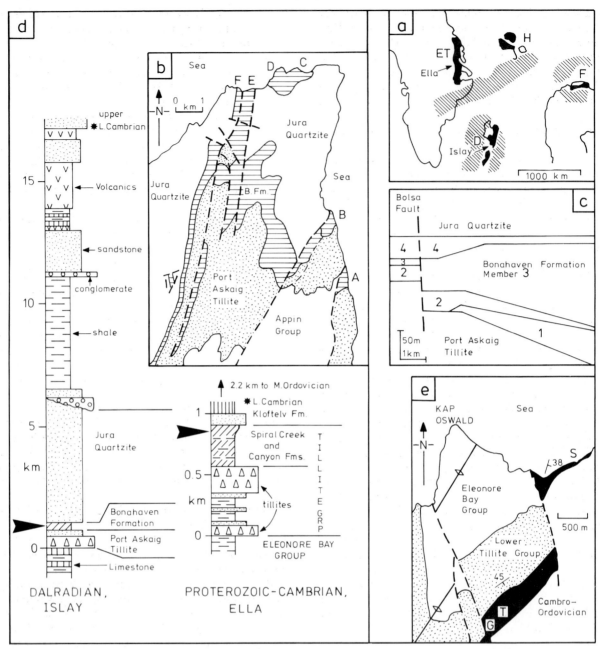

Fig. 1. (*a*) Late Proterozoic reconstruction (after Anderton, 1982; Hambrey, 1983) with preserved sediments of uppermost Proterozoic basins in black and inferred land areas shaded. F = Finnmark, north Norway; H = Hecla Hoek Complex of eastern Svalbard; ET = Eleonore Bay and Tillite Groups of East Greenland, with location of Ella Ø; D = Dalradian Supergroup of Scotland and Ireland with location of Isle of Islay. (*b*) Geological map of northeastern Islay (56° N, 6° W), western Scotland. Outcrop of Bonahaven Formation (B.Fm.) is horizontally shaded. A to F are coastal sections of the Bonahaven Formation, after Spencer and Spencer (1972) and Fairchild (1980b). (*c*) Stratigraphic reconstruction of the Bonahaven Formation of Islay along an east–west line around the level of section A of Fig. 1(*b*). The Bolsa fault also divides sections E and F on the north coast of the island. (*d*) Stratigraphic sections through latest Precambrian sediments of the upper part of the Dalradian Supergroup in Scotland and the Tillite Group and adjacent strata in East Greenland. Arrowheads show the position of the units compared in this paper. Asterisks denote Cambrian fossil horizons. (*e*) Geological map of the northern part of Ella Ø (73° N, 25° W), East Greenland, with outcrop of Canyon and Spiral Creek Formations in black, offset by faults (dashed lines). The position of the axis of the Kap Oswald Anticline is also shown. Studied sections are Solitaerbugt (S), Storeelv Gorge (G) and Storeelv tributary (T).

tional basin was bounded in the southern quadrants by a major landmass (Hambrey, this volume; Moncrieff, this volume). The Dalradian is understood to be a narrow, fault-bounded basin lying southeast of a major landmass, Laurentia (Harland and Gayer, 1972; Anderton 1985).

Stratigraphic columns for the Vendian deposits (Fig. 1(*d*)) draw attention to differences between the two basins. It can be concluded that not only does the Dalradian basin differ in its earlier (Lower Cambrian) cessation of deposition and far more intense deformation and metamorphism than the East Greenland succession, but that it subsided at roughly an order of magnitude greater rate and hence is

likely to represent a completely different tectonic setting from that of East Greenland. The latter conclusion can be drawn from a comparison of the stratigraphic thicknesses from the top of the glacial formations to the base of the Cambrian: that is, the rocks representing the Valdaian (e.g., Vidal, 1985) or Ediacaran (*sensu* Harland *et al.*, 1982) epochs. Harris *et al.* (1978) had argued for positioning of the Precambrian–Cambrian boundary in the Dalradian at a horizon of basin-deepening (at either 6 km or 12.5 km on Fig. 1(*d*)), since the Lower Cambrian in the NW Highlands of Scotland is transgressive. Anderton (1982, 1985) argued that basin-deepening episodes would be controlled by local tectonics and that the boundary would lie within the thick volcanic sequence (the Tay-vallich Lavas, located around the 15 km mark on Fig. 1(*d*)) by comparison with basal Cambrian volcanicity else-where in the North Atlantic region. An Ediacaran strati-graphic thickness of 14–15 km in the Dalradian basin compares with 300–400 m in the East Greenland basin (Fig. 1(*d*)).

The stratigraphic position of the equivalent facies being discussed in this paper is shown on Fig. 1(*d*) by the arrowheads to the left of the columns. It seems unlikely that these two occurrences formed at the same time. The Dalradian unit is member 3 of the Bonahaven Formation (Fig. 1(*c*)), which lies stratigraphically only about 100 m above the last evidence of glacial activity (50 m below the top of the Port Askaig Tillite Formation) in the thickest sections of Fig. 1(*c*) (Spencer, 1971). In contrast, the Greenland unit is near the top of the Canyon Formation, nearly 300 m above the upper glacial unit and 25 m below the disconformable base of probable Cambrian sandstones (the Kloftelv Formation), and less than 100 m below the first occurrence of early Cambrian trilobites.

It should be noted, however, that the Canyon and Spiral Creek Formations of East Greenland compare with the lower 300 m of a 525 m thick succession between the top of the highest tillite and earliest Cambrian deposits in NE Spitsbergen (Fairchild and Hambrey, 1984) and so there is the possibility that a substantial time gap, representing up to half the Ediacaran epoch is represented by the disconformity at the top of the Spiral Creek Formation below the transgressive Kløftelv sandstone.

The duration of the Ediacaran epoch is difficult to assess, but 80 Ma (630–550 Ma) is suggested by limited con-straints provided by radiometric dating (Pringle, 1972; Cope and Gibbons, 1987; Dallmeyer and Gibbons, 1987). Assuming no major undetected breaks in sedimentation, the Bonahven Formation is early Ediacaran and the upper Canyon Formation is either mid Ediacaran or late Edia-caran, depending on the time gap assigned to the dis-conformity between the Spiral Creek and Kløftelv For-mations. The time difference between the two facies associations discussed in this paper is likely therefore to be several tens of millions of years.

DALRADIAN CASE STUDY: BONAHAVEN FORMATION

The Dalradian rocks of northern Islay, despite having undergone high-pressure greenschist facies meta-morphism, are only locally severely strained, and sedi-mentary facies and some diagenetic textures are well preserved (Fairchild, 1985). The Bonahaven Formation is a distinctive assemblage of mixed carbonate–siliciclastic sediments lying above the glaciomarine and shelf sedi-ments of the Port Askaig Tillite (Spencer, 1971; Eyles and Eyles, 1983) and below the Jura Quartzite, a tidal shelf sandstone (Anderton, 1976). Spencer and Spencer (1972) informally divided the Bonahaven Formation into four members that show lateral thickness and some facies changes as depicted on Fig. 1(*c*).

The Formation contains several lines of evidence that the Islay area was at the northwest margin of the Dalradian basin, bordering a major landmass.

(1) Member 1 rests on a thick succession of quartzites at the top of the Port Askaig Tillite Formation, and consists of shoreface, tidal inlet and back-barrier sandstones and mudrocks, becoming dolomitic near

the top. It wedges out to the west, the lower parts disappearing first as a lag conglomerate develops at the top of the Port Askaig Tillite.

(2) The dolomitic member 3 thins in the far southwest of the area and westwards across the Bolsa Fault (Fairchild, 1980b), which is thus interpreted as a syndepositional fault.

(3) Member 3 shows similar facies throughout the area, but the proportion of shallow intertidal sandflat deposits (the sandstone facies, see below) is higher in the northwest (26% of section E) than in the southeast (8% of section B). Subtidal sediments are correspondingly more abundant in the southeast.

(4) Wave ripple asymmetries and preserved ripple lamination suggest a shoreline lay to the north-northwest (Fairchild, 1980a).

(5) Member 4 consists of sandstones and mudrocks with a central pure dolostone that thickens to the west and is absent in the most southeasterly out-crops. The siliciclastic sediments contain some sedimentary structures indicating a tidal flat origin, whereas the dolostone lacks structures apart from rare floating quartz grains of inferred aeolian origin. Such a facies must lie landward rather than seaward of the upper and lower parts of member 4, with which it is partly laterally equivalent.

The sedimentology of member 3 is summarized below (after Fairchild, 1980a). The underlying member 2 is a transgressive open-shelf quartzite with a poorly exposed, but apparently sharp, contact with member 3. This sharp contact is taken to denote a sudden palaeogeographic change to allow the accumulation of muddy dolomitic facies.

Member 3 consists of a restricted range of sedimentary facies (summarized in Fig. 2) that occur interbedded in all the sections studied. Figure 3 illustrates the tabular geometry of the facies units along the north coast of Islay, with some lateral transitions between facies.

Between 60 and 80% of the member is occupied by the layered facies, (Fig. 2C) consisting of centimetre-scale alternations of dolomitic fine sandstones and variably dolomitic mudrocks. The sand layers may be thin graded and/or lenticular (lenticular-graded subfacies, Fig. 2C), or display wave-ripple lamination, locally overlying parallel lamination. The absence of current-related structures indicates that the dominant sedimentation mechanism was wave action during infrequent storms. Rare laterally continuous dolostone conglomerate layers (Fig. 3) probably are also storm-related. The mudrock layers vary from slightly dolomitic mudstones (e.g., carbonaceous dolomitic mudrocks typically of rippled subfacies, Fig. 2C) to pure dolostones (especially scoured dolostones subfacies) and at most horizons display ptyg-matically-compacted shrinkage cracks filled by sand from the overlying layer. The geometry of the cracks, par-ticularly the poor linking in plan, and absence of associated sedimentary brecciation point to a subaqueous origin, possibly by synaeresis induced by salinity changes in the overlying water body. Irregular tops to dolostone layers occur at a number of horizons (especially in the scoured-dolostone subfacies), only some of which can be explained by loading. Most are probably related to scouring of desiccated surfaces, since characteristic profiles of desic-cation cracks are seen at several horizons, commonly associated with intraclasts that are concentrated in depressions on the surface. One horizon was visible in plan and demonstrated the polygonal arrangement of the flake-filled furrows (flake pockets of Fairchild, 1980a)

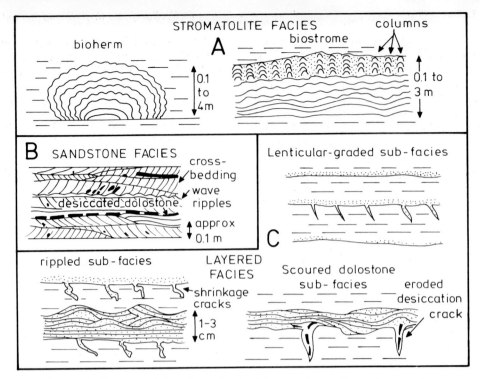

Fig. 2. Sedimentary facies of member 3 of the Bonahaven Formation on Islay.

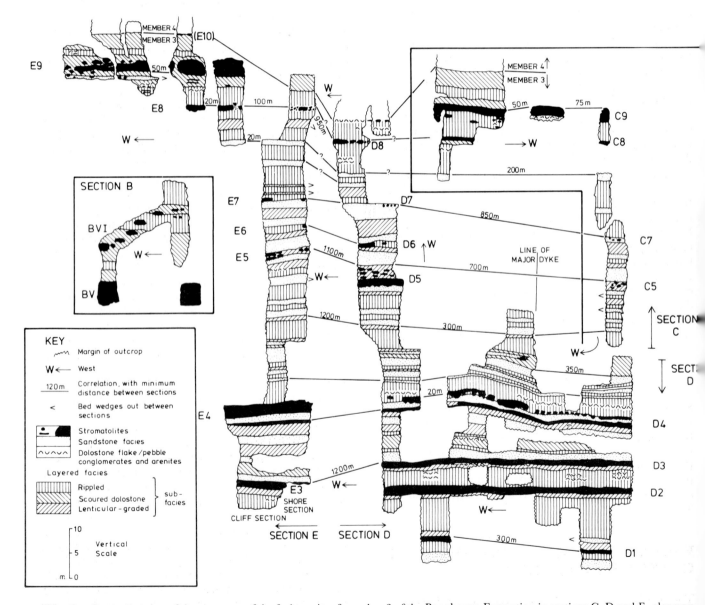

Fig. 3. Reconstruction of the geometry of the facies units of member 3 of the Bonahaven Formation in sections C, D and E, along an east–west line on the north coast of Islay (Fig. 1(b)). The base of the member is unexposed in these sections. Stromatolite horizons are coded by the section letter followed by a number. The inset shows the facies geometry at the top of exposed member 3 at Section B.

and hence an origin as enlarged desiccation cracks. Quartz–calcite-dominated nodules of millimetre to centimetre-scale occur commonly and are probably dominantly anhydrite pseudomorphs (Fairchild, 1985; original suggestion of A. R. MacGregor), although they are not displacive, nor do they occur preferentially in horizons with evidence of desiccation (Fairchild, 1980a).

The layered facies represents the sedimentation of mudrocks accompanied by active carbonate precipitation interrupted by infrequent storms. A broadly lagoonal environment is indicated, with local emergence. More precise interpretation of the subfacies of the layered facies is given later.

The sandstone facies (Fig. 2B) forms beds 0.1 to 3 m thick dominated by dolomitic cross-stratified and parallel-laminated medium-grained sandstone with subordinate pure dolostone interbeds commonly desiccated and incorporated as intraclasts in the sandstones. There is a variable content of dolomitic or calcitic ooids, with one horizon dominantly oolitic. Palaeocurrents are radial overall, but bimodal or multimodal in individual beds. This facies represents the deposits of low intertidal sand flats with variable tidal current direction related to complex local topography around emergent areas. Identical structures are shown by Cambrian oolitic limestones described by Chow and James (1987) and compared with intertidal sandflats bordering small ooid shoals on the windward margin of the modern Bahamas Banks.

Wave ripples in the sandstone and layered facies have wavelengths averaging 7.5 cm, which was used to infer a long wave fetch (greater than 100 km, Fairchild, 1980a).

Just under 10% of member 3 is occupied by dolomitic stromatolites (Fig. 2A) that exist both as continuous beds (biostromes) and as bioherms. Regular-shaped oblate spheroidal and ellipsoidal bioherms occur within layered facies sediments. These typically display centimetre to decimetre-scale growth forms based on an overall convex-up surface that in extreme cases becomes overturned on the margins, demonstrating syndepositional relief (up to 4 m). Bioherms with irregular, partly sandy lamination and ragged margins occur locally within sandstone facies beds. Biostromes are known to persist for at least 3 km in some cases (Fig. 3) and display internal lamination ranging from flat to broadly arched with decimetre-scale relief. Relief tends to diminish upwards. There are rare examples of columnar stromatolites (columns typically 2–3 cm wide) developing at the tops of biostromes corresponding to an increase in grain size of siliciclastic sediment and associated with intraclastic breccias. Although Spencer and Spencer (1972) and Fairchild (1980a) attempted to name them systematically, the thinness of the columnar horizons and the association with medium sand detritus make them unsuitable taxonomic material (M. A. Semikhatov and J. Sarfati, personal communication, 1985). The stromatolites were dominantly subtidal in origin, particularly the bioherms encased in layered facies sediments, but persisted in intertidal environments at times, as demonstrated by their coexistence with sandstone facies and possibly by the upward reduction in growth relief at some horizons that could be explained by shallowing upwards (Grey and Thorne, 1985).

EAST GREENLAND CASE STUDY: UPPER CANYON FORMATION

The stratigraphy of the Tillite Group in East Greenland has recently been summarized by Hambrey and Spencer (1987). The mudrock-dominated Canyon Formation with its dolostone top is known to persist for 150 km in the basin, although the overlying variegated dolomitic sandstone and mudrock unit, the Spiral Creek Formation, is missing in the north. Palaeogeographic evidence from palaeocurrents in the Canyon Formation is as follows.

(1) Subaqueous shrinkage cracks (see below) are aligned WSW–ENE. This direction would have been parallel to slope contours (Donovan and Foster, 1972), presumably parallel to the basin margin.
(2) Gutter marks at the base of storm beds (see below) are aligned north–south.
(3) Inferred offshore-directed cross-lamination in storm beds is orientated to the north–northwest.

This evidence is consistent with the proximity of the southern basin margin (Hambrey, this volume; Moncrieff, this volume) and points to a WSW–ENE-orientation of that margin.

The upper part of the Canyon Formation (Fig. 4) is an excellent example of a shallowing-upwards succession from offshore mudrocks through shoreface, lagoonal and tidal sandflat sediments to playa lakes of the Spiral Creek Formation. Of the several stratigraphic units represented in this interval, just the stromatolitic member of the Canyon Formation and the basal sandstone of the Spiral Creek Formation are comparable with member 3 of the Bonahaven Formation, but since the sequence can be interpreted simply in terms of regression, the nature of the underlying and overlying stratigraphic units helps to constrain the depositional environments. The absence of the Spiral Creek Formation in the north could represent a facies change, although erosion at the base of the overlying Kloftelv Formation is also a possibility.

A detailed sedimentological description is given by Fairchild and Herrington (1988) and is summarized here for comparative purposes. The following text should be read throughout in conjunction with Fig. 4.

Shallow-water sedimentary structures (ripples and desiccation cracks) are present in the basal metre of the Canyon Formation, but are absent in the remaining 250 m of the mudrock member, which consists of variably dolomitic silty shales of monotonous aspect. An offshore environment below wave base can be inferred. Towards the top of the mudrock member occur discrete centimetre to decimetre-scale dolomitic siltstone beds. These beds contain intraclasts of carbonaceous material (seen *in situ* as films in interbedded shales) interpreted as microbial mat fragments. Shallowing to the photic zone is inferred.

The dolomitic siltstone beds coarsen slightly to coarse silt-grade upwards and become dominant in the overlying siltstone member where a storm origin ("tempestites") is demonstrated by the combined presence of gutter casts, parallel to undulatory to hummocky stratification, and absence of current ripple lamination.

The succeeding flakestone member continues the storm-dominated theme, but the background sediments develop thin laminae of pure dolomite, each stromatolitic at their top. These become eroded and incorporated as flexible intraclasts in tempestite beds and isolated gutter-fills. They play the same hydrodynamic role as shell lags in Phanerozoic tempestites (e.g., Aigner, 1985). Basal gutter marks and upper portions with parallel lamination passing up into wave-ripple laminated dolomitic sand/siltstones are typical features of these tempestites. One variety of tempestite has a basal intraclastic layer with an upper diamictite-textured graded layer of intraclast in a silt matrix: these are storm-triggered debris flows. The flakestone member records a further shallowing of water into depths where early carbonate mineralization of microbial

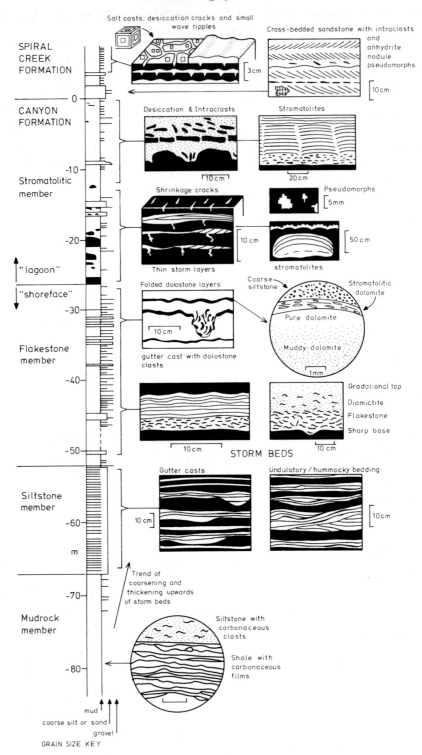

Fig. 4. Summary of the lithologies and sedimentary structures of the upper part of the Canyon Formation and basal Spiral Creek Formation on Ella Ø.

mats could occur to yield the distinctive dolomite layers and intraclasts.

There is then a transition over a 2 m vertical interval to the overlying stromatolitic member. The bulk of the member consists of finely crystalline dolostones, with shaley intercalations that seem to have arisen by pressure-solution in burial diagenesis. At a number of horizons occur repeated centimetre-scale graded, very fine sandstones to siltstones with parallel lamination, undulatory lamination and/or wave-ripple lamination. Incomplete shrinkage cracks are morphologically identical in vertical section to those described from member 3 of the Bonahaven Formation and provide evidence for salinity changes. They

display an alignment on bedding surfaces that is evidence for formation on a slight palaeoslope. In the upper half of the member, there are several discrete desiccated horizons, commonly modified by scour and locally partly covered with an intraclastic horizon. Several decimetre to metre-scale dolomitic arenites occur with a variable content of ooids. The thickest such bed defines the base of the Spiral Creek Formation and contains abundant cross-stratification with apparent bipolar transport directions (interpreted as tidal) and desiccated thin dolostone layers. Stromatolite bioherms and biostromes are conspicuous, being volumetrically more important in the lower half of the member. All transitions are seen from continuous beds with metre-wide broad arches to isolated bioherms with similar characteristics. Growth relief sometimes diminishes upwards in a bed and tends to be less in the upper half of the member. Columnar types are rare. Pseudomorphs of anhydrite crystals (millimetre to centimetre-scale) and nodules (centimetre to decimetre-scale) occur commonly throughout the member. Their present mineralogy is largely quartz and calcite, but anhydrite relics are common.

Shortly above the sandstone at the base of the Spiral Creek Formation, there is a 12 m thick unit with centimetre to decimetre-scale interbedding of dolomitic sandstones and mudstones. The sandstones display wave ripple marks of short (2–3 cm) wavelength and the mudstones are usually cracked by desiccation. Halite pseudomorphs are abundant. The evidence for exposure, limited wave fetch and absence of *sulphate* evaporite pseudomorphs points to a playa lake environment for this facies.

COMPARISON OF THE CASE STUDIES

Although a formal facies analysis has not been made of the stromatolitic member of the Canyon Formation, all the lithologies present can be accommodated within the facies scheme devised for the Bonahaven Formation member 3. Both can be interpreted as the deposits of a storm-influenced evaporitic lagoon that was locally emergent, with spatially-restricted strong tidal currents. Especially distinctive similarities include the almost entirely dolomitized nature of the carbonate; occurrence of thin storm layers associated with incomplete shrinkage cracks; stromatolite bioherms and biostromes of broadly arched morphology; desiccated and eroded dolostone profiles; cross-stratified sandstones with thin desiccated dolostone layers and variable content of ooids; non-displacive evaporite pseudomorphs occur both in strata displaying evidence of desiccation and those

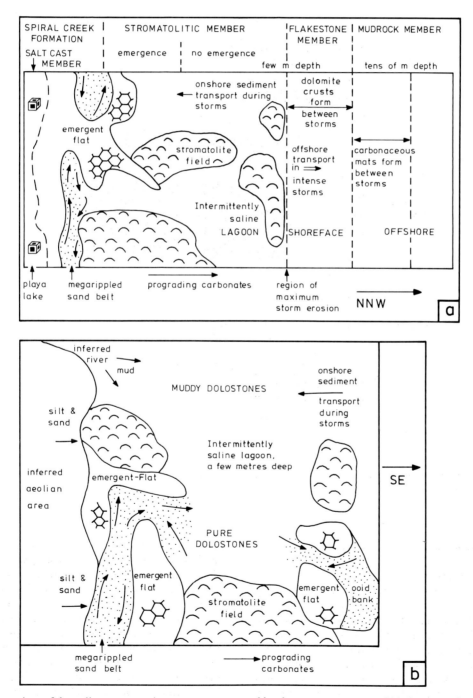

Fig. 5. Reconstructions of the sedimentary environments represented by the present outcrops of (*a*) the upper Canyon Formation and lower Spiral Creek Formation and (*b*) member 3 of the Bonahaven Formation on Islay.

that do not. Similar palaeoenvironmental reconstructions can be made (Fig. 5). The Greenland sections have the advantage of preserving facies inferred to be both onshore and offshore of the lagoon, so that the congruent dolomitic lagoonal facies can be more precisely placed in a bathymetric context.

The differences between the facies occurrences will now be examined with reference to the reconstructions of Fig. 5.

The upper Canyon Formation sediments are generally lower in siliciclastic detritus than member 3. The dolostones contain only minor clay minerals, whereas variable, sometimes high terrigenous mud contents are found in member 3. This can be explained by relative proximity of a fluvial source for mud in the member 3 environment (Fig. 5(b)). The Canyon Formation sediments are also deficient in terrigenous sand and coarse silt, as illustrated by the fact that storm layers are relatively thin or absent at many horizons. This can be related to a deficiency of source sediment supplied either from a nearby aeolian terrain or transported laterally by longshore drift to be redistributed by storms. By comparison, the layered facies of member 3 shows considerable variation in terrigenous sand, silt and clay content. The pure dolostones of the scoured dolostone subfacies occur in emergent areas far from fluvial mud input (Fig. 5(b)) but usually have relatively thick storm layers, perhaps representing redistributed aeolian detritus. The rippled subfacies typically has thick storm layers separating muddy carbonaceous dolostones, clearly proximal to sediment supply (top of Fig. 5(b)). The lenticular-graded subfacies is deficient in coarser detritus (perhaps to the right in Fig. 5(b)).

Beds attributable to the sandstone facies of member 3 are scarce in the Greenland rocks, and are best represented by the basal sandstone of the Spiral Creek Formation. This could be explained by a lower tidal range in the Greenland example. The sandstone facies apparently could only develop where the configuration of bottom topography allowed significant tidal current velocities to develop: perhaps less topographic variations was present across the Canyon Formation lagoon.

The stacking of sedimentary facies is quite different in the two cases. In the upper Canyon and Spiral Creek Formations the evidence points to continuously shallowing water upwards, so that the lateral distribution of facies (Fig. 5(a)) can be predicted from the vertical sequence by application of Walther's rule of facies. In member 3, the facies units were said to be stacked near randomly by carrying our analysis of vertical transitions between facies (Fairchild, 1980a). The correct statistical technique for testing the significance of such vertical transitions only became available later (Powers and Easterling, 1982). Application of the method of Powers and Easterling (1982) to data derived from Fig. 3 indicates that there is an 80% probability that the facies stacking is random. The facies are not therefore strongly cyclic, but stromatolite fields and emergent areas are distributed fairly randomly throughout the area (Fig. 5(b)), in a somewhat similar manner to the "tidal flat island model" of Pratt and James (1986). The seaward margin of the lagoonal system is only represented in the Greenland example. It is depicted on Fig. 5(a) as having a discontinuous stromatolitic rim. The existence of such a rim would help to explain the build-up of hypersaline conditions in thae lagoon. It can only have been discontinuous, however, because anhydrite pseudomorphs and subaqueous shrinkage cracks occur below the first thick stromatolite bioherm in the studied sections.

Some other minor differences are related to the differing burial history of the two successions. The Dalradian metasediments appear to be somewhat lower in organic carbon content, lack anhydrite relics and contain metamorphic minerals, especially phlogopite and phengite (Fairchild, 1985). In contrast, clayey layers resulting from pressure-solution during burial are more prominent in the upper Canyon Formation sediments than in member 3; the reason for this is unknown, although it could be explained by overpressuring of member 3 during burial diagenesis.

SIGNIFICANCE OF THE EQUIVALENT FACIES

It was established in the introduction that the congruent facies were deposited in different basins and at different times. Why, then, are they so similar?

Although not synchronous, both facies occurrences are late Vendian in age, a time when skeletal organisms had not yet evolved and infaunal activity was restricted. Stromatolites are the only record of benthic organic activity. The general similarity of the stromatolites is expected, given that the environments were in other respects comparable, because stromatolite biostratigraphy is imprecise even for distinctive columnar forms (Bertrand-Sarfati and Walter, 1981).

A warm climate seems to be required for these distinctive facies. Not only are there abundant finely crystalline carbonate and horizons of ooids attesting to active evaporation (unlike Cainozoic polar evaporitic sequences; Walter and Bauld, 1983), but anhydrite pseudomorphs are also common. The anhydrite formed by subaqueous early diagenesis from brines that would have represented a several-fold evaporation of normal sea water (Fairchild and Herrington, submitted). Attainment of such salinities in an open coastal setting points to a semi-arid, rather than humid climate. The presence of subordinate siliciclastic material is another distinctive characteristic. This probably reflects both a similar climate and moderate degree of tectonic relief of the hinterland.

A particular combination of physical parameters is required to produce the distinctive facies assemblages. Normal wave activity was low: if it had been significant then a sandbody would have been formed at the seaward margin of the lagoon. Such a sandbody is conspicuous by its absence in the Greenland section. The preserved wave-generated features instead must be attributed to infrequent storms, most of whose energy was dissipated on the leading edge of the lagoon since the thickness of storm units decreases up-section in Greenland, that is, in the shoreward direction. Tidal activity is recognizable only in certain beds: tidal range was almost certainly microtidal, but tidal currents were locally amplified in the vicinity of topographic highs and constrictions.

Elements of similarity are present in various other late Proterozoic carbonate units (e.g., Tucker, 1977; Siedlecka, 1978; Fairchild and Hambrey, 1984), but the combination of infrequent devastating storms, restricted tidal activity and presence of siliciclastic mud impurities is not seen elsewhere. A partial modern physiographic analogue is found in the modern German Bay of the North Sea (Aigner, 1985), where tidal flats (rather than a lagoon) directly border a storm-dominated shoreface. The fact that two non-synchronous examples in different sedimentary basins have been described in this paper suggests that the combination of warm climate, minor siliciclastic influx, low tidal range and low wave energy (yet intense storm action) is one that should be recognized in facies modelling. This ought to lead to the discovery of comparable Phanerozoic examples.

Acknowledgements

Fieldwork in Greenland was supported by N.E.R.C. Grant GR3/5438 awarded to Mr W. B. Harland (Cambridge University). I am grateful for the support of field companions in Greenland, particularly Paul Herrington (Birmingham), with whom I collected data, and Mike Hambrey (Cambridge) who organized the expedition. I thank Carl Burness for draughting the figures, Rodney Gayer for his encouragement and Alan Wright, Mike Hambrey and Maurice Tucker for commenting on the manuscript.

REFERENCES

Aigner, T. (1985). *Storm Depositional Systems*. Springer-Verlag, Berlin, 174pp.

Anderton, R. (1976). Tidal shelf sedimentation: an example from the Scottish Dalradian. *Sedimentology* **23**, 429–58.

Anderton, R. (1982). Dalradian deposition and the late Precambrian–Cambrian history of the N Atlantic region: a review of the early evolution of the Iapetus ocean. *J. Geol. Soc. Lond.* **139**, 421–31.

Anderton, R. (1985). Sedimentation and tectonics in the Scottish Dalradian. *Scott. J. Geol.* **21**, 407–36.

Bertrand-Sarfati, J. and Walter, M. R. (1981). Stromatolite biostratigraphy. *Precambrian Res.* **15**, 353–71.

Chow, N. and James, N. P. (1987). Facies-specific calcitic and bimineralic ooids from Middle and Upper Cambrian platform carbonates, western Newfoundland, Canada. *J. Sediment. Petrol.* **57**, 907–921.

Cope, J. C. W. and Gibbons, W. (1987). The Precambrian–Cambrian boundary in southern Britain: new evidence for the relative age of the Ercall Granophyre. *Geol. J.* **22**, 53–60.

Dallmeyer, R. D. and Gibbons, W. (1987). The age of blueschist metamorphism in Anglesey, North Wales: evidence from $^{40}Ar/^{39}Ar$ mineral dates of the Penmynydd schists. *J. Geol. Soc. Lond.* **144**, 843–52.

Donovan, R. N. and Foster, R. J. (1972). Subaqueous shrinkage cracks from the Caithness flagstone series (Middle Devonian) of North-East Scotland. *J. Sediment. Petrol.* **42**, 309–317.

Eyles, C. H. and Eyles, N. (1983). Glaciomarine model for upper Precambrian diamictites of the Port Askaig Formation, Scotland. *Geology* **11**, 692–6.

Fairchild, I. J. (1980a). Sedimentation and origin of a late Precambrian "dolomite" from Scotland. *J. Sediment. Petrol.* **50**, 423–446.

Fairchild, I. J. (1980b). The structure of NE Islay. *Scott. J. Geol.* **16**, 189–99.

Fairchild, I. J. (1985). Petrography and carbonate chemistry of some Dalradian dolomitic metasediments: preservation of diagenetic textures. *J. Geol. Soc. Lond.* **142**, 167–185.

Fairchild, I. J. and Hambrey, M. J. (1984). The Vendian of NE Spitsbergen: petrogenesis of a dolomite-tillite association. *Precamb. Res.* **26**, 111–167.

Fairchild, I. J. and Herrington, P. M. (1988). A tempestite-stromatolite-evaporite association (late Vendian, East Greenland): a shoreface-lagoon model. *Precambrian Research*, in press.

Grey, K. and Thorne, A. M. (1985). Biostratigraphic significance of stromatolites in upward shallowing sequences of the Early Proterozoic Duck Creek Dolomite, Western Australia. *Precambrian Res.* **29**, 183–206.

Hambrey, M. J. (1983). Correlation of Late Proterozoic tillites in the North Atlantic region and Europe. *Geol. Mag.* **120**, 209–232.

Hambrey, M. J. (1988) The Late Proterozoic sedimentary record of East Greenland: its place in understanding the evolution of the Caledonide Orogen, *this volume*.

Hambrey, M. J. and Spencer, A. M. (1987). Late Precambrian glaciation of central East Greenland. *Meddr. Grønland* **19**, 50pp.

Harland, W. B. and Gayer, R. (1972). The Arctic Caledonides and earlier oceans. *Geol. Mag.*, **109**, 289–314.

Harland, W. B., Cox, A., Llewellyn, P. G., Pickton, C. A. G., Smith, A. G. and Walters, R. (1982). *A Geologic Time Scale*, Cambridge University Press, Cambridge.

Harris, A. L., Baldwin, C. T., Bradbury, H. J., Johnson, H. D. and Smith, R. A. (1978). Ensialic basin sedimentation: the Dalradian Supergroup. In: Bowes, D. R. and Leake, B. E. (eds.) *Crustal Evolution in NW Britain and Adjacent Areas*, pp. 115–138. Seel House Press, Liverpool.

Moncrieff, A. C. M. (1988). The Tillite Group and related rocks of East Greenland: Implications for Late Proterozoic palaeogeography, *this volume*.

Powers, D. W. and Easterling, R. G. (1982). Improved methodology for using embedded Markov chains to describe cyclical sediments. *J. Sediment. Petrol.* **52**, 913–23.

Pratt, B. R. and James, N. P. (1986). The St. George Group (Lower Ordovician) of western Newfoundland: tidal flat model for carbonate sedimentation in shallow epeiric seas. *Sedimentology* **33**, 313–43.

Pringle, I. R. (1972). Rb–Sr age determinations on shales associated with the Varanger Ice Age. *Geol. Mag.* **109**, 465–72.

Siedlecka, A. (1978). Late Precambrian tidal-flat deposits and algal stromatolites in the Båtsfjord Formation, East Finnmark, North Norway. *Sediment. Geol.* **21**, 277–310.

Spencer, A. M., (1971). Late pre-Cambrian glaciation in Scotland. *Mem. Geol. Soc. Lond.* **6**.

Spencer, A. M. and Spencer, M. O. (1972). The late Precambrian/Lower Cambrian Bonahaven Dolomite of Islay and its stromatolites. *Scott. J. Geol.* **8**, 269–82.

Swett, K. and Knoll, A. (1985). Stromatolitic bioherms and microphytolites from the Late Proterozoic Draken Conglomerate Formation, Spitsbergen. *Precambrian Res.* **28**, 327–47.

Tucker, M. E. (1977). Stromatolite biostromes and associated facies in the Late Precambrian Porsanger Dolomite Formation of Finnmark, Arctic Norway. *Palaeogeog. Palaeoclimatol. Palaleoecol.* **21**, 55–83.

Vidal, G. (1985). Biostratigraphic correlation of the Upper Proterozoic and Lower Cambrian of the Fennoscandinavian Shield and the Caledonides of East Greenland and Svalbard. In: Gee, D. G. and Sturt, B. A. (eds.) *The Caledonide Orogen—Scandinavia and Related Areas* pp. 331–338. Wiley, Chichester.

Walter, M. R. and Bauld, J. (1983). The association of sulphate evaporites, stromatolitic carbonates and glacial sediments: examples from the Proterozoic of Australia and the Cainozoic of Antarctica. *Precambrian Res.* **21**, 129–148.

24 The Tillite Group and related rocks of East Greenland: implications for Late Proterozoic Palaeogeography

A. C. M. Moncrieff

Department of Earth Sciences, University of Cambridge,
Downing Street, Cambridge CB2 3EQ, UK

The Late Precambrian (Vendian) Tillite Group of central East Greenland contains well-preserved strata of glacigenic origin. In the fjord region, north of Scoresby Sund, the glacial succession is well developed and rests conformably on the Limestone–Dolomite "series", the uppermost part of the Eleonore Bay Group. The sedimentology and stratigraphy of the glacigenic sequence are briefly described and indicate a depositional environment involving a sub-polar ice sheet of continental proportions. To the west and south of this area are two isolated outcrops of tillite, in Charcot Land and Paul Stern Land. Their relationship to the main outcrops in the fjord region has been unclear, since they rest directly on metamorphic basement and are more extensively deformed. These outcrops are described in detail with the conclusion that they are lateral equivalents of the Tillite Group and lie in the source area for the main outcrops in the fjord region. The western and southern exposures therefore mark the position of a Late Precambrian landmass of considerable size, on which a continental ice sheet formed and flowed northeast to deposit the glacigenic sediments of the Tillite Group.

INTRODUCTION

Vendian sequences worldwide often contain abundant evidence for glaciation (Hambrey and Harland, 1981) and around the North Atlantic–Arctic region the glacial record is particularly well preserved (Hambrey, 1983), with thick glacigenic successions in Scotland (Spencer, 1971), northern Norway (Edwards, 1984) and Svalbard (Fairchild and Hambrey, 1984). The Vendian sequence in East Greenland contains further evidence for glaciation, with good exposures and, in general, well-preserved sedimentary structures (Hambrey and Spencer, 1987). In all these areas, large parts of the sequence are interpreted as being deposited at, or near sea level, and this, combined with their great lateral extent, indicates a major climatic event rather than a series of local upland glaciations. As such, these deposits are useful, not only as stratigraphic marker horizons, but also as indicators of large emergent source areas on which extensive ice sheets could have formed. The East Greenland succession is derived from such a source area, whose identification can make a major contribution to Vendian palaeogeography.

The Tillite Group, comprising up to 1300 m (Higgins,

1981) of glacigenic and non-glacigenic strata, crops out within the Caledonian fold belt immediately north of the Scoresby Sund area (Fig. 1). It rests on a thick sequence of carbonate and clastic rocks, the Eleonore Bay Group, with the entire Late Proterozoic–Early Palaeozoic sequence being involved in north–south trending Caledonian folding and westward-directed thrusting. However, to the west and southwest of the main exposures, two tillite outcrops occur in distinctly different tectonic and stratigraphic settings. In Charcot Land and Paul Stern Land (Fig. 1), thin glacigenic sequences lie directly on metamorphic basement and are exposed in tectonic windows beneath major overthrusts of Caledonian age. Another so-called "tillite" exposure to the southeast of Paul Stern Land, in Gåseland, is deformed to a far greater extent than the others and in contrast contains no evidence of glacial activity. While a correlation with the Paul Stern Land tillite cannot be positively refuted, there is little evidence to support it. The Gåseland exposure is described in more detail by Manby and Hambrey (this volume) and will not be considered further here.

Early opinion correlated the overthrust schists with the base of the Eleonore Bay Group and considered the tillites

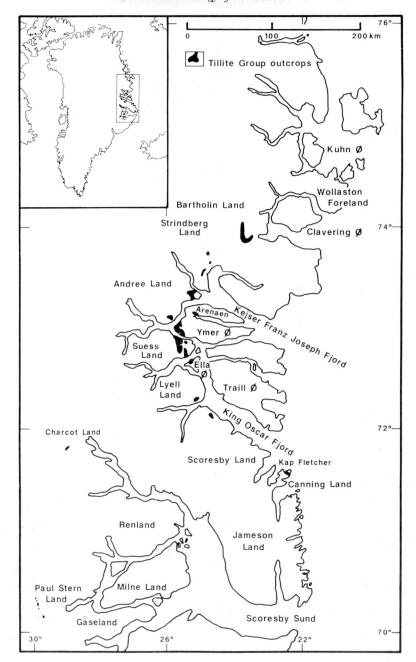

Fig. 1.   Map of central East Greenland showing the location of places mentioned in the text.

as a basal series to this (Wenk, 1961; Haller, 1971). More recent work has recognized the unlikelihood of a link between the tillites and the overthrust schists and a correlation with the Tillite Group farther north was suggested on circumstantial evidence (Phillips and Friderichsen, 1981; Henriksen, 1981). In this paper, more rigorous evidence in support of this correlation is presented, using lithological and palaeogeographical data collected in 1984 and 1985. This work is part of a larger-scale project to test various ideas concerning the Late Proterozoic–Early Phanerozoic evolution of the North Atlantic region, in particular, the timing of the opening of the Iapetus Ocean.

THE TILLITE GROUP IN THE FJORD REGION

Stratigraphy and depositional processes

The Tillite Group conformably overlies a sequence of marine carbonates and sandstones, the Eleonore Bay

Group (Fig. 2). The latter has a cumulative thickness of up to 13 km (Henriksen and Higgins, 1976; Henriksen 1985), although this figure is likely to be exaggerated by thrusting (Manby and Hambrey, this volume). The uppermost part of the Eleonore Bay Group, the Limestone–Dolomite "series" is a carbonate succession that, towards the top, contains signs of tectonic instability in the form of slumps, intraformational breccias and mass flows (Herrington and Fairchild, this volume). The Tillite Group itself comprises five formations, of which only the lower three are relevant to this discussion and will be referred to as the Lower Tillite Group (Fig. 3). Stratigraphic nomenclature is taken from Hambrey and Spencer (1987). At the base, the Ulvesø Formation comprises between 300 m and 15 m of diamictites with thin interbeds of sandstone and conglomerates. The diamictites are dominantly yellowish-grey with a dolomitic matrix, and contain clasts derived almost entirely from the underlying Limestone–Dolomite "series". The bulk of the formation is interpreted as waterlain tillite deposited from floating ice, but grounded ice episodes also occurred. The Arena Formation consists of up to 300 m of

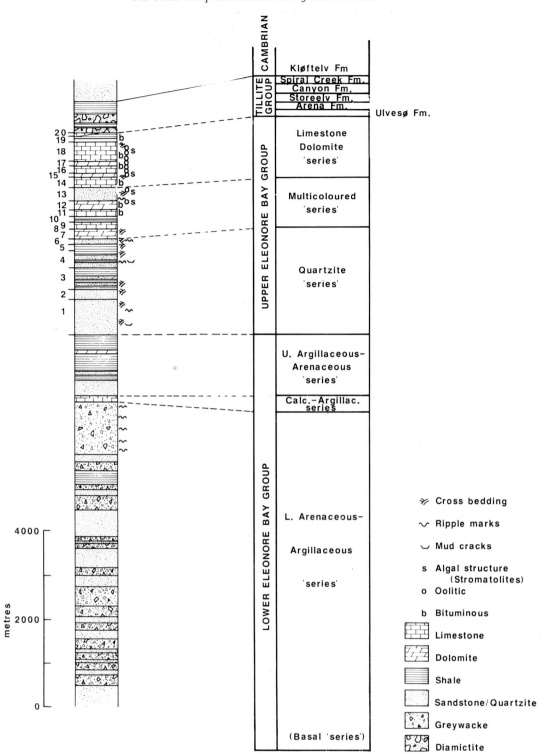

Fig. 2. Summary stratigraphy and sedimentology of the Eleonore Bay Group and Tillite Group. Numbers to the left of the column refer to Bed groups.(Adapted from Henriksen and Higgins, 1976.)

sandstones, siltstones and shales, in which frequent waning-flow deposits occur, and is interpreted as an interglacial episode with dominantly turbiditic deposition. Glacial conditions are represented once again by the Storeelv Formation, in which maroonish-brown and grey diamictites are the characteristic rock types. Although the white- and yellow-weathering carbonates of the Limestone–Dolomite "series" are still the dominant clast type, the Storeelv Formation is distinct in containing up to 30% igneous and metamorphic clasts, dominantly red granite-gneiss. The maroonish-brown colour of the matrix is thought to have been derived from the breakdown of ferromagnesian minerals in these clasts. Sandstone units, associated with diamictites, are more common in the

Storeelv than the Ulvesø Formation, and in part reflect the greater proportion of lodgement tillites with meltwater channel deposits. The remainder of the Tillite Group, comprising the Canyon and Spiral Creek formations, represents a brief interlude of shallow-water deposition followed by a deeper-water, dolomitic mudstone sequence. This shallows towards the top into regressive tide- and storm-dominated facies with mud cracks and halite pseudomorphs (Fairchild, this volume).

The Tillite Group conformably overlies the Limestone–Dolomite "series" and continuous sedimentation can be demonstrated between the Ulvesø Formation and Bed group 19 in the area south of Arenaen on Ymer Ø (Fig. 1). To the north, Bed group 20 occurs above Bed group 19, but

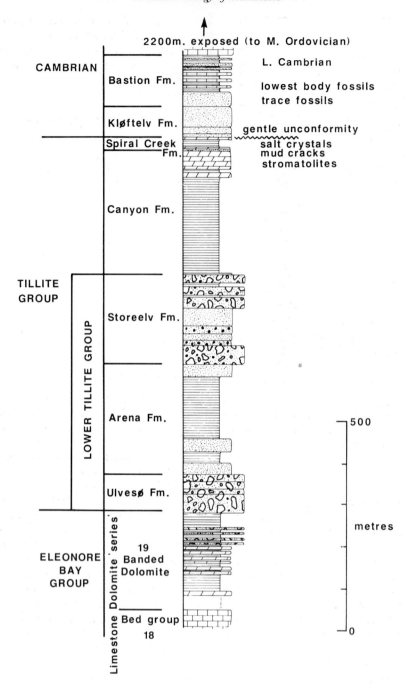

Fig. 3. Summarized stratigraphy and sedimentology of the Tillite Group and Cambrian rocks in East Greenland.

a conformable transition to the Ulvesø Formation persists, indicating that Bed group 20 is a lateral equivalent either of the top of Bed group 19 or of the bottom of the Ulvesø Formation. However, to the southwest at Kap Fletcher in Canning Land, both the Ulvesø and the Arena formations are missing and the Storeelv Formation rests on a lower division of the Limestone–Dolomite "series", probably Bed group 18. The uppermost Eleonore Bay and Tillite Groups have been described by Hambrey and Spencer (1987).

Depositional environment of the Tillite Group

The Lower Tillite Group extends over at least 300 km from north to south and, although lateral facies changes are extensive, individual units can sometimes be traced for up to 50 km. Throughout the Lower Tillite Group there is little angular, supraglacially derived material, indicating that valley sides and nunataks were not significant features. Furthermore, deposition took place on an extensive,

shallow shelf with little pre-existing topography. Thus, an ice sheet, rather than a series of restricted valley glaciers was responsible and a glaciation of widespread extent is indicated, as supported by the occurrence of Vendian Tillites in many other parts of the North Atlantic region.

In the past, the Lower Tillite Group has been attributed to either glaciomarine (Schaub, 1950) or terrestrial deposition (Huber, 1950), or both (Hambrey and Spencer 1987). Both processes occur, the former being more common in the Ulvesø Formation and the latter in the Storeelv Formation. Despite the occurrence of grounded ice deposits, it is unlikely that much of the sequence accumulated above water level, since evidence of emergence is limited to one horizon within the Ulvesø Formation. Instead, the association of deposits from both floating and grounded ice indicates an environment close to the grounding line of the Lower Tillite Group ice shelf.

The sedimentology of the tillites reveals the nature of the ice involved. In present-day Antarctica, the action of meltwater is slight, most of the ice being polar in

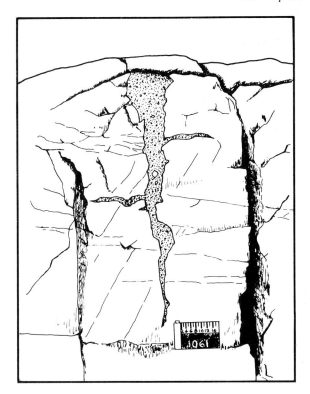

Fig. 4. Drawing from photograph of highly angular sandstone wedge penetrating the top of a well-cemented diamictite unit. The wedge is interpreted as of frost-contraction origin, rather than soft-sediment injection, on the basis of its sharp, angular corners.

nature, that is, below its pressure melting point. However, in the Tillite Group, the influence of meltwater is widespread in the form of sub-glacial meltwater deposits, and rhythmites produced by seasonal variations in meltwater discharge. Polar ice was therefore not responsible and sub-polar or temperate ice, in which parts or the whole respectively, are above their pressure melting point, are indicated. The occurrence of periglacial features in the Lower Tillite Group (Fig. 4) further limits the type of ice to sub-polar, since permafrost conditions are not found in areas with temperate glaciers.

The Lower Tillite Group was therefore deposited by an extensive sub-polar ice sheet, which must have formed on an equally extensive landmass. The sediments suggest that the landmass had only recently formed. The Tillite Group rests conformably on the Limestone–Dolomite "series" and yet contains large numbers of clasts derived from this underlying carbonate unit. This indicates that at the end of Limestone–Dolomite "series" deposition, uplift occurred in the Tillite Group source area, which until a short time before was covered by a shallow shelf sea. The recently lithified carbonates were eroded and transported by the ice into the Tillite Group basin, where deposition had continued unchecked. The Eleonore Bay Group is quoted as being up to 13 km thick in the fjord region (Henriksen, 1985), but is missing from the "foreland" region. However, the discovery of thrusting suggests that this figure is an overestimate. There are two points that suggest that only the Upper Eleonore Bay Group was deposited in the source area. Firstly, by far the most common clast type is a yellow- or white-weathering carbonate derived from the underlying Limestone–Dolomite "series". Quartzites and greywackes, so common in the Lower Eleonore Bay Group, are much rarer as clasts. This suggests that the Lower Eleonore Bay Group was not a major part of the source area. Secondly, metamorphic clasts begin to appear at the top of

the Ulvesø Formation and are then common throughout the Storeelv Formation. This indicates that in some areas, ice eroded swiftly down to the metamorphic basement and that the overlying sedimentary cover was not particularly thick (Fig. 10(a)).

THE "FORELAND" TILLITES

Two tillite outcrops occur to the south and southwest of the fjord region, at Tillit Nunatak in Charcot Land (Henriksen, 1981) and in Paul Stern Land (Phillips and Friderichsen, 1981) (Fig. 1). These tillites are distinct from the fjord region examples in that they rest on metamorphic basement rather than the Eleonore Bay Group and occur within the thrusted parts of the Caledonian fold belt. It has been suggested that these tillites lie on the foreland to the Caledonian thrusts (Henriksen and Higgins, 1976). However, the deformation, and in the case of the Charcot Land tillite, the metamorphism, is rather too intense for this to be true, and a more probable explanation is that they are themselves involved in westward-directed thrusting (Manby and Hambrey, this volume).

Charcot Land

In the northwestern part of the Scoresby Sund region, at the rim of the Inland Ice, an isolated outcrop of tillite occurs on Tillit Nunatak (71°54′ N 29°47′ W). The exposure has been described by Henriksen (1981), but some observations could not be confirmed in the 1985 field work. The name Tillit Nunatak Formation is proposed for the Charcot Land tillite.

The tillite occurs in the Charcot Land window (Henriksen, 1985), where a gneiss and granite complex of probable Archaean to early Proterozoic age is overlain by the Charcot Land supracrustal sequence, comprising about 200 m of basic metavolcanics, marbles, semipelitic and quartzitic metasediments. Both the older gneisses and the supracrustal metasediments are cut by two dioritic and granitic intrusions emplaced at 1900 and 1600–1870 Ma respectively (Steiger and Henriksen, 1972; Hansen *et al.*, 1972, 1973). The tillite lies unconformably on, and is folded in with, the diorite.

The tillite is exposed along the backwall of a westward-facing corrie, over a distance of about 1.5 km. (Fig. 5). A glacigenic origin is inferred from numerous dropstones in the laminated mudstones (Fig. 6). Striated stones were not observed, but the deformation is such that they are unlikely to be preserved. Within the diamictites, the dominant clast type (70% in the lower parts) is the locally derived diorite on which the tillite rests. The basal few metres contain nothing else and include 2 m long, angular slabs, prised from the basement. Other clast types include fine-grained quartzites, pelites and carbonates from the surrounding Charcot Land supracrustal sequence into which the diorite is emplaced. Among the carbonate clasts, accurate identification of metamorphic grade is impossible owing to the mineralogy of the rock. In some cases they are clearly highly deformed with strong fabrics defined by chlorite and elongate domains of recrystallized quartz. In other types, no such fabric occurs, but the calcite is completely recrystallized. In hand specimens, none of the carbonate clasts resemble those from the Eleonore Bay Group, having a distinctive "sugary" recrystallized appearance. It seems likely that these carbonates have experienced metamorphism of a grade far higher than the tillite matrix. Thus all the clasts in the tillite are locally derived.

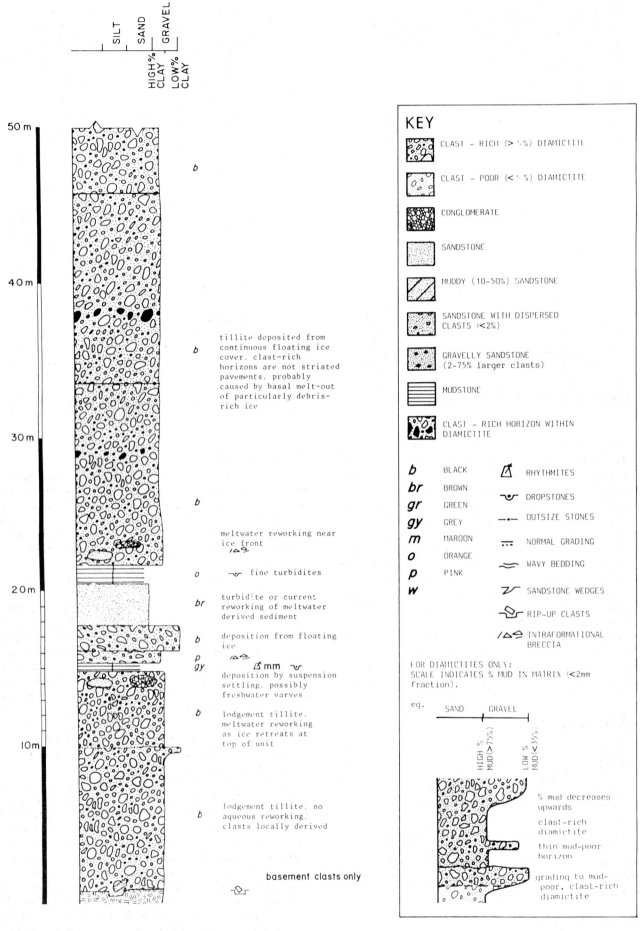

Fig. 5.   Sedimentary log of the Tillit Nunatak Formation. Thicknesses are estimated from photographs.

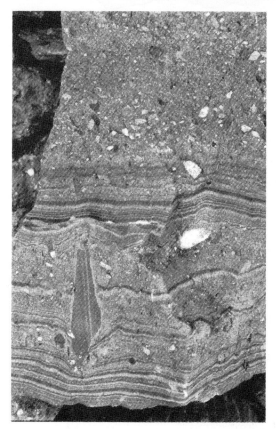

Fig. 6. Dropstone structure in a laminated part of the Tillit Nunatak Formation. The dropstone is 8 cm long.

The tillite is tightly folded with the basement in the form of 150–200 m wavelength cuspate folds, reflecting the large differences in competency. The deep synclinal keels exaggerate the thickness of the tillite to about 200 m in the backwall of the corrie (Manby and Hambrey, this volume). A strong axial planar cleavage is associated with this folding, which is probably Caledonian in age, being roughly parallel to the main Caledonian structural grain. Biotite, mostly retrogressed to chlorite, is abundant in the tillite, but distinguishing between detrital and metamorphic examples is difficult. Examples of definite metamorphic origin occur in the higher-pressure areas around larger quartz and diorite grains.

Although the contact with the underlying diorite is tectonically disturbed in the tighter parts of the folds where strain has been highest, elsewhere there are numerous examples that show that the tillite is in its original position and that the contact is an unconformity.

Paul Stern Land

The Paul Stern Land tillites lie in a structural window, which exposes gneissic basement, considered to be part of an Archaean–Early Proterozoic complex. Above this basement is a parautochthonous sequence of deformed limestones and sandstones, themselves overthrust by micaschists.

A total of six exposures were examined, centred on the main section at Støfvanget, which is described here. The name Støfvanget Formation is suggested for the Tille Paul Stern Land tillites. Two small sections from the northwest and southeast are also described.

Støfvanget main section

The name Støfvanget is given to a small embayment by the north side of the glacier at 70°21′05″ N, 29°44′20″ W. The best exposure is located here and is referred to throughout as the Støfvanget section. The sequence was first described by Phillips and Friderichsen (1981).

Lateral continuity is impossible to assess, owing to the limited nature of the outcrop; thus, the lithological log (Fig. 7) represents a single section up to 15 m across. Seventy five metres to the southeast, across scree, a second exposure comprises 10–15 m of diamictite. This sits with an undisturbed contact directly on the basement, the basement/tillite contact here being topographically higher.

The sandstone units are dominantly quartz, compacted by pressure-solution. However, yellow-weathering carbonate grains also occur, up to 1 cm in diameter. In the diamictite units, clast compositions are as described by Phillips and Friderichsen (1983), the basal diamictite being particularly rich in red-granite clasts, almost certainly derived locally. Dolostone and limestone clasts are unmetamorphosed and are texturally very similar to the yellow- and white-weathering examples from the Limestone–Dolomite "series", found as clasts in the Lower Tillite Group. A glacial origin for the sequence is assumed on the basis of dropstones in the laminated units and the occurrence of a striated dolostone clast.

The section is cut by a spaced cleavage of moderate intensity, thrusting and numerous joints, with beddingparallel deformation increasing upwards. The smaller exposure, 75 m to the southeast, is less affected by deformation and retains a harder and more consolidated appearance.

The Støfvanget section is unmetamorphosed, containing no signs of recrystallization, even to lowest greenschist facies.

Northwestern section

A second, thinner section occurs in a small stream halfway up the hillside, about 400 m northwest of the Støfvanget section. Sixteen metres of varying lithologies, dominantly maroon or green diamictite and gravelly sandstone occur with a considerable amount of tectonic disturbance. Exposure is limited, so lateral continuity cannot be assessed.

Clasts are similar to those found in the main section, but are rarely over 50 mm in diameter.

A steeply dipping, prominent joint pattern occurs, together with bedding-parallel thrusts and associated mylonites, in the lower part. Throughout the maroon diamictite, "bedding" occurs as thin, green, discontinuous and undulating horizons up to 10 cm long. Thin sections show these to be rich in very fine-grained carbonate and clay minerals altering to chlorite and muscovite. The style with which they deform around clasts strongly suggests a tectonic origin for this feature. However, quartz grains are not affected by pressure-solution, suggesting that mica formation is a result of localized pressure close to the thrust planes rather than regional burial.

At this northwestern exposure, the basement is cut by fault planes dipping to the east at 40°. In section, the sense of displacement of the foliation in the granite suggests these to be normal faults. A 0.5 m long pod of a friable maroon aggregate occurs, a few metres down within the basement. Its colour strongly resembles that of the overlying diamictite. In thin section it has a strong fabric defined by dark-brown laminae concentrated on clusters of opaque material, probably iron oxides. The bulk of the rock is a carbonate-rich clayey material, in which float numerous fractured quartz crystals or grains. This pod is interpreted as a piece of the overlying tillite involved in basement faulting.

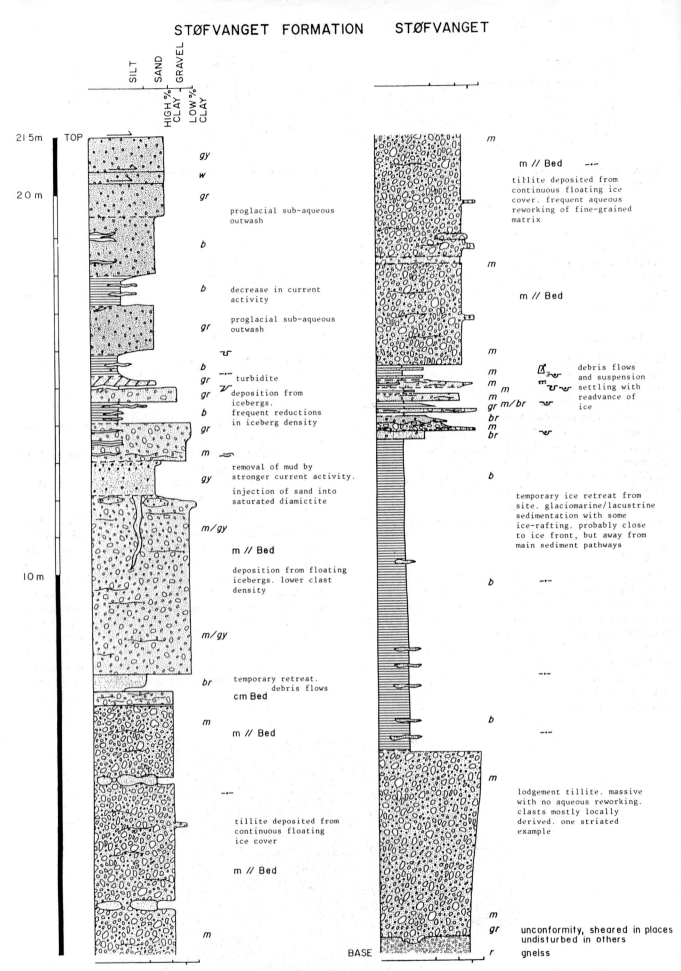

Fig. 7. Sedimentary log of the Støfvanget Formation at the type locality. The key is the same as that in Fig. 5. Thicknesses were measured by tape.

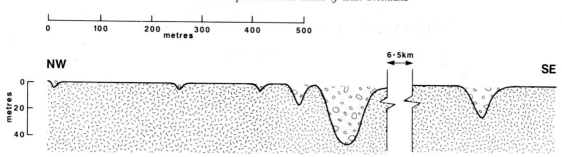

Fig. 8. Outcrop pattern and lateral extent of the Støfvanget Formation in Paul Stern Land. The lower contact of each tillite outcrop is inferred from closely spaced exposures, but cannot be followed continuously.

Three other minor exposures occur to the northwest (Fig. 8). All consist of less than 4 m of hard, maroon diamictite, capped by a sandstone. Thrusted contacts occur within the diamictite, but the only visible contact with the basement, at the westernmost exposure, is undisturbed.

Southeastern section

Six and a half kilometres southeast of Støfvanget is a small water-filled embayment in the cliff face, dammed by the side of the glacier. Exposure at this site is poor, but significant. The tillite comprises 25 m of hard, maroon diamictite, with occasional beds and lenses of sandstone and gravel. No other lithological detail was observed because the diamictite is highly jointed and mostly covered by scree. The exposure is about 75 m wide and thins rapidly to the side. The altitude of the lower contact varies, while the top remains horizontal, so that the tillite fills a basement hollow. The relationship of the tillite to the surrounding rocks is discussed in a later section.

Basement

The underlying basement was examined at each of the above localities and at the base of the cliffs for about 3 km southeast of Støfvanget. The bulk consists of a foliated granite–gneiss, with subordinate granodiorite–gneiss and occasional amphibolite bands up to 1 m thick. Typical compositions are: 40% quartz with undulose extinction and sutured contacts; 35% alkali feldspar, principally microcline with some microperthite; 20% plagioclase generally albitic and usually altered to carbonate and clay minerals; 5% micas and opaque minerals. Alteration of feldspars to carbonate and clay minerals is present in varying degrees, particularly near the contact and always affects the plagioclase crystals far more than the microcline. Basement boulders in the tillite may have all feldspars completely weathered, with only quartz preserved unaltered.

Contact of tillite with basement

The contact between the basement and the overlying tillite is exposed in most of the sections and may be of four types:

(1) knife-sharp, associated with 1–2 cm of black friable soil or mylonite;
(2) imbricated and brecciated basement slices overlain abruptly by diamictite;
(3) brecciated, indurated basement, overlain without a break by diamictite;
(4) undisturbed basement overlain without a break by undisturbed diamictite.

In (1) and (2), where a distinct break is visible, thrust activity is assumed. However, relationships (3) and (4) suggest an undisturbed original contact. Although the nature of the contact may vary rapidly along-strike, type four is the most common. In the smaller exposure east of the main section there is a particularly clear example of such a contact. Here, the basement forms two steps, each about 2 cm high and 2 cm wide. The diamictite overlies these without a break, and in the angle of the steps, coarser material has accumulated, giving a fining-upward sequence over a few centimetres (Fig. 9). Clear examples of both thrusted and undisturbed contacts have therefore been observed along the Paul Stern Land sections. This suggests one of two things.

Fig. 9. Contact of tillite with basement in Støfvanget. Although shearing has occurred within the tillite, the actual contact is undisturbed. The scale is in centimetres.

(1) The contact has been strained, but has only failed in selected places, and to a limited extent.

(2) There has been extensive thrusting, which in places runs along the contact, but in others, penetrates either above, into the diamictites and sandstones, or below into the basement.

Thrusts are visible within the lower parts of the tillite sequence; however, they do not appear to be of any great extent. They do not, for instance juxtapose different lithologies. No thrusts were observed running down to the contact. In view of these observations and the fact that a great deal of strain appears to have been accommodated in the weaker overlying succession, the first of the two possibilities mentioned above is preferred.

The relationship of tillite to basement

The two larger sections, at Støfvanget and 6.5 km to the east, both show rapid lateral thinning by elevation of the lower contact, while the upper contact remains horizontal (Fig. 8). Such a pattern could be explained in two ways.

(1) The tillite was deposited in basement hollows.

(2) The tillite was deposited on a horizontal surface and has been folded or faulted in with the basement.

All evidence points to the former. Bedding within the tillite is not strongly folded and interfaulting of tillite and basement is limited to the small pod at the northwestern exposure. Cleavage development is far weaker than would be expected in such tight folds as those observed in the Tillit Nunatak Formation. However, the contact cannot be followed continuously around the margin of such a valley and post-depositional step-faulting in the unexposed parts, although considered unlikely, cannot be excluded as a means of explaining the outcrop pattern.

The upper contact

The contact between the tillite and the overlying sedimentary succession is only clearly seen in the Støfvanget and eastern sections. At Støfvanget the sequence runs through a number of quartzite beds and into the overlying, finely laminated grey limestone. The cliff immediately northwest of the top of the section has good exposures of most of these lithologies. At all boundaries, some undisturbed contacts and more frequent sheared contacts were observed. This in some ways mimics the lower contact. Strain appears to have been accommodated at different levels and to different degrees. Moving up into the limestone, deformation becomes more intense. A grey, silty limestone is extensively cut by subhorizontal pressure-solution horizons, picked out by residues of insoluble clays. Elongate, recrystallized areas of calcite indicate compression in the plane of the bedding. The most extreme deformation, however, has occurred in several 20–80 cm thick black-shale/siltstone bands, interbedded with the limestone. The exposures are extremely friable, making observation of structures difficult. However, extensive overturned folding and nappe structures occur. Fold axes trend approximately 350°/25° W. A subhorizontal, well-developed fabric, associated with extensive chloritic mylonitization, penetrates each bed. At the southeastern locality, the limestone/diamictite contact is extensively thrust and is associated with a thin mylonite.

The relationship of tillite to cover sequence

A well-defined thrust exists between the tillite and the cover sequence, or the basement and the cover sequence where the tillite does not occur. Deformation within the tillite is in the form of joints rather than a well-defined cleavage. However, much of the deformation in the tillite may be attributable to the westward overthrusting of the cover. The thinner sections are more strongly affected, being closer to the thrust, while bedding parallel deformation increases upwards in the thicker, Støfvanget section. Finally, the possibility exists that the tillite was once considerably thicker. Parts of the sequence that protruded above the basement hollows would have been removed by westward thrusting, while those surrounded by protecting basement have survived, largely undeformed.

PALAEOGEOGRAPHY DURING THE DEPOSITION OF THE LOWER TILLITE GROUP

The location of the landmass on which the ice sheets formed is clearly an important part of Vendian palaeogeography and the two outlying areas of Charcot Land and Paul Stern Land mark a possible site. These tillites are likely to mark a landmass because they rest directly on metamorphic basement, the implication being that any sedimentary cover equivalent to the Limestone–Dolomite "series" has been removed and transported north by glacial erosion. Interpreting these localities as part of the Vendian landmass depends on (1) whether the "foreland" tillites are equivalent to the Lower Tillite Group, and (2) whether they lie in the source area.

Correlation of the "foreland" tillites with the Lower Tillite Group

Chronostratigraphic correlation

Unfortunately, dates available for the "foreland" tillites are of little help in establishing such a correlation between the two areas. The fjord region tillites are dated biostratigraphically as Vendian (Vidal, 1976, 1977, 1979) with a Late Riphean age for the underlying Limestone–Dolomite "series" (Vidal, 1979). However, mild metamorphism of the "foreland" localities makes identification of acritarchs unlikely. The tillites can only be constrained radiometrically, between the age of the basement, 1900 Ma in both Charcot Land (Steiger and Henriksen, 1972) and Paul Stern Land (Haller and Kulp, 1962), and the age of Caledonian deformation that affects both localities, the main phase of which was between the middle Ordovician and middle Devonian periods.

Lithostratigraphic correlation

The maroon coloration, so typical of the Storeelv Formation, also occurs in the Støfvanget Formation, although the colour is probably disguised by the growth of metamorphic chlorite in the Tillit Nunatak Formation. Similar lithologies occur in both the "foreland" and fjord regions. Grounded ice and waterlain deposits occur, together with evidence of meltwater activity in the form of seasonal rhythmites, suggesting a sub-polar or temperate ice sheet in both areas.

The most convincing basis for correlation comes from unmetamorphosed yellow- and white-weathering carbonate clasts found in the Støfvanget Formation. Although they form a small part of the total clast content, they are identical to those derived from the Limestone–Dolomite "series" and as such would alone fix the age of the Støfvanget Formation between Late Riphean and the Caledonian orogeny, leaving little option but to correlate it with the Vendian tillites of the fjord region. In Charcot Land, no such clasts occur and the stones in the Tillit

(a) Limestone – Dolomite "series"

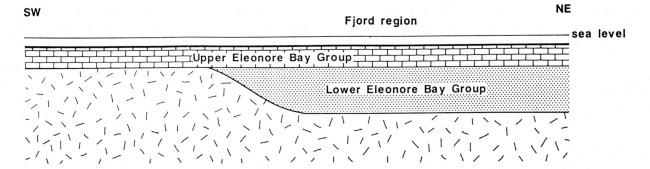

(b) Upper Ulvesø Formation

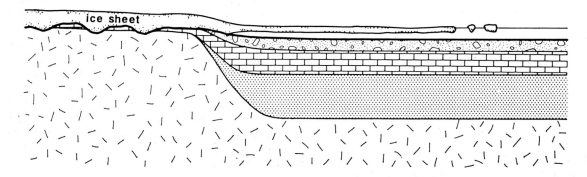

(c) Storeelv Formation

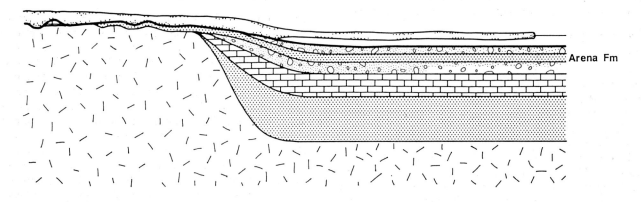

(d) Lower Canyon Formation

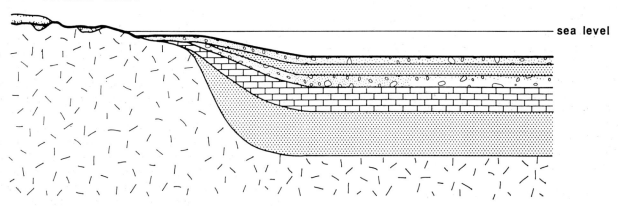

Fig. 10. Schematic cross-sections from the "foreland" tillite area in the southwest, to the Tillite Group basin in the fjord region. Sections show the proposed evolution of the source area and basin from uppermost Limestone–Dolomite "series" time to lowest Canyon Formation time. Sections are not to scale. (*a*) Uppermost Limestone–Dolomite "series". (*b*) Upper Ulvesø Formation. (*c*) Storeelv Formation. (*d*) Lowest Canyon Formation.

Nunatak Formation are all locally derived.

Throughout the fjord region, a distinction can be made between the Ulvesø and Storeelv formations on the basis of metamorphic clast content: very rare in the former and common in the latter. Both the "foreland" localities lie on metamorphic basement and contain a large proportion of metamorphic clasts, and it seems reasonable to assume that the first appearance of metamorphic clasts in the fjord region only marginally post-dates their unroofing in the source area. The "foreland" tillites must therefore be younger than the Ulvesø Formation. At the top of the Storeelv Formation a gradual ice retreat is inferred. Distal glaciomarine and non-glacial facies appear first in the northeast, and the subsequent ice-retreat towards the southwest was much slower than at the top of the Ulvesø Formation. Since deposits left by a slowly retreating ice sheet would be diachronous, the "foreland" tillites are probably equivalent to the very top of the Storeelv Formation or the lower part of the Canyon Formation. No glacial imprint would occur in the fjord region because the ice had retreated on to land and the only sediment input would be by fluvial processes.

However, the "foreland" tillites mark the position of a landmass only if it can be shown that they lie in the source area for the Tillite Group, rather than just another part of the basin. The evidence that suggests that they were in the source area is of three types. Firstly, a strong similarity can be seen between metamorphic clasts contained in the Storeelv Formation and areas of the basement metamorphic complexes in the region of the "foreland" tillites. The most common type is a red or white microcline-rich granite–gneiss, but garnet–mica schists, quartzites, more basic gneisses and amphibolites also occur. An Rb/Sr whole rock isochron age from granitic stones in the Storeelv Formation gave 1830 Ma (Spencer, 1971), significantly different from the 2290 ± 50 Ma age from Rb/Sr whole rock in the basement in Paul Stern Land (Haller and Kulp, 1962). In any case, a direct matching would be simplistic and no systematic comparison of basement and clasts has been undertaken. The point is that the basement metamorphic complex contains all the necessary lithologies and was exposed at the time, as the correlation of the tillites shows. The area around the "foreland" tillites to the west and southwest was therefore the closest exposed source area for the rock types found as metamorphic clasts in the tillites.

The second point suggesting that the "foreland" tillites lie in the source area concerns their contact with the basement. In Paul Stern Land, the tillite appears to lie in basement hollows (Fig. 8). The tillite is not strongly folded and has no axial planar cleavage such as that in the Tillit Nunatak Formation. At two exposures of the Støfvanget Formation, the tillite/basement contact can be followed topographically upwards and outwards from outcrop to outcrop, defining the sides of a palaeovalley. This suggests that the tillite was deposited in eroded basement valleys as would be expected in a glacial source area.

Finally, palaeocurrent data from the Lower Tillite Group suggest a derivation from the southwest. Striation orientations, cross-bedding from non-glacial interbeds and cross-bedding from waning flow deposits in the Arena Formation consistently indicate flow to the northeast, placing the "foreland" tillites upstream of the fjord region outcrops.

Pre-Caledonian configuration

The correlation of the "foreland" tillites with those in the fjord region raises the question of their relative positions at the time of deposition, since the "foreland" tillites have undergone westward thrusting. In Paul Stern Land, the degree of deformation suggests that such movement was limited to a few kilometres, but the Charcot Land rocks are more pervasively deformed and movements of several tens of kilometres are considered likely. However, when considered on the scale of the whole area, restoring these localities to their original areas would have only a relatively minor effect on the reconstruction. This would be further reduced when thrusting in the fjord region, previously thought to be slight (Higgins, 1981), is considered. The discovery of at least one major thrust within the Late Proterozoic sediments shows that westward movement has occurred in both the fjord region and the so-called "foreland". The relative positions of all the tillite localities today is therefore likely to be similar to those at the time of deposition. The relationship between the "foreland" tillites and the Tillite Group is summarized in Fig. 10.

CONCLUSIONS

The two so-called "foreland" tillites in central East Greenland are correlated with the Tillite Group that crops out within the fjord region. This correlation is based on a similarity of facies and depositional conditions and the occurrence of late Riphean Limestone–Dolomite "series" clasts in the "foreland" tillite. A more precise correlation with the top of the Storeelv Formation is suggested, based on the first appearance of metamorphic clasts and the depositional model for the Tillite Group. Palaeocurrents, clast composition and an eroded basement surface indicate that the "foreland" tillites lie in the source area for the Tillite Group and therefore mark the position of a Vendian landmass to the south and southwest of the central fjord region. This Vendian landmass must have been large enough to support a continental ice sheet, but more accurate estimates of its size are, at present, impossible.

Acknowledgements

This study is part of a larger programme on "Vendian stratigraphy and sedimentology of the East Greenland Caledonides", financed by the U.K. Natural Environment Research Council (grant no GR3/5438 to W. B. Harland). I acknowledge the tenure of a N.E.R.C. research studentship. For comments on the manuscript I am grateful to M. J. Hambrey.

REFERENCES

Edwards, M. B. (1984). Sedimentology of the upper Proterozoic glacial record, Vestertana Group, Finnmark, North Norway. *Norges Geol. Unders. Bull.* **394**, 1–76.

Fairchild, I. J. (1988). Dolomitic stromatolite-bearing units with storm deposits from the Vendian of East Greenland and Scotland: a case of facies equivalence, *this volume*.

Fairchild, I. J. and Hambrey, M. J. (1984). The Vendian succession of northeastern Spitsbergen: petrogenesis of a dolomite-tillite association. *Precambrian Res.* **26**, 111–167.

Haller, J. (1971). *Geology of the East Greenland Caledonides.* Interscience, New York.

Haller, J. and Kulp, J. L. (1962). Absolute age determinations in East Greenland. *Meddr. Grønland* **171**(1), 77 pp.

Hambrey, M. J. (1983). Correlation of late Proterozoic tillites in the north Atlantic region and Europe. *Geol. Mag.* **120**(3), 209–320.

Hambrey, M. J. (1988). Late Proterozoic sedimentary record of East Greenland: its place in understanding the evolution of the Caledonian Orogen, *this volume*.

Hambrey, M. J. and Harland, W. B. (eds.) (1981). *Earth's pre-Pleistocene Glacial Record*. Cambridge University Press, Cambridge.

Hambrey, M. J. and Spencer, A. M. (1987). Vendian stratigraphy and sedimentology of central East Greenland. *Meddr. Grønland. Geoscience*. **19**, 50 pp.

Hansen, B. T., Steiger, R. H. and Henriksen, N. (1972). The geochronology of the Scoresby Sund area. *Grønlands Geol. Unders*. Rapport Nr. 48, 105–107.

Hansen, B. T., Frick, U. and Steiger, R. H. (1973). The geochronology of the Scoresby Sund area. *Grønlands Geol. Unders*. Rapport Nr 58: 59–61.

Henriksen, N. (1981). The Charcot Land tillite, Scoresby Sund, East Greenland. In: Hambrey, M. J. and Harland, W. B. (eds.) *Earth's pre-Pleistocene Glacial Record*, pp. 776–777, Cambridge University Press, Cambridge.

Henriksen, N. (1985). The Caledonides of central East Greenland 70°–76°N. In: Gee, D. G. and Sturt, B. A. (eds.) *The Caledonide Orogen—Scandinavia and Related Areas*, pp. 1095–1113. Wiley, Chichester.

Henriksen, N. and Higgins, A. K. (1976). East Greenland Caledonian fold belt. In: Escher, A. and Watt, W. S. (eds.) *Geology of Greenland*. Grøn. Geol. Unders.

Herrington, P. M. and Fairchild, I. J. (1988). Carbonate shelf and slope facies evolution prior to Vendian glaciation, central East Greenland, *this volume*.

Higgins, A. K. (1981). The Late Precambrian Tillite Group of the Kong Oscars Fjord and Kejser Franz Josefs Fjord region of East Greenland. In: Hambrey, M. J. and Harland, W. B. (eds.) *Earth's pre-Pleistocene Glacial Record*, pp. 778–781, Cambridge University Press, Cambridge.

Huber, W. (1950). Geologisch-Petrographische Untersuchungen in der inneren Fjord Region des Kejser Franz Josefs Fjordsystemas in Nordost Grönland. *Meddr. Grønland* **151**(3), 83 pp.

Manby, G. and Hambrey, M. J. (1988). The structural setting of the Late Proterozoic tillites of East Greenland, *this volume*.

Moncrieff, A. C. M. and Hambrey, M. J. (1988). Late Precambrian glacially related grooved and striated surfaces in the Tillite Group of central East Greenland. *Palaeogeog. Palaeoclimatol. Palaeoecol*., **65**, 183–200.

Phillips, W. E. A. and Friderichsen, J. D. (1981). The Late Precambrian Gåseland tillite, Scoresby Sund, East Greenland. In: Hambrey, M. J. and Harland, W. B. (eds.) *Earth's pre-Pleistocene Glacial Record*, pp. 773–775, Cambridge University Press, Cambridge.

Schaub, H. P. (1950). On the Precambrian to Cambrian sedimentation in NE Greenland. *Meddr. Grønland* **114**(10), 50 pp.

Spencer, A. M. (1971). Late Precambrian glaciation in Scotland. *Mem. Geol. Soc. Lond*. **6**, 98 pp.

Steiger, R. H. and Henriksen, N. (1972). The geochronology of the Scoresby Sund area. *Grønlands Geol. Unders*. Rapport Nr 48, 109–114.

Vidal, G. (1976). Late Precambrian acritarchs from the Eleonore Bay Group and Tillite Group in East Greenland. *Grønlands Geol. Unders*. Rapport Nr 78, 19 pp.

Vidal, G. (1977). Late Precambrian microfossils. *Geol. Mag*. **114**, 393–394.

Vidal, G. (1979). Acritarchs from the upper Proterozoic and lower Cambrian of East Greenland. *Bull. Grønlands. Geol. Unders*. **134**, 40 pp.

Wenk, E. (1961). On the crystalline basement and the basal part of the pre-Cambrian Eleonore Bay group in the southwestern part of Scoresby Sund, *Meddr. Grønland*, **168**, 1.

25 The structural setting of the Late Proterozoic tillites of East Greenland

G. M. Manby[1] *and M. J. Hambrey*[2]

[1]Earth Sciences Department, Goldsmiths College, Creek Road,
London SE8 3BU, UK
[2]Department of Earth Sciences, Downing Street, Cambridge CB2 3EQ, UK

The Vendian tillites of the East Greenland fold belt are found in two contrasting tectonostratigraphic settings. In the eastern fjord region, the tillites overlie the Upper Eleonore Bay Group. Although affected by westward-verging folds and thrusts, many not previously recognized, the succession is largely unmetamorphosed. The thrusting has led to truncation of beds and may explain significant differences in stratigraphic relationships throughout the region. Major shortening across this part of the fold belt is inferred in contrast to previous interpretations of minimal shortening. Devonian Old Red Sandstone facies rocks lie with marked unconformity on these rocks and exhibit similarly oriented structures.

To the west, the tillites lie directly on the basement in what have been interpreted as tectonic windows through the main Caledonian thrust front into the underlying undeformed foreland. In the Charcot Land window the tillites that have previously been described as being several hundred metres thick, weakly deformed and metamorphosed have also been found to exhibit large cusp-like folds. These features suggest that the window may not be autochthonous basement and may be underlain by a thrust.

The so-called Gåseland tillites to the south again rest on basement and are described as occurring in a window through the Caledonian thrust sheets. Although grouped together, the "tillites" from the different localities in this area are markedly dissimilar in composition and degree of deformation and metamorphism. The Gnejssø "tillites" are more appropriately described as psephites exhibiting two phases of folding, strong shearing and greenschist facies metamorphism. In contrast, the Paul Stern Land tillites show only limited deformation and no recognizable metamorphic effects. Again, major detachments are required to be present at depth to explain the deformation of these cover–basement sequences.

It is apparent, therefore, that Caledonian deformation extends farther west than previously supposed and that undisturbed foreland may only truly exist beneath the inland ice.

INTRODUCTION

The East Greenland Caledonian fold belt forms part of the western limit of the circum-Atlantic Caledonian Orogen and is a mirror image of the Scandinavian Caledonides, as large-scale tectonic movements were westwards whereas those in Scandinavia were eastwards.

Stratigraphically, however, the East Greenland Caledonides bear a much closer similarity to the Caledonides of NE Svalbard (Kulling, 1930; Harland and Wright, 1979; Hambrey, 1983 and this volume) than to Scandinavia. Nevertheless, there is a distinctive linking feature throughout the Caledonides—the ubiquitous Vendian tillites—that has provided a framework for assessing the pre-Caledonian palaeogeography of the North Atlantic–Arctic region.

Following earlier work in Svalbard, our studies of the stratigraphy and sedimentology of the East Greenland tillite-bearing sequence were carried out in both the main part of the fold belt and on the so-called foreland to constrain Late Precambrian palaeogeographic reconstructions of the whole region. In the field, it became apparent that the structural setting of the tillites was not entirely as described in the literature. In particular, two features that are outlined in this paper are significant: (*a*) the so-called foreland rocks of Archaean–Early Proterozoic basement and tillite cover are together affected by westward-verging folds and thrusts, implying that undisturbed foreland must lie farther west beneath the inland ice; (*b*) the 17 km thick Late Riphean–Early Ordovician sedimentary sequence in the fold belt also shows as well as the simple open folding described in the literature significant westward-directed thrusting. Although the stratigraphic integrity is not generally destroyed by the thrusting, it does explain some of the marked changes of thickness of some rock units over short distances and implies much more crustal shortening than previously estimated.

The Caledonide Geology of Scandinavia © R. A. Gayer (Graham & Trotman), 1989, pp. 299–312.

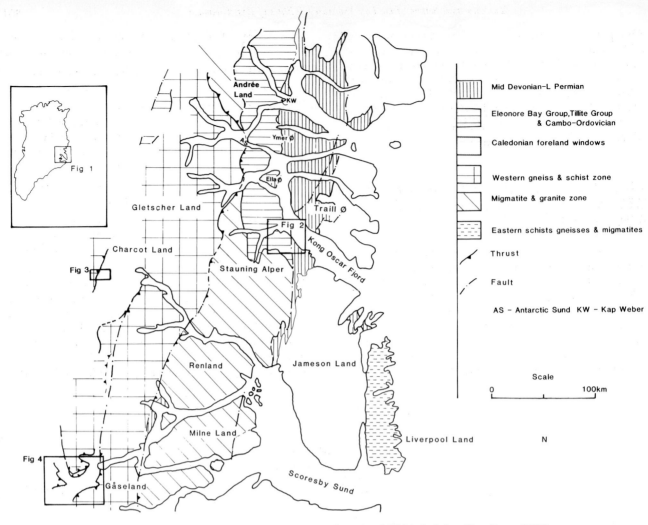

Fig. 1. Outline geological map of the East Greenland fold belt (after Henriksen, 1985).

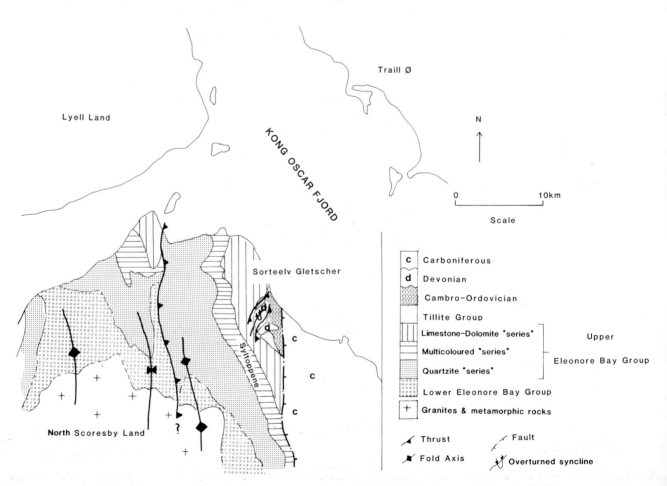

Fig. 2. Sketch map of North Scoresby Land (modified after Hambrey and Spencer, 1987).

Our present knowledge of the East Greenland Caledonides is largely the result of the Danish expeditions led by Lauge Koch (1928–1958) and the work of the Greenland Geological Survey (G.G.U.) since 1968. This work has led to the publication of the outstanding monograph by Haller (1971) on the entire fold belt and to a series of review papers (Henriksen and Higgins, 1976; Higgins and Phillips, 1979; Henriksen, 1985).

STRUCTURAL AND STRATIGRAPHIC SETTING OF THE TILLITE-BEARING SEQUENCES

The main development of tillite occurs within the Vendian Tillite Group exposed in the fjord region between 72° and 76° north (Figs. 1 and 2), which in most localities lies conformably above the Late Riphean Eleonore Bay Group and is succeeded unconformably (low-angle) by the Cambro–Ordovician sediments. Hambrey and Spencer (1987) have summarized previous research on the Tillite Group and the sequence is now defined as consisting of five formations (see also Hambrey, this volume).

Spiral Creek Formation (25–55 m)	Sandstone, mudstone and limestone breccia
Canyon Formation (250–300 m)	Dolostone, mudstone, marl and limestone
Storeelv Formation (up to 200 m)	Diamictite (mainly interpreted as lodgement or waterlain tillite), sandstone and conglomerate
Arena Formation (150–310 m)	Sandstone and dolomitic mudstone
Ulvesø Formation (up to 320 m)	Diamictite (as above), sandstone, conglomerate and mudstone.

Although the Tillite Group and the upper parts of the Eleonore Bay Group are unmetamorphosed, the latter become progressively more metamorphosed downwards (to kyanite/sillimanite grade: Higgins *et al.*, 1981) and then pass into high-grade schists and gneisses of the Caledonian crystalline complexes. The nature of this change has been much debated. Haller (1971) considered it to be a gradational contact, but recent work (Henriksen and Higgins,

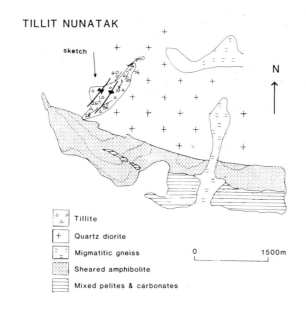

TILLIT NUNATAK

sketch

N

△ Tillite

+ Quartz diorite

Migmatitic gneiss

Sheared amphibolite

Mixed pelites & carbonates

0   1500m

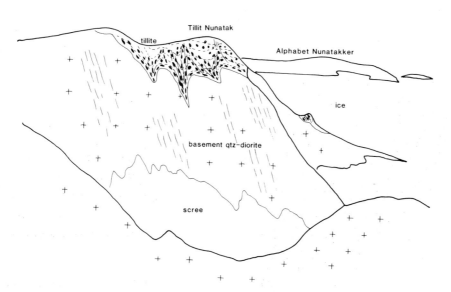

Tillit Nunatak

tillite

Alphabet Nunatakker

ice

basement qtz–diorite

scree

Fig. 3. Outline map and field sketch of Tillit Nunatak, Charcot Land. The cuspate folds in the tillite sequence trend NE–SW and plunge mostly to the southwest. The sketch is an oblique view from the northern spur of the corrie looking southeast.

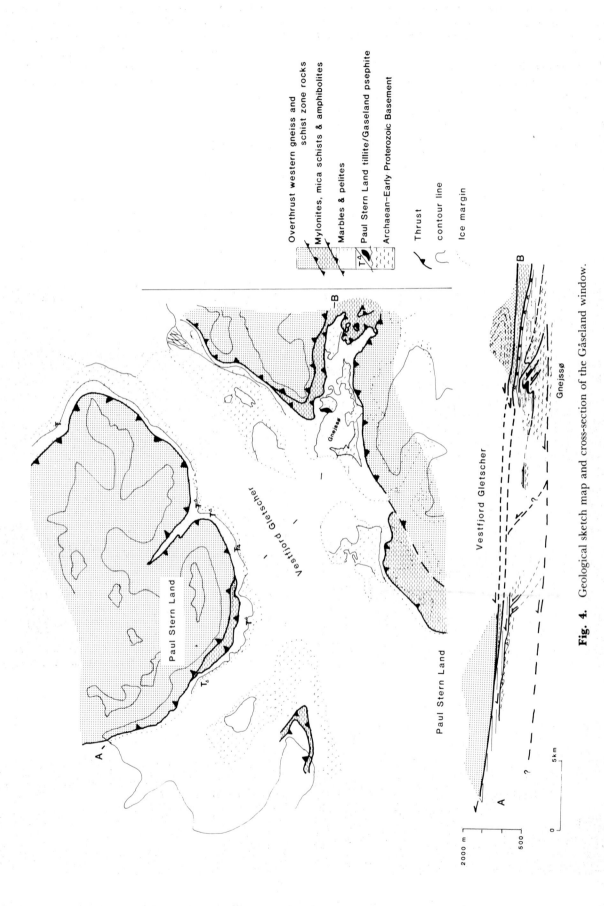

Fig. 4. Geological sketch map and cross-section of the Gåseland window.

1976) has revealed a tectonic décollement plane or zone of intense shearing deformation. Henriksen (1985), summarizing more recent thinking, considered the metamorphic crystalline complexes to represent reactivated pre-Caledonian basement that has been affected by older episodes of reworking between 3000–2300, 2000–1800 and 1200–1000 Ma.

Tillites resting directly on the Archaean–Proterozoic basement occur in tectonic windows beneath the overthrust western gneiss and schist zone rocks in a belt extending from Gåseland–Paul Stern Land (Figs. 1 and 4) north to Charcot Land (Figs. 1 and 3). The tillites in the Gåseland window were first described and referred to as the Basal Psephite Series by Wenk (1961), who equated them with the lower part of the Eleonore Bay Group. More recently Henriksen and Higgins (1976), Phillips and Friderichsen (1981) and Henriksen (1985) have argued that the Gnejssø and Paul Stern Land tillites are more likely to be equivalent to the Tillite Group. However, our more detailed lithological and structural data do not entirely support this suggestion.

It is argued later in this paper that the Gnejssø tillite (preferred term psephite) is so deformed that it cannot confidently be equated with the Charcot Land and Paul Stern Land tillites as previously supposed. In a companion paper, Moncrieff (this volume) demonstrates that the latter two areas can probably be correlated with the Storeelv Formation tillite of the fjord region.

In all of the localities noted above, the "tillites" show significantly more deformation than reported in the literature.

FJORD REGION

Eastern Andrée Land

The Tillite Group and bounding formations are well-exposed as flat-lying units on the three sides of the triangular mountain complex to the west of Kap Weber (Fig. 1). Previous authors attributed the marked thickness changes in Bed nos. 19 and 20 directly below the Tillite

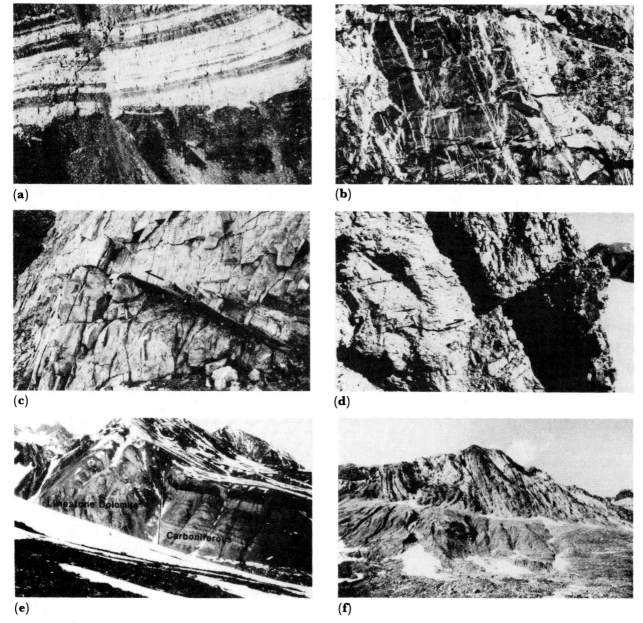

(a)

(b)

(c)

(d)

(e)

(f)

Fig. 5. (*a*) Kap Weber, thrust contact between Beds nos. 19 and 20 of the Limestone–Dolomite "series", looking north. (*b*) Kap Weber, calcite veining in Bed no. 20 below thrust. (*c*) Kap Weber, thrust in basal sandstones of the Ulvesø Formation; Bed no. 20 is below. Viewed north; height of exposure is 3–4 m. (*d*) Kap Weber, thrust and small fold in dolostone of Canyon Formation, looking south; height of exposure is 2–3 m. (*e*) North Scoresby Land, normal faulted contact between Carboniferous and isoclinally folded Limestone–Dolomite "series". (*f*) Isoclinally folded Limestone–Dolomite rocks (see (*e*)); both views looking north.

Group and the disappearance of Bed no. 20 southwards to facies changes. Earlier authors (Poulsen, 1930; Fränkl, 1953; Katz, 1954; Haller, 1971) made no mention of tectonic modification, but our observations on the upper flank of the Kap Weber mountain complex (summit 1410 m) indicated that the base of the Tillite Group is thrust-faulted (Hambrey and Spencer 1987). Indeed, high-level thrusting typical of foreland-propagating fold and thrust belts is in evidence throughout the exposed sequence, with well-developed ramp-flat trajectories. Although the shortening was not determined, major tectonic repetition of the Tillite Group has not occurred. Locally developed small-scale recumbent folds produce limited (ca. 10%) overthickening of the Tillite Group succession.

In the Limestone–Dolomite "series", directly beneath the Tillite Group, westward-directed thrusting in Bed no. 19 is associated with marked calcite veining (Figs. 5A and B) and slickensided surfaces. The most clearly defined and universally developed thrust cuts up-sequence through Bed no. 20 westwards to its boundary with the Ulvesø Formation (Fig. 5C). Slickensides, small shear zones, basal fault gouge and many small back-thrusts have developed with this thrust. Well-developed spaced pressure-solution cleavages are found throughout the sequence, particularly in the Ulvesø diamictite. The Arena Formation dolomitic shales also show small-scale thrusting and in places take on an "S–C" fabric. Small folds and crenulations on bedding surfaces are common, most of them verging to the northwest.

The Arena Formation–Sorteelv Formation junction is also a thrust fault with local shearing and fracturing of the diamictites. At the base of a ca. 8 m dolostone unit (the Canyon Formation), capping the clast-poor diamictite and sandstones of the Sorteelv Formation, strongly sheared to mylonitic fabrics are again found. In laminated horizons of the dolostone, northwest- and southwest-verging folds and small-scale thrusts locally double the thickness of this unit (Fig. 5D).

Despite the degree of thrusting and folding, deformation is inhomogeneously distributed throughout the sequence. Cleavages in particular tend to be confined to thrust-faulted sections where significant layer-parallel shortening and shearing has occurred. Consequently, original sedimentary fabrics such as those that characterize waterlain tillites can still be recognized in other parts of the sections away from the high-strain zones. Shortening of the stratal units appears to have been accommodated along extensive bedding plane parallel detachments with low to moderate angled, eastward-dipping linking thrusts. The amount of shortening has not been determined here, but the structures described above would suggest more than the 15% estimated by earlier workers (Eha, 1953; Higgins *et al.*, 1981; Henriksen, 1985). The age of the thrusting has not yet been determined, but it probably pre-dates the deposition of the Devonian sediments. Whilst some late Devonian reactivation of the thrusts may have occurred, their generation was almost certainly associated with the main phase of open north–south folding in Silurian time (Haller, 1971).

Antarctic Sund

The Tillite Group is exposed on both sides of Antarctic Sund in a large open syncline (Hambrey and Spencer, 1987). The lowest exposed parts of the succession, particularly the rocks of Bed no. 17 and the Ulvesø Formation, have a strong penetrative cleavage associated with chlorite-grade metamorphism. In these rocks, clast fabrics are strongly influenced by the deformation and clasts in

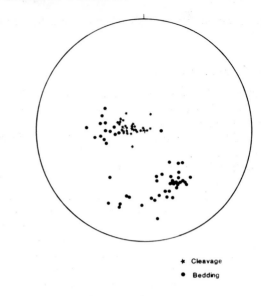

★ Cleavage
● Bedding

Fig. 6. Kap Oswald (Ella Ø) anticline; photograph taken looking north.

waterlain tillites have a marked NNW–SSE preferred orientation parallel to small fold axes and bedding cleavage intersections. Small-scale westward-directed thrusts with some overturning of beds are common and layer-parallel shearing is indicated by the frequent occurrence of sigmoidal tension gashes. On the north side of the sound the Arena Formation rocks have a well-developed slate cleavage associated with asymmetric chevron folds that verge to the west. Higher in the succession, many low-angle to bedding-parallel thrusts are apparent and these, with the slaty cleavage and minor folds, pre-date the large synclinal structure. The low (15%) crustal shortening estimates by earlier workers were undoubtedly derived from the open nature of the larger folds like the syncline at this locality. However, the many thrusts, folds, slaty cleavage and degree of metamorphism would be more consistent with greater shortening (>30%) during the Caledonian Orogeny.

Ella Ø

This is the best-known locality for the Tillite Group (Fig. 1), the rocks being well-exposed and easily accessible in a tight northward-plunging anticline (Fig. 6; Hambrey and Spencer, 1987). The fold is associated with an axial planar pressure-solution cleavage (Fig. 6), most strongly developed in the hinge region. Apart from some boudinage of a massive diamictite bed, set in a less competent shaly diamictite, deformation features are limited to small-scale accommodation thrusts in the core of the fold. This structure is thought to be Caledonian in age although mid-Devonian, westward-directed stress movements may have been responsible for tightening the fold (cf. Haller, 1971).

North Scoresby Land

Except for a small area in Canning Land described by Caby (1972), North Scoresby Land contains the most southerly tillite outcrop in the fjord region of the fold belt (Figs. 1 and 2). It is evident from aerial photographs and published 1:250 000 geological maps (Koch and Haller, 1971) that the Eleonore Bay Group rocks are repeated in North Scoresby Land by several large-scale open folds (cf. Henriksen and Higgins, 1976) and at least one important thrust (Fig. 2). Two localities were visited, one north and the other south of the mountain range Syltoppene. The Tillite Group was not exposed in the southern locality as depicted on the 1:250 000 geological map (Koch and Haller, 1971), but highly folded, thrust-faulted and weakly metamorphosed representatives of the Limestone–Dolomite "series" were found (Fig. 5E). To the west, these rocks overthrust much less obviously deformed limestones and other rocks belonging to the Multicoloured "series". To

the east, they have a normal faulted contact with the Carboniferous conglomerates and sandstones (Fig. 5F). On the northern slopes of Syltoppene and south of the Sorteelv Gletscher, an almost complete sequence from the Limestone–Dolomite "series" through the lower part of the Tillite Group is exposed. This succession is overlain by pure cross-bedded quartz sandstones that have no counterpart in the northern sequences (Hambrey and Moncrieff, 1985). However, the upper parts of Beds nos. 19 and 20 are missing and Tillite Group has a distinct waterlain aspect with massive and stratified diamictite, dolomitic mudstone, sandstone, conglomerate, rhythmites and dolostones (Hambrey and Spencer, 1987, Figs. 9 and 16). Except for the larger-scale structures noted above, previous workers have considered the sequence to be largely undeformed. However, many small-scale thrusts have been observed by us producing repetitions in the measurable stratigraphy on the ground, which if not recognized would lead to an overestimation in thicknesses of the affected rock units. The various rock types have responded differently to the deformation. The sandstone component of the Arena Formation and the quartz sandstones above

Fig. 7. (*a*) North Scoresby Land, low-angled thrust in tillites; looking south. (*b*) Small-scale box fold in the Ulvesø Formation rhythmites, North Scoresby Land. (*c*) Thin section of rhythmite from Tillit Nunatak. Bedding (S0) cut by spaced S2 cleavage. Long edge of photograph 2.5 mm. (*d*) Photomicrograph of tillite/basement contact, Tillit Nunatak, S2 is refracted from basement (left) to tillite (right). Long edge of photograph is 2.5 mm. (*e*) Tillit Nunatak peak looking northwest. Crest of anticlinal dome (cuspate fold) with S2 parallel to tillite/basement contact. Higher in the section S2 swings parallel to the axial surface of the fold. (*f*) Deformed skolithus pipes in quartzite, Tillit Nunatak (erratic block).

the tillite levels display mostly brittle deformation, and low-angle thrusting. Imbrication (Fig. 7A) and duplex structures are common. The tillites with a carbonate-rich matrix in contrast have responded in a more ductile manner with folds (Fig. 7B), boudinage and cleavage development; thrust-faulting is also observed in these rocks. The cleavage orientation varies according to lithology, so that in fine-grained rocks it tends to be subparallel to bedding and is close-spaced. In more competent sandy rocks it is less obvious and develops as a wide-spaced fracturing and does not significantly modify the waterlain character of the clast fabric. Again there appear to be no major repetitions in the varied assemblage of rocks comprising the Tillite Group. The tillites and associated rocks are overlain with marked angular unconformity, with deep-red weathering, by predominantly fluviatile Devonian Old Red Sandstone sediments (cf. Yeats and Friend, 1978). The rocks include red sandstones, breccias and conglomerates deposited in steep-sided palaeovalleys that have been partially reexcavated by recent erosion. The bulk of the sediments were probably deposited by fast-flowing streams, although the angular debris could represent rock-fall material (Yeats and Friend, 1978). Clasts in the conglomerate include locally derived representatives from the Limestone–Dolomite "series" and quartzites of the type lying directly above the uppermost tillite formation. It has long been known that the Devonian rocks in

the main basin to the north are strongly folded and thrusted (e.g. Haller, 1971; Friend *et al.*, 1983), but in the accounts on these rocks no mention is made of their deformation in the Syltoppene area. Here, the sandstones and conglomerates are found in three northeast-plunging synclinal folds. The northwest-dipping limbs are flat lying, whereas the southeast-dipping limbs are steep to overturned. Associated with these folds is a divergent fanning, pressure-solution cleavage (with strong suturing of clasts) that is best-developed in the sandstones and finer-grained rocks. The steep to overturned limbs are replaced by thrusts. It is not clear whether these structures are the result of the reactivation of earlier Caledonian thrusts or simply Devonian in origin. It is suggested here that the Devonian rocks accumulated in a series of half-grabens generated by the reactivation on Caledonian thrusts within the Tillite Group and associated rocks. Renewed NW–SE compression (or transpression) in the Late Devonian caused further reactivation of the thrusts and resulted in the folding of the Devonian rocks (Fig. 8).

THE "FORELAND" REGION

Charcot Land

The Charcot Land window (Fig. 3) exposes an Archaean–Proterozoic basement complex with a small isolated cover

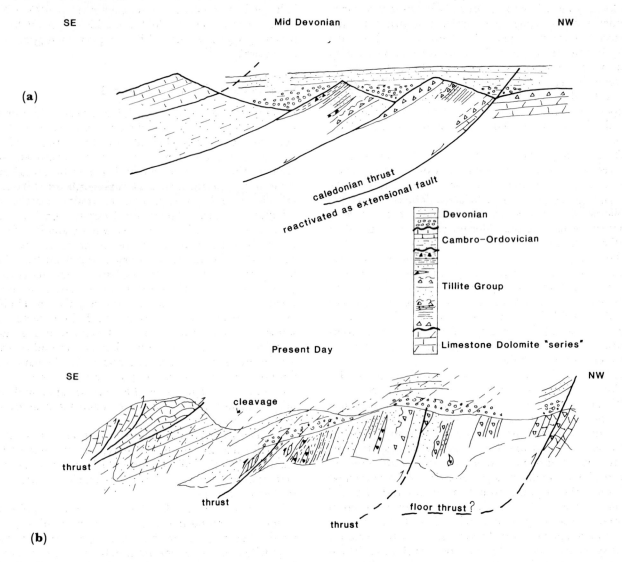

Fig. 8. Schematic evolution of the Old Red Sandstone south of Sorteelv Gletscher, North Scoresby Land. (*a*) Old Red Sandstone initially deposited in half-grabens generated by extensional relaxation on Caledonian thrusts. (*b*) Present-day compiled section. Devonian transpression has reactivated earlier Caledonian thrusts and folded/overthrust the Old Red Sandstone, rocks.

of diamictites on the peak of Tillit Nunatak (2060 m). The window occurs below a major Caledonian thrust mass that has been transported at least 40 km westwards (Henriksen and Higgins, 1976). The basement rocks are described as constituting the autochthonous Caledonian foreland. At Tillit Nunatak, the basement consists predominantly of a post-kinematic quartz diorite from which zircon ages of 1900 Ma (Steiger and Henricksen, 1972) have been obtained. This intrusion cuts earlier gneisses and migmatites as well as the adjacent supracrustals and highly sheared amphibolites. Implicit in this relationship is that the supracrustals are older (cf. Henriksen and Higgins, 1976). The shearing in the amphibolites developed during amphibolite facies conditions as the amphiboles have in places recrystallized, with a reduction in grain size in the mylonite fabric, whilst other, larger grains show semibrittle deformation. This deformation clearly pre-dates that seen in the diamictite. The surface of the unconformity between the diamictite and the basement is irregular on the metre scale. At its base, the diamictite incorporates large fragments (up to 1.7 m across) of the underlying basement that have been displaced little from their point of origin. These clasts appear to have been produced by glacial plucking of the basement associated with the deposition of the massive diamictite as a lodgement till. Clasts in the overlying diamictite often show some degree of rounding, although original sedimentary clast fabrics are obliterated by a tectonic foliation (S2). Other facies at this locality include weakly bedded, green-grey to black diamictite of the same composition but with a distinct waterlain character. Varve-like rhythmites with graded beds, dropstones and gravelly horizons containing rip-up clasts are an important component of the middle part of the sequence and are thought to be glacially influenced. Minor stratified sandstones, with a dispersed clast and grit component, appear at the top of the exposed section and they display good cross-lamination and convolute structures.

The distinct green coloration of the diamictites is due to the presence of chlorite in the matrix, which in thin section appears to be predominantly metamorphically recrystallized and defines an early S1 foliation. The occurrence of stilpnomelane reported by Steck (1971) has not been observed by us, but the chlorites are often oxidized in weathered sections.

Flattening of quartz grains and deformation of early calcite veins are also a product of this synmetamorphic deformation. Overprinting this fabric is a more obvious post-metamorphic, axial planar pressure-solution cleavage (S2) described below and seen penetrating the basement (Figs. 7C and D). Large flakes of chlorite (rhomboidal) truncated by the S2 may be metamorphic, but a detrital origin cannot be excluded.

Although earlier workers reported the diamicite at this locality to be several hundred metres thick (Henriksen, 1981), our observations show the maximum undisturbed thickness to be 50–60 m. The diamictite/basement rocks are folded together into a series of northwest-vergent textbook 150/200 m wavelength/amplitude cuspate folds. On the broad anticlinal domes, the diamictite does not exceed 50–60 m in thickness, whereas in the tight synclinal keels the rocks are pinched-out to an apparent thickness of 200 m. Cleavage (S2) and clast flattening orientations in the diamictites change around the folds such that they are strongly axial plane-parallel and most intensely developed in the synclines. Close to the hinge regions of the anticlinal domes, the S2 in the diamictite is parallel to the layer boundary as are the long axes of the flattened clasts. Up profile, however, S2 and the long axes of the clasts swing into the axial plane orientation through a finite neutral point indicated by the lack of deformation in the clasts at that point (Fig. 7E).

The combination of the Caledonian (411 Ma) mineral ages (biotite Rb/Sr; Hansen *et al.*, 1973a,b) obtained from the basement rocks, the extent of deformation and low-grade metamorphism of the diamictite argue strongly against the Charcot Land Windon representing truly autochthonous foreland. The shortening represented by the folding of the diamictites/basement rocks and the arching of the overthrust sheet of supracrustal rocks (Henriksen and Higgins, 1976) would need to be accommodated at depth by some kind of ductile and/or brittle detachment zone(s) involving several tens of kilometres transport from east to west.

At this locality, in common with others (Haller, 1971) along the margin of the inland ice, erratics of skolithus quartzite (pipe rock) have been observed. Significantly, and apparently overlooked by other authors, some of the pipes are deformed and are flattened in the plane of the cleavage cutting the rock (Fig. 7F). The quartzites are not seen anywhere in outcrop but are thought to originate from beneath the inland ice (Haller, 1971). The deformation exhibited by these erratics suggests that undisturbed Caledonian (?) foreland must lie further to the west than previously suggested. The occurrence of metamorphosed conodont-bearing Ordovician sediments as erratics at Cecilia Nunatak, which have yielded conodont alteration index values indicative of temperatures of 190–300° C (compared with only 50–90° C from Ella Ø), (M. P. Smith personal communication, 1987) appear to support this conclusion.

Gnejssø and Paul Stern Land

In the Gåseland window (Fig. 4) is a segment of presumed autochthonous or parautochthonous Caledonian foreland beneath the westerly-directed, overthrust masses of the western gneiss and schist zone rocks (Henriksen and Higgins, 1976). The foreland consists of Archaean to Early Proterozoic gneissic basement rocks with amphibolites, migmatites and augen gneisses of granitic composition (Henriksen, 1985). Above the basement gneisses and preserved in isolated pockets in the Gnejssø and Paul Stern Land areas, respectively, are low-grade to unmetamorphosed deposits including psephites and diamictites first interpreted as tillites by Wenk (1961). These rocks were named the Basal Psephitic Series by Wenk, who correlated them with the base of the Eleonore Bay Group. Phillips and Friderichsen (1981) re-examined these rocks, described them informally as the Gåseland tillites, and correlated them along with the Charcot Land tillites with the (Varangian) Tillite Group above the Eleonore Bay Group in the fjord region. Implicit in these earlier works are that the psephites of the Gnejssø area and the unambiguous tillites of Paul Stern Land are stratigraphically equivalent.

Although the psephite clasts have been described as being predominantly similar in type to the underlying basement (Wenk, 1961), their range is limited to quartzites and granite–gneiss types, unlike the diamictites elsewhere in East Greenland.

The clasts account for 30–70% of the psephite by volume, the proportion declining stratigraphically upwards. The majority of the clasts appear to have been originally rounded to sub-rounded, but near the base a few large sub-angular types are found. The matrix of the psephite is quartzose to pelitic, with widely varying proportions of quartz, feldspar, chlorite and carbonate.

The psephite–basement junction is obscured by the strong shearing deformation that accompanied both D1 and D2 events. Overlying the psephite with sharp contact is a weakly stratified psammite with graded bedding indicating inversion of the succession at the main locality. Clasts in the psammite occur either as dispersed stones or else in thin layers (Fig. 9A). In both the psephite and the psammite gradations into thin pelitic horizons are found. Stratigraphically upwards, the pelites become dominant and they also contain dispersed clasts or distinct pebble layers. Sedimentary structures are not preserved and the pelite consists of metamorphically recrystallized quartz, chlorite, muscovite, feldspar and carbonate. Above this sequence are intensely folded and highly sheared blue-grey marbles, that are interbanded with pelitic layers. The pelitic layers increase in abundance up section (structurally).

Although Wenk (1961) strongly favoured a glaciogenic origin for the psephite sequence, we believe that the degree of deformation precludes any certain assessment of the depositional environment. It appears that the psephite was deposited on an uneven surface on the basement, conceivably by a grounded glacier but equally likely by debris flowage or even sheet flood. Psammites and pelites with dispersed stones can also be produced by any of the above processes. In the absence of any positive criteria for deposition by ice, it is unlikely that unequivocal correlations with other areas can be made.

Structurally, the psephite sequence is found in an overturned to recumbent synclinal fold core (Fig. 9B and Fig. 4 cross-section A–B). This fold is clearly F2, as the clasts in the psephite have been flattened by an earlier F1 (D1) deformation event (Fig. 9C). Significantly, the psephite is cut by a similarly deformed quartz–feldspar pegmatite (Fig. 9D), which would lend itself to isotopic age determination to constrain the age of the psephite more

(a)

(b)

(c)

(d)

(e)

(f)

Fig. 9. (*a*) Psammite with thin pebble layers. (*b*) Gåseland psephite in core of recumbent syncline, basement is to the right, marbles to the left. Looking north. (*c*) Basal part of the psephite, clasts are flattened in S1 fabric and refolded by F2 folds. S2 cleavage is parallel to hammer shaft. (*d*) Pegmatite sheet cutting psephite and deformed by F1 and F2 folding. (*e*) Ductile to brittle shearing in the basement, parallel to S2 fabric in the psephite, Gnejssø. (*f*) Strong pressure-solution cleavage with shearing in basement in core of F2 syncline seen in psephite (see (*b*)).

closely. Wenk (1961) and Phillips and Friderichsen (1981) have also drawn attention to the metamorphism of the psephite. Mineralogically the pelitic fraction consists of metamorphically grown chlorite (occasionally seen replacing biotite), muscovite and albite, indicating greenschist facies conditions with some slight retrogression during the deformation of these rocks. Similar deformation and metamorphic effects can also be recognized in the adjacent basement gneisses, such that the basement and psephite were folded together. Also cutting the earlier gneissic fabrics and related to these fold structures are a number of ductile to brittle shear surfaces (Figs. 9E and 9F). The presence of these structures and the intrusive pegmatite strongly suggest that the so-called foreland is more likely allochthonous and has been transported on an underlying (unexposed) thrust, possibly from the more distinctly metamorphic Caledonian belt to the east.

The complexity and intensity of the deformation in these rocks make it difficult to estimate the extent of shortening within this window, but 50% would be conservative. Although strict stratigraphic and structural constraints are lacking, the extent of transport on the underlying thrust must be in excess of 40 km. This amount of transport is required to bring the essentially greenschist-facies psephite sequence of Gnejssø into proximity to the unmetamorphosed Paul Stern Land tillites (see below).

The marbles and intermixed pelitic sequences that structurally overlie both the psephites of Gnejssø and the tillite of Paul Stern Land are equated in the literature on positional grounds (Henriksen and Higgins, 1976). They are infolded with the psephites and basement rocks at Gnejssø and in some places lie in direct contact with the basement. Rootless F1 folds are common in the marbles and the F2 folds are asymmetric with strong shearing associated with the westerly-vergent overturned limbs, these rocks are pervaded by mylonitic fabrics (Fig. 10A).

(a)

(b)

(c)

(d)

(e)

Fig. 10. (*a*) Deformation in marbles above psephite/basement, Gnejssø. Note development of S–C fabric in pelitic layers and westward-vergent folds in the marbles. (*b*) S–C fabric in quartz phyllites above the marbles, Gnejssø. (*c*) Cleaved (pressure-solution) tillite from Paul Stern Land. (*d*) Støfvanget tillite outcrop, Paul Stern Land, Typical pocket-like outcrop. This shows strong deformation in upper part (several metres); the lower part is only deformed in the vicinity of the contact. (*e*) Basement/marble contact, Paul Stern Land. Eastward-dipping fractures in the basement are possibly imbricate faults.

In more silica-rich horizons, typical upper-greenschist calc–silicate assemblages are found, including albite, epidote and pale green amphiboles. It is our view therefore that the marbles, being of higher metamorphic grade, overthrust the psephites and the Paul Stern Land tillite/ basement succession and are not part of a normal stratigraphy as has been implied.

Above the marbles in both localities is a sequence of quartzite phyllonites, again with a strong mylonitic fabric (Fig. 10B). These are succeeded by quartz–mica schists of apparently higher grade than those rocks immediately below, which also display two generations of folds, an early set of isoclines overprinted by asymmetric crenulations indicating transport to the west. Garnet–mica schists and magmatitic amphibolites are found in this overthrust sequence, which together are believed to be equivalent to the Proterozoic Krummedal supracrustal sequence (cf. Henriksen, 1985). Finally, overthrusting these rocks are high-grade rocks of the gneiss and schist zone (Fig. 4).

The Paul Stern Land tillites outcrop discontinuously above the basement (Phillips *et al.*, 1973; Phillips and Friderichsen, 1981; Moncrieff this volume) and are said to occupy hollows in the basement, a position that protected them from the thrust-related deformation. However, some repetition due to imbrication during thrusting has been identified from aerial photographs along the slopes of Paul Stern Land facing Vestfjord Gletscher. Unlike the psephites, the Paul Stern Land tillites are of undoubted glacial origin, being unmetamorphosed and internally undeformed. At one of these localities (Støfvanget) the lower contact with the basement gneisses is foliated (Fig. 10C) and locally sheared, but according to Moncrieff (personal communication), undeformed tillite/basement contacts are found at other localities . The upper contact with the marbles is more pervasively deformed and can be described as mylonitic (Fig. 10D) in character (see Moncrieff, this volume). This deformation of the upper parts of the tillites is almost certainly the result of overthrusting by the marbles and the high pore fluid pressures generated along the zone of thrusting. The deformation in the tillites and the high-angle, eastward-dipping brittle fractures observed in the underlying basement rocks (Fig. 10E) would indicate the presence of a thrust beneath the present level of exposure (Fig. 4 cross section A–B). Although this thrust is necessary to accommodate the obvious shortening in the tillite/basement, the amount of transport that has occurred on it need not be great. This floor (or sole) thrust could have been generated off the toe of the footwall ramp of the thrust that carried the Gnejssø psephites west (Fig. 4).

SUMMARY AND CONCLUSIONS

This study has demonstrated that the tillites and associated rocks of both the eastern fjord region and in the "foreland" windows have been subjected to considerably more deformation than is apparent from the literature.

In North Scoresby Land south of Syltoppene, the occurrence of isoclinally folded representatives of the Limestone–Dolomite "series" and low-grade slates suggests that Caledonian thrusting must have involved significant westward displacement. Their present position above the much less-deformed lower Limestone–Dolomite and Multicoloured "series" and the more open folding affecting the whole stack of rocks is thought to be the result of the less-intense Devonian deformation.

In contrast to other parts of the fjord region, the Eleonore Bay and Tillite Group rocks of Antarctic Sund

are metamorphosed to chlorite grade but all areas display significant thrust-related shortening.

Although it has not been possible to assess fully the amount of thrust transport in the Kap Weber exposures, local overthickening and duplication of parts of the sequence are due to folding and thrusting. However, major repetitions of the stratigraphy are not apparent. Deformation is restricted to incompetent or more ductile horizons, but on the whole original sedimentary structures are preserved and the sequence is distinctly non-metamorphic.

The Eleonore Bay and Tillite Groups on Ella Ø are also folded principally by Caledonian movements but are not detectably metamorphosed and other than fold-related cleavages show little internal deformation.

The "tillitic" rocks of the forland region (Charcot Land and Gåseland/Paul Stern Land) windows have also been affected by deformation related to the Caledonian thrusting. In none of the three localities can the basement on which the tillitic rocks rest be regarded as unmoved foreland. At Tillit Nunatak, Charcot Land, the tillite and adjacent basement has been subjected to chlorite-grade metamorphism accompanying the formation of an early S1 foliation and this is overprinted by a later pressure-solution fabric (S2) axial planar to the large cuspate folds. This deformation cannot be due simply to the overthrusting of the Proterozoic supracrustals; a further thrust at depth is required to accommodate the shortening reflected by the cuspate folds and the arching of the supracrustal thrust sheet. The presence of deformed Skolithus pipe rock erratics along this part of the margin of the inland ice suggests that stable foreland must be located some unknown distance farther west.

The Paul Stern Land tillites are usually correlated with those in Charcot Land, but the former are distinctly non-metamorphic and separated north–south by 180 km. The discontinuous outcrop pattern of the Paul Stern Land tillites may, in part, be due to preservation in hollows in the basement, but tectonic imbrication has also occurred. Shearing fabrics in the Støfvanget section and eastward-dipping fractures in the basement imply the presence of thrusts above and below tillite.

Correlation of the Paul Stern Land tillite with the Gnejssø psephite (western Gåseland) as suggested by Wenk (1961) and Phillips and Friderichsen (1981) is based on their similar structural position above the basement. However, a greater amount of shortening and thrust displacement is evident in the Gnejssø section. The psephites are more deformed with two phases of folding, flattening and shearing and have been metamorphosed to greenschist facies. The psephites cannot be shown to have been tillites originally; other interpretations are possible. Similar ductile to brittle structures can also be recognized in the basement and imply the presence of a (floor) thrust at depth from which they were generated. This floor thrust must be structurally above the Paul Stern Land tillite/ basement sheet. It is assumed that the floor thrust to the Paul Stern Land sheet is linked to the westwards propagation of the Gnejssø floor thrust.

The difference in metamorphic grade between the Paul Stern Land tillites and the Gnejssø psephites suggests a greater separation (possibly be as much as 40 km) of these rocks prior to thrusting.

Isotopic dating of the pegmatite cutting the psephite is needed to clarify the age relations between these rocks and the Paul Stern Land tillites.

The marbles above the tillites and the psephites are highly sheared and certainly in thrusted contact with both sequences. It is possible that these highly ductile rocks have acted as a major thrust zone carrying the overlying thrust sheets westwards. The F2 folding of the marbles and psephites together at Gnejssø was the ductile response of these rocks to deformation ahead of more discrete shear zones propagating up through the basement from the floor thrust. The thrust contact between the marbles and tillites in Paul Stern Land has only a shallow easterly dip and appears to be unaffected by the imbrication of the tillite/basement rocks.

The thrust carrying the western gneiss and schist zone rocks over both sequences truncates the intervening mylonites, garnet–mica schists and amphibolites above the marbles in Paul Stern Land but is virtually planar.

The extent of deformation recorded in the rocks described here implies that very little if any of the successions are autochthonous and most have been transported by varying degrees westwards. For the fjord region rocks, the displacements appear to have taken place predominantly during Caledonian time but with some (variable) Devonian modifications. In all fjord region localities described here, deformation of the Tillite Group and adjacent rocks can be attributed to a thin-skinned, foreland-propagating, fold-and-thrust mechanism. In the deeper, stratigraphically older and higher-grade sections, deformation is more pervasive and thick-skinned thrusting mechanisms are more appropriate. Simply, it appears that during the westwards thrusting of the whole succession the younger parts (Tillite/Eleonore Bay Groups) have been carried in a relatively passive mode on the backs of the far-travelled basal (Krummedal Supracrustal) sheets.

Caledonian deformation of the Charcot Land and Gåseland foreland windows can be explained in much the same way, although in the former area the evidence is somewhat circumstantial. The case for the Gåseland window is stronger, with progressively higher-grade/deeper, more intensely deformed rocks occurring in successively higher thrust sheets.

POSTSCRIPT

Finally, this contribution would be incomplete without some consideration of the Devonian deformation in central East Greenland. To understand how this might be important, some observations on the main Devonian Old Red Sandstone basin to the north are needed. The rocks are essentially bounded by north–south trending faults that form a graben-like structure 80 × 140 km. In the northern part of the trough, the Devonian rocks are affected by NE–SW-trending folds and thrusts (Henriksen and Higgins, 1975). The basement rocks to the west also exhibit WNW–ESE and NNW–SSE fault systems. Taken together, these structures would indicate that, whatever the origin of the trough, the rocks have been subjected to sinistral strike-slip with an element of transpression. The Devonian rocks responded by folding and thrusting, the adjacent basement outside the parent faults by brittle failure along Reidel fractures. This type of mechanism would account for some of the high-level deformation of the Late Precambrian–Early Palaeozoic rocks west of the graben. It is of interest to note that similarly displaced Devonian grabens exist in Svalbard and together support the operation of sinistral strike-slip plate motions in the N Atlantic–Arctic Ocean region during Late Devonian–Carboniferous time.

Acknowledgements

This contribution was made possible by N.E.R.C. grant GR3/5438 awarded to W. B. Harland 1984/5. Drs N. Henriksen and A. K. Higgins of the G.G.U. are sincerely thanked for their geological and logistical advice in preparing the field work. We thank M. P. Smith, A. C. M. Moncrieff and the referees for their comments on earlier drafts of the manuscript.

REFERENCES

Caby, R. (1972). Preliminary results of mapping in the Caledonian rocks of Canning Land and Wegener Halvø, East Greenland. *Rapp. Grønlands Geol. Unders.* **48**, 21–38.

Eha, S. (1953). The pre-Devonian sediments on Ymers Ø, Suess Land, and Ella Ø (East Greenland) and their tectonics, *Meddr. Grønland* **111**(2), 105pp.

Fränkl, E. (1953). Geologische Untersuchung in Ost-Andrées Land (NE Grønland) *Meddr. Grønland* **113**(4), 160pp.

Friend, P. F., Alexander-Marrack, P. D., Allen, K. C., Nicholson, J. and Yeats, A. K. (1983). Devonian sediments of East Greenland VI, Review of results. *Meddr. Grønland* **206**(6), 96pp.

Haller, J. (1971). *Geology of the East Greenland Caledonides.* Wiley Interscience, London.

Hambrey, M. J. (1983). Correlation of Late Proterozoic tillites in the North Atlantic region and Europe. *Geol. Mag.* **120**, 209–232.

Hambrey, M. J. (1988). Late Proterozoic sedimentary record of the Arctic–North Atlantic region; implications for the evolution of the Caledonide Orogen, *this volume*.

Hambrey, M. J. and Moncrieff, A. C. M. (1985). Vendian stratigraphy and sedimentology of the East Greenland Caledonides. *Rapp. Grønlands Geol. Unders.* **125**, 88–94.

Hambrey, M. J. and Spencer, A. M. (1987). Late Precambrian Glaciation of Central East Greenland, *Meddr. Grønland Geoscience*, **19**, 50pp.

Harland, W. B. and Wright, N. J. R. (1979). Alternative hypothesis for the pre-Carboniferous evolution of Svalbard. *Skr. Norsk Polarins.* **161**, 90–117.

Hansen, B. T., Oberli, F. and Steiger, R. H. (1973a). The geochronology of the Scoresby Sund area. 4: Rb/Sr whole rock and mineral ages. *Rapp. Grønlands Geol. Unders.* **58**, 55–58.

Hansen, B. T., Frick, U. and Steiger, R. H. (1973b). The geochronology of the Scoresby Sund area. 5: K/Ar mineral ages. *Rapp. Grønlands Geol. Unders.* **58**, 59–61.

Henriksen, N. (1981). The Charcot Land tillite, Scoresby Sund, East Greenland In: Hambrey, M. J. and Harland, W. B. (eds.) *Earth's pre-Pleistocene Glacial Record*, pp. 776–77. Cambridge University Press, Cambridge.

Henriksen, N. (1985). The Caledonides of central East Greenland 70°–76° N. In: Gee, D. G. and Sturt, B. A. (eds.). *The Caledonian Orogen—Scandinavia and Related areas*, pp. 1095–1113. Wiley, Chichester.

Henriksen, N. and Higgins, A. K. (1976). East Greenland Caledonian fold belt. In: Escher, A. and Watt, W. S. (eds.) *Geology of Greenland*, pp. 182–247. Grønlands Geol. Unders., Copenhagen.

Higgins, A. K. and Phillips, W. E. A. (1979). East Greenland Caledonides, an extension of the British Caledonides. In: Harris, A. L., Holland, C. H. and Leake, B. E. (eds.) *The Caledonides of the British Isles Reviewed*. Geol. Soc. Lond. Spec. Publ. **8**, 19–32.

Higgins, A. K., Friderichsen, J. D. and Thyrsted, T. (1981). Precambrian metamorphic complexes in the East Greenland Caledonides (72°–74° N)—their relationships to the Eleonore—Bay Group, and Caledonian orogenesis. *Rapp. Grønlands Unders.* **104**, 5–46.

Katz, H. R. (1954). Eine Bemerkungen zur Lithologie und Stratigraphie der Tillitprofile im Gebiet des Kjeser Franz Josephes Fjord, Ostgrönland, *Meddr. Grønland* **72**(4), 63pp.

Koch, L., Haller, J. (1971). Geological map of East Greenland, 72°–77° N Lat. (1 : 250,000) *Meddr. Grønland* **183**, 26pp + 13 maps.

Kulling, O. (1930). Stratigraphic studies of the geology of Northeast Greenland. *Meddr. Grønland*, **74**, 317–346.

Moncrieff, A. C. M. (1988). The Tillite Group and related rocks of East Greenland: implications for Late Proterozoic Palaeography, *this volume*.

Phillips, W. E. A. and Friderichsen. (1981). The Late Precambrian Gåseland tillite, Scoresby Sund, East Greenland. In: Hambrey, M. J. and Harland, W. B. (eds.) *Earth's pre-Pleistocene Glacial Record*, pp. 773–775. Cambridge University Press, Cambridge.

Phillips, W. E. A., Stillman, C. J., Friderichsen, J. D. and Jemelin, L. (1973). Preliminary results of mapping in the western gneiss and schist zone around Vestfjord and inner Gåsefjord, south-west Scoresby Sund. *Rapp. Grønlands Geol. Unders.* **58**, 17–32.

Poulsen, C. (1930). Contributions to the stratigraphy of the Cambro-Ordovician of East Greenland, *Meddr. Grønland* **74**, 297–316.

Steiger, R. M. and Henriksen, N. (1972). The geochronology of the Scoresby Sund area. Progress Reports: zirconages. *Rapp. Grønlands Geol. Unders.* **48**, 109–114.

Steck, A. (1971). Kaledonische metamorphose der praekambrischen Charcot Land serie, Scoresby Sund, Ost-Grönland, *Bull. Grønlands Geol. Unders.* **97**, 69pp. [also *Meddr. Grønland* **192**(3).]

Wenk, E. (1961). On the crystalline basement and the basal part of the Pre-Cambrian Eleonore Bay Group in the southwestern part of Scoresby Sund. *Meddr. Grønland* **168**(1), 54pp, XI plates.

Yeats, A. K. and Friend, P. F. (1978). Devonian sediments of East Greenland IV The western sequence, Kap Kolthoff Supergroup of the western areas. *Meddr. Grønland* **206**(4), 112pp.